CRC Handbook of Clinical Chemistry

Volume IV

Editor

Mario Werner, M.D.
Professor of Pathology
(Laboratory Medicine)
The George Washington University
Medical Center
Washington, D.C.

CRC Press, Inc.
Boca Raton, Florida

Library of Congress Cataloging-in-Publication Data
(Revised for volumes 4 and 5)

CRC handbook of clinical chemistry.

(CRC series in clinical laboratory science)
Vols. 2- lack series statement.
Includes bibliographies and indexes.
1. Chemistry, Clinical. I. Werner, Mario, 1931-
II. C.R.C. Handbook of clinical chemistry. III. Handbook of clinical chemistry.
RB40.C73 1982 616.07'56 83-135797
ISBN 0-8493-7081-7 (v. 1)
ISBN 0-8493-7082-5 (v. 2)
ISBN 0-8493-7083-3 (v. 3)
ISBN 0-8493-7084-1 (v. 4)
ISBN 0-8493-7085-0 (v. 5)

Direct all inquiries to CRC Press, Inc., 2000 Corporate Blvd., N.W., Boca Raton, Florida, 33431.

International Standard Book Number 0-8493-7081-7 (Volume I)
International Standard Book Number 0-8493-7082-5 (Volume II)
International Standard Book Number 0-8493-7083-3 (Volume III)
International Standard Book Number 0-8493-7084-1 (Volume IV)
International Standard Book Number 0-8493-7085-0 (Volume V)

Library of Congress Number 83-135797
Printed in the United States

PREFACE

Some branches of laboratory medicine, such as microbiology, are defined according to pathology; others, such as hematology, are defined according to a functional system. Clinical chemistry, on the other hand, is defined by a technology, and so is capable of cutting across the conceptional structures and boundaries defined by pathology or organ systems. Such protean applicability is an obstacle to structuring a text which is comprehensive yet free of redundancy. To overcome this difficulty, the Handbook has been divided into three parts which represent each a fundamental but conceptually different aspect of clinical chemistry: Elements of Methodology; Analytes; and Clinical Applications. This arrangement should allow the reader to find information intuitively at the most plausible place. The first part, Elements of Methodology, comprises ten sections which make up Volumes I, II, and III.

The second part, Analytes, comprises five sections which make up Volumes IV and V.

The material presented has been selected and edited to render it succinct and relevant from a practical viewpoint. Unavoidably, this necessitated omission of all information which may be of historical interest but no longer has practical use. It has been my good fortune to find support from the best talent in making the required critical judgments and decisions. The success of this editorial effort should be judged by the frequency this text is used as a convenient source of information.

It remains for me to thank two individuals who crucially contributed to my undertaking. Dr. David Seligson conceived this Handbook as the original Editor-in-Chief of the *CRC Series in Clinical Laboratory Sciences*. Mr. B. J. Starkoff, CRC Press, supported the expanding project from its modest beginnings as the publisher.

Mario Werner

THE EDITOR

Mario Werner, M.D., is Professor of Pathology (Laboratory Medicine) at the George Washington University Medical Center Washington, D.C.

Dr. Werner obtained his medical training in Lausanne (Switzerland), Hamburg (West Germany), Paris (France), and Zurich (Switzerland). He was awarded his medical license in 1956 and a Doctorate of Medical Science, University of Zurich, Switzerland, in 1960. After specializing in internal medicine at the University of Basel, Switzerland, he worked at the National Institute of Arthritis and Metabolic Diseases, Bethesda, Maryland and at the Rockefeller University, New York, as a Fellow of the Swiss Academy of Medical Sciences. In 1964 he was called by the newly formed Medical School of Essen, West Germany to organize and head the Central Laboratory. Subsequently, he served as Assistant Professor at the University of California in San Francisco, as Associate Professor at Washington University, St. Louis, Missouri, and, since 1972, at his present position.

Dr. Werner is a Fellow of the American Society of Clinical Pathology, the College of American Pathologists, and the National Academy of Clinical Biochemistry, and has held either elective office or committee appointments in these societies. Among other memberships, he belongs to the American Association for Clinical Chemistry, the Academy of Clinical Laboratory Physicians and Scientists, the Society for Experimental Biology and Medicine, the American Federation for Clinical Research, the German Society for Clinical Chemistry, and the International Society for Clinical Enzymology.

Dr. Werner has been a member of the Pathology, a study section of the National Institutes of Health, is a Director of the American Board of Clinical Chemistry and has chaired its Examination Committee, has been a member of the Clinical Chemistry Panel of the Food and Drug Administration, and the Chairman of the Committee on Clinical Pathology, College of American Pathologists.

Dr. Werner is the author of approximately 100 scientific papers and has edited 9 books. He sits on the editorial boards of *Clinical Chemistry, Clinica Chimica Acta, Clinical Biochemistry, Clinical Chemistry Lookout, American Journal of Clinical Pathology, Clinical Physiology and Biochemistry, Pathologist, Medical Laboratory Observer (MLO), and LAB (Italy).* His major current research interests relate to clinical enzymology, therapeutic drug monitoring, and the mathematical modeling of medical decision making.

ADVISORY BOARD

William C. Purdy, Ph.D.
Professor of Analytical Chemistry and Associate in Medicine
McGill University
Montreal, Quebec, Canada
Editor: Electrochemistry

Robert Rej, Ph.D.
Wadsworth Center for Laboratories and Research
New York State Department of Health
Albany, New York
Co-editor: Enzymes

Robert Ritchie, M.D.
Foundation for Blood Research
Scarborough, Maine
Editor: Ligand Assays

William H. C. Walker, M.B.
Professor of Pathology
McMaster University Medical Center
Hamilton, Ontario, Canada
Editor: Apparatus and Instrumentation

CONTRIBUTORS

VOLUME IV
VOLUME V

Helmut V. Ammon, M.D.
Division of Gastroenterology
Henry Ford Hospital
Detroit, Michigan

Walter Appel, Ph.D.
Professor
Central Laboratory
St. Vincentius Hospital
Karlsruhe, Federal Republic of Germany

Onno W. van Assendelft, M.D., Ph.D.
Chief, Clinical Medicine Branch
Division of Host Factors
Center for Infectious Diseases
Centers for Disease Control
Atlanta, Georgia

Andrew D. Baines, M.D.
Biochemist-in-Chief
Toronto General Hospital
and
Professor
Department of Clinical Biochemistry
University of Toronto
Toronto, Ontario, Canada

Alan Belfield, Ph.D.
Head
Department of Biochemistry
Wexham Park Hospital
Slough, England

Michael G. Bennett, Ph.D.
Principal Biochemist
Children's Hospital
Western Bank
Sheffield, England

Ernest Beutler, M.D.
Head
Divison of Hematology/Oncology
and
Chairman
Department of Basic and Clinical Research
Research Institute of Scripps Clinic
La Jolla, California

Angelo Burlina, M.D.
Professor and Chairman
Institute of Clinical Chemistry
University of Padova
and
Laboratorio Centrale
Ospedale Civile
Padova, Italy

John W. Callahan, Ph.D.
Division of Neurosciences
The Hospital for Sick Children
and
Associate Professor
Departments of Biochemistry and Pediatrics
University of Toronto
Toronto, Ontario, Canada

Amos Cohen, Ph.D.
Division of Immunology/Rheumatology
The Hospital for Sick Children
and
Associate Professor
Department of Pediatrics
University of Toronto
Toronto, Ontario, Canada

Jean-Pierre Colombo, M.D.
Professor and Director
Department of Clinical Chemistry
University of Berne
Berne, Switzerland

Basil T. Doumas, Ph.D.
Professor of Pathology
Medical College of Wisconsin
Milwaukee, Wisconsin

Graham Ellis, Ph.D.
The Hospital for Sick Children
and
Associate Professor
Department of Clinical Biochemistry
University of Toronto
Toronto, Ontario, Canada

John W. Fenton II, Ph.D.
Research Scientist
Wadsworth Center for Laboratories and Research
New York State Department of Health
Albany, New York

Margaret M. Fletcher, Ph.D.
Mercy Catholic Medical Center
Darby, Pennsylvania

Jiri Frohlick, M.D., Ph.D.
Associate Professor
Department of Pathology
University of British Columbia
Vancouver, British Columbia, Canada

U. Gerlach, M.D.
Professor und Direktor
Medizinische Klinik und Poliklinik
Universität Münster
Münster, Federal Republic of Germany

David M. Goldberg, M.D., Ph.D.
Biochemist-in-Chief
The Hospital for Sick Children
and
Professor and Chairman
Department of Clinical Biochemistry
University of Toronto
Toronto, Ontario, Canada

Robert A. Goyer, M.D.
Professor and Chairman
Department of Pathology
University of Western Ontario
London, Ontario, Canada

F. Heinz, M.D.
Professor
Physiologische Chemie
Zentrum Biochemie
Medizinische Hochschule Hannover
Hannover, Federal Republic of Germany

J. Thomas Hindmarsh, M.D.
Chief, Department of Laboratory Medicine
Ottawa General Hospital
and
Associate Professor
Department of Biochemistry and Pathology
University of Ottawa
Ottawa, Ontario, Canada

Liliana Konarska, Ph.D.
Research Associate
Department of Clinical Biochemistry
Medical School of Warsaw
Warsaw, Poland

Winston W. K. Koo, M.D.
Research Scholar
Children's Hospital Medical Center
University of Cincinnati Medical Center
Cincinnati, Ohio

Norman C. Kramer, M.D.
Professor Emeritus
Department of Medicine
The George Washington University
Washington, D.C.

Jürgen Kusche, M.D.
Assistant Professor and Head
Biochemical and Experimental Division
Second Department of Surgery
University of Cologne
Federal Republic of Germany

William K. W. Lam, Ph.D.
Professor
Department of Opthalmology
University of Texas Health Science Center
San Antonio, Texas

Corey Largman, Ph.D.
Associate Research Biochemist
Departments of Internal Medicine and Biological Chemistry
University of California, Davis
School of Medicine
and
Director
Biochemistry Research Laboratory
Martinez VA Medical Center
Martinez, California

Clifford A. Lingwood, Ph.D.
The Hospital for Sick Children
and
Assistant Professor
Department of Clinical Biochemistry
University of Toronto
Toronto, Ontario, Canada

Gillian Lockitch, M.D.
Head, Pathophysiology
British Columbia Children's Hospital
and
Assistant Professor
Department of Pathology
University of British Columbia
Vancouver, British Columbia, Canada

Jack W. London, Ph.D.
Director
Computer Laboratory
Fox Chase Cancer Center
Philadelphia, Pennsylvania

Donald Mahuran, Ph.D.
The Hospital for Sick Children
and
Assistant Professor
Department of Clinical Biochemistry
University of Toronto
Toronto, Ontario, Canada

Dieter Maruhn, M.D.
Professor of Clinical Chemistry
University of Essen
and
Isotopenlabor Biokinetik
Institut für Pharmakokinetik
Bayer AG
Wuppertal, Federal Republic of Germany

Robert B. McComb, Ph.D.
Associate Director
Clinical Chemistry Laboratory
Hartford Hospital
Hartford, Connecticut

Ross F. McCurdy, Ph.D.
Associate Professor
Department of Pathology
Dalhousie University
Halifax, Nova Scotia, Canada

Matthew J. McQueen, M.B., Ph.D.
Head, Department of Clinical Chemistry
Hamilton General Hospital
and
Professor
Department of Clinical Biochemistry
McMaster University Medical Center
Hamilton, Ontario, Canada

Karen J. McVety, Ph.D.
Department of Pharmacology and Toxicology
College of Pharmacy
University of Rhode Island
Kingston, Rhode Island

Lindsay G. Morgan, M.D.
Inco Europe Ltd.
Clydach, Swansea, Wales

Donald W. Moss, Ph.D.
Professor
Department of Chemical Pathology
Royal Post-Graduate Medical School
Hammersmith Hospital
University of London
London, England

Michael A. Moss, M.D.
Director, Division of Clinical Chemistry
Victoria General Hospital
and
Assistant Professor
Department of Pathology
Dalhousie University
Halifax, Nova Scotia, Canada

Robert L. Murray, Ph.D.
Lutheran General Hospital
Park Ridge, Illinois

Ethan A. Natelson, M.D.
Professor
University of Texas Medical School
and
St. Joseph Hospital
Houston, Texas

Samuel Natelson, Ph.D.
Adjunct Professor
Department of Environmental Practice
College of Veterinary Medicine
The University of Tennessee
Knoxville, Tennessee

Daniel A. Nealon, Ph.D.
Wadsworth Center for Laboratories and Research
New York State Department of Health
and
Assistant Professor
Department of Biomedical Sciences
School of Public Health Sciences
State University of New York at Albany
Albany, New York

Jerome S. Nisselbaum, Ph.D.
Associate Attending Clinical Chemist
Memorial Sloan-Kettering Cancer Center
New York, New York

Paul O'Brien, Ph.D.
Department of Chemistry
Queen Mary College
University of London
London, England

Joel G. Parkes, Ph.D.
Assistant Professor
Department of Clinical Biochemistry
University of Toronto
Toronto, Ontario, Canada

John Patrick, M.D.
Professor
Department of Biochemistry
University of Ottawa
Ottawa, Ontario, Canada

Sherry L. Perkins, Ph.D.
Division of Biochemistry
Ottawa Civic Hospital
Ottawa, Ontario, Canada

Billy W. Perry, Ph.D.
Assistant Professor of Pathology
Medical College of Wisconsin
Milwaukee, Wisconsin

William H. Porter, Ph.D.
Associate Director
Clinical Chemistry Laboratory
and
Professor
Department of Pathology
University of Kentucky Medical Center
Lexington, Kentucky

Jennifer Potter, Ph.D.
Department of Biology
The Open University
Milton Keynes, England

Hayden Pritchard, Ph.D.
Assistant Professor
Department of Pathology
University of British Columbia
Vancouver, British Columbia, Canada

K. B. Rajani
Department of Pathology
Valley Children's Hospital
Fresno, California

Robert Rej, Ph.D.
Wadsworth Center for Laboratories
and Research
New York State Department of Health
Albany, New York

Alan D. Rinker, Ph.D.
Department of Clinical Chemistry
University of Kentucky
Albert B. Chandler Medical Center
Lexington, Kentucky

Thomas J. Ryan, Ph.D.
Research Scientist
The Wadsworth Center for Laboratories
and Research
New York State Department of Health
Albany, New York

B. Sarkar, Ph.D.
Research Institute
The Hospital for Sick Children
and
Professor
Department of Biochemistry
University of Toronto
Toronto, Ontario, Canada

Edward A. Sasse, Ph.D.
Associate Professor of Pathology
Medical College of Wisconsin
Milwaukee, Wisconsin

John Savory, Ph.D.
Professor
Department of Pathology
University of Virginia Medical Center
Charlottesville, Virginia

Paul H. Scott, Ph.D.
Senior Biochemist
Selly Oak Hospital
Birmingham, England

Ellen Schmidt, Ph.D.
Professor
Division of Gastroenterology
and Hepatology
Department of Internal Medicine
and Dermatology
Hannover Medical School
Hannover, Federal Republic of Germany

Friedrich Schmidt, M.D.
Professor
Division of Gastroenterology
and Hepatology
Department of Internal Medicine
and Dermatology
Hannover Medical School
Hannover, Federal Republic of Germany

Morton K. Schwartz, Ph.D.
Professor and Chairman
Department of Clinical Chemistry
Memorial Sloan-Kettering Cancer Center
New York, New York

H. Schweisfurth, M.D.
Privatdozent
Department of Internal Medicine
University of Essen Medical School
Essen, Federal Republic of Germany

L.E. Sendelbach, Ph.D.
Department of Pharmacology
and Toxicology
University of Kansas Medical Center
Kansas City, Kansas

Zahir A. Shaikh, Ph.D.
Professor and Chairman
Department of Pharmacology
and Toxicology
College of Pharmacy
University of Rhode Island
Kingston, Rhode Island

Leslie M. Shaw, Ph.D.
Associate Professor
Department of Pathology and
Laboratory Medicine
University of Pennsylvania
Philadelphia, Pennsylvania

John E. Sherwin, Ph.D.
Department of Pathology
Valley Children's Hospital
Fresno, California

Marie-Anne Skomorowski, Ph.D.
Research Institute
The Hospital for Sick Children
Toronto, Ontario, Canada

Alistair F. Smith, M.B., Ph.D.
Department of Clinical Chemistry
The Royal Infirmary
Edinburgh, Scotland

Richard Spooner, Ph.D.
Principal Biochemist
Department of Pathological Biochemistry
Gartnavel General Hospital
Glasgow, Scotland

Patrick J. St. Louis, Ph.D.
Staff Biochemist
The Hospital for Sick Children
and
Lecturer, Department of
Clinical Biochemistry
University of Toronto
Toronto, Ontario, Canada

John E. Stone, M.D.
Assistant Professor of Gastroenterology
Medical College of Wisconsin
Milwaukee, Wisconsin

Paula M. Strasberg, Ph.D.
Research Associate
Division of Neurosciences
The Hospital for Sick Children
and
Department of Clinical Biochemistry
University of Toronto
Toronto, Ontario, Canada

F. William Sundermann Jr., M.D.
Northeast Utilities Professor of
Toxicology
University of Connecticut School of
Medicine
Farmington, Connecticut

C. Brian Taylor, Ph.D.
Reader
Department of Biochemistry
University of Sheffield
Sheffield, England

Norbert W. Tietz, Ph.D.
Professor and Chairman
Department of Clinical Chemistry
Albert B. Chandler Medical Center
Lexington, Kentucky

Reginald C. Tsang, M.D.
Director
Division of Neonatology
Children's Hospital Medical Center
and
Professor
Departments of Pediatrics, Obstetrics, and
Gynecology
University of Cincinnati Medical Center
Cincinnati, Ohio

Raymond E. Vanderlinde, Ph.D.
Director
Clinical Chemistry
and
Professor
Department of Pathology and Laboratory
Medicine
Hahnemann University Hospital
Philadelphia, Pennsylvania

Myron M. Warshaw, Ph.D.
Department of Pathology
Northwest Community Hospital
Arlington Heights, Illinois

B. Weisner, M.D.
Universitäts-Krankenhaus Eppendorf
Hamburg, Federal Republic of Germany

P. Willnow, M.D.
Boehringer Mannheim GmbH
Biochemical Research Center
Tützing, Federal Republic of Germany

Michael R. Wills, M.D., Ph.D.
Professor
Department of Pathology
University of Virginia Medical Center
Charlottesville, Virginia

Jo Ellen Welsh, Ph.D.
Associate Professor
Department of Biochemistry
University of Ottawa
Ottawa, Ontario, Canada

D. C. Wigfield, Ph.D.
Professor
Department of Chemistry
Carleton University
Ottawa, Ontario, Canada

H. P. Witschi, Ph.D.
Biology Division
Oak Ridge National Laboratory
Oak Ridge, Tennessee

Paul Wolf, M.D.
Professor
Department of Pathology
University of California, San Diego
San Diego, California

Tai-Wing Wu, Ph.D.
Professor
Department of Clinical Biochemistry
University of Toronto
and
Toronto General Hospital
Toronto, Ontario, Canada

TABLE OF CONTENTS

VOLUME I

TABLE OF CONTENTS

VOLUME II

TABLE OF CONTENTS

VOLUME III

TABLE OF CONTENTS

VOLUME IV

TABLE OF CONTENTS

VOLUME V

ENZYMES

CRC HANDBOOK OF CLINICAL CHEMISTRY

Mario Werner, Editor

Volume I

Nomenclature, Quantities, and Units
Physical and Chemical Data
Reagents, Specimen Collection, Calibrators,
Reference Material and Standards
Quality Assurance
Apparatus and Instrumentation

Volume II

Optical Methods of Analysis
Electrochemistry
Separation Methods

Volume III

Ligand Assays
Statistical and Mathematical Methods

Volume IV

Regulation of Water and Electrolyte Balance
Trace Elements
Nutrients and Vitamins

Volume V

Enzymes
Miscellaneous Compounds

Regulation of Water and Electrolyte Balance

Samuel Natelson and Ethan A. Natelson, M.D.

WATER, ELECTROLYTES, AND BLOOD GASES

Samuel Natelson and Ethan A. Natelson

HISTORICAL BACKGROUND AND FUNDAMENTAL CONCEPTS

Prior to the systematic study of electrolyte and acid-base problems, fluids were administered empirically to patients who were obviously dehydrated, with markedly elevated temperatures characteristic of certain illnesses such as malaria and yellow fever. The patient was sponged with alcohol or ice water to reduce the temperature, and water was then administered orally. Fluids were also administered rectally, by enema. Usually, small amounts of baking soda ($NaHCO_3$) were added to the enema fluid. This served to combat the acidosis evidenced by the strong odor of acetic acid and acetone in the patient's sweat. In these patients, it was noted that the amount of water in the enema administered was far greater than what was then returned. In many cases the deep breathing (Kussmaul respiration) was substituted by normal respiration as the acidosis was corrected. During epidemics the survival rate was low, often not due to the underlying disease, but consequent to inadequate fluid therapy.

Toward the end of the 19th and early into the 20th century, C. Bohr (1855—1911) reported on his studies of oxygen and carbon dioxide in blood, using an apparatus developed by J.S. Haldane (1860—1936). It was during this period that Bohr noted the effect of p_{CO_2} and, thus, pH on the hemoglobin oxygen dissociation curve. The studies of Bohr and Haldane inspired D. D. Van Slyke (1883—1967) and his students G. E. Cullen and A. B. Hastings to carry out their classical studies on the carbon dioxide content of blood and its significance in acid-base balance.

S.P.L. Sørensen (1868—1939) had proposed the use of the concept of pH (in 1909) describing buffer systems. His equation was applied to the bicarbonate buffer system by L.J. Henderson (1878—1942) and blood by K.A. Hasselbalch (1874—1962).[1,2] With this equation, Hastings designed a nomogram in 1934, relating pH and total CO_2 to p_{CO_2} and hematocrit (as a measure of Hb concentration).[3] The nomogram was designed for the purpose of the diagnosis of respiratory or metabolic acidosis, or alkalosis. The major objective was to obtain an estimate of the "alkali reserve". This meant the base excess (mainly Na^+) over chloride, which was available to neutralize the organic acids, so as to guide fluid therapy. At that time, sodium could not be measured readily. Subsequently, Astrup and Siggard-Anderson simplified and modified this nomogram to be used for similar purposes.[4,5]

A major stride was taken in the 1930s, when the sealed glass electrode was introduced to clinical medicine. This had been originally devised by Fritz Haber (1868—1934). Thus, the state of acidemia or alkalemia could now be measured directly.

While the measurement of potassium or chloride could be performed accurately and quickly by titration, no rapid practical method for Na determination was available until the introduction of the flame photometer at the end of World War II.[6]

The availability of direct, rapid, and accurate measurements of the electrolytes, blood pH, total CO_2, P_{CO_2}, P_{O_2}, and lactic acid, due in part to the dramatic developments in the electronic industry, is a major factor in the success of complicated surgical procedures, treatment of acute respiratory failure, neonatal survival, and medicine in general.

It is one objective of this chapter to present the fundamental chemistry and physiology underlying the rationale for fluid therapy. Another objective is to present the manner in which these data are used for the diagnosis of disease and the management of the patient.

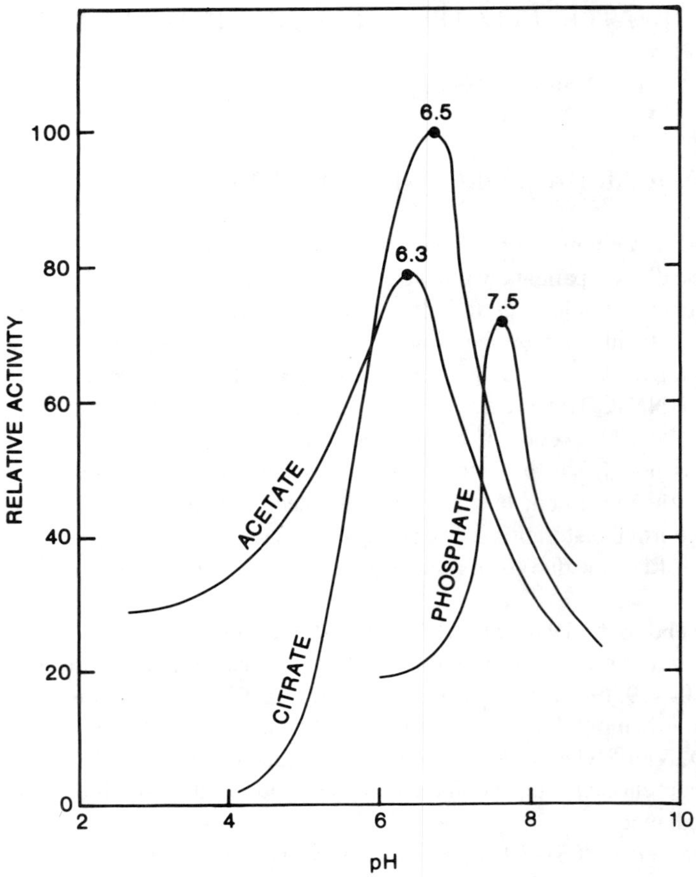

FIGURE 1. Effect of pH and buffer composition on the optimum pH for urease.

MAINTENANCE OF CONSTANT BLOOD pH

There are at least 40 enzymes which have been prepared with a high degree of purity and which are involved in the intermediate metabolism of the nutrients. These enzymes differ in activity, depending upon the pH and ionic strength of the medium (Figure 1). These enzymes act in sequence, as the metabolic intermediates are produced to eventually generate energy for survival and growth. At a fixed pH their rate of activity can be designed so that there is no unnecessary pileup or deficiency of intermediates, so that the metabolic chain can proceed smoothly. For example, if the blood pH is lowered from 7.4 to 7.0, cell metabolism is in chaos, lactate accumulates, and cell metabolism is brought to a standstill.

Another reason for maintaining a constant pH at 7.4 is the fact that at this pH the CO_2 in the blood is available for rapid exchange at the lungs without changing the blood pH. This is practicable because of the very nature of the manner in which blood pH is maintained constant. This can be seen in a simplified model in Figure 2.

The blood contains about a 0.04 *M* excess of sodium over chloride. The container in Figure 2 contains a 0.05 *M* sodium biocarbonate solution. This has a pH of about 8.5. If the air valve of Figure 2 is opened, then CO_2-free air will be blown through the solution in fine bubbles. Mixing is by the magnetic stirrer. This will cause the solution to lose CO_2, and the pH will move to about 10.2 after about 5 min. The added hemolyzed blood provides the carbonic anhydrase needed to bring about equilibrium quickly. The reaction may be written:

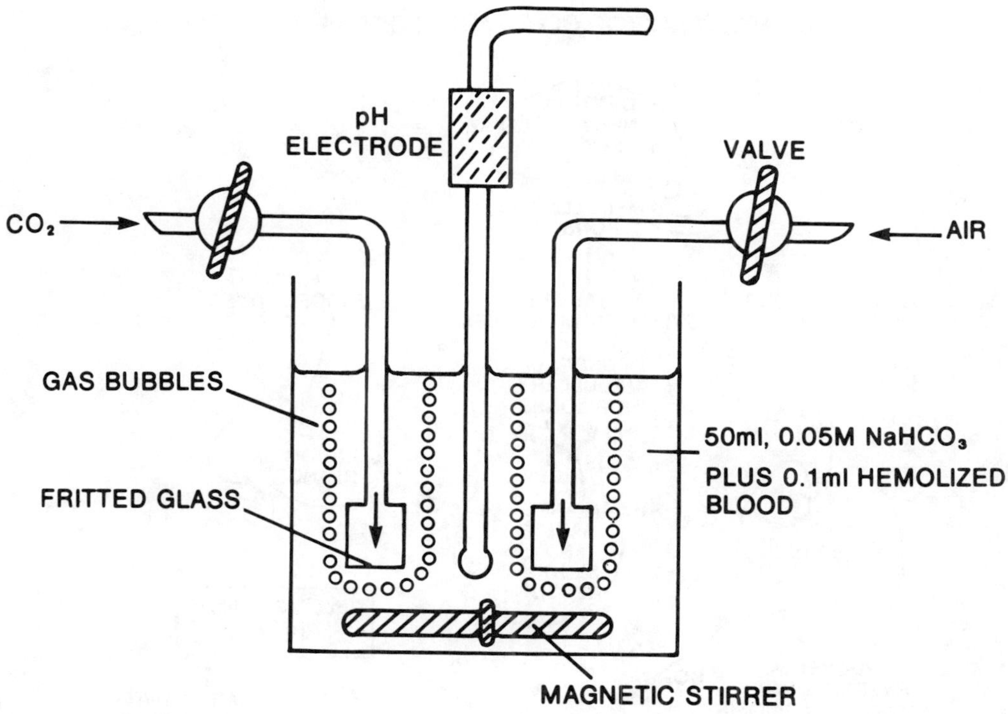

FIGURE 2. Mechanical model of the mechanism for maintaining constant blood pH.

$$2NaHCO_3 \rightarrow Na_2CO_3 + H_2O + \uparrow CO_2$$

If CO_2-free air continues to be blown through the solution, then eventually more CO_2 can be blown off, bringing the pH to about 11.

$$H_2O + Na_2CO_3 \rightarrow 2NaOH + \uparrow CO_2$$

This results from the generation of some OH ion, replacing the original carbonate ion. The overall reaction may be written:

$$HCO_3^- \rightarrow OH^- + \uparrow CO_2$$

This is the reaction taking place at the lungs.

If the air valve is closed and CO_2 is blown in vigorously, then the pH will move over to about 6.2 in about 1.5 min. The reverse reaction is taking place:

$$(1) \quad OH^- + CO_2 \rightarrow HCO_3^-$$

$$(2) \quad CO_3^{2-} + H_2O + CO_2 \rightarrow 2HCO_3^-$$

If both valves are opened partially, the pH can be adjusted readily to any pH from 6.9 to 7.7 (the physiological range) and maintained at exactly 7.40 pH units. How the body performs a similar task is shown in Figure 3.

In Figure 3, the pulmonary ventilation rate (with air) is controlled by two factors: one is the mechanical action of the intercostal muscles and the diaphragm; the other is the blood flow from the heart to the lungs. These need to be matched. For example, with vigorous

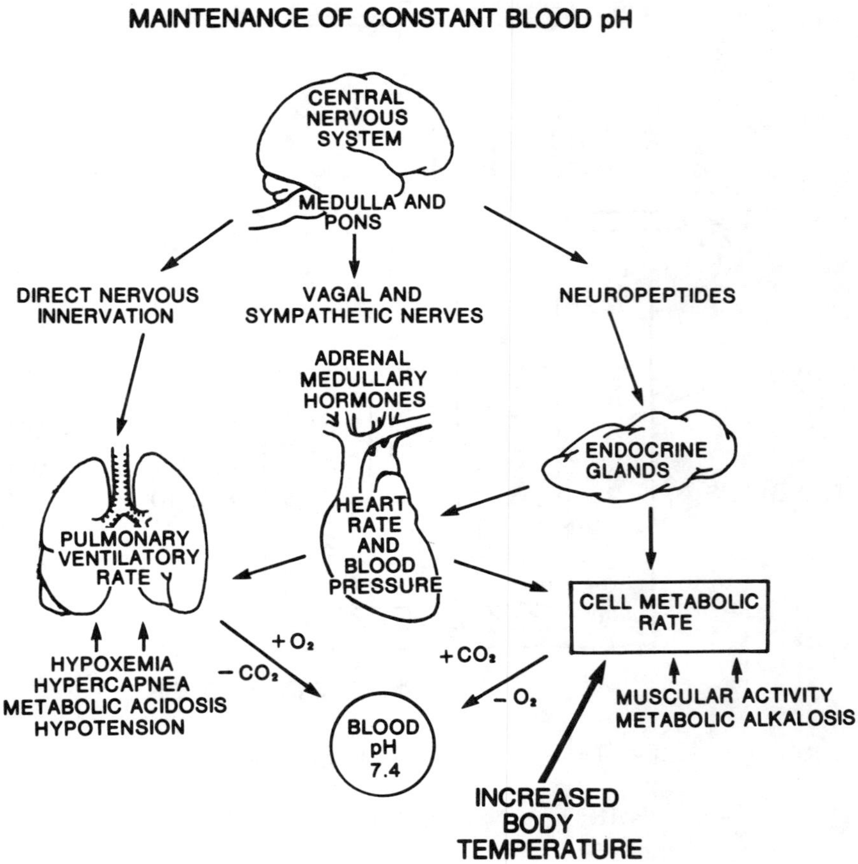

FIGURE 3. Blood pH is maintained constant at about pH 7.4 by a balance between the flow of CO_2 to the blood and the rate of excretion at the lungs. This balance is maintained by varying the breathing rate and metabolism rate. The central nervous system responds to minute changes in $[H^+]$, and to stimuli via arterial receptors sensitive to oxygen tension and blood pressure, to maintain this balance. This is an example of a servo-mechanism.

exercise, a period of mismatch exists at first when a person has breathing difficulties. As the blood flow from the heart catches up, individuals get their ''second wind'' and breathe easier.

The heart rate is controlled by its own feedback system. Thus, the heart will continue to beat even if removed from the body. Superimposed upon this system are outside controls (Figure 3). Secretion of adrenaline in stress will cause an increase of the heart rate. This is balanced by the action of the vagus nerve (parasympathetic innervation) which prevents the heart from beating too rapidly. Carotid and aortic receptors sense the blood pressure and react on the medulla, to adjust the movement of the diaphragm and intercostal muscles through the phrenic and intercostal nerves, so as to match the blood flow through the lungs. This system permits adjustment of the pulmonary ventilation rate to match the blood flow rate. *The net result is the blowing off of CO_2 from the blood.*

The amount of CO_2 blown off is substantial. At rest, each exchange represents about 500 ml of air at 38°C containing about 5% CO_2. This represents 25 ml of pure CO_2. An individual at rest respires about 16 times per minute. Thus, about 400 ml of anhydrous CO_2 is expelled per minute. At standard conditions (0°C, 760 mmHg), this represents about 350 ml, or dividing by 22.2 ml (the CO_2 gram molecular volume/1000), *about 15.8 mmol CO_2 is excreted per minute.*

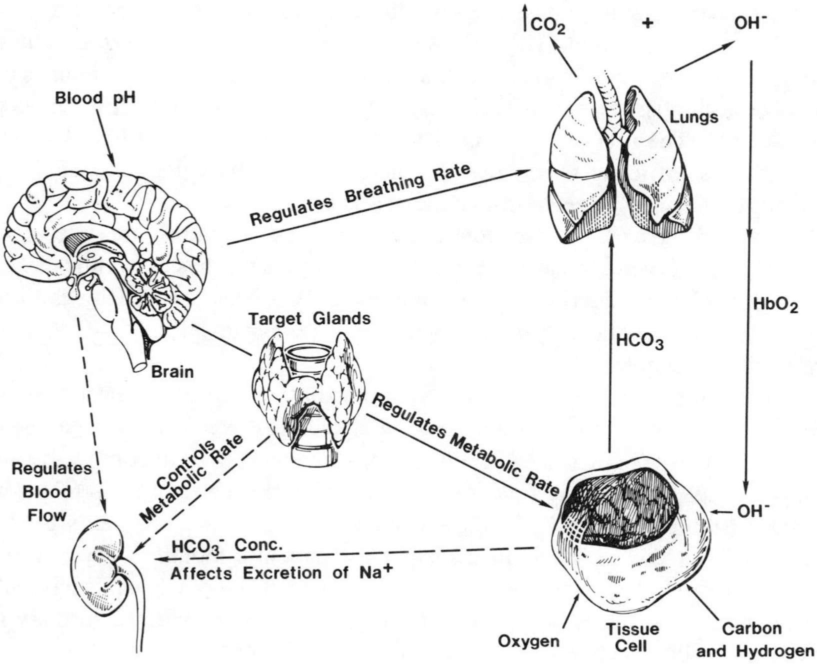

FIGURE 4. Maintenance of constant pH at the lungs and cells. CO_2 generated at the cells unites with OH ion to form bicarbonate. This reduction in pH is balanced by conversion of the acid HbO to basic Hb. At the lungs, the bicarbonate is decomposed to liberate CO_2 and regenerate the OH ion, whose effect in raising the pH is balanced by the conversion of basic Hb to acidic oxy-Hb. The balance is controlled by the central nervous system by its action on the ventilation rate and rate of cell metabolism and, thus, CO_2 generation. (From Natelson, S. and Natelson, E. A., *Principles of Applied Clinical Chemistry,* Vol. I, Plenum Press, New York, 1975, 73.)

The CO_2 derives from the metabolism of the nutrients by the cells. The rate of metabolism of the cells is controlled by the release of certain hormones from the pituitary gland. The secretion of the hormones of the pituitary gland is controlled, in turn, by the secretion of low molecular weight neuropeptides from the hypothalamic tract. The pituitary hormones stimulate the release of hormones from the target glands. These hormones from the target glands regulate cell metabolism and, thus, the rate of CO_2 production. As an example of this process, thyrotropin-releasing hormone (TRH) from the hypothalamus stimulates the release of thyrotropin from the pituitary, which causes, in turn, the release of thyroxin from the thyroid to increase metabolic rate and, thus, CO_2 production. Muscular activity also has a direct influence on both the rate of CO_2 production and pulmonary ventilation rate. *The net result is a regulated flow of CO_2 into the lungs which expels it at the same rate that it is produced in order to maintain a constant blood pH.* How this is done, without producing an acidemia at the tissues and alkalemia at the lungs, is illustrated in Figure 4.

In Figure 4 it can be seen that the OH ion shuttles between the lungs and tissues. At the tissues it combines with CO_2 to form bicarbonate ion, as indicated in the figure. At the lungs bicarbonate decomposes to OH ion as CO_2 is blown off:

$$\text{at tissues:}\quad HO^- + CO_2 \rightarrow HCO_3^-$$

$$\text{at lungs:}\quad HCO_3^- \rightarrow OH^- + CO_2$$

It is important to note that while bicarbonate is a base, according to the Brönsted theory, hydroxyl ion is a stronger base. *Thus, in Figure 2, an increase in bicarbonate concentration (addition of CO_2) acidifies the solution.*

At the lungs, hemoglobin, a base (pK = 7.85), combines with oxygen to form oxyhemoglobin (pK = 6.60), an acid. Thus, the OH ion generated at the lungs is neutralized by oxyhemoglobin. On the other hand, the OH ion generated, when oxyhemoglobin is converted to hemoglobin at the tissues, reacts with CO_2 to form bicarbonate ion. This serves to buffer any change in pH. The role of the kidney, referred to in Figure 4, will be discussed later. At this point, it is sufficient to say that the blood flow to the kidneys is regulated by the central nervous system by mechanisms indicated in Figure 3.

Figures 2 and 4 indicate that the central nervous system controls the rate of breathing and, thus, CO_2 and oxygen exchange. Figure 4 also indicates that by controlling cell metabolic rate, the rate of CO_2 production is also controlled. Feedback, sensed by the blood pH, regulates the balance of these systems to maintain a constant pH at 7.40. Compare this system with that of Figure 2.

The concepts illustrated in Figure 4 have immediate practical applications to clinical medicine. From the figure it can be seen that in alkalemia the cells are encouraged to raise their metabolic rate to a higher level in order to produce more CO_2 to combat the alkalemia. In so doing, the cell's metabolic functions, in general, are accelerated. Secretory cells will secrete greater volumes of fluid. On the other hand, in acidemia, the cell's production of CO_2 will be slowed down. Thus, in acidemia the patient's mouth is dry, since salivary secretion is decreased. Urine output is also decreased. On the other hand, with alkalemia, the mouth is moist and the patient salivates. If not severely dehydrated, urinary output is increased. This is illustrated with a typical patient[7] in Figure 5 and Table 1.

An appendectomy was performed on a 12-k female child. In the immediate post-operative period, intestinal obstruction occurred. A nasogastric suction tube was inserted, and aspiration of fluids was initiated to reduce distension. This was not completely effective and an ileostomy was performed on the 11th post-operative day. Fluids were given intravenously to replace the K, Na, Cl, and water losses. In spite of increased i.v. fluids and electrolytes, the rising voluminous output could not be controlled. The blood hematocrit was maintained with small amounts of blood. B-complex vitamins and Vitamin C were given daily. Table 1 lists the volume of fluids and their composition of the 17th day.

Administration of NH_4Cl (10 g) to the fluids brought the blood pH down to 7.48 from 7.65. This resulted in a dramatic decrease in fluid output. Addition of NH_4Cl (2 g) on the 18th day brought the pH into the normal range, and the fluids were then readily controlled. The patient's condition improved so that she could be returned to surgery where intestinal continuity was restored. Recovery was uneventful.

When NH_4Cl, KCl, or NaCl is administered to a patient with alkalemia, the cations are excreted, and the chloride ion serves to correct the alkalemia. In severe cases, like that of Figure 5, NH_4Cl will act more quickly since its pH is at 4.5, while NaCl and KCl are neutral. The NH_4 ion is readily excreted as such, conserving Na and especially K ion. It is also converted to urea and excreted.

THE BUFFERING ACTION OF BLOOD

Buffers

If 0.5 mmol of acetic acid is dissolved in 20 ml of water, and the solution is titrated with 0.05 *M* NaOH, it will take 10 ml of the NaOH to neutralize the acetic acid to form sodium acetate to reach a pH of about 9. If the pH is measured repeatedly as the NaOH is slowly added, a curve such as in Figure 6 will be obtained. Sørensen obtained this type of curve by plotting the negative logarithm of the hydrogen ion concentration against the milliliters of NaOH added. He used a hydrogen electrode to measure hydrogen ion concentration. The shape of the curve of Figure 6 derives directly from the equation for the ionization of an acid.

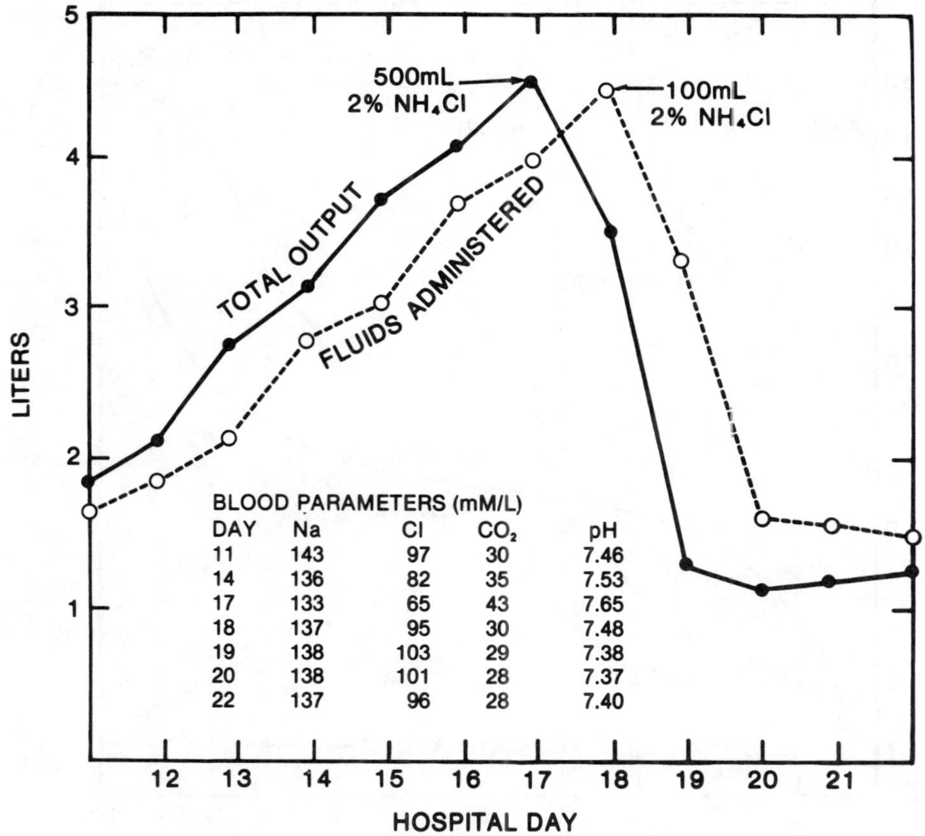

FIGURE 5. Effect of blood pH on rate of secretion and excretion of body fluids. In alkalosis cells are encouraged to metabolize more rapidly and generate more CO_2. They thus generate more fluids. In acidosis cells are inhibited from normal CO_2 production and, thus, secrete less fluids. Adjustment of blood pH to normal encourages normal urine flow and normal flow of drainage fluids. (From Natelson, S., *Clin. Chem.*, 4, 32, 1958.)

Table 1
VOLUME AND COMPOSITION OF FLUIDS EXCRETED ON THE 17TH DAY[7] BY THE PATIENT DESCRIBED IN FIGURE 5

		Composition (m*M*/l)		
Fluid	**Volume (ml)**	**Na**	**Cl**	**K**
Urine	900	84	105	11.0
G I drainage	2070	126	134	3.5
Ileostomy	1230	145	130	4.0
		Total lost (mmol)		
Total	4200	514	531	22.[illegible]

$$\frac{[H][Ac^-]}{[HAC]} = K$$

Sørensen rearranged this equation to solve for $[H]^+$, to obtain

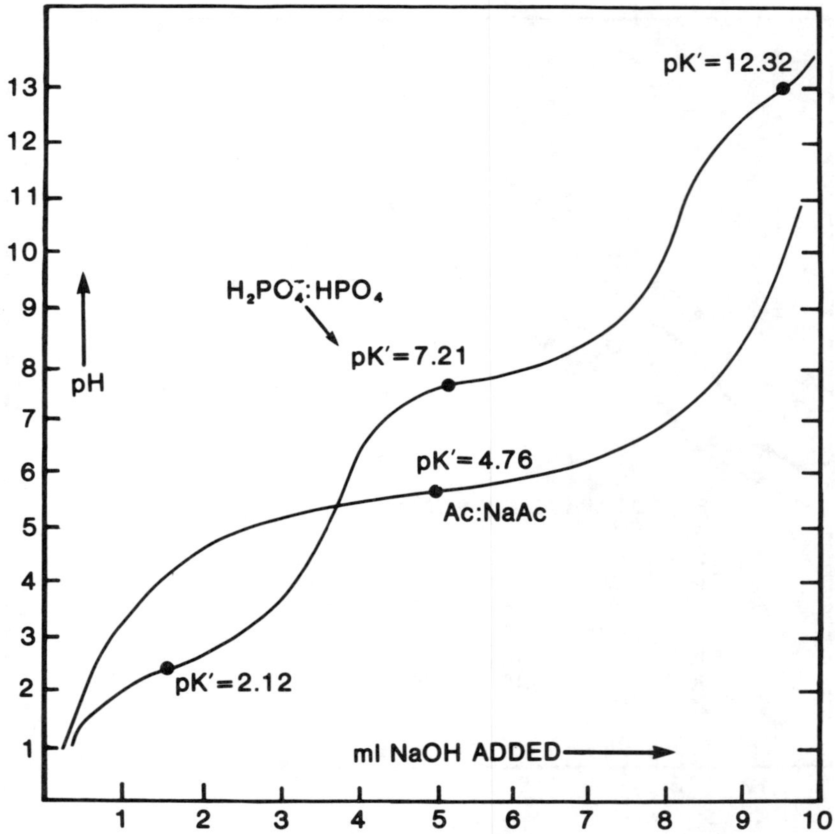

FIGURE 6. Plot of the equation pH = pK′ + log (Ac^-/HAc) illustrating the action of a buffer. A similar buffering effect will be noted if the equation $[H^+] = K' \, HAc/Ac^-$ is plotted. For phosphoric acid there are three pK values, one for each H^+.

$$[H^+] = K \times \frac{[HAC]}{[Ac^-]}$$

Taking the negative logarithm of both sides of the equation, we obtain

$$-\log[H^+] = -\log K - \log \frac{[HAC]}{[Ac^-]}$$

Replacing the negative logarithm of the hydrogen concentration with pH, and the negative logarithm of the ionization constant with pK′, and inverting the fraction to obtain a positive sign for the logarithm, we obtain

$$pH = pK' + \log \frac{[Ac^-]}{[HAC]} \qquad (1)$$

If the components of the fraction are expressed in terms of concentration instead of activities, then pK becomes pK′.

This is the Sørensen buffer equation. If plotted for acetic acid it will yield the curve of Figure 6. From Figure 6 it can be seen that between the addition of 3 to 7 ml of alkali, there is very little change in pH. This explains the buffering action of the titration curves

Table 2
DISSOCIATION CONSTANTS IN WATER (ACIDS)

Acid	pK'_1	Acid	pK'_1
Acetic	4.75	Malic	
Boric		(1)	3.40
(1)	9.14	(2)	5.11
Carbonic		Oxalic	
(1)	6.37	(1)	1.23
(2)	10.25	(2)	4.19
Citric		Phenol	9.89
(1)	3.08	Phosphoric	
(2)	4.74	(1)	2.12
(3)	5.40	(2)	7.21
Glycine	9.87	(3)	12.67
Hepes (*N*-2-hydroxyl ethyl-) piperazine-*N*′-2-ethane-sulfonic acid	7.5	Tartaric	
		(1)	2.98
		(2)	4.34

Table 3
DISSOCIATION CONSTANTS IN WATER (BASES)

Base	pK_{H^+}	pK_{OH^-}	Base	pK'_{H^+}	pK'_{OH^-}
Ammonia	9.25	4.75	Triethanol-amine	7.76	6.24
Ethanolamine	9.44	4.56	Triethyl-amine	10.76	3.24
Picoline	6.48	7.52	Tris(hydroxymethylaminomethane)	8.08	5.92
Piperazine					
(1)	9.81	4.19			
(2)	5.57	8.43			
Piperidine	11.21	2.79			
Pyridine	5.23	8.77			

of weak acids with strong bases, or weak bases with strong acids. If KOH solution is titrated with dilute HCl, a linear plot downward will be obtained until a pH of about 2 is reached, before there is some flattening of the curve and buffering action. For this reason, strong acid-strong base buffering systems are of lesser value in biological applications.

An ideal buffer for many biological applications would have its pK′ value at about pK′ = 7.4. The Na_2HPO_4/NaH_2PO_4 system meets this criterion, since its pK′ value is at about 7.2 (Figure 6). The pK′ values of various acids are shown in Table 2. For bases, the pK_{H^+} values are those for the quaternary ammonium ion (e.g., $[C_2H_5]_3NH^+$), since they are Brönsted acids. Subtracting from 14 gives the pK_{OH} value (Table 3).

From the Sørensen buffer equation it is apparent that when the concentration of $[Ac^-]$ is equal to that of [HAc] in the fraction, then the fraction has a value of 1. Since the logarithm of 1 is 0, then pH = pK′. The pK′ is, thus, a point of inflection from a concave to a convex curve and is the point where pH change is at a minimum, yielding maximum buffering action. The pK values shown in Tables 2 and 3 thus serve as a guide in selecting a buffer for a particular purpose. Handbooks of chemistry and physics have extensive tables for this purpose. Catalogs of commercial sellers of chemicals also list the buffer ranges of selected acids and bases (e.g., Sigma Chem. Co.).

The Henderson-Hasselbalch Equation

Hasselbalch applied the buffer equation to the blood bicarbonate system and wrote[2]

$$pH = pK' + \frac{[NaHCO_3]}{[H_2CO_3]} \qquad (2)$$

It was assumed that the ionization of $NaHCO_3$ was 100%. The ionization of $NaHCO_3$ at pH 7.4 in a 0.05 *M* solution is closer to 80%. The degree of ionization of H_2CO_3 is only 0.17% at 0.1 *M* concentration as compared to 82% for HNO_3 at 18°C. The activity coefficient for a 0.1 *M* NaCl solution is about 0.72.[8] Thus, it acts as though it were ionized to the extent of 72%. The assumption of 100% ionization did not invalidate the equation. It merely changed the value of pK'.

For aqueous solutions of $NaHCO_3$, the $[H_2CO_3]$ is proportional to the dissolved CO_2 concentration, which, in turn, is proportional to the p_{CO_2}. This can be represented as follows:

$$[H_2CO_3] = K_1 \times (\text{dissolved } CO_2) = K_2 \times p_{CO_2}$$

$$[H_2CO_3] = K_2 \times p_{CO_2}$$

K_2 is actually the millimoles of CO_2 which dissolve in 1 l of water at 1mmHg pressure, when p_{CO_2} is expressed in millimeters of mercury and concentrations are expressed in millimoles/l. K_2 is represented by S and has the value of 0.0323 for water, and this can be substituted in the Henderson-Hasselbalch equation. HCO_3^- is substituted for $NaHCO_3$, assuming 100% ionization. This is usually written as

$$pH = pK_1 + \log \frac{[HCO_3^-]}{S' \, p_{CO_2}} \qquad (3)$$

where S is the solubility of CO_2 in the medium expressed in millimoles CO_2 per liter at 1 mmHg and 37°C. For water it is 0.0323 and for normal human serum it is 0.0302.

Van Slyke et al. measured the partition coefficient α of CO_2 in water at 38°C and 760 mmHg pressure.[9] They found that 0.5455 ml CO_2 dissolved in 1 ml H_2O, or 545.5 ml/l. Dividing by 22.2, this calculates to 24.56 mmol/l at 760 mmHg. Dividing by 760 we obtain 0.0323 mmol/l/mmHg pressure. For plasma, the solubility is less and $\alpha = 0.51$. The solubility (S) then calculates to 0.0302 mmol/l/1 mmHg. Since the Henderson-Hasselbalch equation is intended to apply to blood, this value is usually seen substituted for S in the equation.

With the hydrogen electrode, pH could be measured with CO_2 gas added, at a pressure of 40 mm, to the flowing hydrogen of the electrode. With a known p_{CO_2} of 40 mm, aqueous solutions of $NaHCO_3$ could be examined for the value of pK'. Total CO_2 content at equilibrium was measured gasometrically. The $[HCO_3^-]$ could be calculated from the total CO_2 and the value for S.

$$[HCO_3^-] = CO_2 \text{ content} - 0.0323(40) = CO_2 \text{ content} - 1.292$$

With this information, pK_1' for water at 38°C for the bicarbonate system was found to be 6.37. The subscript 1 refers to the first dissociation constant of the carbonic acid system.

The First Dissociation Constant of the Blood Bicarbonate Buffering System (pK_1')

Using the technique described above, with the hydrogen electrode simultaneously tonometering the blood at $p_{CO_2} = 40$ mm, Hasselbalch[2] calculated in 1916, a value of $pK_1' = 6.22$ on one blood and 6.145 on another. Warburg restudied this problem in 1922 by essentially the same technique, and obtained $pK_1' = 6.15$ at 38°C for human blood with tonometered blood.[10] Van Slyke used horse serum and determined the pK'_1 as 6.13, and found 5.93 for hemolyzed cells at 38°C.[11] In 1937 the glass electrode was used by Dill, who found a pK'_1 of 6.11 for serum and 5.97 for hemolyzed cells. These studies were done with tonometered hemolyzed cells or serum.[12]

In 1925 Cullen, using tonometered blood with the hydrogen electrode, came to the conclusion that consistent results could not be obtained with whole blood, and went over to tonometered serum and plasma.[13] It was difficult to measure the pH at that time. For example, Cullen found 7.276, 7.317, and 7.371 for the pH of healthy dogs blood. These are in significant error. He found values of pK'_1 in tonometered plasma ranging from 6.007 to 6.133.

In 1928 Hastings, using the hydrogen electrode with tonometered serum, found a pK'_1 in patients who were chronically ill, ranging from 6.098 to 6.134. He also used tonometered serum to avoid the need for measuring p_{CO_2}. Hastings' pH values for normal human blood were also in error. In six healthy controls he found pH values ranging from 7.225 to 7.510.[11]

Cullen restudied this problem with the glass electrode, and again obtained the wrong pH for serum from healthy humans and dogs. For example, he found pH values ranging from 7.227 to 7.416 on tonometered serum (p_{CO_2} = 40 mm) from a single healthy dog over a 12-d period.[14]

It is apparent from the discussion above and especially Figure 4 that circulating blood does not qualify as an equlibrium system, but is maintained in a steady state, arbitrarily at pH = 7.4 where the human can best function. Another pH could have been chosen. For example, in a squid which is found in Japanese waters, the blood pH is less than 5.0.

Even if blood is removed to a test tube, the glucose continues to be metabolized. When this is exhausted, K ion moves out of the cell. *Thus, equilibrium laws do not apply to such a system. An equilibrium is one in which there is one and only one point of equilibrium at constant temperature.* Systems like a constant temperature bath are in a *steady state* or in "homeostasis", since the temperature is under the control of an outside mechanism. If the plasma or serum is separated and equilibrated with a gas of constant p_{CO_2}, this qualifies as an equilibrium system. *However, this p_{CO_2} is not the p_{CO_2} of the blood, and the pK'_1 is not the pK'_1 of the blood drawn originally.*

Van Slyke introduced the concept of "*CO_2 combining power*" for plasma, tonometered at p_{CO_2} of 40 mm, where the total CO_2 content was then measured. Usually, the residual gas from a technician's lungs was blown through the plasma, on the assumption that it had a P_{CO_2} value of 40 mm. The objective was to find a practical way of measuring blood bicarbonate ($NaHCO_3$) concentration or "buffer base".

The bicarbonate and, thus, residual sodium bicarbonate available for buffering was

$$HCO_3^- = CO_2 \text{ combining power} - (0.03 \times 40)$$
$$HCO_3^- = CO_2 \text{ combining power} - 1.20$$

The pH could then be calculated as follows:

$$pH = 6.1 + \log \frac{CO_2 \text{ combining power} - 1.2}{1.2} \quad (4)$$

Even if pK'_1 was constant, there was a fallacy in this approach. The pH calculated was not that of the original blood, but that of the tonometered plasma. For example, if the original p_{co_2} was 15 or 60 mm, the calculated pH and HCO_3^- levels would be in significant error.

Van Slyke and students set about designing experiments to explain the variation in the pK'_1 in blood and its difference from that of the bicarbonate system in water. The factors influencing the pK'_1 value of tonometered plasma of serum were then studied.

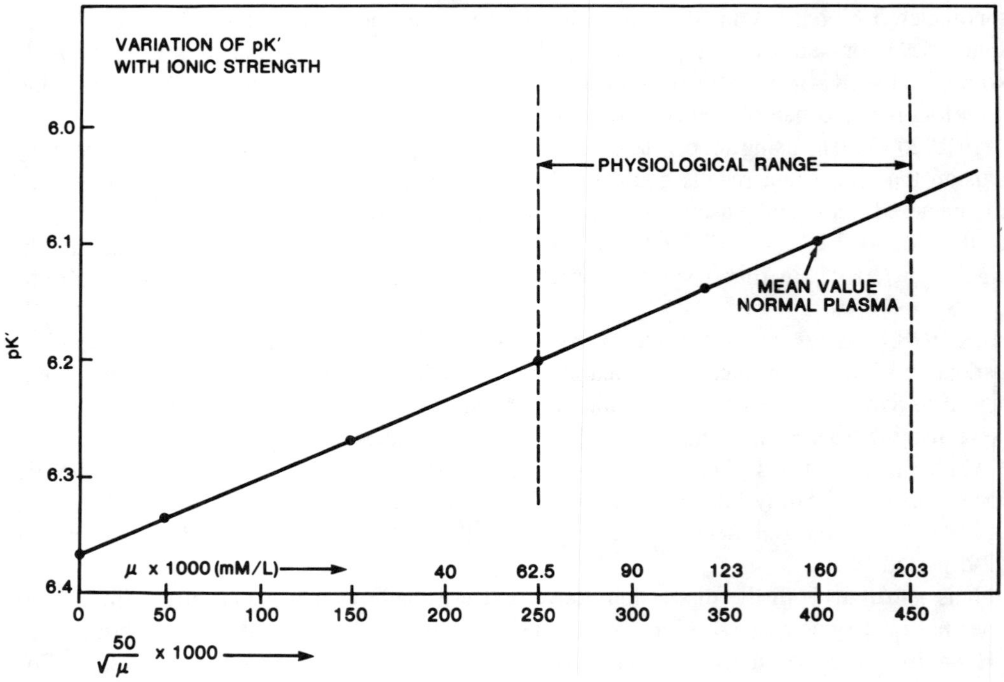

FIGURE 7. Variation of pK′ with ionic strength.[15]

Effect of Ionic Strength on pK_1'

Several studies have been carried out measuring the variation of pK_1' of the bicarbonate system with the ionic strength of the solution (μ).[15-17] The results are in general agreement and are summarized in Figure 7.

The ionic strength (μ) of serum is about 0.167 *M*. At this ionic strength, it can be seen from the figure that pK_1' has a value of about 6.1. Water has an ionic strength of 0, and in water the bicarbonate system has a pK_1' of about 6.37.

The values for ionic strength observed in the clinical laboratory for plasma range from about 0.06 to about 0.2 *M*. At these extremes the pK_1' varies from about 6.06 to about 6.21. The high ionic strength is seen in "Neurogenic Hyponatremia and Hypochloremia".[18] The low ionic strength is seen with adrenal insufficiency as examples.

From Table 4 it can be seen that if the bicarbonate concentration is calculated using 6.1 as the pK_1', then for the "high salt syndrome" an error of about 26% in calculated bicarbonate would be obtained. This would translate to an error of about 26% in calculated base excess and repair fluid requirements.

Effect of pH on the Value of pK_1'

Figure 8 illustrates the variation of pK_1' with blood pH as observed in dogs who were given $NaHCO_3$ or NH_4Cl to create an alkaloses or acidosis, respectively.[19] Similar observations have been made by others in *in vitro* experiments.[16,17,20] The extreme pH values seen in patients' blood in the clinical laboratory range from 6.9 to 7.7. Extrapolating from the data reported, a pK_1' of about 6.35 would be observed in severe acidosis and about 6.0 in severe alkalosis. Referring to Table 4, an error of as much as 80% in the bicarbonate concentration would be calculated if a pK_1' of 6.1 were assumed in the calculation in severe alkalosis.[21,22]

The variation of pK_1' with pH of the bicarbonate system is probably due to the change of activity coefficient with pH. If the dissociation of $NaHCO_3$ were 100% as assumed in the

Table 4
EFFECT OF CHANGE OF pK'_1 FROM 6.1 ON THE % CHANGE OF CALCULATED BICARBONATE CONCENTRATION USING THE HENDERSON-HASSELBALCH EQUATION

pK'_1	% Variation of $[HCO_3]^-$	pK'_1	% Variation of $[HCO_3^-]$
6.34	+86.0	6.08	−4.5
6.32	+66.0	6.06	−8.8
6.30	+58.5	6.04	−12.9
6.28	+51.4	6.02	−16.8
6.26	+44.4	6.00	−20.6
6.24	+38.0	5.98	−24.1
6.22	+31.8	5.96	−27.6
6.20	+25.9	5.94	−30.8
6.18	+20.2	5.92	−33.9
6.16	+14.8	5.90	−39.7
6.14	+9.1	5.88	−42.5
6.12	+4.7	5.86	−45.0
6.10	0.0	5.84	−47.5

Adapted from Natelson, S. and Nobel, D., *Clin. Chem.*, 23, 767, 1977.

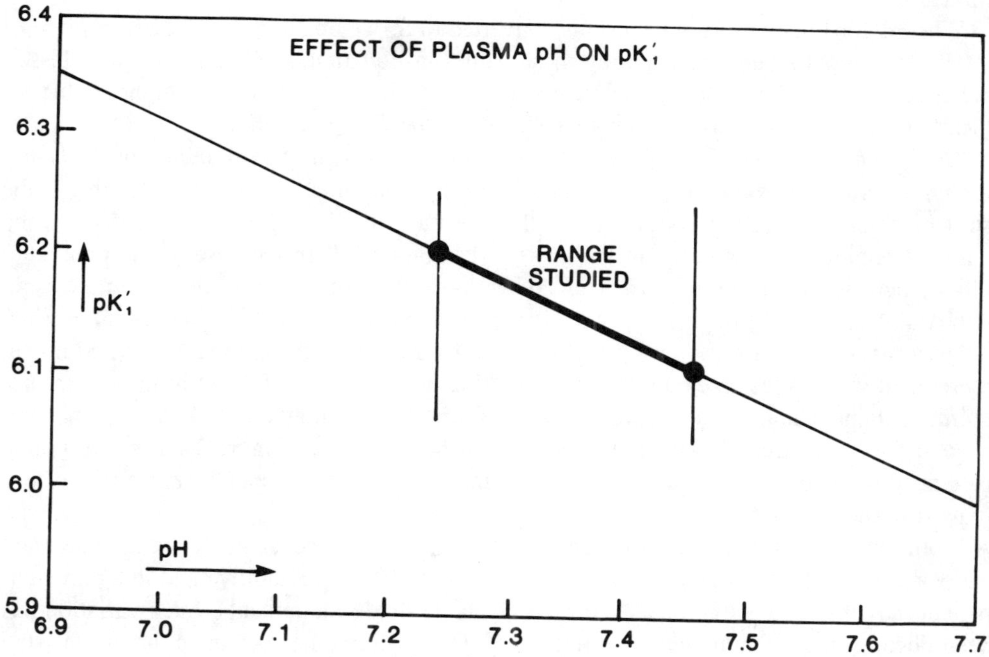

FIGURE 8. Effect of plasma pH on pK'_1 *in vivo.*[19]

derivation of the Henderson-Hasselbalch equation, the pH should not have a significant effect. However, the activity coefficient of the bicarbonate ion at a concentration of 0.05 is 0.82; that for the sodium ion is also 0.82.[8] The ionic activity of $NaHCO_3$ is then $\sqrt{(0.82)\ (0.82)} = 0.82$. Thus, $NaHCO_3$ at the approximate concentration that it is in serum acts as though it is 82% dissociated.

Table 5
THE EFFECT OF DIFFERENT SALTS ON THE SOLUBILITY OF CARBON DIOXIDE IN WATER AT A CONCENTRATION OF 0.15 *M*

Compound	Alpha (α) at 38°C	% Depression of Solubility from 0.5455 ml/ml H_2O
NaCl	0.4345	20.7
KCl	0.4545	15.9
Na lactate	0.3905	28.4
NaH_2PO_4	0.3275	40.2

Adapted from Van Slyke, D. D., Sendroy, J., Jr., Hastings, A. B., and Neill, J. M., *J. Biol. Chem.*, 78, 765, 1929.

Effect of Other Factors on the Value of pK_1' of the Bicarbonate System

The *solubility coefficient (α)* for CO_2 in normal human plasma or serum at 38°C is 0.506.[9,23,24] This varies with the salt concentration and other factors (Table 5). Lipids in serum will tend to increase the solubility of CO_2 in plasma and, thus, increase the solubility coefficient (α).

In one lipemic serum studied, the value of α was 0.551, the value of which is significantly higher that the α value even for water (0.5455). The value of 0.506 for serum is higher than what would be expected from its salt composition. This is due mainly to its normal lipid content.

When α for lipemic serum (0.551) is converted to S, we obtain S = 0.0327, in place of 0.030 for normal serum. Substituting these values in Equation 4 of the Henderson-Hasselbalch equation, and calculating the bicarbonate concentration, we obtain a higher value for bicarbonate in lipemic serum of about 10% due to the change in pK_1'.

The hematocrit value, hemoglobin concentration, and extent of oxygenation of the hemoglobin all affect the value of pK_1' as calculated for whole blood. One reason for this is the shift of chloride from the erythrocyte to the plasma as the blood is oxygenated when the basic hemoglobin is converted to the acid oxyhemoglobin.[25] In the case of the premature infant, hematocrit values of 65% and higher have been observed while the infant is in hypoxia. Thus, this factor is of special significance in the acutely ill premature infant.[26,260]

Urea levels in serum vary from as low as about 2 mmol/l (5 mg urea N per 100 ml) in a pregnant woman, as an example, to about 70 mmol/l (200 mg/100 ml) in the serum of a uremic patient. Urea acts as a cation, since it forms salts with anions and can significantly change the ionic strength and, thus, the pK_1' of the serum (see Figure 7). For the uremic patient, this calculates to an underestimate of the bicarbonate of about 13% if a pK_1' of 6.1 is used in the calculations.

From the discussion above, it is apparent that pK_1' is not a constant in human blood and can vary substantially in an acutely ill patient. These effects are additive and in a particular patient may result in a significant error in calculated fluid requirements. Such variation has been documented by numerous investigators.[19,21,22,25-30] Some have claimed that usually the error in calculated bicarbonate from the Henderson-Hasselbalch equation is small enough to be disregarded[31-35a] if one averages results. Unfortunately, the composition of a particular blood sample from an acutely ill patient may be significantly out of the range of the averaged value. It is not acceptable procedure to depend on average values for the management of a particular acutely ill patient. See especially Reference 26a.

Fortunately, methods for rapid and direct measurement of total CO_2 in serum (an approximation of the bicarbonate level) are now readily available. The original motive for

Table 6
VARIATION OF VAPOR PRESSURE OF WATER WITH TEMPERATURE

Temp (°C)	VP (mmHg)	Temp (°C)	VP (mmHg)	Temp (°C)	VP (mmHg)	Temp (°C)	VP (mmHg)
16	13.6	23	21.1	30	31.8	37	47.1
17	14.5	24	22.4	31	33.7	38	49.7
18	15.5	25	23.8	32	35.7	39	52.4
19	16.5	26	25.2	33	37.7	40	55.3
20	17.5	27	26.7	34	39.9	41	58.3
21	18.7	28	28.3	35	42.2	42	61.5
22	19.8	29	30.0	36	44.6	43	64.8

calculating base excess from the Henderson-Hasselbalch equation is now moot, since Na, K, Cl, pH, p_{CO_2}, p_{O_2}, and lactate levels can be measured rapidly, and fluid requirements can be calculated from these data. These procedures are more reliable and reproducible than those obtained by calculation from pH and p_{CO_2} levels in the Henderson-Hasselbalch equation.

PARTIAL PRESSURE

P_{CO_2} and P_{O_2}

The total pressure of a gas mixture is the sum of the partial pressures of the component gases. Thus, air is composed of 78.08% nitrogen, 20.95% oxygen, 0.033% CO_2, and 0.934% argon. Other gases, such as the rare gases xenon, neon, and krypton, are present in trace amounts. Water is present in variable amounts. The partial pressure of water in air saturated with water depends upon the temperature (Table 6). At 37°C, this is 47.1 mm. The hygrometer reads the *relative humidity,* which is the % saturation of air with water. Thus, when the relative humidity is 60%, then the partial pressure of H_2O in air at 37°C is 28.26 mm, when the total air pressure is 760 mm.

Neglecting the rare gases, the total pressure (P_{air}) of air may be represented as follows:

$$P_{air} = p_{N_2} + p_{O_2} + p_{CO_2} + p_A + p_{H_2O}$$

The *vapor pressure* of an aqueous solution is the partial pressure of water in air, usually in millimeters of mercury. The *vapor tension* is the escaping tendency of water from a solution, in millimeters of mercury. In a sealed system, such as a solution in a stoppered bottle, the vapor pressure is numerically equal to the vapor tension. When blood is injected into the chamber of an instrument which measures the vapor tensions for CO_2 and O_2, the results are reported as the partial pressures of these gases. In scientific journals, the partial pressure is put in lower case, while the component referred to is the subscript (viz., p_{O_2}). However, in the clinical chemical literature, the partial pressures are often indicated by a lower case p, but the component gases are written for convenience, at the same level as the lower case p (viz., pCO_2, pO_2). This has the disadvantage in that this type of designation has been usurped to mean the negative logarithm of a parameter, such as pK or pH.

In the alveoli, the air is saturated with water, p_{O_2} is lowered to about 100 mm, and the p_{CO_2} is increased to about 40 mm at 37°C. A comparison of the composition of inhaled air (760 mmHg, 25°C, 60% humidity, as an example) with air which has been held in the lungs for a few minutes and then expelled may be represented as follows:

Table 7
PARTIAL PRESSURES OF BLOOD AND ALVEOLAR GASES AT 37°C (mmHg)

	Alveolar gas	Arterial blood	Venous blood
p_{O_2}	104	100	40
p_{CO_2}	40	40	46
p_{H_2O}	47	47	47
$p_{(N_2+A)}$	569	573	573
p (total)	760	760	706

Adapted from Arndt and Dolle, W., *Klin. Wochenschr.* 44, 511, 1966.

$$\text{(inhaled)} \quad P_{air} = 760 \text{ mmHg}$$
$$= p_{(N_2+A)}(589) + p_{O_2}(156) + p_{CO_2}(0.25) + p_{H_2O}(14.3)$$
$$\text{(exhaled)} \quad p_{air} = 760 \text{ mmHg}$$
$$= p_{(N_2+A)}(569) + p_{O_2}(104) + p_{CO_2}(40) + p_{H_2O}(47)$$

Venous blood contains less oxygen than arterial blood and the differences are shown in Table 7.

The Bohr Effect

If the amount of gas dissolved in an inert liquid is plotted against varying pressures, then a straight line is obtained, indicating proportionality between the amount of gas dissolved and its partial pressure. This phenomenon is known as Henry's Law. If a reaction occurs between the solvent and the gas, this law does not apply. This is the case for blood and oxygen, since the oxygen reacts with the hemoglobin of the blood. Thus, an S-shaped curve is obtained (Figure 9). There are four equilibrium constants involved, since there are four heme residues in hemoglobin. As each heme is converted to oxyheme, it affects the other constants. A simplified equation was devised by Hill in 1910, which is still in general use today, to approximate this curve using a single constant:[3,5]

$$y = \text{fraction Hb saturated} = \frac{\%\ HbO_2}{100} = \frac{Kp^n}{1 + p^n} \tag{5}$$

The partial pressure of oxygen (p_{O_2}) is designated by p. The % HbO_2 represents the % Hb saturated with O_2. The value of n varies from 1.5 to 3, depending upon the Hb concentration and other factors. In 0.1 *M* phosphate buffer at the normal blood pH (7.4), it is about 2.7 to 2.8.

From Table 7 it can be seen that at p_{O_2} = 100 mm (the partial pressure of O_2 in the alveoli), the blood hemoglobin is almost completely saturated. On change in pH, the shape of the curve is altered. This is the "Bohr effect". The curve is characterized by the partial pressure of oxygen required to result in 50% of the Hb being saturated. This is known as the P_{50} value of the curve. This value will vary not only with pH, but with other factors as well, and the significance of the P_{50} value will now be explained.

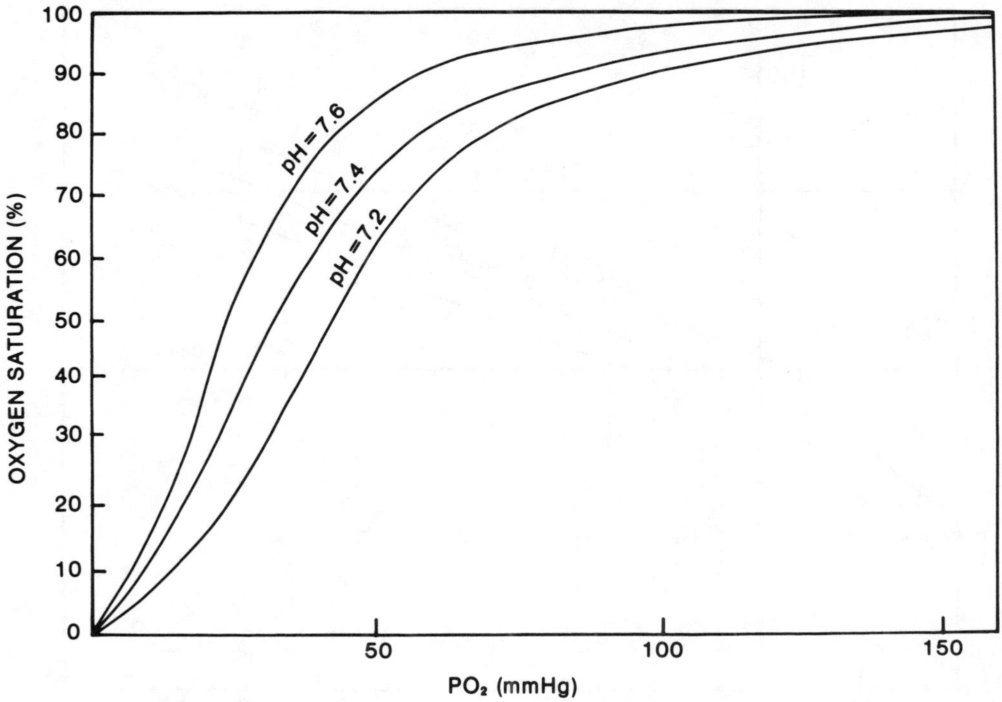

FIGURE 9. Plot of the % oxygen saturation of hemoglobin with change in p_{O_2}. Effect of pH on the shape of the curve (the Bohr effect).

The P_{50} Value

Equation 5 can be rearranged to yield:

$$\frac{y}{K - Ky} = p^n \tag{6}$$

At 50% saturation y = 0.5 and P_{50} is the partial pressure of oxygen at which half-saturation occurs. Thus, in Equation 6 we get

$$\frac{0.5}{K - 0.5K} = \frac{1}{K} = P_{50} \tag{7}$$

Dividing Equation 6 by Equation 7, and taking the logarithm of both sides of the equation, we obtain Equation 8:

$$\log \frac{y}{1 - y} = n \log \frac{P}{P_{50}} \tag{8}$$

P_{50} is a constant, and thus if we plot the log of $[y/(1-y)]$ against log p, we get a straight line.

When y, the fraction of Hb saturated is 0.5, then p is the P_{50} value. However, at this point, the log $(y/1-y)$ is the logarithm of 1 or 0. The intercept at the ordinate is, thus, the P_{50} value. Figure 10 illustrates how this is calculated.

The P_{50} value will be different for some abnormal hemoglobins and this value will serve to aid in the characterization of the hemoglobin. The P_{50} value is also controlled by the

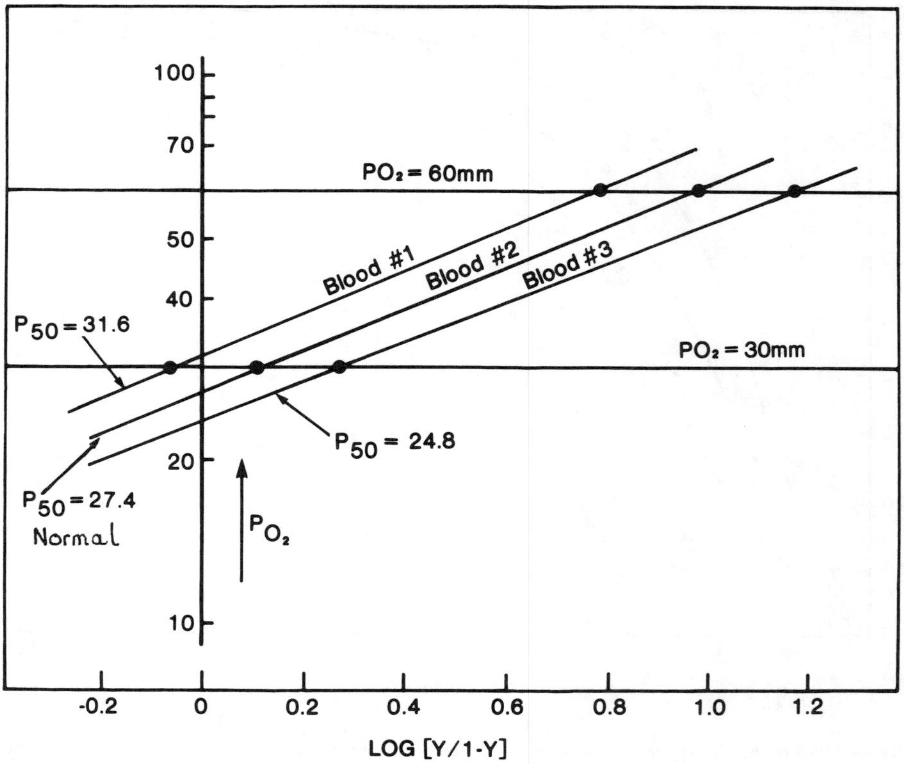

FIGURE 10. Plot on semilog paper of the logarithm of p_{O2} vs. log [y/(1−y)] to obtain the P_{50} value. Blood is equilibrated with gas at 30 and 60 mmHg of oxygen, and the fraction oxygenated (y) is determined. (From Natelson, S. and Natelson, E. A., *Principles Applied Clinical Chemistry*, Vol. 1, Plenum Press, New York, 1975, 48.)

concentration of 2,3-diphosphoglycerate (DPG) in the erythrocyte. Figure 11 shows the effect of change in DPG level on the P_{50} value. An elevated DPG level results in a higher P_{50} value. This results in a greater release of oxygen at the tissues in conditions tending toward anoxia. This is of importance to the cardiac patient who is able to release greater amounts of oxygen to the tissue with a lower rate of blood flow and, thus, put less stress on the heart.

ABNORMAL BLOOD pH

The blood pH value ranges, normally, from 7.38 to 7.42. If respiration is impaired, CO_2 will accumulate in the blood and the pH will fall. If the air is cut off completely, as in asphyxiation, the pH will drop to as low as 6.7 in less than 5 min. This is called a *respiratory acidosis or respiratory acidemia*. The terms respiratory acidosis and respiratory acidemia (referring to acid blood pH) are now used interchangeably to mean an acid blood pH induced by an error in ventilation.

An acidosis may also be induced by loss of sodium ion and retention of chloride ion. This is the case in various disease states, especially in diabetes mellitus. In diabetes, the inability to burn fatty acids completely results in an accumulation of acetic acid in the blood. This reacts with itself to form acetoacetic acid which cuts down the acid present by a factor of two. On reduction of acetoacetic acid to β-hydroxy butyric acid, the degree of ionization of the acid is reduced further, ameliorating the condition. Some of the acetoacetic acid loses CO_2 so that acetone appears in the urine. Acetoacetate, hydroxybutyrate, and acetone are

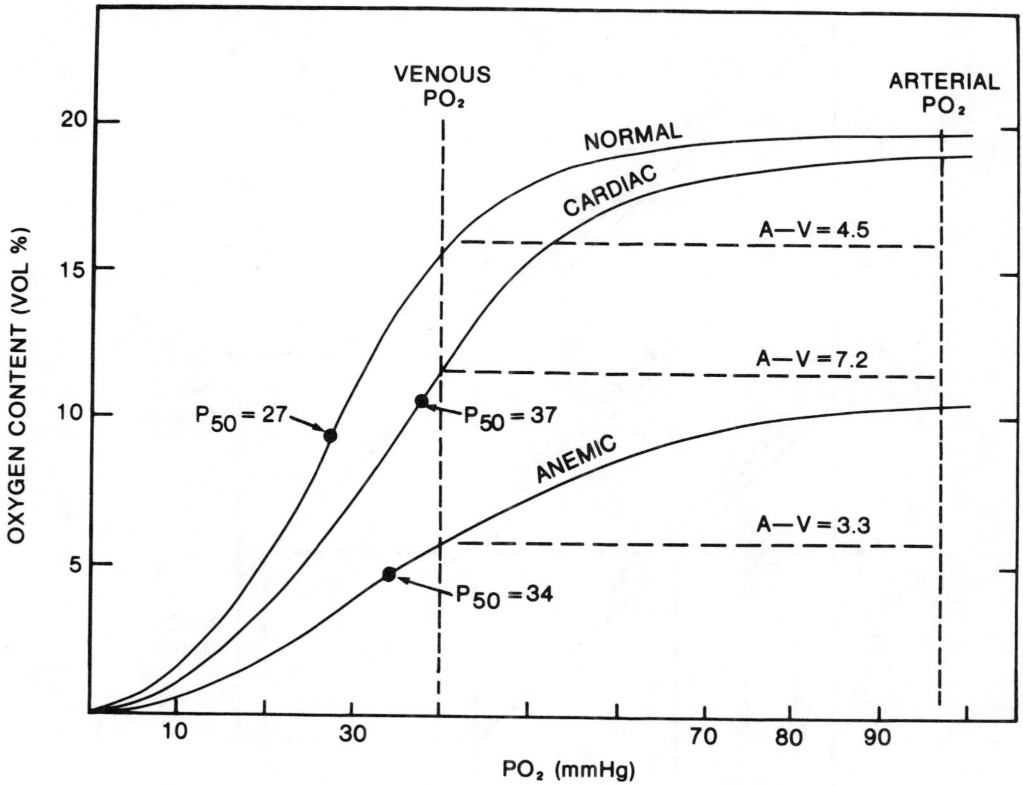

FIGURE 11. Blood oxygen dissociation curve of the normal (P_{50} = 27 mmHg) compared with that of the patient with heart failure and low cardiac output (P_{50} = 37 mmHg), and the anemic patient (P_{50} = 34 mmHg). The arterial venous difference (A − V) for oxygen is shown, which indicates the advantage gained by an increase in the P_{50} value. Thus, a higher % O_2 is released at the tissues in the low cardiac output patient. This increase in P_{50} value is caused by an increase of erythrocyte DPG. (From Natelson, S. and Natelson, E. A., *Applied Clinical Chemistry*, Vol. I, Plenum Press, New York, 1975, 43.)

referred to as the *ketone bodies,* and along with lactic acid are usually elevated at acid blood pH.

The excretion of hydroxy butyric acid and acetoacetic acid occurs in combination with sodium, ammonium, and potassium ions. Chloride ion is forced to compete with the organic acids for excretion sites and thus accumulates in the body.

This results in a disparity between sodium and chloride levels in the blood. Where Na plasma concentration is normally at about 140 meq/l and Cl concentration is at 103 meq/l, we now find a sodium level of 120 and chloride at 110 meq/l in a typical example. The pH of the blood is lowered, causing an increase in the rate of ventilation to attempt to adjust the pH to normal levels. This condition is called *metabolic acidosis.*

The sodium and chloride concentrations in the stomach are about the same, with an excess of chloride concentration of about 0.01 mol/l in gastric juice (pH = 2). When a child vomits repeatedly due to an intestinal obstruction, then the chloride loss is only slightly higher than the sodium loss. However, the relationship between Na and Cl in the blood is normally 140 to 103. Thus, vomiting causes a disproportionate loss of chloride and the patient goes into an alkalosis. This is a *metabolic alkalosis,* distinct from the respiratory alkalosis discussed above.

It should be stressed that a respiratory acidosis or alkalosis cannot be corrected by fluid therapy. The cause for CO_2 retention or hyperventilation needs to be removed. For example, anxiety can cause an individual to hyperventilate involuntarily; this patient continues to gasp

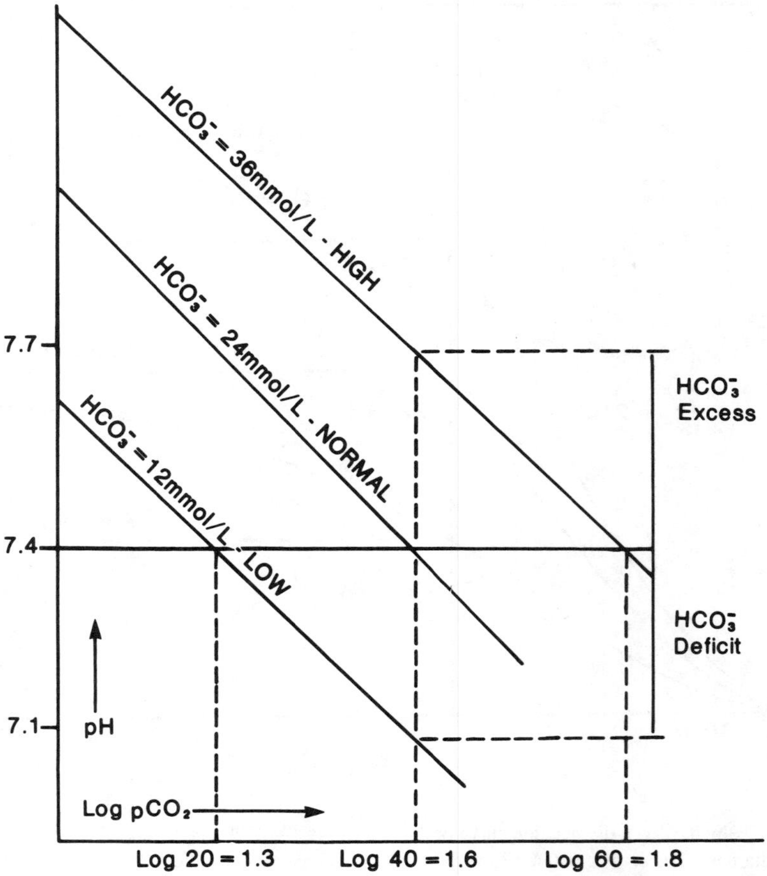

FIGURE 12. Relationships between pH and the logarithm of p_{CO_2} illustrating base excess and deficit.[4]

for air even though he is aggravating the alkalosis. If a bag is placed over his head and he rebreathes the expelled air, the blood CO_2 level returns to normal and the blood pH is corrected. Another example is the individual with food lodged in the trachea. If the obstruction is removed or a tracheotomy is performed, the excess CO_2 is expelled and the blood pH returns to normal. Table 8 illustrates this point. Attempts to correct the acidosis with alkali (NH_4OH) served no purpose. Only increased ventilation would bring the pH to the normal range.

On the other hand, a metabolic acidosis or alkalosis can be corrected by fluid administration, and this will be addressed now.

ESTIMATION OF FLUID AND ELECTROLYTE REQUIREMENTS FOR REPAIR AND MAINTENANCE

The objectives of fluid and electrolyte therapy may be summarized as follows:

1. Correction of a metabolic acidosis or alkalosis promptly
2. Fluid administration for maintaining the patient on a 24-h basis until the patient can be nurtured orally
3. Administration of trace elements and vitamins to meet daily requirements
4. An attempt to approach the caloric needs of the patient

Each of these points listed will now be explored.

Table 8
DEMONSTRATION THAT ALKALI ADMINISTRATION IS NOT EFFECTIVE IN CORRECTING RESPIRATORY ACIDOSIS, ONLY INCREASING CO_2 RETENTION[a]

	Initial	After 500 ml 0.8 *M* NH_4OH	After 7 min assisted ventilation	After addition 500 ml 0.8 *M* NH_4OH	After 7 min assisted ventilation	After 30 min
Blood pH	6.80	6.81	7.30	6.82	7.35	6.80
CO_2 content (mmol/l)	60	77	31	95	36	95

[a] Bullet wound in head. Patient breathing spontaneously and maintained by oxygen insufflation with an endotracheal tube. NH_4OH was administered over 30 min. NH_4OH was dissolved in a solution, 0.4% with respect to NaCl and 2.5% with respect to glucose.

Adapted from Natelson, S. and Barbour, J. H., *Am. J. Clin. Pathol.*, 22, 426, 1952.

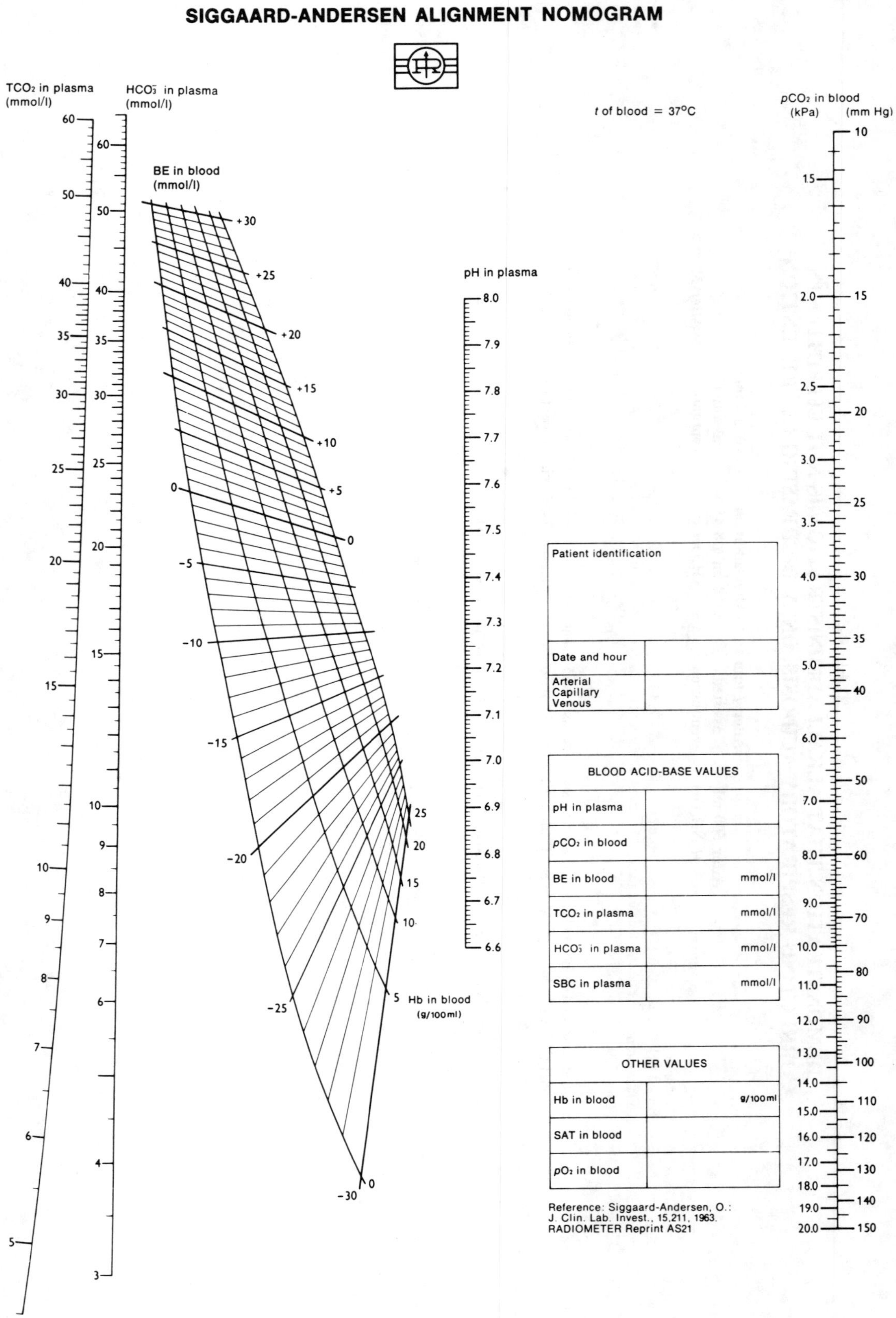

FIGURE 13. Nomogram for relationship between pH, p_{CO_2}, HCO_3^-, total CO_2, and hemoglobin concentration.[5] ©1962 Radiometer AS, Copenhagen, Denmark.

Table 9
APPROXIMATE SERUM ELECTROLYTE COMPOSITION IN THE NORMAL ADULT

Cations	(meq/l)	Anions	(meq/l)
Na	140	Cl	103
K	4.6	Bicarbonate	26
Ca	5.0	Phosphate	1.8
Mg	2.0	Protein	16
		Sulfate	0.6
	151.6	R (organic acids)	4.2
		Total	151.6

Note: Values with flame photometer for Na and K, and atomic absorption for Ca and Mg. The value for phosphate is estimated from the $HPO_4^{2-}/H_2PO_4^-$ ratio at pH 7.4.

Correction of an Alkalosis or Acidosis

Blood pH values of less than 7.3 or more than 7.5 are life threatening. For this reason, they need to be corrected promptly.

For severe metabolic acidosis, the preferred substance to be administered is sodium bicarbonate. For severe alkalosis, 2% ammonium chloride solution is used commonly. *Thus, the acidosis is being corrected with sodium ion, and the alkalosis with chloride ion.*

With a mild acidosis, some prefer to use sodium lactate. Excess lactate is metabolized to generate CO_2. This material has the advantage that there is less danger of overtreatment, since excess lactate can be metabolized or excreted into the urine as such. However, below pH = 7.2, lactate cannot be metabolized effectively. Thus, lactate may serve to aggravate the acidosis. This is so because excess lactic acid is added to sodium lactate solutions, to a pH below 7.0, to act as a preservative.

For a mild alkalosis, some prefer to use saline solutions. In NaCl, the Na:Cl ratio is 1:1, and in plasma it is about 1.4:1. NaCl is, thus, an acidifying solution, since it supplies an excess of chloride ion. If the pH change is not too severe and urine flow is maintained, the excess sodium can be excreted with other anions, particularly the organic acids. NaCl has also been used when the acidosis in diabetes is mild. In this case, reliance is placed on the ability of the kidneys to conserve sodium while excreting chloride ion.

There are several methods which have been recommended for estimating the amount of sodium or chloride ion to be given to the patient to correct an acidosis or alkalosis. These will now be explored.

Correction of Acidosis and Alkalosis by Calculation of $[Na^+]$ and $[Cl^-]$ Difference[37]

In Table 9, the reference values for the anions and cations are listed. Sodium ion is at 140 and chloride ion at 103 meq/l in the healthy individual. In a particular patient with diabetic acidosis or an acidosis associated with dehydration, a chloride level of 110 meq/l and sodium level of 120 meq/l may be observed. The question arises as to how much sodium bicarbonate is needed to correct the acidosis.

It is apparent that 20 meq/l of $NaHCO_3$ is required to correct the sodium deficit. An additional 7 meq/l of Na is required to neutralize the chloride excess. Thus, a total of 27 meq/l of $NAHCO_3$ is required. We need to know how many liters of fluid are to be corrected. This can be calculated from the weight of the patient and the data in Table 10. Since the interstitial and intravascular fluids are in communication with each other, we need to correct

Table 10
DISTRIBUTION OF BODY WATER IN HUMANS

Compartment	Mean (% of body weight)	Range
Total body water		
Male	70	60—76
Female	60	50—70
Infant	78	70—83
Intracellular fluid		
Male	45	36—50
Female	35	26—40
Infant	48	45—50
Extracellular fluid		
Male	25	20—30
Female	25	20—30
Infant	30	25—33
Interstitial fluid		
Male	21.5	18—23
Female	21.5	18—22
Infant	25.9	22—27
Intravascular fluid		
Male	4.5	3.4—5.8
Female	4.3	3.2—5.0
Infant	4.1	3.2—5.0

Adapted from Weisberg, H. T., *Water, Electrolyte and Acid Base Balance,* 2nd ed., Williams & Wilkins, Baltimore, 1962.

Table 11
ADDITIVES TO MAINTENANCE FLUIDS AVAILABLE IN AMPOULES

Solution	Ampoule volume (ml)	Content	Total content
Sodium bicarbonate (7.5%)	50	$NaHCO_3$	44.6 meq
Potassium chloride	10	KCl	20 meq
Calcium gluconate (10%)	10	Ca^{2+}	4.47 meq
Calcium chloride (10%)	10	$CaCl_2$	13.4 meq
Magnesium sulfate (4 meq/ml)	20	$MgSO_4$	80 meq
Ascorbic acid	2	Ascorbic acid	500 mg
Vitamin B complex (e.g., MVI-12®)	10	B-complex vitamins, including folic acid	
Potassium phosphate	10	K_2HPO_4	3 m*M* P/ml 4.4 meq K/ml
Sodium phosphate	10	Na_2HPO_4	3 m*M* P/ml 4 meq Na/ml
Trace elements (e.g., MTE-4®)	10	$ZnSO_4$	1 mg/ml
		$CuSO_4$	0.4 mg/ml
		$CrCl_3$	4 μg/ml
		$MnSo_4$	0.1 mg/ml

the sum of these (the extracellular fluid) as a minimum. Traditionally, we assume that the extracellular fluid is about 20% of the body weight. This is a convenient underestimate and permits fluid administration without the danger of overtreatment (see Table 10).

For a 70-kg man, the extracellular fluid volume is about 14 kg or 14 l. The deficit in this individual is, thus, $27 \times 14 = 378$ meq. An ampoule of $NaHCO_3$ contains 44.6 meq (Table 11). Thus, about 8.4 ampoules of $NaHCO_3$ need to be added to the fluids. This is

Table 12
PARENTERAL FLUIDS USED IN REPAIR AND MAINTENANCE

Solution	Na (meq/l)	Cl (meq/1)	K (meq/l)	PO_4 (meq/l)	Ac (meq/l)	CHO[a] (g/l)	kcal	mOsm/l
Normal saline (0.9%)	154	154						280
Normal saline plus 5% dextrose	154	154				50	170	
Half-normal saline + dextrose	77	77				50	170	
Hypertonic saline (3%)	513	513						
Isotonic dextrose (5%)						50	170	252
Dextrose (10%)						100	340	505
Dextrose (20%)						200	680	1010
Sodium lactate (1/6 *M*)	167						45	
Ringer's lactate	130	109	4				8	
Ammonium chloride (2.14%)		400						
Amigen® (5%)[b]	35	22	19	30			370	
Aminosyn®[c] (3.5% *M*)	40	49	18.4		68		146	460
Aminosyn®[c] (5%)			5.4		60		128	500
Freamine® III[b] (8.5%)	10	2		10	74		324	810
Intralipid® (10%)							1100	280
Lyposyn® (20%)							2000	330

[a] Carbohydrate.
[b] Casein hydrolysate with 50 g protein equivalent.
[c] Crystalline amino acids to 1 l.

usually administered, diluted with hypotonic glucose solution. It is administered slowly, measuring the patient's pH from fingertip blood at frequent intervals to monitor the correction of the acidosis. The hypotonic fluids permit the body a latitude to excrete or retain water, so as to maintain a constant osmotic pressure in the extracellular fluid.

In bringing the patient from an acidosis to normal pH, the ionic activity of Ca ion is lowered. This may result in tetany. For this reason, calcium gluconate (Table 10) is added to the fluids if symptoms of tetany are observed.

Similar calculations are performed for a metabolic alkalosis. For example, the patient of Figure 5 shows a Na value of 133 and a Cl value of 65 meq/l on the 17th day. By adding 7 meq to the Na and Cl values, we obtain 140 and 72, respectively. This indicates a chloride deficit of $103 - 72 = 31$ meq/l. From Table 12 it can been seen that an NH_4Cl solution available contains 400 meq/l; 31 meq require 77.5 ml of this solution. The child weighed 12 kg and, thus, there were 2.5 l of extracellular fluid. This suggested the use of 194 ml of NH_4Cl to correct the alkalosis. In addition, 7 meq/l of NaCl was required as indicated above. Thus, $7 \times 2.5 = 17.5$ meq/l of NaCl is required. This corresponds to about 42 ml of normal saline (Table 12). In actual practice, in this particular case, additional NH_4Cl and NaCl were given to compensate for anticipated losses in the urine and drainage fluids, and to replenish stores (see Table 1).

In patients of this type which have been depleted of sodium and chloride ions over a long period of time (which was several weeks in this case), body stores of these ions are severely depleted. There is about ten times the amount of NaCl stored in the cartilage and bone as

Table 13
DISTRIBUTION OF Na, K, AND Cl AMONG VARIOUS COMPARTMENTS OF THE BODY IN A 70-kg MAN

		Plasma	Interstitial fluid[a]	Connective tissue, cartilage, and transcellular	Bone	Intracellular	Total
Na	meq	455	1246	511	1750	98	4060
	meq/kg	6.5	17.8	7.3	25	1.4	58
K	meq	14	35	49	287	3381	3786
	meq/kg	0.2	0.5	0.7	4.1	48.3	53.8
Cl	meq	322	861	497	350	287	2317
	meq/kg	4.6	12.3	7.1	5.0	4.1	33.1

[a] From Natelson, S. and Natelson, E. A., *Applied Clinical Chemistry*, Vol. 1, Plenum Press, New York, 1975, 114. Includes lymph, cerebrospinal fluid, etc.

is circulating in the extracellular fluid (Table 13). Thus, extra Na and Cl administered are redeposited in these tissues. This is the case with animals in the wild which periodically visit ''salt licks'' to replenish losses from their body stores.

It is important to note that the bicarbonate (most of total CO_2) of the plasma, while it is affected by the state of acidosis or alkalosis in a particular patient, does *not* enter into the calculation of Na, Cl, and water requirements. The reason for this is that a patient in severe acidosis will be hyperventilating (Kussmaul respiration). On correction of the Na/Cl ratio, resulting in the pH approaching the normal range, the ventilation rate will be reduced and the total CO_2 level will increase. Thus, the CO_2 content and bicarbonate level are variables which adjust themselves as the pH approaches the normal range. This is equivalent to what has been stressed above, namely, that a respiratory acidosis or alkalosis cannot be corrected for by fluid therapy.

The organic acid levels which have been ignored above will increase the requirement for Na ion. Thus, if the patient is titrated with $NaHCO_3$ to normal pH in the presence of increased amounts of organic acids, the result will be an apparent imbalance in the sodium/chloride ration, with an excess of sodium. This is illustrated in Table 14 by two children, both in severe acidosis, who were titrated to normal pH with $NaHCO_3$. In Case 1 when normal pH was reached, sodium was at 137 meq/l and chloride at 83 meq/l, an apparent excess of 17 meq/l of Na. This actually was an indication of the elevated organic acid concentration. In the second case, Na was at 138 and chloride at 88 meq/l, a sodium excess of 13 meq/l. It became apparent that as the organic acids were metabolized and excreted, an alkalosis would ensue in both cases. This was countered by NaCl administration. In spite of this, the alkalosis did demonstrate itself and adjustment had to be made by further administration of NaCl.

The Anion Gap

In Table 9, the cations and anions are aligned and R (organic acids) is adjusted in value so that the sum of the anions and cations becomes equal. This difference due to R plus the protein concentration as an anion was originally referred to as the *anion gap*. Subsequently, the ions in lesser concentration were ignored and the sum of the $[HCO_3]$ and $[Cl^+]$ was subtracted from the sum of the $[K^+]$ and $[Na^+]$ concentrations to yield an anion gap, which was supposed to represent the acid base status of the patient. This was further simplified by subtracting the sum of $[HCO_3^-]$ and $[Cl^-]$ from the Na level. Finally, the total CO_2 was substituted, by some, for the $[HCO_3^-]$ yielding the equation:

$$[Na^+] - [Cl^-] - [T\ CO_2] = \text{anion gap} = 12 \pm 5$$

Table 14[39]

DEMONSTRATION THAT CORRECTION OF SODIUM DEFICIT DUE TO BOTH CHLORIDE AND ORGANIC ACID EXCESS WILL EVENTUALLY RESULT IN ALKALOSIS DUE TO METABOLISM AND EXCRETION OR ORGANIC ACIDS: ELECTROLYTE LEVELS BEFORE AND AFTER TREATMENT WITH $NaHCO_3$

	pH	CO_2 content (mmol/l)	Cl (meq/l)	Na (meq/l)	K (meq/l)	Protein (g/100 ml)	Hematocrit (%)	Urea (mg/100 ml)	P (mg/100 ml)	pCO_2 (mmHg)
Case 1 (13.7 kg)										
Initial	7.04	23.5	92	134	4.4	8.1	47	69	8.0	70
After 6.9 g $NaHCO_3$[a]	7.42	38.3	83	137	4.2	5.8	36	63	6.8	61
After 8 h[b]	7.52	31.1	94	139	2.5	4.8	33	19.2	5.0	40
Case 2 (5 kg)										
Initial	6.94	8.3	85	118	5.0	5.2	37.2	88	—	33
After 3.5 g $NaHCO_3$[c]	7.37	21.1	88	138	4.6	4.9	33.0	60	—	37
After 12 h[d]	7.48	18.9	89	131	2.2	4.6	43.0[d]	24	—	26

[a] Administered in 200 ml of 5% glucose in water mixed with 200 ml of ''normal saline''.
[b] Received 50 meq of sodium chloride, made up to 600 ml with 5% glucose in water. Excreted 500 ml of urine.
[c] Administered in 100 ml of ''normal saline'' mixed with 100 ml of 5% glucose in water.
[d] Received 150 ml of ''normal saline'' mixed with 300 ml of 5% glucose in water followed by 140 ml of whole blood. Excreted 380 ml of urine.

Table 15
ANION GAP OBSERVED IN FIVE ACUTELY ILL PATIENTS

meq/l					
Na^+	Cl^-	Total CO_2	Anion gap	pH	Remarks
120	112	10	−2	7.15	Diabetic coma (organic acidemia)
139	94	31.1	+13.9	7.52	Status asthmaticus treated with $NaHCO_3$
140	102	53	−15	7.10	Mechanical obstruction in trachea (respiratory acidemia)
141	103	18	+20	7.56	Hyperventilation due to apprehension (respiratory alkalemia)
134	92	23.5	+18.5	7.04	Asthmatic attack untreated (organic acidemia and respiratory acidemia)

If the anion gap were less than 12, then the anion gap was negative and suggested decreased [Na^+] and excess of organic acids and chloride ion, and we were dealing with an acidosis and possible lactic acidemia. An increased anion gap would then suggest an alkalosis.

The ineffectiveness of this approach is readily apparent from Table 15. In Table 15, it can be seen that increased anion gaps can occur with an acid or alkaline pH. Severe acidosis can occur with a normal anion gap or a severely increased anion gap.

The fallacy of this approach is that it does not take into account the rate of ventilation of the individual and its effect on the blood pH. Thus, it attempts to evaluate the extent of respiratory acidosis or alkalosis from the measurement of electrolyte concentrations.

The computation of the anion gap involves the addition and subtraction of ionic components, each with a different activity coefficient. This is not acceptable chemical practice. The term also implies an electrical imbalance, which does not exist. For the above reasons, the concept of anion gap is of little value in estimating acid or base requirements in the patient, and is not acceptable as a descriptive term from a theoretical point of view.

Calculation of Electrolyte Requirements from the Henderson-Hasselbalch Equation[3-5]

In the Henderson-Hasselbalch equation (Equation 3), the [HCO_3^-] is approximately equal to the [Na^+] (140 meq/l) minus the sum of the [Cl^-] (103 meq/l) and weak acid concentration (protein + organic acids, about 19 meq/l (see Table 9), if we assume $NaHCO_3$ to be 100% ionized. This amount of $NaHCO_3$ is available to neutralize excess organic acids, which may accumulate under abnormal conditions. since this $NaHCO_3$ is available for resisting changes in pH, it has been referred to as "buffer base", "bicarbonate base", and other terms by various investigators. Any deviation from the normal value suggested the amount of $NaHCO_3$ or NH_4Cl to be administered to correct the blood pH.

As presented earlier, in the absence of convenient methods for Na, P_{CO}, or pH measurements, the total CO_2 was estimated on tonometered plasma (at p_{CO_2} = 40 mmHg) and used to estimate the HCO_3^- level with the Henderson-Hasselbalch equation. A nomogram was then devised for this calculation by Hastings in 1934. This took into account, in addition, the effects of hemoglobin concentration on the system.[3] The extent of metabolic acidosis or alkalosis was also indicated in the nomogram. If the pH was less than 7.4 and the bicarbonate level was lowered, then a lowered sodium-to-chloride difference existed and suggested a metabolic acidosis. It has been suggested that the difference between the strong cations (Na^+ and K^+) and anions (CL^-, phosphate and sulfate) be referred to as the "strong ion difference" (SID).[38] This is mainly the difference between [Na^+] and [Cl^+] concentration in milliequivalents per liter. If the total CO_2 level were elevated with an acid pH, then we had a respiratory acidosis. If the total CO_2 was elevated and an alkaline pH was demonstrated, we had a respiratory alkalosis.

With the development of practicable glass electrodes, the pH could now be measured directly. Astrup introduced the method of measuring the whole blood pH before and after tonometering the blood with gas mixtures at p_{CO_2} of 20 and 60 mmHg. The pH was then plotted against the log of the p_{CO_2}. The original pH of the blood permitted the estimation of p_{CO_2} from the curve (Figure 10), and the Henderson-Hasselbalch equation permitted the calculation of the HCO_3^- level, so as to estimate [HCO_3^-] excess or deficit and, thus, the $NaHCO_3$ or NH_4Cl requirement to bring to normal pH. A nomogram was devised, then, for calculating the base excess,[4,5] correcting for the effect of hemoglobin. This is shown in Figure 11.

In these studies, the p_{CO_2} estimated on the assumption of a constant pK_1' was assumed to be the p_{CO_2} of the original patient's blood. *This was not necessarily so.* In a case report using this technique, the alkali requirements of a patient were overestimated, resulting in an alkaline blood pH.[4] Another factor which was ignored with this procedure was the fact that *the rate of ventilation* changes as the pH of the blood is brought from an alkalosis or an acidosis to a pH = 7.4. This can be corrected for and is presented in the next section.

Calculation of $NaHCO_3$ Required to Bring to Normal pH and Correction for Change in Ventilation on Treatment[39]

For our purpose, the Henderson-Hasselbach equation is best written where pH is that of the blood of the patient to be treated:

$$pH = pK_1 + \log NaHCO_3/(0.03 \cdot p_{CO_2}) \tag{8}$$

This is also to indicate clearly that [HCO_3^-] represents the $NaHCO_3$ available for buffer action.

Let us assume that we now add an amount of $NaHCO_3$ (Q) to the numerator of the fraction so as to correct to pH = 7.4. Then the equation will read:

$$7.4 = 6.1 + \log[NaHCO_3 + Q]/(0.03 \cdot p_{CO_2})$$

$$\text{and} \quad 1.3 = \log[NaHCO_3 + Q]/(0.03 \cdot p_{CO_2})$$

$$\text{or antilog } 1.3 = 19.95 + [NaHCO_3 + Q]/(0.3 \cdot p_{CO_2}) \tag{9}$$

The original equation before treatment, Equation 8, can be presented in the same form:

$$\text{antilog } (pH - 6.1) = [NaHCO_3]/(0.03 \cdot p_{CO_2}) \tag{10}$$

In addition, the original $NaHCO_3$ can be replaced by the total CO_2 as follows:

$$\text{total } CO_2 = [NaHCO_3] + (0.03 \cdot p_{CO_2}) \tag{11}$$

We can now solve for Q by combining Equations 9, 10, and 11: thus,

$$\text{Base deficit or excess} = Q = \frac{[\text{total } CO_2]\ [19.95 - \text{antilog } (pH\ 6.1)]}{1 + \text{antilog } (pH\ 6.1)} \tag{12}$$

With two measurements, patient's pH and total CO_2, Q, the $NaHCO_3$ required in mmoles/liter of extracellular fluid to bring to pH = 7.4 can be calculated. Table 16 uses this equation to permit the ready estimation of the $NaHCO_3$ required to bring the patient's blood to normal pH after measuring his blood pH and total CO_2 level.

To see how this table is used, let us assume a patient weighing 50 kg with a blood pH of 7.15 and a total CO_2 level of 10 mmol/l. From Table 16, the patient requires 7.14 mmol of $NaHCO_3$ per liter of extracellular fluid. Assuming that the extracellular fluid is 20% of the body weight (Table 10) we are dealing with 10 l of extracellular fluid. The patient then requires 71.4 mmol of $NaHCO_3$ or 1.6 ampoules (Table 11). *If this amount is administered, the pH will remain low.* The reason for this is that the patient has been hyperventilating (Kussmaul breathing): as his pH is corrected the depth and rate of his breathing will be decreased, and the total CO_2 will increase. Finally, his total CO_2 will reach about 25 mmol/l. From Table 16, with this total CO_2 and a pH of 7.15, the patient requires 17.9 mmol of $NaHCO_3$. Thus, his hyperventilation was equivalent to 17.9 − 7.4 = 10.5 mmol of $NaHCO_3$ per liter of extracellular fluid. This advantage is lost as the patient begins to breathe normally. With more CO_2 retention, more alkali needs to be administered. In actual practice, it was found that the patient required the 17.9 mmol/l × 10 = 179 meq of $NaHCO_3$. From Table 11, this meant 4 ampoules of $NaHCO_3$ to bring to normal pH. From the above, it is apparent that it is necessary to measure the pH and total CO_2 repeatedly to adjust the calculations to compensate for the change in ventilation rate.

In testing the validity of Table 16, over 100 patients were titrated successfully. Generally, the $NaHCO_3$ was administered as a 3% solution in 2.5% glucose. Repeated measurements of pH and total CO_2 were taken at regular intervals to monitor the process. Table 17 shows the values obtained for blood pH with two typical patients during the titration. Table 18 illustrates the initial and final results on an additional 11 patients.[39]

The theoretical titration curve, using Equation 13, is shown in Figure 14. Study of this diagram shows that when 90% of the $NaHCO_3$ has been administered, the pH will reach only 7.25. However, from that point, pH correction is rapid because of the steep curve. Thus, if care is not taken, even a small excess of $NaHCO_3$ will result in an alkalosis. For this reason, it is important to add the $NaHCO_3$ slowly, monitoring the pH at frequent intervals.

Equation 13 serves the additional purpose of estimating the status of the patient. If a patient is moribund, or if the acidosis is due mainly to a respiratory problem, then Equation 13 will not apply. In a typical case of diabetic acidosis, a patient was treated with fluids containing NaCl to correct a severe acidosis over a 48-h period. At that time the blood pH had not changed and was at 7.0. $NaHCO_3$ administration was then started, calculated on the basis of Table 16. After the calculated amount of $NaHCO_3$ had been administered, the pH had moved to only 7.05. Even though reflex respiration continued, it became apparent that the patient's metabolism had ceased to function normally, and the cells were secreting (not utilizing) the organic acids. Equation 13 also suggests when further addition of $NaHCO_3$ will not serve a useful purpose.

With this procedure all of the acid substances are being neutralized, including the excess organic acids. When the patient is brought to normal pH, a strong ion imbalance will then be evident (Table 14). Note that in case 1, Na^+ = 137 and Cl^- = 83 (SID = 54). In Case 2, Na^+ = 138 and Cl^- = 88 (SID = 50). Both of these patients were anuric. As the pH was brought to the normal range, the patient began to urinate normally (see the discussion of Figures 4 and 5). With the SID elevated (normal value is about 37), these patients will develop an alkalosis, unless additional fluids are administered which are designed to neutralize the impending alkalosis.

CONSIDERATION OF [H$^+$] IN PLACE OF pH IN ACID-BASE STUDIES[37,39]

Since pH is the negative logarithm of the hydrogen ion concentration, substantial changes in [H$^+$] are masked when presented as pH changes. For example, Table 19 lists the pH values in the physiological range and the corresponding [H$^+$]. In severe alkalosis (pH = 7.7) the hydrogen ion is lowered to 50% of the value of that at pH = 7.4. In severe acidosis, the [H$^+$] will rise to as high as 100 mmol/l from a norm of 39.8 for a 250% increase. This is not as evident for a change from a pH of 7.4 to 7.0. It has been proposed, therefore, that

Table 16
RELATIONSHIP BETWEEN pH, CO_2 CONTENT, AND BASE DEFICIT OR EXCESS IN ACIDOSIS OR ALKALOSIS[a]

	CO_2 content (mmol/l)											
pH	**5**	**10**	**15**	**20**	**25**	**30**	**35**	**40**	**45**	**50**	**55**	**60**
6.90	9.35	18.7	28.11	37.4	46.8	56.1	65.5	74.8	84.2	93.5	102.9	112.2
6.95	7.86	15.7	23.6	31.4	39.3	47.2	55.0	62.8	70.7	78.6	86.5	94.3
7.00	6.70	13.4	20.1	26.8	33.5	40.2	46.9	53.6	60.3	67.0	73.7	80.4
7.05	5.57	11.1	16.7	22.3	27.9	33.4	39.0	44.5	50.1	55.7	61.3	66.8
7.10	4.53	9.05	13.6	18.1	22.7	27.1	31.7	36.1	40.7	45.3	49.8	54.3
7.15	3.57	7.14	10.7	14.3	17.9	21.4	25.0	28.5	33.3	35.7	39.3	42.8
7.20	2.71	5.42	8.13	10.8	13.6	16.3	19.0	21.6	24.3	27.1	29.8	32.5
7.25	2.06	4.12	6.18	8.24	10.3	12.4	14.4	16.4	18.5	20.6	22.7	24.7
7.30	1.22	2.43	3.66	4.86	6.08	7.29	8.51	9.72	10.9	12.2	13.4	14.6
7.35	0.580	1.16	1.74	2.32	2.90	3.48	4.06	4.64	5.22	5.80	6.38	6.96
7.40	0	0	0	0	0	0	0	0	0	0	0	0
7.45	−0.52	−1.04	−1.56	−2.08	−2.60	−3.12	−3.64	−4.16	−4.68	−5.20	−5.72	−6.24
7.50	−0.99	−1.98	−2.97	−3.96	−4.95	−5.94	−6.93	−7.92	−8.91	−9.90	−10.9	−11.9
7.55	−1.41	−2.82	−4.23	−5.64	−7.05	−8.46	−9.87	−11.3	−12.69	−14.4	−15.5	−16.9
7.60	−1.79	−3.58	−5.37	−7.16	−8.95	−10.7	−12.5	−14.3	−16.11	−17.9	−19.7	−21.5
7.65	−2.13	−4.26	−6.39	−8.52	−10.7	−12.8	−14.9	−17.0	−19.2	−21.3	−23.4	−25.6
7.70	−2.44	−4.87	−7.31	−9.74	−12.2	−14.6	−17.1	−19.5	−21.9	−24.3	−26.8	−29.2
7.75	−2.71	−5.42	−8.13	−10.8	−13.5	−16.2	−18.9	−21.7	−24.4	−27.0	−29.8	−32.4

Note: Table based on the Equation: Base deficit or excess $= \dfrac{T[19.95 - \text{antilog}\ (pH - 6.1)]}{[1 + \text{antilog}\ (pH - 6.1)]}$

[a] Values represent meq of sodium bicarbonate needed to be added to bring to pH 7.4 in acidosis, per liter of extracellular fluid (5 kg of body weight). Negative values represent the meq of acid (NH_4Cl) needed to be added to bring to normal pH in alkalosis. Interpolate for CO_2 content from the adjacent value on the pH line. Example: for CO_2 content of 18 mmol/l and pH 7.20, base requirement is $8.13/15 \times 18 = 9.76$ mmol of $NaHCO_3$ per 5 kg body weight.

Adapted from Natelson, S. and Natelson, E. A., Eds., *Principles of Applied Clinical Chemistry,* Vol. I, Plenum Press, New York, 1975, 29.

Table 17
IN VIVO TITRATION OF BLOOD OF PATIENTS WITH SEVERE ACIDOSIS AND ANURIA[39]

Case 1 (age 2 years, weight 13.7 kg)						
pH (measured)	7.04	7.08	7.16	7.20	7.30	7.42
$NaHCO_3$ (added) (g)	—	1.0	2.6	3.4	5.1	6.9
$NaHCO_3$ (calculated)[a] (g)	—	1.1	2.9	3.7	5.5	6.8
Case 2 (age 10 months, weight 5 kg)						
pH (measured)	6.94	7.12	7.25	7.32	7.37	
$NaHCO_3$ (added) (g)	—	2.0	2.8	3.1	3.6	
$NaHCO_3$ (calculated)[a] (g)	—	1.9	2.8	3.3	3.4[b]	

[a] Calculated to bring to the pH measured, assuming 20% extracellular fluid volume or 1 l of extracellular fluid per 5 kg of body weight.
[b] Calculated to bring to pH 7.40.

with the use of computers it is not necessary to use the form of the buffer equation designed by Sørensen, but to return to the equations used before the development of the concept of pH. All of these equations have already been presented above. These include the following:

$$[H^+][OH^-] = K_W \tag{14}$$

$$H_2CO_3/p_{CO_2} = K_P \tag{15}$$

$$[H^+][HCO_3]/[H_2CO_3] = K_{H_2CO_3} \tag{16}$$

$$[H^+][CO_3^{2+}]/HCO_3 = K_{HCO_3^-} \tag{17}$$

$$[H^+]\ [A^-]/[HA] = K_{HA} \tag{18}$$

In Equation 18, HA represents the weak acids, mainly[41] proteinate.[41] This author includes the organic acids normally present and assumes a normal value of 19 meq/l for R + HA.[37]

In addition to these equations, there must be electrical neutrality and the sums of the cations minus the sum of the anions must be equal to zero. The cations include Na^+, K^+, Ca^{2+}, and Mg^{2+}, all expressed in milliequivalents per liter. The strong anions are Cl^-, SO^{2-}, HPO_4, and H_2PO_4. The difference between the strong anions and cations is represented by SID^+, as discussed above. For electrical neutrality, A^- represents the weak acids:

$$[SID^+] + [H^+] - [HCO_3] - [A^-] - [CO_3^2] - [OH^-] = 0 \tag{19}$$

For Equations 14 to 18, the equilibrium constants are all available in the literature and can be inserted into the computer to calculate the relationships between the various parameters. $[SID^+]$, P_{CO}, and $[HA^-]$ are considered the independent variables to be measured in the equations.

In Figure 15, the change in $[H^+]$ with P_{CO} is plotted. This is seen to be a straight line instead of a logarithmic relationship as it would be with pH. Normally, p_{CO_2} = 40 mm and $[H^+]$ = 39.6 nmol/l (Table 19). This is the upper point of the solid triangle in Figure 15.

The results by this procedure are necessarily about the same as from the Henderson-Hasselbalch equation. Compare this set of curves with that of Figure 12. Variations in SID

Table 18
EFFECT OF SODIUM BICARBONATE ADMINISTERED TO PATIENTS IN ACCORDANCE WITH EQUATION 13 ON BLOOD pH AND CO_2 CONTENT[39]

			CO_2 content (mmol/l)		pH		$NaHCO_3$ (g)	
Case no	Diagnosis	Weight (kg)	Initial	Final	Initial	Final	Administered	Calculated[a]
1	Diarrhea	7	20.3	31	7.03	7.43	3.6	3.7
2	Partial respiratory obstruction	68	33.5	41	7.23	7.36	15.0	13.5
3	''Status asthmaticus''	60	29.3	34	7.25	7.41	10.0	10.0
4	Nephritis	65	14.4	23	7.20	7.35	18.0	15.3
5	Salicylism	5	8.1	25	7.28	7.45	0.65	0.65
6	Diarrhea	15	18.9	26	7.15	7.39	4.5	4.7
7	Cardiac decompensation	72	20.3	25	7.23	7.36	16.3	14.5
8	Salicylism	53	18.0	20	7.10	7.23[b]	11.3	10.7
9	Diarrhea	11	7.7	23	7.23	7.40	2.54	2.24
10	Diarrhea	13	12.6	27	7.24	7.45	2.34	2.34
11	Diarrhea	1.75	1.0	22	6.92	7.42	1.15	1.30

[a] Calculated to bring to pH 7.40.
[b] Further treatment with $NaHCO_3$ discontinued. Calculated to bring to pH 7.23

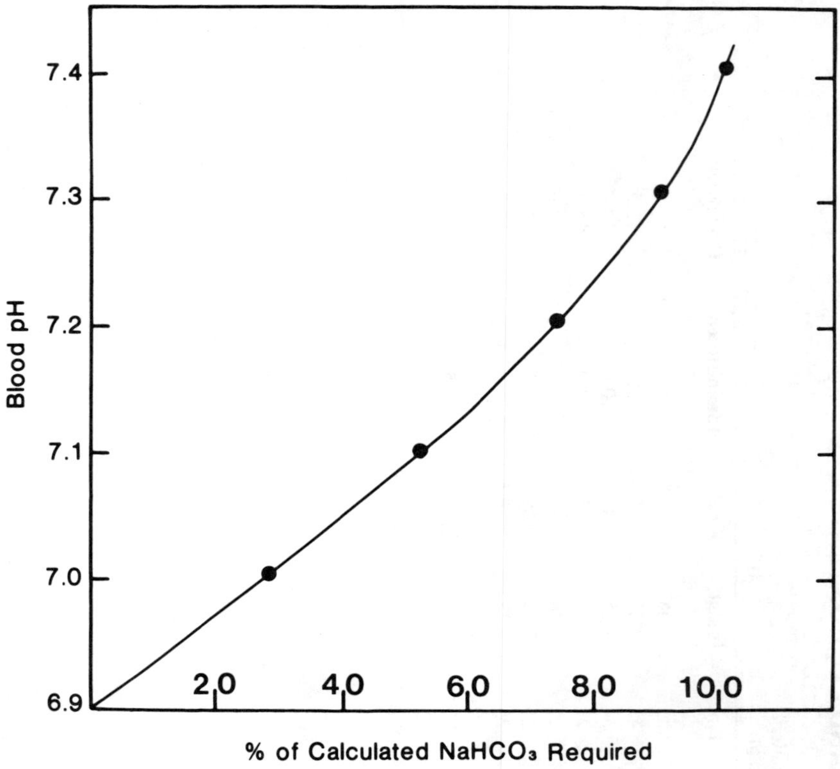

FIGURE 14. Plot of the equation Q = T [19.95 − antilog (pH 6.1)] ÷ [1 + antilog (pH 6.1)] relating the amount of $NaHCO_3$ (Q) required to bring to pH = 7.4 in an acidosis. T represents the total CO_2; pH is the initial pH of the patient.

Table 19
CONVERSION TABLE FROM pH TO HYDROGEN ION CONCENTRATION $[H]^+$ IN nmol/l, pH = −LOG $[H^+]$

pH	$[H^+]$ (nmol/l)	pH	$[H^+]$ (nmol/l)
6.70	199.5	7.25	56.2
6.75	177.8	7.30	50.1
6.80	158.5	7.35	44.7
6.85	141.3	7.40	39.8
6.90	125.9	7.45	35.5
6.95	112.2	7.50	31.6
7.00	100.0	7.55	28.2
7.05	89.1	7.60	25.1
7.10	79.4	7.65	22.4
7.15	70.8	7.70	20.0
7.20	63.1	7.75	17.8

in Figure 15 actually represent variations in bicarbonate as shown in Figure 9, since the authors assume that the weak acid concentration is constant. Thus, SID = HCO_3^- + 19 and the slope of the lines would remain the same in Figure 15. Thus, for the SID concentration, the sum of $[HCO_3^-]$ plus a constant can be substituted.

The presentation of Figure 15 has an advantage in that it clearly shows a direct increase of $[H^+]$ with an increase in p_{CO_2}. More CO_2 added to the serum results in an increase in

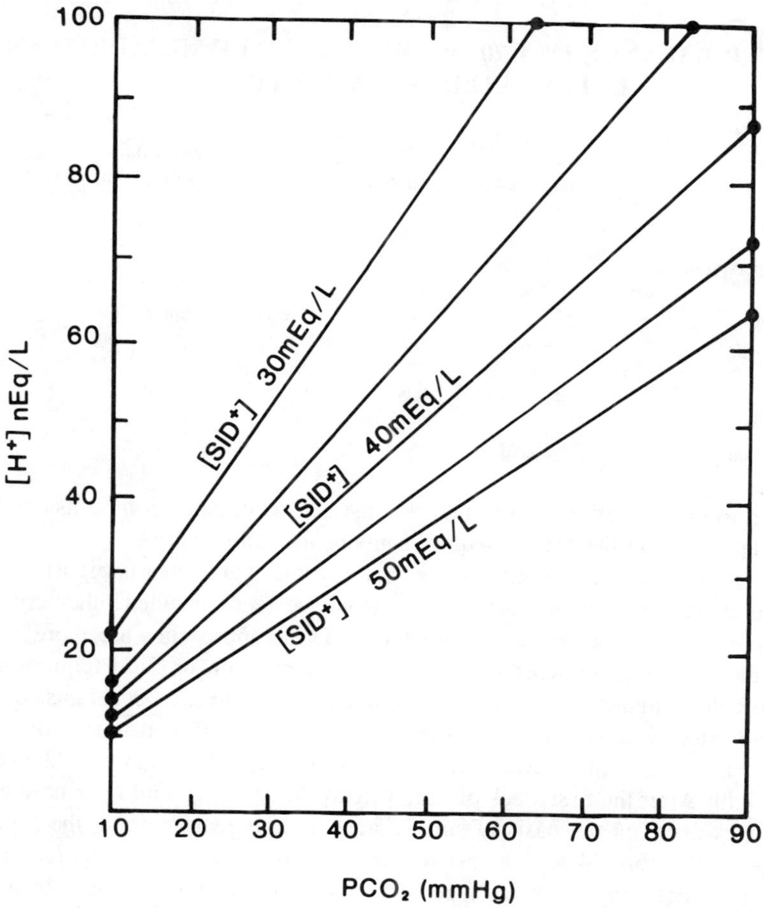

FIGURE 15. Plot of relationship between $[H^+]$ and pCO_2. SID is the strong ion difference and is approximately equal to $[K^+] + [Na^+] - [CL^-]$.

HCO_3^- and an increase in $[H^+]$. Thus, bicarbonate is acidic and not a base. This is also shown in Figure 9 but in reverse, since pH and $[H^+]$ are inverse functions of each other. Sodium bicarbonate should be substituted for the expression ''bicarbonate base''.

Nomograms have also been designed to estimate ''base excess'' using the equations listed above.[41]

MAINTENANCE FLUIDS

It is necessary to plan the maintenance fluids, on a 24-h basis, while correcting a metabolic acidosis or alkalosis. The volume of fluids administered must be adequate to compensate for losses in the urine, sweat, drainage fluids (if any), the feces, and from the lungs (insensible loss). A typical study is summarized in Table 20 for a healthy 70-kg male. The water volume in the urine and stool was measured. The losses through the skin and lungs were estimated from published data. The volume of the liquid intake and water in the food was also measured. That due to oxidation of the hydrogen in the food was calculated. Thus, the volume of fluids, if administered to this 70-kg man, would be about 3 l to meet his daily losses.

These fluids must replace the losses in electrolytes, nitrogen, and phosphorus. These are summarized in Table 21. The losses of the electrolytes (Na, K, Cl, Ca, and Mg) are given in milliequivalents per kilogram of body weight per 24 h. Nitrogen and phosphorus are given in milligrams per kilogram of body weight per 24 h.

Table 20
WATER BALANCE IN A 70-kg HEALTHY MALE MEASURED AND CALCULATED FOR A 24-h PERIOD

	Water loss		Water intake (ml/70 kg/24 h)	
	ml/kg/24 h	ml/70 kg		
Via the skin (insensible perspiration)	11	770	As liquid	1700[a]
Via the lungs	7	490	With food	900[a]
Via the kidney	22[a]	1540[a]	Due to oxidation of food	305
Via the feces	1.5[a]	105[a]		
Total	41.5	2905		2905

[a] Measured, other values were calculated.

In addition to consideration of the fluid volume and its composition, consideration must be given to approaching the caloric requirements of the individual.

In presenting the fluid requirements above, only one parameter (weight) was used to calculate the fluid requirements. Actually, some have recommended other criteria, most commonly *surface area* and *caloric expenditure*. These approaches are more effective in calculating the fluid requirements of infants. The calculation of fluid requirements from surface area, caloric intake, and weight are compared in Table 22. For infants, these values need to be adjusted. For example, an infant weighing 4 kg will consume about 700 ml of formula and water over a 24-h period, while from Table 22, only 160 ml would be calculated based on weight. After the first week of life, a figure of 125 to 160 ml/kg is used commonly for infants. However, on the basis of surface area, infants require about the same amount as adults — 1800 ml/m^2/24 h. For this reason some prefer to use the surface area as an estimate of fluid requirements, since it remains about the same at all stages of growth. On the basis of caloric expenditure, a figure of about 125 ml/100 cal is used for adults. This approach is recommended for those who are to engage in vigorous activity. Note that Calories (upper case C) is used for kilocalories.

A major problem is that of meeting the caloric requirements of the individual with i.v. fluids. Referring to Table 12, it can be seen that a liter of 5% glucose (isotonic) yields only 170 cal. Thus, if the 70-kg man receives 3 l of such a solution, he would only receive 510 Cal. His daily requirements are about four to five times that amount.

Protein hydrolysate also yields about the same number of Calories per gram as glucose (glucose = 3.4 Cal/g; protein = 4 Cal/g). Thus, a liter of 5% glucose plus 5% protein hydrolysate yields only 370 Cal. Further, there is a limit to the concentration of fluids to be used, since solutions which are more than 600 mOsm are inhibiting to the function of the peripheral veins and may cause thrombosis. Some have used 10% glucose, supplemented with fat emulsions (9 Cal/g) and added crystalline amino acids (Table 12). Such combinations can approach the caloric requirement of patients who cannot be fed orally.

To these fluids must be added an ampoule of B-complex vitamins, an ampoule of ascorbic acid (500 mg), and KCl in accordance with Table 18 and in accordance with the patient's plasma potassium level. Fluids containing potassium in excess of 5 meq/l should be given slowly, and only if the patient's serum potassium level is not in excess of 5 meq/l. Na, K, and Cl requirements are best estimated by analyzing the patient's urine and drainage fluids for these elements.

Generally, with patients receiving fluids for only a few days, it is not necessary to meet caloric requirements, since there are, usually, adequate stores of body fats. Long range, it is necessary to supply adequate nutrients to maintain an electrolyte and caloric balance. This

Table 21
EXCRETION OF ELECTROLYTES, NITROGEN, AND PHOSPHORUS IN THE NORMAL ADULT

	meq/kg/24 h					mg/kg/24 h	
Pathway	Na	K	Cl	Ca	Mg	Nitrogen	Phosphorus
Urine	1.10—4.10	0.40—1.45	1.10—5.10	0.03—0.42	0.035—0.20	120—250	9—15
Feces	0.05—0.10	0.10—0.20	0.05—0.15	0.25—0.50	0.13—0.30	11—36	0.01—0.02
Sweat[a]	0.10—0.60	0.05—0.30	0.10—0.60	0.005—0.20	0.001—0.003	2—6	0.001—0.005
Total	1.25—4.8	0.55—1.95	1.25—5.75	0.285—1.12	0.166—0.503	133—292	9.01—15.025

[a] Calculated, assuming excretion of 11 ml sweat per kilogram per hour. From Natelson, S. and Natelson, E. A., *Principles Applied Clinical Chemistry*, Vol. I, Plenum Press, New York, 1975, 250.

Table 22
MAINTENANCE FLUID AND NUTRIENTS FOR THE NORMAL ADULT FOR 24 h (APPROXIMATE VALUES)

Component	Per kg	Per m^2 [a]	Per 100 Cal
Water (ml)	40	1800	100
Salt (NaCl)(meq)	2	75	5
Potassium (meq)	1.0	37	2.5
Protein (g)	1.0	27	1.5
Caloric requirement	40	1800	—

[a] Surface area correlates with height and weight. For 70-kg male, 6 ft tall, surface area = 1.90 m^2.

Adapted from Natelson, S. and Natelson, E. A., *Principles Applied Clinical Chemistry,* Vol. I, Plenum Press, New York, 1975, 325.

can only be met with hypertonic solutions. This process is generally referred to as *intravenous hyperalimentation*. In these cases, the solutions need to be infused slowly, typically through a central venous catheter. Several commercially available solutions and additives are available (Tables 11 and 12).

REFERENCES

1. **Henderson, L. J.,** The theory of neutrality regulation in the animal organism, *Am. J. Physiol.,* 21, 427, 1908.
2. **Hasselbalch, K. A.,** The calculation of blood pH from free and bound CO_2 as a function of pH, *Biochem. J.,* 78, 112, 1916; *Biochem. J.,* 78, 251, 1917.
3. **Hastings, A. B. and Shock, N. W.,** Studies of the acid base balance of blood. Nomogram for calculation of acid-base data for blood, *J. Biol. Chem.,* 104, 578, 1934.
4. **Astrup, P.,** A new approach to acid base metabolism, *Clin. Chem.,* 7, 1, 1961.
5. **Siggard-Andersen, O.,** The acid base status of the blood, *Scand. J. Clin. Lab. Invest.,* 15 (Suppl.), 211, 1963.
6. **Natelson, S.,** Routine use of the Perkin Elmer flame photometer in the clinical laboratory, *Am. J. Clin. Pathol.,* 20, 463, 1950.
7. **Natelson, S.,** Chronic alkalosis with damage to the central nervous system, *Clin. Chem.,* 4, 32, 1958.
8. **Dean, J. A., Ed.,** Acitivity coefficients, in *Lange's Handbook of Chemistry,* McGraw-Hill, New York, 1985, 53.
9. **Van Slyke, D. D., Sendroy, J., Jr., Hastings, A. B., and Neill, J. M.,** The solubility of CO_2 at 38°C in water, salt solution, serum and blood cells, *J. Biol. Chem.,* 78, 765, 1928.
10. **Warberg, E. J.,** Studies on carbonic acid compounds and hydrogen ion activities in blood and salt solutions, *Biochem. J.,* 16, 311, 1922.
11. **Hastings, A. B., Sendroy, J., Jr., and Van Slyke, D. D.,** The value of pK′ in the Henderson-Hasselbalch equation for blood serum, *J. Biol. Chem.,* 79, 183, 1928.
12. **Dill, D. B., Daly, C., and Forbes, W. H.,** The pK′ of serum and red cells, *J. Biol. Chem.,* 117, 569, 1937.
13. **Cullen, G. E., Keller, H. R., and Robinson, H. W.,** The pK′ of the Henderson-Hasselbalch equation for hydrion concentration of serums, *J. Biol. Chem.,* 66, 301, 1925.
14. **Robinson, H. W., Price, J. W., and Cullen, G. E.,** The value of pK′ in the Henderson-Hasselbalch equation for human and dog serum determined with the Sims electrode, *J. Biol. Chem.,* 106, 7, 1934.
15. **Hastings, A. B. and Sendroy, J., Jr.,** The effect of variation in ionic strength on the apparent first and second dissociation constants of carbonic acid, *J. Biol. Chem.,* 65, 445, 1925.
16. **Tibi, L., Bhattacharya, S. S., and Flear, C. T. G.,** Variability in pK_1 of human plasma, *Clin. Chim. Acta,* 121, 15, 1982.

17. **Tibi, L., Bhattacharya, S. S., and Flear, C. T. G.,** Actual or standard bicarbonate, *Lancet,* p. 1139, November 24, 1979.
18. **Natelson, S. and Alexander, M. O.,** Marked hypernatremia and hyperchloremia with damage to the central nervous system, *Arch. Intern. Med.,* 96, 172, 1955.
19. **Van Stekelenburg, G. J., Valk, C., and Donekerwolke, A. M. G.,** Variation of carbonic acid pK′ in blood and urine during $NaHCO_3$ infusion and NH_4Cl loading. A study of two renal acidotic patients, *Clin. Chem.,* 26, 60, 1980.
20. **Siggard-Anderson, O.,** The first dissociation exponent of carbonic acid as a function of pH, *J. Clin. Lab. Invest.,* 14, 587, 1962.
21. **Natelson, S. and Nobel, D.,** Effect of the variation of pK_1 of the Henderson-Hasselbalch equation on values obtained for total CO_2, calculated from P_{CO_2} and pH values, *Clin. Chem.,* 23, 767, 1977.
22. **Natelson, S. and Nobel, D.,** More on blood bicarbonate measurement, *Clin. Chem.,* 24, 1082, 1978.
23. **Harned, H. S. and Davis, R. Jr.,** The ionization constant of carbonic acid in water and the solubility of carbon dioxide in water and aqueous salt solutions from 0—50°C, *J. Am. Chem. Soc.,* 65, 2030, 1943.
24. **Harned, H. S. and Bouner, F. T.,** The first ionization of carbonic acid in aqueous solutions of NaCl, *J. Am. Chem. Soc.,* 67, 1026, 1945.
25. **Van Slyke, D. D., Hastings, A. B., Murray, C. D., et al.,** The distribution of hydrogen, chloride, and bicarbonate ion in oxygenated and reduced blood, *J. Biol. Chem.,* 65, 701, 1925.
26. **Rosan, R. C., Enlander, D., and Ellis, T.,** Unpredictable error in calculated bicarbonate homeostasis during pediatric intensive care. The delusion of fixed pK_1, *Clin. Chem.,* 29, 69, 1983.

26a. **Kost, G. J., Treat, K. T., Saeed, D.,** Indications for measurement of total carbon dioxide in arterial blood, *Clin. Chem.,* 34, 1650, 1988.

27. **Trenchard, D., Noble, M. I. M., and Guz, A.,** Serum carbonic acid pK′. Abnormalities in patients with acid-base disturbances, *Clin. Sci.,* 32, 189, 1967.
28. **Tietz, N. W.,** Comparison of calculated and experimental p_{CO_2} values, *Ann. Clin. Lab. Sci.,* 3, 341, 1973.
29. **Wills, M. R. and Lacte, P. A.,** Apparent change in pK′ of carbonic acid in plasma in response to acute metabolic acidosis in normal subjects, *Clin. Chim. Acta,* 35, 514, 1971.
30. **Hyde, T. A.,** The inconstant constant pK_1, *Can. Med. Assoc. J.,* 120, 1486, 1979.
31. **Austin, W. H.,** Further comments on the variation of pK_1 *Clin. Chem.,* 74, 1081, 1978.
32. **DeRaedt, M., Vandenbergh, E., and Van de Woestyne, K. P.,** Direct and indirect determination of partial pressure of CO_2 in arterial blood of patients with respiratory insufficiency, *Clin. Sci.,* 347, 1968.
33. **Austin, W. H., Terrante, V., and Anderson, C.,** Evaluation of whole blood pK′ in the acutely ill patients, *J. Lab. Clin. Med.,* 72, 129, 1968.
34. **Dean, N. and Smith, H. W.,** The apparent first dissociation constant pK′ of carbonic acid in the human erythrocyte, *J. Biol. Chem.,* 227, 101, 1957.
35. **Maas, A. H. J., Rispens, P., Siggard-Anderson, O., and Zjlstia, W. G.,** On the reliability of the Henderson-Hasselbalch equation in routine acid-base chemistry, *Ann. Clin. Biochem.,* 21, 26, 1984.

35a. **Masters, P., Blackburn, M. E. C., Henderson, M. T., Barrett, J. F. R., and Dear, D. R. F.,** Determination of plasma bicarbonates of neonates in intensive care, *Clin. Chem.,* 34, 1483, 1988.

36. **Hill, A. V.,** The possible effects of the aggregation of the molecules of haemoglobin on its dissociation curves, *J. Physiol.,* 40, IV, 1910.
37. **Natelson, S. and Natelson, E. A.,** Calculation of base excess and base deficit from Na^+, K^+ and chloride concentrations, in *Applied Clinical Chemistry,* Vol. 1, Plenum Press, New York, 1975, 219.
38. **Stewart, P. A.,** Independent and dependent variables of acid base control, *Respir. Physiol.,* 33, 9, 1978.
39. **Natelson, S. and Barbour, J. N.,** Equation and nomogram for approximation of alkali requirements in acidosis, *Am. J. Clin. Pathol.,* 22, 426, 1952.
40. **Stewart, P. A.,** Modern quantitative acid-base chemistry, *Can. J. Physiol. Pharmacol.,* 61, 1442, 1983.
41. **Stewart, P. A.,** *How to Understand Acid-Base,* Elsevier, New York, 1981, 174.

GENERAL REFERENCES

Natelson, S. and Natelson, E. A., *Principles of Applied Clinical Chemistry,* Vol. 1, Maintenance of Fluid and Electrolyte Balance, Plenum Press, New York, 1975.

Siggard-Anderson, O., *Acid Base Status of the Blood,* 3rd ed., Williams & Wilkins, Baltimore, 1974.

Suramicz, B., *Potassium and the Heart,* Lea & Febiger, Philadelphia, 1969.

Stewart, P. A., *How to Understand Acid-Base,* Elsevier, New York, 1981.

Weil, W. B. and Bailie, M. D., *Fluid and Electrolyte Metabolism in Infants and Children,* Grune & Stratton, New York, 1977.

Brenner, B. M. and Steen, J. H., *Sodium and Water Homeostasis,* Churchill Livingstone, Edinburgh, 1978.

Olszowka, A. J., Rahn, H., and Farlie, W. E., *Blood Gases, Hemoglobin, Base Excess and Maldistribution,* Lea & Febiger, Philadelphia, 1973.

Halpern, M. L. and Fields, L. A., Lactic acidosis — emphasis on the carbon precursors and buffering of the acid load, *Am. J. Med. Sci.,* 289, 154, 1985.

Frommer, J. P., Lactic acidosis, *Med. Clin. North Am.,* 67, 815, 1983.

DuBose, T. D., Jr., Clinical approach to patients with acid-base disorders, *Med. Clin. North Am.,* 67, 799, 1983.

Brueckner, P. J., Water, electrolyte and hydrogen ion disorders in *Applied Biochemistry of Clinical Disorders,* Gornall, A. G., Ed., Harper Medical, Lippincott, New York, 1981, 65.

Kaehny, W. D., Respiratory acid-base disorders, *Med. Clin. North Am.,* 67, 915, 1983.

Knoben, B. E. and Anderson, T. O., *Handbook of Clinical Drug Data,* 5th ed., Hamilton Press, Hamilton, IL, 1983.

Goldforb, I. W. and Yates, A. P., *Total Parenteral Nutrition. Nutritional Assessment and Parenteral Nutrition,* Synapse Publishers, Pittsburgh, 1978, 208.

Arndt and Dolle, W., *Klin. Wochenschr.,* 44, 511, 1966.

Weisberg, H. T., *Water, Electrolyte and Acid Base Balance,* 2nd ed., Williams & Wilkins, Baltimore, 1962.

FLUID MANAGEMENT OF THE PREMATURE AND NEWBORN INFANT

J. E. Sherwin and K. B. Rajani

INTRODUCTION

Improvements in prenatal, perinatal, and neonatal care have all contributed to decreased neonatal mortality. These improvements have been aimed not only at the immediate survival of the infant, but also at the prevention of long-term morbidity and developmental handicaps.

The immediate postnatal period is attended by growth and developmental changes which adapt the infant to extrauterine life.[1-4] The ability of the infant to adapt is a function of the intrauterine environment and the infant's developmental maturity. The normal infant adapts with little difficulty. On the other hand, a 26-week-gestation infant weighing 600 g will require intensive care. Fluid balance in these infants is precarious and requires careful monitoring.

The neonate has a relatively high surface-to-volume ratio.[3] This leads to increased insensible water loss (Figure 1). Renal function is immature and glomerular filtration rates are low.[5] Additionally, water balance is complicated by the fact that the body water content is normally being reduced by contracting the extracellular fluid volume at this time. In fact, for the most part, this accounts for the observed normal weight loss following birth.[2] Electrolyte balance is also precarious in these infants, not only because of fluid volume changes, but because of immaturity in renal filtration and reabsorption of electrolytes.[5]

DEVELOPMENTAL CHANGES IN BODY FLUID AND ELECTROLYTE COMPOSITION DURING THE NEONATAL PERIOD

The total body water content of the normal newborn is about 80%.[4] Approximately 55% of this is intracellular fluid, with the remainder as extracellular. The extracellular fluid is about 20% plasma water and 80% interstitial fluid. During the first weeks of postnatal life, the total body water content decreases to about 60% primarily at the expense of interstitial fluid.

Renal Function

Renal function is immature even in the normal neonate.[5] The glomerular filtration rate is only about 25% of the adult and matures slowly during the first 6 months of life (Table 1). The maximal concentrating ability of the kidney is also reduced by 25%. This is due to the fact that the loop of Henle is shorter and does not penetrate the medulla as deeply as in the adult kidney. Response to antidiuretic hormone in the neonatal period appears to be normal, but is limited by the renal concentrating ability.

Insensible Water Loss

Insensible water loss is inversely proportional to birth weight or gestational age.[4] This is due to increased permeability of the skin epidermis to water, larger body surface per unit of weight, and greater skin blood flow relative to metabolic rate. The higher respiratory rate of infants may also be a contributing factor. Elevation of ambient temperature by 1°C results in a three- to fourfold increase in insensible water loss. Use of a radiant warmer and phototherapy also increases insensible water loss by as much as 50%. In sick newborn infants, these considerations are very important as the infant's fluid intake is solely derived from parenteral sources. Fluid overload may lead to congestive heart failure secondary to patent ductus arteriosis with a left-to-right shunt. Factors that may reduce insensible water

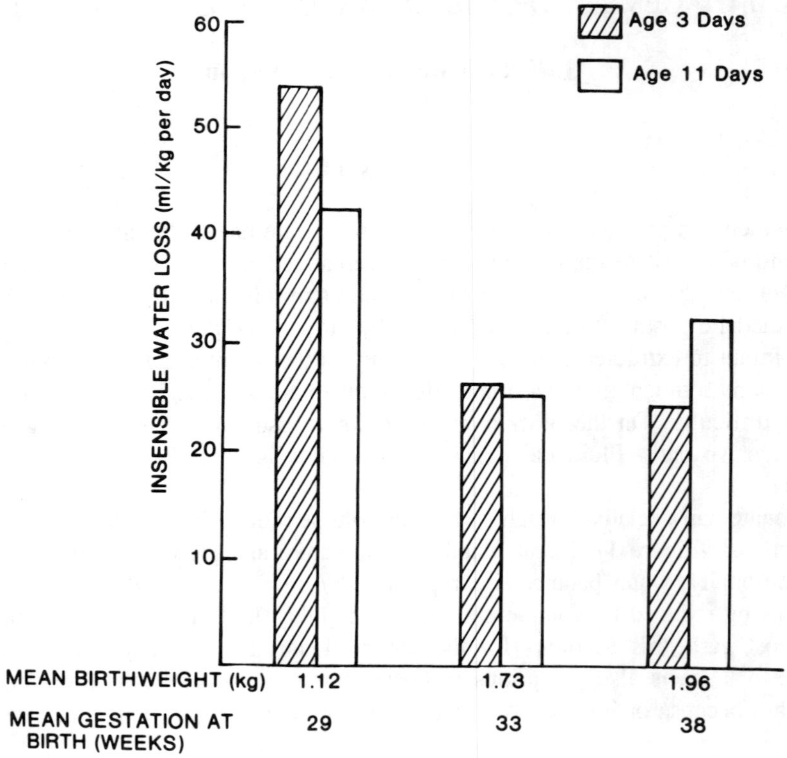

FIGURE 1. A comparison of insensible water loss in immature infants. (Reprinted from Evans, S. E. and Durbin, G. M., *Ann. Clin. Biochem.*, 20, 193, 1983. With permission.)

Table 1
GLOMERULAR FILTRATION RATES OF INFANTS IN THE FIRST 3 MONTHS OF LIFE

Age (days)	Glomerular filtration rate (ml/min/1.73 m^2)	
1	24	3—38
2—8	38	17—60
10—22	50	32—69
37—95	58	30—86

Reprinted from Evans, S. E. and Durbin, G. M., *Ann. Clin. Biochem.*, 20, 193, 1983. With permission.

loss include use of head shields, high relative humidity in the ambient environment, maintaining an infant in a thermoneutral environment, and keeping the infant in an isolette.

Fluid Administration

In those infants receiving enteral fluids, water and electrolytes are absorbed in the proximal jejunum.[2,3,6,7] Intestinal disorders resulting in diarrhea can rapidly result in dehydration in the neonate. Administration of hyperosmotic enteral fluids also results in an increased rate of water secretion by the mucosa of the small intestine. This may also result in necrotizing enterocolitis. Administration of parenteral fluids does not ordinarily alter the secretion of water by the gastrointestinal tract in the infant.[8]

Table 2
FASTING (4 H) REFERENCE VALUES FOR THE NEONATE (PLASMA FROM FINGERTIP BLOOD) DURING THE FIRST 48 H OF LIFE

Sodium	133—139 mmol/l	Urea N	8—14[a]
Potassium	4.9—5.9 mmol/l	pH	7.35—7.41
Chloride	96—102 mmol/l	pCO_2	30—40 mmHg
Calcium	8.0—9.8 mg/100 ml	pO_2	80—93 mmHg
Magnesium	1.6—2.1 mg/100 ml	T. CO_2	18—24 mmol/l
Glucose	55—90 mg/100 ml	Hematocrit	48—58%
Total protein	5.0—5.6 g/100 ml	T. Bilirubin	1.8—5.1 mg/100 ml
Osmolality			
Plasma	275—290 mmol/ KgH_2O		
Urine	60—450 mmol/ KgH_2O		

[a] On breast milk or synthetic formula simulating breast milk. Higher values will be obtained on a higher protein milk.

Electrolyte Losses

The electrolytes of concern here are the four major ones: sodium, potassium, chloride, and bicarbonate. The reference ranges for these electrolytes are nearly the same as for the adult, although the concentrations are less rigidly controlled in the neonate (Table 2). Serum sodium concentrations change most commonly in response to alterations in water homeostasis.[9,10] Hypernatremia is usually associated with water loss, but it may be accompanied by an increased extracellular fluid. It is particularly common in the pre-term infant because of high insensible water loss, a reduced ability to respond to shifts in sodium balance (Figure 2), and a physical inability to replace water loss *ad libitum.* Iatrogenic hypernatremia can result from osmotic diuresis due to hyperosmolar fluid or glucose solution administration. Hypernatremia can also result from excessive sodium intake from formula, parenteral fluids, or sodium bicarbonate administration. These sodium overload states are often accompanied by increased extracellular fluid volume and the infant becomes edematous.

Hyponatremia with hypovolemia largely results from the fact that the kidneys of premature infants have a low capacity to retain sodium when they are in negative sodium balance.[5,9] In this instance, the urinary sodium will be high. It is important to distinguish this renal immaturity from congenital adrenal hyperplasia which may present similar electrolyte values. Enteric fluid loss by vomiting or diarrhea due to systemic infection or other cause can also result in hyponatremia with a decreased extracellular volume.[11] Hyponatremia accompanied by edema due to increased extracellular volume is generally due to inappropriate antidiuretic hormone secretion or administration of hyperosmolar or dextrose solutions. Hyperkalemia can result from cell destruction, impaired renal function, or excessive intake. Symptoms of hyperkalemia include muscle weakness and cardiac conduction changes manifested in the electrocardiogram. Clinical manifestations of hyperkalemia are apparent at 6.5 meq/dl. Hypokalemia may be due to decreased intake, increased urinary losses due to renal tubular acidosis or hyperaldosteronism, or increased gastrointestinal losses due to vomiting, diarrhea, or diuretics. Hypokalemia results in muscle weakness, loss of abdominal sounds, and abdominal distention, and can be accompanied by polyuria.

Under normal circumstances sodium and chloride are excreted in equimolar amounts. In most instances, sodium is absorbed as the neutral sodium chloride across the intestinal brush border. Hypochloremia can accompany states of hyponatremia, e.g., inappropriate ADH syndrome, but may also selectively occur in infants who vomit excessively. This hypo-

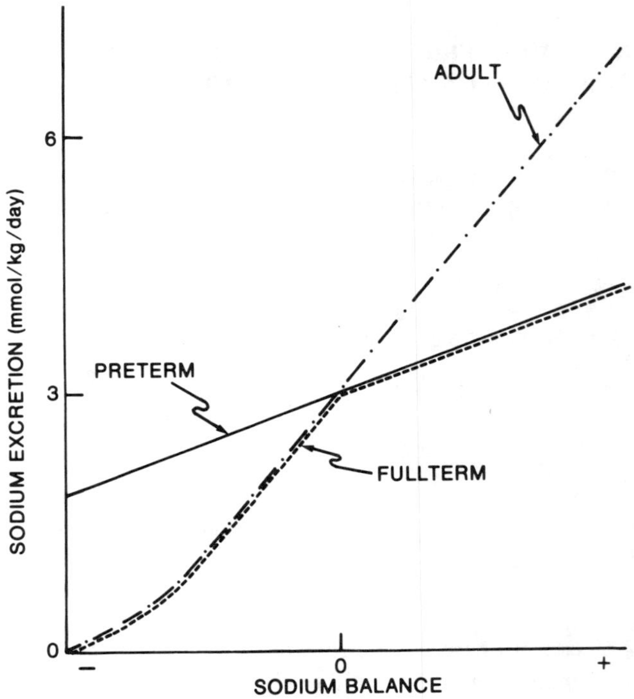

FIGURE 2. Model of the relationship between sodium balance and sodium excretion in full- and pre-term newborn infants compared with adults. Fullterm newborn infants have the same ability as adults to retain sodium when in negative balance, but have a low capacity to excrete a sodium load when in positive balance. Preterm infants are not only unable to excrete a sodium load rapidly when in positive balance, but have also a low capacity to retain sodium when in negative balance. (Redrawn from Aperia et al.[9])

chloremia is due to loss of chloride in the stomach acid. Hyperchloremia occurs primarily due to increased intake and is associated with hypernatremia.

Newborn infants are very susceptible to acidosis because of a limited capacity to excrete hydrogen ion.[4,5,9] Renal reabsorption of bicarbonate appears to be only slightly reduced in the neonate, which is reflected in the somewhat lower plasma bicarbonate concentration of 22 mmol/l. This approaches the adult normal of 24 to 27 mmol/l by 1 to 2 months of age. The proximal tubule reabsorbs the majority of bicarbonate and any remaining bicarbonate is filtered by the distal tubule. Therefore, two types of renal tubular acidosis can occur, one involving the proximal tubule and the other more common distal tubular acidosis. Treatment of neonatal acidosis is usually done with sodium bicarbonate.

Laboratory Findings in the Blood of the Low-Weight Neonate

The following observations may be made when examining the blood of the low-weight neonate. (See Reference Values, Table 1.)

Hematocrit Values

Hematocrit values of the low-weight neonate will tend to be high and continue to be elevated for several weeks post-partum. Subsequently, a steady drop of the hematocrit will be observed. In some, the fall will be to anemia levels. These infants may require packed erythrocyte administration to return the hematocrit to the normal range. This tendency toward anemia may persist in some infants for several months until the erythropoietic system matures.[13,14]

Sodium and Chloride

Sodium and chloride concentrations will tend to be low even with an apparently adequate intake of formula in the low-weight neonate. The lowered sodium and chloride levels may not be apparent until the infant is several days old and is well hydrated.

This phenomenon was first observed by Natelson et al.[13] and reported in 1952. He used the term "adrenal immaturity" for this phenomenon and stated, " . . . this term is used here to denote a condition resulting in the inability of the infant to hold water and sodium and chloride ion in the face of the administration of salt in what would be more than adequate amounts for a normal infant of equal weight. There is no unequivocal histological or pathological evidence to support the thesis that the adrenals are involved."

In reaction to this report, a symposium was held by the leading pediatricians of that time, in which this finding was questioned seriously.[15] In a response to this symposium, Natelson et al. reaffirmed the accuracy of their findings and cited a study undertaken by others to evaluate their findings which confirmed their observations.[16-18] Natelson et al. again stressed that *replacement fluids should be hypotonic.*

Since that time, several reports have appeared which confirm the fact that low-weight neonates tend to lose water, Na, and Cl, resulting in depressed plasma Na and Cl levels.[9,19] Aperia et al. state, "Very early born babies not being supplemented with extra sodium run a great risk of developing hyponatremia . . . "

Potassium

Potassium concentrations tend to be high, even in non-hemolyzed serum during and after the period of physiological jaundice has passed in untreated infants. This suggests that the potassium is probably derived by leakage rather than rupture of the erythrocyte.

Serum Glucose

Serum glucose concentrations are generally lower in the low-weight premature neonate as compared to the full-term infant of normal weight at birth. This may affect the Na-K pump and account for the tendency of K to diffuse into the plasma. Lactose is better absorbed from breast milk, as compared to pasteurized milk. Thus, with breast milk, blood sugar levels are higher.[20]

Urea

Urea concentrations fluctuate depending upon the state of hydration of the infant and the formula administered. Human milk is about 1% protein or somewhat less. Cow's milk is about 3% protein. If the protein in the formula is higher than that found in breast milk, urea and non-protein nitrogen concentrations will be higher, and there will be a tendency to develop an acidemia. This was first shown in 1952 and has been confirmed.[13,21-25]

Blood pH and Blood Gases

The low-weight neonate may have bouts of dyspnea. In this case, the pCO_2 *will rise* and the pH and pO_2 will be lowered, resulting in a respiratory acidosis. As pointed out above, this is distinct from the metabolic acidosis produced by high protein intake. With high casein intake, there will be a tendency to excrete organic acids in combination with Na ion and to retain chloride ion. In this case, pH and pCO_2 levels will be *lower.*[13]

The hazard in an isolette is a tendency toward high blood pO_2, unless the circulating gases are carefully controlled and monitored. Chronic exposure to gases of high pO_2 concentration may result in retrolental fibroplasia. High pO_2 levels also result in dyspnea, high pCO_2 levels, and acid pH.[26,27]

High Plasma Bilirubin

High plasma bilirubin concentrations will persist with the low-weight neonate for longer periods of time than with the normal infant. When an Rh or A-O incompatibility occurs, it needs to be treated as with the normal weight and full-term infant. If untreated, kernicturus may develop associated with alkaline blood pH.[28]

Fluid and Electrolyte Management

Maintenance of fluid and electrolyte balance requires assessment of three parameters: (1) estimate the quantity of the fluid and electrolyte deficit, (2) calculate the amounts required to replace losses and correct the abnormal deficit, and (3) implement therapy and monitor (and modify, if necessary) this therapy.[2-4] If an accurate measure of weight loss is available, this can be used as a direct reflection of fluid loss. If not, clinical assessment of dehydration can be used. The physical signs of skin turgor, blood pressure, activity, and mucous membrane hydration are used to differentiate hypotonic (serum sodium less than 130 meq/l), isotonic, and hypertonic (serum sodium greater than 150 meq/l) dehydration. Laboratory measurements of hematocrit, blood urea nitrogen, creatinine, and total protein are useful in conjunction with serum electrolytes for assessment of dehydration. Replacement of fluids is ordinarily done in the same manner as the fluid was lost, e.g., if dehydration is of rapid onset, fluid replacement should also be rapid. The one exception is in hypertonic deydration where rapid rehydration may cause convulsions. The assessment of electrolyte deficit is based upon physical examination and laboratory measurements of serum electrolytes.[3,4] Electrolyte disorders depend upon the underlying cause of the disorder. Severe diarrhea leads to isotonic, isonatremic dehydration.[11] Diarrhea with fever which has been treated by administration of milk formula can result in hypertonic dehydration; and hypotonic dehydration can result from inappropriate parenteral fluid administration, central nervous system disease, and fluid overload.

Repair Fluids

Calculation of the amount of electrolyte required for replacement is based upon the difference between the current total body solute and the normal body solutes.[12] Replacement of fluids is ordinarily done over at least 48 h, with 50% of the fluids administered in the first 8 h, except in the case of hypertonic dehydration in which replacement should be done evenly over 3 to 5 d. Calculations are done to return the plasma osmolality to normal, 280 mOsm/kg. It must be remembered that 50% of the osmolality deficit is due to cations and 50% due to anions. The percentage distribution of cations between sodium and potassium is a function of the underlying disorder. Potassium requirements can be estimated, but they may exceed the amount which can be safely administered rapidly so therapy must be prolonged. Potassium should not be administered to neonates without at least some urinary flow.

Once the fluid and electrolyte deficit has been estimated and maintenance requirements considered, therapy can be initiated using appropriate combinations of the readily available fluids. Calcium and magnesium may also be required, but are beyond the scope of this presentation. Maintenance therapy in the neonate requires 75 to 80 ml water per kilogram per 24 h. This fluid should contain 2 to 3 meq of potassium and sodium. As long as the sodium bicarbonate is greater than 12 mmol/l, the cations can be given as the chloride salts. Obviously, this fluid therapy will also require the administration of calories as glucose. Abnormal losses are added to the maintenance requirements for the total amount required.

Monitoring therapy in the neonate is crucial to an appropriate outcome, since the neonate has higher insensible water loss due to increased permeability of the epidermis, greater skin blood flow, and a higher respiratory rate, as well as a larger surface-to-volume ratio than the adult. In sick newborns, these differences are very important, because the infant's fluid intake is restricted to the parenteral route. Fluid overload in these patients may lead to congestive heart failure secondary to a patent ductus arteriosis with a left-to-right shunt.

The Environment

As necessary, infants should be kept in a thermoneutral environment, an isolette, or with a head shield to reduce insensible water loss. The composition of the air flow should be monitored continuously including pO_2, pCO_2, temperature, and humidity. Therapy should be monitored regularly and accurate records kept. Monitoring therapy includes maintaining a record of fluid input and output, monitoring specific gravity of urine, an accurate weight, and periodic reassessment of needs. Laboratory tests include serum sodium, potassium, chloride, and osmolality. These parameters should be monitored periodically until they return to normal. Other laboratory tests which may be required in specific disorders include blood urea nitrogen, serum creatinine, and urinary electrolytes.

REFERENCES

1. **Natelson, S. and Natelson, E. A.,** *Principles of Applied Clinical Chemistry: Maintenance of Fluid and Electrolyte Balance,* Vol. 1, Plenum Press, New York, 1975, 109.
2. **Levin, D. L. and Perkin, R. M.,** Abnormalities in fluids, minerals and glucose, in *A Practical Guide to Pediatric Intensive Care,* 2nd ed., Levin, D. L., Morriss, F. C., and Moore, G. C., Eds., C. V. Mosby, St. Louis, 1984, 93.
3. **Philip, A. G. S.,** *Neonatology: A Practical Guide,* Medical Examination Publishing, Flushing, NY, 1977, 129.
4. **Oli, W.,** Fluid and electrolyte management, in *Neonatology,* Avery, G. B., Ed., Lippincott, Philadelphia, 1975, 471.
5. **Evans, S. E. and Durbin, G. M.,** Aspects of the physiological and pathological background to neonatal clinical chemistry, *Ann. Clin. Biochem.,* 20, 193, 1983.
6. **Lebethal, E. and Hietlinger, L. A.,** Impact of development of the gastrointestinal tract on infant feeding, *J. Pediatr.,* 102, 1, 1983.
7. **Olsson, K.,** Central control of vasopressin release and thirst, *Acta Paediatr. Scand. Suppl.,* 305, 1983.
8. **Seashore, J. H.,** Metabolic complications of parenteral nutrition in infants and children, *Surg. Clin. North Amer.,* 60, 1239, 1980.
9. **Aperia, A., Broberger, O., Herin, P., Thodenius, K., and Zetterstrom, R.,** Post-natal control of water and electrolyte homeostatis in pre-term and full-term infants, *Acta Paediatr. Scand. Suppl.,* 305, 61, 1983.
10. **Milla, P. J.,** Disorders of electrolyte absorption, *Clin. Gastroenterol.,* 11, 31, 1982.
11. **Hirschhorn, N.,** Oral rehydration therapy for diarrhea in children — a basic primer, *Nutr. Rev.,* 20, 97, 1982.
12. **Natelson, S. and Natelson, E. A.,** Water, Electrolytes, and Blood Gases in *Handbook of Clinical Chemistry, Vol. IV.* Werner, M., Ed., CRC Press, Boca Raton, FL, 1989.
13. **Natelson, S., Crawford, W. L., and Munsey, F. A.,** Correlation of Clinical and Chemical Observations in the Immature Infant, Dept. Preventive Medicine Publ., Illinois Department of Public Health, Springfield, 1952.
14. **Steinberg, A. G.,** The hematocrit of capillary blood, *N. Engl. J. Med.,* 253, 308, 1955.
15. Forum: does the premature infant need adrenal cortical (DOC) support? A critical evaluation of the Rockford Program, *Q. Rev. Pediatr.,* 8, 63, 1953.
16. **Natelson, S. and Crawford, W. L.,** The author reappraise their work on the problem: does the premature infant need adrenal cortical (DOC) support?, *Q. Rev. Pediatr.,* 9, 59, 1954.
17. **Burk, E. C.,** Failure of premature infant to gain weight due to low Na feeding, *Proc. Staff Meet. Mayo Clin.,* 29, 90, 1954.
18. **O'Brien, D., Hansen, J. D. L., and Smith, C. A.,** Effect of supersaturated atmospheres on insensible water loss in newborn infants, *Pediatrics,* 13, 126, 1954.
19. **Engelke, S. C., Shah, B. L., Vasan, U., and Raye, J. A.,** Sodium balance in very low birth weight infants, *J. Pediatr.,* 93, 837, 1978.
20. **Natelson, S., Sherman, M., and Kramer, B.,** Carbohydrate supplements (lactose and starch hydrolysates) in infant feeding, *J. Pediatr.,* 34, 443, 1949.
21. **Natelson, S., Penniall, A., Crawford, W. L., and Munsey, F. A.,** Noncasein protein to casein ratio of feeding formulas, *Am. J. Dis. Child.,* 89, 1, 1955.
22. **Naudie, S. P., Prinsloo, J. G., and Haupt, C. E.,** Comparison between a humanized cow's milk and a soy product for premature infants, *S. Afr. Med. J.,* 55, 982, 1979.

23. **Davos, D. P., Saunders, R., and Gray, O. P.,** Weight gain and chemical changes in low birth weight infants after change from human milk to modified cow's milk, *Arch. Dis. Child.*, 47, 946, 1972.
24. **Svenninsen, N. W., Lindroth, M., and Lindquist, B.,** A comparative study of varying protein intake in low birth weight infant feeding, *Acta Paediatr. Scand,* 96, (Suppl.), 28, 1982.
25. **Atkinson, S. A., Bryan, M. H., and Anderson, G. H.,** Human milk feeding in premature infants, protein, fat and carbohydrate balances in the first two weeks of life, *J. Pediatr.*, 99, 617, 1981.
26. **Fujikura, T.,** Retrolental fibroplasia and prematurity in newborn rabbits induced by maternal hyperoxia, *Am. J. Obstet. Gynecol.*, 90, 854, 1964.
27. Symposium on retrolental fibroplasia, *Opthalmology,* 86, 1985, 1979.
28. **Natelson, S.,** Chronic alkalosis with damage to the central nervous system, *Clin. Chem.*, 4, 32, 1950.
29. **Natelson, S.,** Correlation of Clinical and Chemical Findings in the Immature Infant, Publication Division of Preventitive Medicine, Illinois Department of Public Health, Springfield, 1952.

BOOKS ON APPLICATION OF FLUID THERAPY TO THE NEONATE

Natelson, S. and Natelson, E., *Applied Clinical Chemistry,* Vol. 1, Plenum Press, New York, 1975.
Winters, R. W., Ed., *The Body Fluids in Pediatrics,* Little, Brown, Boston, 1973.
Meites, S., Ed., *Pediatric Clinical Chemistry,* American Association Clinical Chemistry Publishers, Washington, D.C., 1977.
Young, D. M. and Hicks, J. M., Eds., *The Neonate,* John Wiley & Sons, New York, 1976.
Faulkner, W. R. and Meites, S., Eds., *Selected Methods for the Small Clinical Laboratory,* American Association Clinical Chemistry Publishers, Washington, D.C., 1982.

CALCIUM AND MAGNESIUM METABOLISM

Winston W. K. Koo and Reginald C. Tsang

INTRODUCTION

The divalant cations, calcium (Ca) and magnesium (Mg) are closely interrelated in their effects on cellular metabolism and cell function. Their actions often mimic each other, but at times may be antagonistic. The physiology and metabolism of these cations will be compared and contrasted in health and disease.

BODY CONTENT

Body Content of Calcium

Ca is the most abundant mineral in the body. In adults, the total body Ca is approximately 1.3 kg. The bone of the infant is less densely mineralized than that of adults, there being approximately 4 and 19 g of Ca per kilogram of body weight in the infant and adult, respectively.[1-3] In contrast, the amount of Ca outside bone in extracellular fluids and soft tissues does not exceed 10 g in adults.[4] The intracellular concentration of Ca is minute, being 10^{-5} to 10^{-8} *M*.[5,6]

At all ages, 98% of the body's Ca is in bone, where it is held in a crystalline form of calcium phosphate that resembles the mineral hydroxyapatite, and in a non-crystalline or amorphous calcium phosphate form, within a soft, fibrous organic matrix. It appears that amorphous material is predominant in early life and is superseded by crystalline apatite in later life.[7] The skeleton plays an important role in the maintenance of Ca homeostasis, since bone Ca may be mobilized when serum Ca concentration is low[2,8] and can act as a reservoir in response to acute Ca loading. In the dog, for example, the predominant means of maintaining Ca homeostasis after an i.v. Ca load is thought to be due to skeletal uptake, with less than 20% of the load excreted via the kidney.[8]

Body Content of Magnesium

Mg is the fourth most abundant metal in the body. In adults, the total body Mg is reported to be between 21 and 28 g.[1-3,9] The body Mg content also increases between infancy and adulthood, there being approximately 0.22 and 0.39 g of Mg per kilogram body weight in the infant and in the adult, respectively. About 60% of the body's Mg is in bone, 20% is in muscle, and most of the remainder is found in intracellular space of other tissues. The majority of bone Mg is intimately associated with the apatite crystal, whereas the muscle Mg pool is largely bound to lipoproteins, nucleoproteins, ribonucleic acids, and adenosine triphosphate.[10] Only 30% of bone and 20 to 30% of muscle Mg is free or is in a readily exchangeable pool[9]

In contrast to the low concentrations of intracellular Ca, Mg is the second most common intracellular electrolyte in the body. In most tissues, the concentration of Mg within the cell is estimated to be 10 mmol/l (1 mmol/l = 2.4 mg/dl), but most of it is bound to cellular constituents, such as nucleic acids and proteins. The free intracellular Mg concentration may be 0.1 to 1 mmol/l, as compared with 0.8 mmol/l in plasma.[11]

The release of both Ca and Mg from bone are apparently stimulated by parathyroid hormone and decreased by calcitonin.[2] In addition, a heteroionic bone exchange of Mg for Ca has been demonstrated *in vitro*.[12] Mg deficiency may decrease bone Ca release *in vitro*[12] and increase the percentage of bone Ca, compared to Mg-replete rats *in vivo*.[13] This supports the hypothesis that less bone to extracellular Ca flux with hypomagnesemia could contribute to the development of hypocalcemia. The skeleton, therefore, is important in maintenance of homeostasis for both Ca and Mg.

SERUM CONCENTRATIONS

The circulating fractions of both Ca and Mg are <1% of the total body content. Differences between plasma and serum values are minimal. Total serum Ca is maintained within narrow limits at 9.9 mg/dl ± 5% (2SD);[14,15] the diurnal range is <±3%.[16] Forty to 45% of serum Ca is bound to protein. Of the protein-bound Ca, about 81% is bound to albumin and 19% to globulin. The Ca binding capacity of human serum ranges from 0.4 to 1 mg/g of protein, varying with hydrogen ion concentration, temperature, and ionic strength of other serum constituents. Total serum Ca concentration is decreased by 0.8 to 1.0 mg/dl for every gram-per-deciliter decrement in serum albumin from normal.[17,18]

Of total serum Ca, 55 to 60% is ultrafilterable; 85 to 90% of this fraction exists in an ionized form and 10 to 15% in a complexed form.[15,19,20] Serum-ionized Ca concentration is the best indicator of physiologic Ca activity. A circadian rhythm for ionized Ca has been reported[21] with a peak occurring at 1000 h; maximal change between mean peak and trough values is small and averages 0.3 mg/dl. Inasmuch as the concentration of ionized Ca is critical to many important biological functions, it is not surprising that there is a finely tuned regulation of the extracellular concentration of this ion. Both serum total Ca and total Mg concentrations are normally distributed.[14,15,22,23]

In a population survey,[23] serum total Mg apparently is higher in infants and early childhood than adults, being 2.2 ± 0.3 mg/dl (mean ± 2 S.D.), compared with approximately 2.1 ± 0.3 mg/dl (mean ± 2 S.D.). Differences between mean serum total Mg values for males and females are negligible; mean serum total Mg concentrations in whites are slightly, but consistently, greater than blacks by approximately 0.1 mg/dl.

About 30% of serum Mg is protein bound. The ultrafilterable portion of serum Mg is greater than that for serum Ca, being approximately 70 to 80%.[24,25] Of the filterable Mg, only 70 to 80% is in the ionic form, the remainder being complexed to anions, particularly phosphate, citrate, and oxalate.[19,25]

INTAKE AND INTESTINAL ABSORPTION

Calcium Intake

The recommended daily intake in infancy and early childhood[4] is about 60 mg Ca per kilogram, and is greater for pre-term infants;[3,26] in adults[4] it is about 800 mg, and higher for women (by an additional 400 mg/d) during pregnancy and lactation;[4] the requirement may be higher for post-menopausal woman, currently recommended to be 1 g elemental Ca per day.[27]

Calcium Absorption

The proportion of Ca absorbed from the intestinal tract may be inversely related to dietary Ca intake in infants[28-30] and adults.[31-33] Intraindividual variations in Ca absorption may contribute to the stabilization of plasma Ca concentrations.[33] Ca absorption occurs by active transport primarily in the duodenum;[2,34,35] there is also absorption by passive diffusion. Normally, the colon plays no significant role in Ca absorption; significantly increased colonic Ca absorption may occur in patients with long-segment small-bowel resection.[36] Active Ca transport may be the saturable component, while the passive transport may be the nonsaturable component of duodenal Ca transport.[34,35] In humans, absorbed Ca may be linearly related to Ca intake; there appears to be a constant passive, concentration-dependent (''slope'' of regression line of absorbed vs. ingested Ca) and constant active (''Y-intercept'') transport of Ca over a wide range of Ca intakes from a mean intake of 65 to 176 mg Ca per kilogram per day.[37]

Vitamin D metabolites, in particular, 1,25-dihydroxyvitamin D (1,25[OH]$_2$D), appear to affect active transport and passive diffusion of Ca.[2,34,35] Inhibitors of protein and RNA synthesis also inhibit vitamin D_3 and 1,25(OH)$_2$D$_3$-dependent Ca absorption in rachitic chick[38] and cultured embryonic chick duodenum.[39] Two categories of vitamin D-dependent intestinal mucosal proteins have been reported: (1) soluble proteins, for example, the soluble intestinal Ca binding protein (Ca BP);[40,41] and (2) proteins associated with membranes at the brush border, mitochondria, or basal lateral cell membrane, for example, brush border Ca BP,[42] alkaline phosphatase,[38] and Ca-stimulated ATPase.[38,43] However, as yet there is no unified model of intestinal absorption, transport, and efflux of Ca, vitamin D-dependent proteins may play a role in intestinal Ca absorption,[40,44] but increase in intestinal Ca BP[45,46] and duodenal alkaline phosphatase activity[46] lag behind the onset and peak rate of intestinal Ca transport. From *in vitro* studies using ligated intestinal loop, there is an apparent dissociation between Ca BP concentration and Ca absorption.[42] However, appearance of at least two brush border membrane proteins following *in vivo* administration of 1,25(OH)$_2$D to rachitic chicks correlate well with Ca transport.[47] Mitochondria are thought to play an important role in the transport of Ca ion from apex to the basal region of the cell where entry into the capillaries takes place.[48]

Lipids also may play a role in the vitamin D-dependent Ca transport process, since an essential fatty acid (EFA)-deficient diet fed to rats for 2 weeks is sufficient to negate the effect of 1,25(OH)$_2$D$_3$ on Ca transport across the gut.[49]

Passive intestinal Ca absorption is directly related to Ca intake. In the suckling rat, this appears to be the dominant mechanism for intestinal Ca absorption until the time of weaning after 2 weeks.[50] 1,25(OH)$_2$D receptors are not found in newborn rat pups until 2 weeks or older.[51] In human infants, Ca absorption may also be linearly related to Ca intake and not vitamin D dependent, particularly after correction for endogenous fecal losses of Ca.[29,52] Others have shown that Ca absorption and retention may be related to vitamin D intake.[53]

Ignorance of endogenous intestinal mineral secretion may result in a significant underestimation of the true mineral absorption. Earlier studies using Ca isotopes[29,54] reported variable endogenous intestinal Ca loss. The endogenous fecal loss of Ca was calculated to range from 4 to 150 mg/d (1.2 to 52.4% of Ca intake) in 13 infants who weighted 1.5 to 2.7 kg.[29] These infants received formulas with three different Ca/P ratios. A recent report of two pre-term infants using stable isotopes of Ca, and three-compartmental model analysis, demonstrated much greater endogenous fecal loss of Ca (approximately 25 times) than urinary Ca loss.[55] Similar findings were found in infant rhesus monkeys.[55]

Since variations in endogenous Ca losses markedly affect results of Ca absorption studies, the true percent Ca absorption is probably higher than the 29 to 80% in earlier reports based on standard balance studies.[3,26,28,53] The percentage of Ca absorption is higher in human milk fed infants, but the absolute quantity of Ca absorbed is lower, since Ca content of human milk is lower when compared to cow milk formula. In a more recent study with extrinsically tagged ^{46}Ca stable isotope, Ca absorption appears very high and may reach 90% of intake in preterm infants fed their own mother's milk.[56] Whether absorption of the ^{46}Ca label is a true reflection of dietary Ca absorption remains to be confirmed.

In infants, maturational changes, particularly length of gestation and postnatal age, appear to directly favor Ca absorption, regardless of whether infants are fed full cream or modified cow milk, or human milk.[2,3] Intestinal Ca absorption decreases in the elderly human.[31,57]

Specific dietary items may affect Ca absorption. Lactose enhances Ca absorption.[58,59] However, in lactase-deficient adults[59] and in rats on a prolonged high lactose diet,[60] lactose lowers the intestinal Ca absorption. In the high lactose rat model, a decrease in intestinal Ca BP is noted.[60]

The effect of dietary fat on Ca absorption, and vice versa, is controversial. Ca absorption appears high with fats low in stearate and palmitate, but high in oleate.[61] Use of easily absorbed fat in the form of medium-chain triglyceride[62] increases intestinal Ca absorption,[63] although others have reported no difference in Ca absorption between the use of medium- and long-chain triglyceride formula in premature infants, with standard (42 mg Ca per deciliter)[64] or high (>128 mg Ca per deciliter)[65] Ca content. Some investigators have found no correlation between Ca and fat absorption over normal ranges of fat intake,[3,66] though excess fat intake is reported to decrease Ca absorption.[67] High Ca intake decreases fat absorption in adults,[68] children,[69] and term infants.[70] In pre-term infants, high Ca intake decreases fat absorption,[52,71] in particular, saturated fatty acids of long-chain length, such as C16:0 (palmitate) and C18:0 (stearate);[72] others show no effect of high Ca intake on fat absorption.[3,73]

The quantity and type of protein ingested may affect Ca metabolism. Increased protein (from mixed sources, including pure L amino acid mixtures) intake, regardless of amount of Ca ingested, leads to increased urinary Ca loss.[74-76] This is thought to be related to increased glomerular filtration with possible inhibition of renal tubular Ca reabsorption. Using meat as the only source of high protein intake, there is no increase in urinary Ca excretion with a daily intake of Ca 2.5 times normal.[77] This is thought to be related to concurrent high phosphorus intakes in the meat diet, which decreases urinary Ca loss through its effect on bone and calciotropic hormones (see later). Phosphorus supplement to high protein diets decreases urinary Ca loss.[78,79]

Increased dietary phosphate increases Ca retention in human adults[78,80] and in term[81] and preterm[53] infants fed human milk, predominantly by a reduction in urine Ca excretion; there is a tendency to increase intestinal Ca absorption in adults[78] and in infants fed low Ca containing milk, such as human milk.[53,81] A high dietary Ca intake impairs phosphate absorption in animals,[82,83] human adults,[78,84] and pre-term infants.[52] In a recent study in term infants over a mean Ca intake 65 to 176 mg/kg/d, phosphate absorption was lowered[37] with high Ca intake; since urinary phosphate excretion also decreased, there was no significant effect on phosphate retention.

High dietary intake of Ca in animals reduces blood and bone zinc[85] and reduces zinc bioavailability.[86] In human adults[87,88] and in infants,[37] zinc balance and retention are not affected by dietary Ca. It appears that with satisfactory zinc nutritional status and ample zinc intake, a modest increase in Ca intake is unlikely to affect zinc absorption. In humans, absorption and retention of copper and nitrogen are not influenced by dietary Ca.[37,87,88] In human adults, intestinal fluoride absorption is decreased when Ca and fluoride supplementation are given concurrently,[89,90] but fluoride balance is not affected.[89] Ca diminishes inorganic iron absorption in a dose-related manner in animals[91] and in humans.[88]

Magnesium Intake

Mg occurs widely in foods — particularly those of vegetable origin.[92] Based on balance studies and usual dietary intakes, the allowance recommended in adults for males is 350 mg/d and for females 300 mg/d. An additional allowance of 150 mg is recommended for pregnant and lactating women. In infants, the allowance is 50 mg/d for the young infant and 70 mg/d for the older infant. The rapid growth rate during infancy, particularly of the skeleton, increases Mg needs per body weight compared to adults.[4] Human milk contains approximately 35 mg mg/l[93] and appears adequate for healthy infants, even those of low birth weight.[94]

Magnesium Absorption

Mg absorption occurs uniformly throughout the small intestine.[95,96] Although fractional absorption of Mg varies inversely with intake,[95-97] there is no convincing evidence that it is physiologically regulated. The intestinal transport mechanism for Mg may be different from that for Ca.[98] Mg absorption may involve two separate processes: a diffusion component, and a facilitated diffusion or saturable component.[96] Intestinal Mg absorption is always down an electrochemical gradient in the usual range of Mg intake; the absorption is a linear function of intake in adults[96] and infants,[28] consistent with diffusion. However, decreased fractional absorption with increasing concentrations is consistent with a carrier-mediated process.

Mg absorption, as with Ca absorption, also varies with age. In adults receiving the usual range of Mg intake, percent Mg absorption is generally between 25 and 60% of intake.[95,97,99,100] Normal infants have greater intestinal Mg absorption in the range of 46 to 73%.[28,37,63,64,81,101] Mg absorption does not appear to vary with gestational age in low-birth-weight infants between 29 and 34 weeks gestation,[64] and there is no increase in Mg absorption with increasing post-natal age during the first month after birth.[63]

Vitamin D or its metabolites in modest doses appear to have little effect on Mg absorption in the growing rat,[102] but may be important for the growing pig.[103] Vitamin D and its metabolites have little effect on intestinal Mg absorption in humans, based on Mg absorption before and after vitamin D supplementation in normal individuals, and in patients with osteoporosis, osteomalacia, and hyper- and hypoparathyroidism;[104] in uremic patients given physiological doses of $1,25(OH)_2D$;[105] and in healthy adult volunteers given supraphysiologic doses of $1,25(OH)_2D$.[106] In healthy adults and anephric patients, there is no correlation between plasma $1,25(OH)_2D$ and intestinal absorption of Mg and phosphorus.[99]

The endogenous fecal excretion of Mg for adults appears small,[107-109] with a small obigatory secretory Mg loss of approximately 0.36 mg/kg/d.[110]

Mg absorption may be improved with use of easily absorbed fat in the form of medium-chain triglyceride.[63] Protein intake may increase Mg absorption.[111] Increased phosphorus uptake may decrease Mg absorption owing to formation of less-soluble phosphate-Mg complexes. However, in full-term breast-fed infants, phosphate supplements to human milk improve Mg absorption and retention.[81] In lambs, an increase in dietary potassium was associated with a dose-dependent decrease in Mg absorption and retention; whereas dietary Mg did not affect the amount of potassium excreted, absorbed, or retained.[112] High oral uptake of Mg may cause substantial phosphate depletion[113] as well as an increase in urinary Ca.[114] Hypokalemia and acidosis have been reported with massive ingestion of magnesium oxide.[115] Selective defects in intestinal Mg absorption has been reported for humans and this condition may be familial.[116,117]

Interaction of Calcium and Magnesium Absorption

Variable Ca intakes have little or no influence on Mg absorption in human adults[96] or infants,[30,37] although Mg depresses Ca absorption to a modest extent in human adults.[96] Ca and Mg ions at physiologic concentrations increase the *in vitro* intestinal transport of fatty acid and monoglyceride in bile salt solution.[118] The soluble chlorides of both Ca and Mg depress intestinal water absorption in humans.[119] Energy-dependent, active, intestinal transport of sodium, potassium, Ca, and Mg may share common mechanisms, since variations in intestinal absorption of these elements can occur with varying doses of ouabain, an inhibitor of sodium potassium-dependent ATPase.[120]

RENAL REGULATION

Calcium Reabsorption in the Nephron

Ca transport occurs in most segments of the nephron, except the descending and ascending thin limbs of the loop of Henle.[121-124] Approximately 80% of the filtered Ca is reabsorbed in the proximal convoluted and straight tubules.[123] Ten to 20% is reabsorbed in the thick ascending limb of loop of Henle and distal nephron.[122,124] Normally, urine Ca excretion is

100 to 200 mg/d during calcium intake of <1 g/d. The fraction of filtered Ca that is excreted is approximately 1%. In the proximal convoluted tubule and loop of Henle, Ca transport is primarily passive by electrochemical gradient, occurs in association with sodium chloride absorption, and may proceed through paracellular electrodiffusive pathways.[121,123] Active transport against both an electrical and concentration gradient occurs particularly in the distal tubule and collecting duct.[121,122,124-126] Although a small fraction of the filtered load is reabsorbed in the distal nephron, the transport process is more selective, allowing specific regulation of urinary Ca excretion independent of alterations in excretion of other ions.

Factors Influencing Renal Regulation of Calcium

Factors that alter nephron Ca transport may be related to effects on altering sodium excretion, secondarily affecting Ca excretion through the transport relationship between these ions; inhibition of transport in the distal nephron; altering cellular mechanisms of Ca transport. Acute increases in glomerular filtration rate,[127] acetazolamide,[128] infusion of renal vasodilators,[129] and saline infusion[130] result in increased delivery of sodium and Ca to distal portions of the nephron and increased urinary Ca excretion. Large increases in urine Ca excretion occur only with the latter two effects, presumably because of simultaneous inhibition of sodium and Ca transport in distal portions of the nephron.

Clinical factors increasing urinary Ca excretion include: increased dietary sodium intake with resultant positive sodium balance[131] and extracellular fluid expansion;[127] i.v.[74] and oral glucose loading,[74,132] and phosphate depletion;[113,133,134] induced[132] or disease-related[135] metabolic acidosis; hypercalcemia;[136,137] and parenteral nutrition.[2,138]

Potent loop-active natriuretic agents such as furosemide cause hypercalciuria.[139] chronic benzothiazide use can produce hypocalciuria,[140] related in part to extracellular fluid volume contraction; replacement of sodium loss and prevention of negative sodium balance markedly blunts the fall in Ca excretion.[140] Ca excretion is decreased with alkali administration.[141] The hypocalciuria of nephrotic syndrome may relate to contraction of extracellular fluid volume.[142]

Various hormones may affect urine Ca excretion. Insulin increases Ca excretion.[143] The parathyroid hormone (PTH) effect is complex. Micropuncture studies in the dog demonstrate that PTH inhibits proximal tubular reabsorption of sodium and Ca.[144] However, PTH increases Ca reabsorption in the cortical thick ascending limb of Henle's loop[145] and distal nephron.[146] The net result is a decrease in urinary Ca excretion.[144-148] The hypocalciuric effect of PTH may be countered by any concomitant hypercalcemia and resultant increase in glomerular filtered Ca load, as seen in hyperparathyroidism. Calcitonin (CT) in pharmacologic doses also may decrease tubular Ca reabsorption in humans,[149] but increases tubular Ca reabsorption in the rat[150] if there is a simultaneous decrease in plasma Ca concentration. Renal effects of both PTH and CT may be mediated via cyclic AMP.[124,144,151]

In the rat, $1,25(OH)_2D_3$ receptors are present in both proximal and distal nephron.[152] In the dog, vitamin D_3, $25(OH)D_3$, and $1,25(OH)_2D_3$ enhance tubular reabsorption of Ca, as well as sodium and phosphate.[153-155] Vitamin D metabolite ($25[OH]D_3$, $1,25[OH]_2D_3$-enhanced Ca reabsorption is independent of sodium reabsorption in the distal nephron and independent of PTH.[155] Vitamin D metabolites also stimulate renal RNA biosynthesis,[156] and a vitamin D-dependent Ca BP (MW 28,000) is present in rat[157] and human kidney tubule.[157,158] Nevertheless, hypercalciuria occurs often in patients on high-dose $1,25(OH)_2D_3$ treatment.[159]

Magnesium Reabsorption in the Nephron

In contrast to Ca reabsorption in the nephron, most (50 to 60%) of the filtered Mg is reabsorbed in the loop of Henle, mostly in the thick ascending limb. Only 20 to 30% of filtered Mg is reabsorbed in the proximal tubule, while approximately 10% of filtered Mg is delivered into the distal nephron.[123,160] The reabsorptive rate of Mg changes in parallel with, but is quantitatively less than that of, sodium and water transport;[160-162] Mg reabsorption is reduced when intraluminal Mg concentration is increased,[163,164] probably associated with inhibition of sodium transport.[164] Mg reabsorption in the ascending limb may be voltage dependent and secondary to active sodium chloride reabsorption.[165]

Factors Influencing Renal Regulation of Magnesium

In humans, the major excretory pathway for Mg is renal, and both oral and i.v. loads are rapidly eliminated.[166-168] Normally, Mg excretion is approximately 5% of the filtered Mg load.[168-170] Mg excretion primarily reflects dietary intake and intestinal absorption. Pronounced renal conservation of Mg is possible in Mg-deficient states, with urinary Mg excretion less than 12 mg/d.[110]

Factors increasing Mg excretion include Mg[166,168] and sodium loading;[171,172] osmotic diuresis;[171] dietary phosphate restriction in humans[134] and rats,[173] correctable with phosphate supplementation;[134,174] readily metabolizable substrates such as glucose, galactose, fructose, protein, and alcohol when given in excess;[74,175] loop diuretics such as furosemide,[139,176] osmotic diuretics such as mannitol[177] and urea;[178] and pharmocological agents such as *cis*-platinum[179] and gentamicin.[180] A primary defect in renal Mg reabsorption has been reported in humans[181,182] and this condition may be familial.

Mg and potassium are both major intracellular cations, and disturbances in potassium balance often are seen in Mg wasting diseases; for example, hypokalemia generally is associated with hypomagnesemia, of either dietary[110] or acquired origin;[179,180,183] in addition, during continuing Mg depletion, intracellular potassium depletion cannot be restored to normal until Mg is also replaced, even with potassium supplementation.[110,184] Conversely, in primary potassium depletion, intracellular Mg is also reduced;[9,185] in animals, a high potassium diet results in greater Mg retention than a normal potassium diet.[186] However, Mg and potassium are handled by different transport mechanisms in the nephron, potassium being extensively reabsorbed in the proximal tubule and secreted in the more distal portions of the nephron.[187]

Hypercalcemia inhibits Mg reabsorption primarily in the loop of Henle and results in magnesuria.[2,188,189] Conversely, Mg infusion can result in increased Ca as well as Mg excretion[167,190] not corrected by PTH.[190] The effect of Mg infusion is primarily due to a decrease in Ca reabsorption in Henle's loop.[191] These *in vivo* studies support the presence of Mg-Ca interaction at a common reabsorptive site. However, *in vitro* studies with isolated rabbit tubules show that elevated perfusate Ca had no effect on Mg absorption, and perfusate Mg had no effect on Ca transport.[165]

Hormones may also affect Mg excretion. PTH increases Mg reabsorption by a mechanism unrelated to changes in plasma Ca and Mg.[147,148] CT may increase loop Mg reabsorption either directly[192] or indirectly through a fall in plasma Ca concentration.[150] Mineralocorticoids, thyroid hormone, vitamin D metabolites, and growth hormone have been associated with increased urinary Mg excretion.[2,187]

HORMONAL REGULATION

The hormonal regulation of Ca and Mg homeostasis is complex. Three major hormones have a predominant role: PTH, CT, and $1,25(OH)_2D$. Some of the interactions between hormone and target organs have been discussed in the sections on intestinal absorption and renal excretion of Ca and Mg. This section provides an overview of hormonal regulation in the maintenance of normal circulating Ca and Mg concentrations.

Parathyroid Hormone

PTH is an 84-amino acid polypeptide with a molecular weight of 9500. It is derived from its precursors, pre-pro- and pro-PTH, through cleavage by specific proteases. The 1-34 fragment at the N termin of PTH appears to be the potent fraction for biological activity. Circulating PTH concentrations may have a circadian rhythm, with the highest concentration of serum PTH reported during the period 2:00 to 4:00 a.m. Other biological factors such as age and pregnancy may influence serum PTH concentrations.[2]

Physiologically, PTH is the most important regulator of extracellular Ca circulation.[2] Serum PTH concentrations rise within 20 s after the beginning of EGTA infusion (which lowers ionized Ca) and elevation of PTH may occur with a lowering of basal serum Ca by as little as 0.1 mg/dl.[193] PTH restores extracellular Ca concentration by stimulating mobilization of mineral from bone, increasing renal tubular Ca reabsorption, and increasing intestinal Ca absorption secondary to stimulating 1,25$(OH)_2$D production.[194] Although PTH action on bone results in an increase in serum phosphate, its potent phosphaturic effect overwhelms the bone effect and results in lowered serum phosphate. The net effect of PTH action is elevation of serum Ca and lowered serum phosphorus concentration.

Extracellular Ca, in turn, is the most potent regulator of PTH secretion.[2] PTH secretion is most pronounced when serum Ca is in the mildly hypocalcemic range, with a relatively constant basal secretion of PTH in normocalcemia and little or no additional PTH secretion when serum Ca is <7.5 mg/dl. There is, thus, a sigmoidal type of PTH secretion response to serum Ca concentrations.[195] Cytosolic Ca concentration is directly correlated with extracellular Ca concentrations;[5] together with adenylate cyclase activity[2,196] it may be important in the direct regulation of PTH secretion.

An acute decrease in serum Mg concentration also can stimulate PTH secretion,[197] although chronic hypomagnesemia inhibits PTH secretion.[198,199] Other factors such as catecholamines, growth hormone, CT, cortisol, and prostaglandin may modulate PTH secretion; however, their role in regulation of Ca metabolism under physiologic conditions is not clear. The effect of PTH on Mg metabolism appears to include mobilization of Mg from bone along with the mobilization of bone Ca, and increasing renal Mg reabsorption (see "Factors Influencing Renal Regulation of Magnesium").

Hypercalcemia and hypermagnesemia are both able to block PTH release.[200,201] Hypermagnesemia has been reported to result in hypocalcemia, probably due, in part, to its effect on suppression of PTH secretion.[201]

Mg and Ca interrelationships remain close with respect to PTH. Hypomagnesemia may lead to hypocalcemia through several mechanisms: decreased secretion of PTH, target organ resistance to PTH, decreased intestinal absorption of Ca (probably via decreased 1,25$[OH]_2$D production); and decreased heterionic exchange of Ca for Mg at the bone surface, with reduction of Ca release from bone.[198,202]

Calcitonin

CT monomer is a 32-amino acid peptide with a molecular weight of 3500. There are biosynthetic CT precursors consisting of short sequences of basic amino acids linked to the amino and carboxyl terminals of CT. The most potent fraction for biological activity of human CT (hCT) appears to reside in the 8- to 32-amino acid region. In contrast to PTH, the linear carboxyl-terminal portion of the CT molecule is of prime importance in its biological function. As with PTH, serum CT concentrations may also follow a circadian rhythm, but unlike PTH, the peak serum CT concentration occurs around midday. This and other biological factors such as sex, age, and pregnancy should be taken into account when interpreting serum CT concentrations.[2,203]

The biological action of CT is frequently opposite to that of PTH; CT probably modulates the effect of PTH on target organs (see previous sections). However, *in vitro* studies demonstrate that the actions of both hormones may be mediated via production of adenosine 3′5′-monophosphate.[204]

In experimental situations, CT and phosphorus are two physiologic inhibitors of bone resorption.[205,206] CT decreases renal tubular reabsorption of Ca, Mg, phosphorus, and sodium in humans. Free water clearance is also increased as a consequence of decreased proximal reabsorption of sodium.[149] However, CT has been shown to increase renal Ca and Mg reabsorption in the rat[150,192] which may be related to the hypocalcemic effect of CT,[150] with secondary elevation of PTH and its effects on renal reabsorption of Ca, Mg, and phosphorus. Infusion of salmon CT induces intestinal secretion of water, sodium, potassium, and chloride ions in humans[207] and may decrease intestinal Ca and phosphorus absorption, secondary to net secretion of water and ions.[208] Elevated serum Ca and Mg concentrations increase CT release. However, the physiologic role of CT in the regulation of Ca and Mg homeostasis is not well defined and there are no major difficulties with Ca or skeletal homeostasis in humans with congenital or acquired CT deficiency.[203]

Vitamin D

Vitamin D is produced endogenously (vitamin D_3) in the skin or ingested (vitamin D_3 or D_2). Conversion to 25-hydroxyvitamin D (25OHD) occurs in the liver and subsequent conversion to 1,25$(OH)_2$D occurs in the kidney. 1,25$(OH)_2$D is the most active vitamin D metabolite. Unlike serum PTH and CT, serum 1,25$(OH)_2$D in the human does not undergo large fluctuations during the day, but is maintained within approximately 20% of its overall 24-h mean.[209] However, serum 1,25$(OH)_2$D concentrations vary with postnatal age[210] and may increase, as an apparently direct consequence of rise in plasma 25OHD in normal individuals,[211,212] and in pathological situations including primary hyperparathyroidism.[213]

1,25$(OH)_2$D has important biological actions in the three classic target organs of vitamin D: bone, where Ca and phosphorus are mobilized; kidney, in which tubular reabsorption of Ca and phosphorus is increased; and intestine, where absorption of dietary Ca is stimulated. There are other target tissues for metabolites of vitamin D_3; for example, insulin secretion from the pancreatic B cell may be affected by 1,25$(OH)_2$D,[214] and hematolymphopoietic stem cell differentiation and growth also may be affected by 1,25$(OH)_2$D.[215]

1,25$(OH)_2$D contributes to maintenance of Ca homeostasis predominantly by its ability to increase intestinal Ca and phosphorus absorption. In bone, 1,25$(OH)_2$D and PTH probably function as physiologic synergists in stimulating resorption by acting at separate sites in bone-resorbing cells.[216] In kidney, production of 1,25$(OH)_2$D is increased when there is an increased Ca or phosphorus need. Hypocalcemia stimulates 1,25$(OH)_2$D production via increased serum PTH concentrations.[217] A Ca-dependent and PTH-independent regulation of 1,25$(OH)_2$D formation also may be present.[218] Dietary phosphorus restriction is a potent stimulus for production of 1,25$(OH)_2$D.[219] Studies using bovine parathyroid cells in tissue culture have shown a dose-dependent, reversible, suppressive effect of 1,25$(OH)_2D_3$ on PTH secretion, as well as levels of pre-pro-PTH mRNA occurring at 24 to 48 h of culture, whereas acute exposure (30 to 120 m) to 1,25$(OH)_2D_3$ does not affect PTH secretion.[220] In humans with hyperparathyroidism secondary to chronic renal failure, use of 1,25$(OH)_2D_3$ when administered intravenously can markedly decrease serum PTH concentrations.[221] Whether there is a direct physiologic regulatory influence of 1,25$(OH)_2D_3$ on parathyroid gland function in normal individuals is not certain.

In humans, there is a delay of 18 to 48 h in the change of plasma 1,25$(OH)_2$D concentrations in response to oral Ca or Ca deprivation,[222] administration of PTH[223] mithramycin-induced hypocalcemia,[217] and after parathyroidectomy.[224] The delayed response to changes in serum Ca is supported by animal studies.[225] Plasma concentrations of 1,25$(OH)_2$D remain decreased or elevated for $>$12 h after normalization of plasma total and ionized Ca, and PTH concentrations.[225] Thus, the role of 1,25$(OH)_2$D in regulation of acute changes in Ca status is probably limited. However, changes in 1,25$(OH)_2$D that persist beyond the period of hyper- or hypocalcemia suggest that it may have a greater role in longer-term Ca homeostasis.

Under physiologic conditions, the role of vitamin D and its metabolites in the regulation

Table 1
CAUSES OF HYPERCALCEMIA

Malignancy — with or without bone destruction by malignant tissue
Hyperparathyroidism
Other endocrine disorders
Hyperthyroidism, acromegaly, pheochromocytoma, acute adrenal insufficiency
Familial hypocalciuric hypercalcemia
Granulomatous disorders
Renal failure (including renal transplant recipient)
Therapeutic excess or abuse
Hypervitaminosis A and D, drugs (thiazide diuretic, lithium, milk alkali syndrome, parenteral nutrition
Usually occurs in infancy
Infantile hypercalcemia, infantile hypophosphatasia, s.c. fat necrosis
Miscellaneous
Immobilization, Paget's disease

of serum Mg concentrations is probably of minor importance. However, Mg is an essential cofactor for the 25 hydroxylation reaction in the formation of 25OHD.[226,227]

DISTURBANCES OF SERUM CONCENTRATIONS

Hypercalcemia

Causes

Hypercalcemia may be defined as a serum total Ca concentration >11 mg/dl and/or an ionized Ca concentration >5.8 mg/dl.[228] Hypercalcemia is detected with increasing frequency with the advent of routine blood autoanalyzer "screening". Asymptomatic hypercalcemia may occur in up to 0.1% of the general population.[229] In pathologic situations of hypercalcemia, elevated serum ionized Ca concentration usually occurs simultaneously with elevated total Ca. However, elevated serum total Ca may occur without elevation of ionized Ca, related to increased protein available to bind Ca.[15,18,230,231] Conversely, reduced binding of albumin for Ca may result in normal serum total Ca, but elevated ionized Ca.[232,233]

Hypercalcemia is the result of imbalance between fluxes of Ca into and out of extracellular fluid. An increase in net Ca mobilization from skeleton is the primary cause of hypercalcemia. Increased mobilization of Ca from bone, increased intestinal Ca absorption, increased renal tubular reabsorption of Ca, and decreased renal tubular reabsorption of phosphate occur with excessive PTH secretion and may all contribute to hypercalcemia. However, where hypercalcemia is associated with non-PTH-related causes, serum PTH concentrations can be normal or suppressed by elevated serum Ca concentrations.

Malignancy and hyperparathyroidism account for almost 90% of cases of hypercalcemia (Table 1). Malignancy is the preponderant cause in adults, especially in elderly populations, hyperparathyroidism is the second most common cause in adults and is the commonest cause in younger and asymptomatic populations.[233-238]

Malignancies, in particular those from breast, lung, and kidney, and, multiple myeloma and lymphoma[234,239] are frequently associated with hypercalcemia. Bone dissolution due to direct osteolytic action of metastases,[234] as well as humoral factors such as PTH-like hormone,[240] osteoclast activity factor,[241] prostaglandin E_1 and E_2,[241,242] contribute in varying degrees to mobilization of minerals from bone, and hypercalcemia.

Primary hyperparathyroidism is estimated to have an adult average age-adjusted incidence of 27.7/100,000; the incidence increases with age, being highest above 60 years and greater in females.[243] Primary hyperparathyroidism is due to a single parathyroid adenoma in >80%, multiple parathyroid adenomas in 5%, and parathyroid hyperplasia in 10 to 15% of cases.[244] Carcinoma of the parathyroid glands is rare and probably accounts for <1% of primary hyperparathyroidism.[245]

Parathyroid hyperplasia may be familial with an autosomal-dominant inheritance and a high degree of penetrance.[246,247] It may manifest as part of the multiple endocrine neoplasia (MEN) syndrome.[247,248]

Hyperparathyroidism may occur *in utero* in the fetus of the hypocalcemic mother with inadequately controlled hypoparathyroidism.[249] Neonatal hyperparathyroidism with parathyroid hyperplasia may be transmitted as an autosomal dominant[250] or recessive trait[251] and may occur in families with familial hypocalciuric hypercalcemia.[252] The latter also is known as ''benign familial hypercalcemia'' and inherited as an autosomal dominant disorder with high penetrance at all ages.

Primary hyperparathroidism can be associated with thyroid, renal metabolic, and neoplastic disorders; low-dose radiation of the head and neck; and carcinoid tumors and other non endocrine malignant tumors.[235-238] In some patients with primary hyperparathyroidism the concentration of serum Ca may be normal; however, in all cases of normocalcemic primary hyperparathyroidism, serum Ca concentration always falls after surgical therapy for the underlying condition.[237]

Non-parathyroid endocrine disorders such as hyperthroidism,[253] acromegaly,[254] pheochromocytoma,[255] and acute symptomatic adrenal insufficiency[230,256] may be associated with hypercalcemia of varying severity. Some of these endocrinopathies may be a part of the MEN syndrome with simultaneous hyperparathyroidism. If the hypercalcemia is directly the result of the non-parathyroid endocrinopathy, successful treatment of the underlying disorder should normalize serum Ca concentrations.

Sarcoidosis is the granulomatous disease best described as associated with hypercalcemia. Increased production of $1,25(OH)_2D$ in these patients[257] may be important in the production of hypercalcemia via increased intestinal absorption[258] and enhanced bone resorption.[259] Other granulomatous diseases including tuberculosis,[260] berylliosis, [261] disseminated coccidioidomycosis,[262] and histoplasmosis[263] have all been reported to be associated with hypercalcemia, although the mechanisms producing hypercalcemia are not well defined.

Renal diseases may be associated with hypercalcemia. Hypercalcemia may occur during the oliguric[264] and diuretic[265,266] phase of acute renal failure. Remobilization of intracellular Ca to extracellular spaces associated with non traumatic rhabdomyolysis is thought to be primarily responsible for the hypercalcemia in this condition.[267] However, increased circulating PTH and $1,25(OH)_2D$ in patients during the hypercalcemic phase also has been reported.[268] There may be multiple contributory causes to hypercalcemia in patients with chronic renal failure (CRF). These include secondary hyperparathyroidism[269] or therapeutic maneuvers in the treatment of CRF, for example, using dialysates with high Ca content,[270] Ca containing phosphate-binding antacids,[271] Ca exchange resin,[272] thiazide diuretics,[273] or vitamin D and its metabolites.[274,275] Hypercalcemia may occur in approximately 10% of renal transplant recipients.[276] It may appear soon after transplantation or several months later. It may persist for years. Phosphate depletion, persistent secondary hyperparathroidism, high doses of corticosteroids, and associated malignancies may be contributors to prolonged hypercalcemia in such patients.[276,277]

Excessive use or abuse of medications may be associated with hypercalcemia. Vitamin D intoxication occurs with prolonged use of high doses of vitamin D and its metabolites in therapeutic situations[278,279] or with its indiscriminate use.[280] Hypercalcemia in these patients is associated with elevated serum 250HD concentrations, but normal or reduced serum $1,25(OH)_2D$ concentrations.

Vitamin A intoxication[281] results generally from self-medication, vitamin faddism, and prolonged high-dose treatment for acne. Excessive resorption of bone is thought to be responsible for hypercalcemia.

Overt hypercalcemia associated with thiazide diuretics usually occurs only if thiazides are used in conditions of increased bone turnover, such as vitamin D therapy,[282] hyperparathyroidism,[283] juvenile osteoporosis,[283] and phosphate depletion.[284] Thiazide diuretics may contribute to production of hypercalcemia via its hypocalciuria effect,[282] natriuresis and extracellular volume contraction,[140] direct osteolytic effect,[285] and potentiation of skeletal action of PTH.[286]

Chronic lithium therapy is thought to alter the "set point" of Ca action at the cellular level; it depresses the suppressive effects of Ca upon PTH secretion *in vitro,* decreases renal excretion of Ca, and induces reversible hypercalcemia and increased secretion of PTH *in vivo* in humans.[287] Milk alkali syndrome has been noted in patients with duodenal ulcers receiving therapy with sodium bicarbonate and large amounts of milk.[288] Parenteral nutrition may be associated with hypercalcemia.[289] Phosphate deficiency contributes to hypercalcemia in the latter situation. Ingestion of large amounts of calcium carbonate alone may cause hypercalcemia, as in Munchausen's syndrome.[290]

Phosphate depletion has been well documented to be associated with hypercalcemia in humans[113,291] and animals.[133,292] Phosphate depletion results in increased bone resorption and inhibition of bone formation[206,292] and development of rickets in growing infants,[293] which is reversed with phosphate therapy alone. The effect of hypophosphatemia on mobilizing skeletal Ca apparently is greater than the simultaneous increased urinary loss of Ca,[133] with the net effect of hypercalcemia. Hypercalcemia in patients receiving total parenteral nutrition is frequently associated with hypophosphatemia[3,289] and hypercalciuria,[2,3,289,293] consistent with phosphate depletion as the contributing cause of hypercalcemia.

Immobilization leads to increased bone resorption. Hypercalcemia under these circumstances occurs usually in patients with high bone turnover prior to immobilization, such as young children and teenagers, and adults with Paget's disease.[294] The parathyroid 1,25$(OH)_2$D axis is suppressed in these patients,[295] but the mechanism leading to increased bone resorption is unknown.

Other causes of hypercalcemia are found in infants with no demonstrated specific defects of vitamin D or PTH physiology. These include idiopathic infantile hypercalcemia (IIH),[296] severe infantile hypophosphatasia,[297] and extensive s.c. fat necrosis.[298] In a case report of an infant with IIH, the serum concentration of 250HD was normal and 1,25$(OH)_2$D was appropriately suppressed with hypercalcemia.[296] Exaggerated response to pharmacologic doses (18,000 to 100,000 U) of vitamin D_2 given orally,[299] and a blunted CT response to i.v. Ca loading[300] may contribute to the pathogenesis of hypercalcemia in IIH.

Clinical and Laboratory Features

Hypercalcemia adversely affects functioning of all organ systems. The severity of signs and symptoms in a given patient is a function of the degree and rapidity of onset of hypercalcemia. The most common presentation is as a coincidental finding on biochemical "screening".[236] Symptoms may be non-specific and include anorexia, fatigue, weight loss, and, in infants, failure to thrive; more specific symptoms may include constipation, thirst and polyuria, deafness, pruritus, myalgia, bone pain, and mental changes, e.g., confusion and depression. Renal calculi, hypertension, peptic ulcer, and acute pancreatitis are less common presentations. Signs of hypercalcemia are few and may include s.c., corneal, conjunctival, arterial, and cardiac calcification, and the "red eye syndrome".[228,236,238]

Manifestations of underlying disease processes are often present, e.g., anatomic anomalies ("elfin faces") associated with the IIH syndrome. There may be a family history of disturbed Ca metabolism. Maternal dietary and drug history during pregnancy are especially important if hypercalcemia is noted in infancy.[228]

Laboratory investigations include serum total and ionized Ca, phosphorus, alkaline phosphatase, and Ca-regulating hormones; urinary Ca, phosphorus and cyclic AMP concentrations; and radiographs of hands and wrists for evidence of hyperparathyroidism. Phosphate loading test (5 g elemental phosphorus over 5 d) may result in a tubular reabsorption of phosphorus (TRP) of <60% in patients with primary hyperparathyroidism, whereas other hypercalcemic patients display little change; and in normal subjects after phosphate loading, the TRP usually remains above 60%.[301] The ratio of serum chloride in millimoles per liter to phosphate in milligrams per deciliter of <33 may be used as supportive evidence against the diagnosis of primary hyperparathyroidism.[236] Earlier reports suggested the use of Ca infusion as a diagnostic parathyroid suppression test; it was felt that PTH secretion was not suppressed in primary hyperparathyroidism. In more recent reports, there appears to be a high degree of variation from one patient to another in parathyroid suppressibility in response to i.v.[302] or dietary Ca load.[303]

A careful history, physical examination, and a relatively small series of tests will, in the majority of patients, lead to correct diagnosis of hypercalcemia. ''Screening'' of family members for hypercalcemia also may be indicated.

Therapy

Therapy of hypercalcemia includes removal of specific underlying causes, e.g., discontinuation of large vitamin D intake. However, patients with marked hypercalcemia (>13 mg/dl) may require therapy to lower serum Ca concentrations while its hypercalcemia is being investigated. Nausea, vomiting, and polyuria that may be present in hypercalcemic patients may require fluid volume repletion.

Symptomatic treatment for hypercalcemia may include enhancement of urinary Ca excretion; increase of net movement of Ca into bone; inhibition of bone resorption; and rarely peritoneal dialysis and hemodialysis with Ca-free dialysate, which may be needed as a life saving procedure in a patient with multiple additional complicating factors such as renal failure and congestive heart failure. Other measures may be helpful in long-term management including restriction of dietary vitamin D and Ca, as well as minimizing sun exposure to lower endogenous vitamin D production. These measures are aimed at reducing circulating vitamin D metabolite concentrations and the renal filtered load of Ca which may be important in reducing complications of hypercalciuria and nephrolithiasis.[304]

For short-term treatment of acute hypercalcemic episodes, enhanced urinary Ca excretion can be accomplished by expansion of the extracellular compartment with 10 to 20 ml i.v. (0.9%) sodium chloride per kilogram followed by i.v. injection of a potent loop diuretic such as 1 to 2 mg furosemide per kilogram. Further i.v. fluids should replace urine losses of water, sodium, and potassium, which are measured frequently at 2- to 4-h intervals. Furosemide may be repeated at 2- to 4-h intervals. Prolonged diuresis requires replacement of Mg losses. With this therapy, Ca excretion may increase to 1 to 2 g, and serum Ca concentration may decrease by 2 to 4 mg/dl. This therapy usually is not effective in congestive heart failure or severe renal failure because of inability to tolerate large i.v. fluid load and/or a decreased Ca excretory response.[228,248,305] This therapy also may not be effective in some patients, if the rate of Ca mobilization from bone skeleton is equal to or exceeds urinary losses.

Shifting of Ca out of extracellular fluid into bone[306,307] and possibly inhibition of bone resorption[205,206] may occur with inorganic phosphate therapy. Ten to 15 mg of elemental phosphorous per kilogram infused with saline over 8 to 16 h or 15 to 45 mg of elemental phosphorus per kilogram given in three or more divided doses may be effective in lowering serum Ca concentrations.[306-309] Further dose adjustment will depend on serum concentrations of phosphorous and Ca. Certain patients with primary hyperparathyroidism may display an escape from the effect of oral phosphorus, and hypercalcemia may recur: such patients may respond to increase in the dosage.[306]

Overtreatment may result in hypocalcemia especially if the initial serum phosphorus or creatinine is elevated. Metastatic calcification is another potential complication with phosphorus therapy. Oral phosphorus therapy may be associated with diarrhea and gastrointestinal upset.[306,308] High-dose prolonged phosphorus supplementation in the dog may lead to increased bone loss (probably associated with secondary hyperparathyroidism) and soft tissue calcification.[309]

Agents that inhibit bone resorption also may be effective in lowering serum Ca concentration. I.v. or i.m. administration of 4 to 8 IU of salmon CT or equivalent per kilogram every 8 to 12 h may be effective to lower serum Ca concentration.[310,311] As little as 10 U of CT may lower serum Ca concentration in healthy adults.[312] However, hypercalcemia recurs within hours after discontinuation of therapy; "escape" from hypocalcemic effects during therapy is a major disadvantage associated with CT use.[203,311] Glucocorticoids, in the form of 0.5 to 1 mg prednisone or its equivalent per kilogram, alone or when used in conjunction with CT, may be effective in reducing serum Ca.[228,307] The major disadvantages of their use are the relatively slow hypocalcemic effect, which may be apparent only after 5 to 10 d of therapy, and multiple side effects of prolonged high-dose glucocorticoid therapy. I.v. infusion of 20 to 25 μg mithramycin per kilogram is effective in the therapy of hypercalcemia.[217,313] Mithramycin is a highly potent toxic antibiotic that blocks bone resorption through inhibition of RNA synthesis in bone cells. Its use is accompanied by multiple side effects and should be restricted to patients resistant to other forms of treatment. Therapeutic trial with prostaglandin synthetase inhibitors (indomethacin, 1 mg/kg/d and/or aspirin, 20 to 40 mg/kg/d may be justified in hypercalcemic patients, since some malignant tumors may be associated with prostaglandin production. Therapy should be stopped after 5 to 7 d if there is no response.

Hypocalcemia

Causes

Hypocalcemia may be defined as a serum total Ca concentration of <8 mg/dl in the adult[238] and in the neonate.[314] Serum-ionized Ca concentrations theoretically are the indicators of physiologic Ca activity; "true" hypocalcemia should be reflected by decrease in serum, ionized Ca. With newer, more precise, and stable ion-selective electrodes, the lower limit of normal serum ionized Ca concentrations is 5.2 mg/dl in the adult and 4.8 mg/dl in term healthy infants in the first 18 months of life.[315]

Unlike hypercalcemia which occurs predominantly in adults, hypocalcemia occurs predominantly in infants, particularly in neonates. The causes of hypocalcemia are listed in Table 2.

Specific endocrine dysfunction involving disturbances in PTH and vitamin D metabolism are the major causes of hypocalcemia. PTH deficiency may result from failure of parathyroid gland function (hypoparathyroidism) and inhibition of PTH secretion (for example, from hypomagnesemia and Mg deficiency, which can cause hypocalcemia via this and other mechanisms [see "Parathyroid Hormone"]). Hypoparathyroidism may be congenital or acquired. Congenital hypoparathyroidism may be due to agenesis, dysgenesis, or hypoplasia of the parathyroid gland. It may occur as a sporadic abnormality or as a familial disorder,[316,317] and may be transmitted as an X-linked recessive trait.[316] It may be associated with other anomalies such as Di George's syndrome (thymic dysplasia, aortic arch anomalies, immune defect, congenital heart disease);[318,319] Kearn-Sayre's syndrome (myopathy, ataxia, ophthalmoplegia, and retinal degeneration);[238] and the complex of congenital nephrosis and deafness, gonadal dysgenesis, and ring chromosome 18.[237] Acquired hypoparathyroidism may occur as a result of cervical radiation[320] or surgery,[321] hemosiderosis after multiple transfusions,[322] and copper deposition in Wilson's disease or infiltrative or metastatic diseases.[237] Acquired hypoparathyroidism may be part of the autoimmune polyglandular failure syndrome[323,324] which may include other endocrine disorders such as diabetes mellitus, Hashimoto thyroiditis, hypoadrenalism, oophoritis, alopecia, vitiligo, chronic active hepatitis, moniliasis, and pernicious anemia.

Table 2
CAUSES OF HYPOCALCEMIA

- Diminished ionized and total Ca
  - Parathyroid hormone
    - Deficiency
    - Defective hormone
    - Inhibition of release
    - End organ (skeletal) resistance
  - Vitamin D
    - Deficiency
    - Defective metabolism
    - End organ (skeletal) resistance
  - Increased flux of Ca into bone
    - Osteoblastic skeletal metastases
    - Healing phase of metabolic bone disease
  - Ca deposition in necrotic soft tissues
    - Rhabdomyolysis
    - Acute pancreatitis
  - Insufficient Ca
    - Decreased intake
    - Increased loss (malabsorption)
  - Excess phosphate load
    - Excess intake
    - Excess production (increased tissue destruction)
    - Decreased excretion
  - Chronic Mg deficiency
  - Miscellaneous
    - Drugs—mithramycin, CT, EDTA, loop diuretic, fluoride
    - ? Endogenous hypercalcitonemia—birth asphyxia
- Diminished ionized Ca without decrease in total Ca
  - Citrate administration with donor blood
  - Long-chain fatty acid (i.v. lipid infusion)
  - Alkalosis
- Diminished total Ca without decrease in ionized Ca
  - Diminished protein-bound Ca (hypoalbuminemia)

Functional hypoparathyroidism may occur as a result of secretion of a biologically defective PTH molecule (pseudoiopathic hypoparathyroidism) or diminished end-organ responsivity to PTH. The latter consists of two major groups of patients, type I and type II pseudohypoparathyroidism (PHP).[325] The classic (type I) PHP[326] is present when there is bone and renal unresponsiveness to PTH; serum concentrations of Ca are low, phosphorus concentrations are high, and $1,25(OH)_2D$ concentrations are low; urinary phosphate excretion and urinary cyclic AMP remain low in spite of having very high circulating endogenous PTH concentrations. These patients also have a typical phenotype — short stature, round face, brachymetacarpals and brachymetatarsals, s.c. deposits of Ca, and intellectual dullness. Patients with the typical phenotype but with normal serum Ca, phosphorus, and $1,25(OH)_2D$ concentrations are considered to have pseudopseudohypoparathyroidism (pseudo PHP).[327] Pseudo PHP may evolve from PHP type I in the same patient; these two disorders also may occur in the same family, with a sex-linked dominant inheritance and variable penetrance.[328] In many, but not all, patients with PHP type I, there are marked reductions in the activity of the stimulatory coupling protein (N_s) of the adenylcyclase system in red blood cell membranes,[329] platelets,[330] cultured skin fibroblasts,[331] and renal cortical membranes.[332] Deficiencies of this protein in kidney and bone could impair adenylcyclase system responsiveness to PTH. Such a defect could also explain the target organ resistance to other hormones such as thyroid-stimulating hormone, gonadotrophins, and glucagon, which is found in many patients with PHP and low N_s.[324] There is no abnormality in the activity of the related adenylate cyclase inhibitory coupling protein (N_i) in erythrocyte membrane of patients with PHP regardless of their N_s protein status.[333] Type II PHP patients have normal phenotype and normal urinary cyclic AMP response to PTH, but have the other biochemical and hormonal features similar to type I PHP.[334,335] Thus, this condition is a postreceptor abnormality of cell function, possibly involving a cyclic AMP-dependent protein kinase. The inheritance pattern is not known.[363] In addition to the classic types I and II PHP, there are multiple clinical presentations featuring various combinations of the clinical and biochemical manifestations of the disease.[325,328,336]

Skeletal resistance to the calcemic action of PTH which leads to hypocalcemia may occur in acute[337] and chronic[338] renal failure, acute pancreatitis with renal failure[339] and vitamin D deficiency.[340,341] Although the exact mechanism of the diminished end organ responsiveness to PTH is not well defined, several factors may contribute: for example, in chronic renal failure, secondary hyperparathyroidism may be followed by "down regulation" of PTH receptors; and decreased synthesis of $1,25(OH)_2D$ in renal failure may affect the expression of PTH effect on bone. In addition, reduced renal excretion of phosphate can further contribute to the lowering of serum Ca concentration. While acute Mg deficiency can stimulate PTH secretion,[197] chronic Mg deficiency can decrease PTH secretion[198,199] and increase end organ resistance to PTH.[199]

Disturbances in vitamin D metabolism could result in hypocalcemia. These include: (1) marked reduction in vitamin D ingestion (e.g., in nursing vegetarian mothers and food faddism) in conjunction with lack of sun exposure; (2) any major defect in vitamin D absorption (e.g., biliary obstruction or decreased hepatic function, intestinal resection, or bypass); (3) defect in vitamin D metabolism, e.g., decreased hepatic 25 hydroxylation, accelerated metabolism of 25OHD (anticonvulsant, alcohol), increased loss of 25OHD (nephrotic syndrome); (4) decreased renal 1-α hydroxylation of 25OHD (chronic renal disease, congenital deficiency of 1-α hydroxylase, e.g., vitamin D dependency type I); or (5) end organ resistance to $1,25(OH)_2D$ (in the newborn human, vitamin D dependency type II with alopecia or with decreased $24R,25[OH]_2D$). The molecular basis of the syndrome of hereditary target organ resistance to vitamin D accompanied by alopecia appears to be due to a spectrum of derangements in the hormone receptor pathway, including defects in receptor binding, nuclear uptake, or post-receptor steps.[342,343] A normal steroid binding domain, but a defective nuclear binding domain of $1,25(OH)_2D$ receptors from fibroblasts of these patients has been postulated as an additional mechanism involved in the clinical manifestations of this syndrome.[344] With the exception of severe vitamin D deficiency and vitamin D dependency, symptomatic hypocalcemia is relatively uncommon in the vitamin D disturbances mentioned earlier.

Increased flux of Ca into bone may be responsible for hypocalcemia that may occur during vitamin D therapy for vitamin deficiency rickets. A similar phenomenon occurs after parathyroidectomy in patients with hyperparathyroidism and osteitis fibrosa cystica and in patients with osteoblastic metastases (the "hungry bone" syndrome).[235,345] Ca deposition in necrotic soft tissues such as in rhabdomyolysis[266] and in acute pancreatitis[339] also may result in hypocalcemia. Potassium therapy has been reported to precipitate tetany in asymptomatic hypocalcemic patients.[346]

CT secretion may be acutely increased in a variety of stress situations (e.g., birth asphyxia) and might theoretically contribute to the hypocalcemia of such illnesses. However, chronic hypercalcitonemia, as in patients with medullary carcinoma of the thyroid, is not associated with hypocalcemia.[203] In theory, any drug used in the treatment of hypercalcemia, as described in the section "Therapy" on therapy for hypercalcemia, is liable to cause hypocalcemia.

Inadequate Ca intake may play a role in the development of neonatal hypocalcemia, and, in theory, malabsorption of Ca from any cause (see the section "Calcium Absorption") may aggravate the degree of hypocalcemia. Hypocalcemia can occur from (1) increased dietary phosphate intake;[314,347] (2) an acute phosphate load which may occur following i.v. phosphate infusion,[308] phosphate enema,[348] massive tissue destruction during chemotherapy for leukemia and malignant tumors,[349] or hyperpyrexia and reduced renal excretion of phosphate.[237] The mechanisms by which hyperphosphatemia induces hypocalcemia are not well defined any may involve sequestration of Ca in extravascular sites,[350,351] and impairment of bone resorption and/or increased bone formation.[352]

Neonatal hypocalcemia occurs with two peaks in frequency: an "early" form, which typically occurs during the first few days of life, with the lowest concentrations of serum Ca reached at 24 to 48 h of age. This condition occurs primarily in preterm infants. The frequency of hypocalcemia varies inversely with birth weight and gestational age. Stressed infants (e.g., infants of diabetic mothers or infants who have suffered birth asphyxia) are also prone to develop hypocalcemia. Early neonatal hypocalcemia appears to be related to decreased calcium intake and possibly other factors. There is some controversy as to whether the parathyroid gland is functionally adequate during the neonatal period.[228,353] Hypercalcitonemia may play a role in stressed infants[354] who developed hypocalcemia, and, in some instances, disordered vitamin D metabolism may play a role, including end organ (skeletal) resistance to $1,25(OH)_2D$ effect.[355] In rare instances, maternal hyperparathyroidism may be associated with congenital hypoparathyroidism, thought to be related to suppression of fetal parathyroids through accentuation of fetal hypercalcemia. The mechanisms underlying the apparent hypocalcemia in neonates receiving phototherapy, and in neonates with *in utero* exposure to anticonvulsants and narcotics, are ill understood.[314]

"Late" neonatal hypocalcemia occurs towards the end of the first week after birth. The condition probably is related to the ingestion of milks with relatively high phosphate contents, acting through production of hyperphosphatemia.[347] Maternal disturbance of vitamin D function has been proposed as an additive mechanism, possibly related to poor dietary intake of vitamin D during pregnancy and compounded by seasonal variations in sunlight exposure. Associated factors include increased maternal age and parity, and poor maternal socioeconomic class. Other potential causes of late neonatal hypocalcemia include ingestion of high phosphate-containing cereals, intestinal malabsorption of Ca, hypomagnesemia, and hypoparathyroidism (see later).[314]

Decrease in serum ionized Ca concentrations can occur without decrease in total Ca concentrations. The classic example is the effect of citrate contained in acid citrate dextrose (ACD) or citrate phosphate dextrose (CPD) donor blood, especially during an "exchange" blood transfusion. Citrate complexes Ca and results in decreased ionized Ca. Similarly, increases in long-chain free fatty acids also can complex Ca and should be watched for during i.v. infusion of lipid-containing solutions. Alkalosis from overtreatment with alkali, or hyperventilation, can result in shifts of Ca from the ionized state to the protein-bound fraction.[314,356] In contrast, a fall in the concentration of serum total Ca may occur without a change in serum ionized Ca, for example, in hypoalbuminemia from whatever cause.[15,17,18]

Clinical and Laboratory Features

The signs and symptoms of hypocalcemia[238,314,357] depend on the rapidity with which it develops, its magnitude, its duration, and the age of the affected individual, but they are usually independent of the cause. Chronic and moderate hypocalcemia may be asymptomatic, but a precipitous and marked fall in serum ionized Ca will be symptomatic. Pre-term infants with hypocalcemia, particularly in "early onset" cases, are usually asymptomatic if circulating ionized Ca concentrations are >2.5 mg/dl.[355,358] However, the less mature the infant, the more subtle and varied are the manifestations and these may be easily confused with other neonatal disorders such as hypoglycemia, sepsis, meningitis, anoxia, intracranial bleeding, and narcotic withdrawal in the sick neonate.

Non-specific manifestations of hypocalcemia may include weakness, fatigue, vague ill health, and emotional disturbances such as irritability and emotional lability. The decrease in serum ionized Ca concentration increases neuromuscular excitability[359] and is responsible for a variety of signs and symptoms related to this phenomenon.

The clinical features of twitching of muscles and extremities, high-pitched cry, laryngospasm, Chvostek's sign (facial muscle twitching on tapping the side of the face), and Trousseau's sign (carpopedal spasm on constriction of limbs) may be present. A positive Chvostek sign or Trousseau sign may be elicited even in asymptomatic hypocalcemic patients; many normal subjects may have positive Chvostek signs. Epileptiform and more often Jacksonian seizures may occur, but usually they are not associated with aura, loss of consciousness, or incontinence.

Patients with chronic hypocalcemia may have papilledema, elevated cerebrospinal fluid pressure, and neurologic signs simulating a cerebral tumor;[360] calcification of basal ganglia;[361] and subcapsular cataracts.[362] The cataracts do not resolve after correction of hypocalcemia.[362] Conjunctivitis and keratits may also be seen.[360,363]

Children with hypoparathyroidism[364] and pre-term very low birth weight infants on follow-up examination[365] are frequently noted to have teeth abnormalities such as defective enamel and root formation, dental hypoplasia, or failure of adult teeth to erupt. However, hypocalcemia may not be the only predisposing cause of enamel hypoplasia.[366] Very severe hypocalcemia may cause congestive heart failure in the neonate[367] and the adult[368,369] and may be associated with resistance to digitalis therapy.[368] In pregnant women with inadequately controlled hypoparathyroidism, severe skeletal demineralization may occur in their fetuses related to secondary fetal hyperparathyroidism.[249]

In patients with idiopathic hypoparathyroidism, the skin may be dry and scaly, eyelashes and eyebrows may be scanty; axillary and pubic hair may be absent.[360,363] Some of these manifestations, including immunologic defects in some patients, may be part of autoimmune processes.[323]

A family history of Ca related disorders or a maternal history suggestive of hyperparathyroidism (nausea, vomiting, polyuria, hypertension, and renal stones) may be present, although the maternal signs and symptoms may be confused with pregnancy-related complications. A history of drug ingestion, perinatal asphyxia, and pre-term gestation may indicate neonates at risk for hypocalcemia.

Measurement of serum total Ca concentration is freely available and should be carried out in all subjects suspected to have hypocalcemia. Determination of serum ionized Ca, phosphorus and Mg concentrations, urinary contents of these elements, and acid base status are important in the differential diagnosis of the causes of hypocalcemia. Serum total and ionized Ca concentrations are correlated; however, one may not be predictive of the other. Electrocardiographic changes include inversion of T waves and prolongation of QT intervals (Qt_c, or QT interval corrected for heart rate, >0.04 s, or Q_oT_c, the beginning of Q to the beginning of T, corrected for rate, >0.2 s,[228,314] which are of some value in assessing the course of hypocalcemia, but do not predict ionized Ca. Electroencephalographs may show a nonspecific increase in fast waves in infants with hypocalcemia.[369] However, there is no correlation between serum Ca concentration and electroencephalographic abnormalities in adults.[370]

Assays of Ca regulating hormones and diagnostic investigations such as urinary Ca, phosphorus, and cyclic AMP response to exogenous PTH and $1,25(OH)_2D$ administration are also important. However, in view of the heterogeneity of circulatory forms of PTH and CT, and differences in specificity of available antisera, multiple serial determinations of these hormone concentrations may be needed for an accurate diagnosis. Serum 25OHD concentrations below 10 ng/ml may be indicative of vitamin D deficiency. Various tests to confirm the presence or absence of intracranial calcifications, cataracts, thymic aplasia (Di

George syndrome), immunologic defects, and other associated autoimmune disorders also may be of use in the differential diagnosis of the causes of hypocalcemia.

Therapy

The major approach to management of hypocalcemia is treatment of the underlying disorder, for example, Mg therapy is needed for hypocalcemia related to Mg deficiency. The major symptomatic treatment consists of Ca supplementation and use of vitamin D or its metabolites. Adequate dietary Ca intake is a prerequisite in management of any hypocalcemic state.

Administration of i.v. Ca salts is the effective and rapid means of elevating serum Ca concentrations. One half to 1 ml of 10% Ca gluconate per kilogram for an infant, or 10 to 20 ml of the same solution for an adult, administered over 10 min, usually reverses the symptoms and signs of hypocalcemia. Complications associated with the use of Ca salts include local tissue necrosis following extravasation, and bradycardia or cardiac standstill following rapid infusion of Ca.

Oral Ca salts may be used for patients with asymptomatic hypocalcemia, following initial reversal of hypocalcemia with i.v. Ca salts, or in conjunction with treatment using vitamin D metabolites in disorders such as hypoparathyroidism.[228,248,328,371] The amount of elemental Ca available may vary depending on the preparation used; for example, elemental Ca constitutes, by weight, 40% of Ca carbonate, 36% of Ca chloride, 13% of Ca lactate, and 9% of Ca gluconate. Neocalglucon syrup is another preparation well accepted for use; each 5 ml contains 115 mg of elemental Ca. In the neonate and young infant, oral therapy may be commenced at 75 mg elemental Ca per kilogram per day and further adjustment of the doses will depend on the measured serum Ca concentration. In the adult, 2 to 4 g of elemental Ca per day may be needed to maintain normocalcemia.

Vitamin D and its potent metabolite $1,25(OH)_2D_3$ are important in the management of hypocalcemia from specific endocrine disorders and in the long-term management of hypocalcemia. Thus, in the management of patients with idiopathic or acquired hypoparathyroidism, large doses of parent vitamin D (50,000 to 100,000 IU/d or 2000 IU/kg/d)[328,372] or small doses of its potent metabolite $1,25(OH)_2D_3$ (0.5 to 2 μg/d or 0.025 to 0.05 μg/kg/d[328,371,373] are needed.

Thiazide diuretics (chlorthalidone, 50 mg/d) and sodium restriction (<1 meq/kg/d) may be useful as adjuncts in the management of hypocalcemia of hypoparathyroidism, since these therapies may result in hypocalciuria and possibly increased intestinal Ca absorption.[374]

Close monitoring of the serum Ca concentration is essential to minimize the complications of therapy. Hypercalcemia may lead to hypercalciuria, nephrocalcinosis, and renal dysfunction with its associated complications.

Neonatal hypocalcemia especially in small pre-term infants tends to resolve spontaneously,[228,355,358] and some investigators have suggested that this condition may be left untreated.[358] However, in view of the major physiologic importance of Ca in all cellular systems and the multiple problems confronting sick preterm infants,[228] it would seem reasonable to prevent or treat hypocalcemia in the neonate.

Hypermagnesemia

Causes

Based on the data from a population survey, the mean plus two standard deviation value for serum Mg is 2.4 mg/dl for adults and 2.5 mg/dl for infants and early childhood.[23] Hypermagnesemia is thus defined as a serum Mg concentration of >2.5 mg/dl.

Hypermagnesemia is a normal accompaniment of hibernation[375] and hypothermia.[376] However, most causes of hypermagnesemia usually are iatrogenic in origin. Diminished renal function is the most common clinical prerequisite for development of hypermagnesemia.

Some degree of hypermagnesemia almost invariably accompanies acute renal failure [377] and serum Mg concentrations may reach extremely high levels with exogenous Mg intake.[378] In CRF, as creatinine clearance falls with progressive renal failure, fractional excretion of Mg increases, and serum Mg concentrations usually remain normal until glomerular filtration rate falls below 30 ml/min.[379] Symptomatic hypermagnesemia in patients with impaired renal function usually involves concomitant administration of a Mg-containing compound by mouth[380,381] or by infusion.[381,382] Both excess and depletion of Mg are inducible by dialysis[383] and excessive dialysate Mg also has caused symptomatic hypermagnesemia.[270]

Acute oliguria accompanied by acidosis as occurs in diabetic ketoacidosis also may result in hypermagnesemia.[384] Various neoplasms with skeletal involvement may be associated with hypermagnesemia.[385] Mild hypermagnesemia may be present in patients with familial hypocalciuric hypercalcemia.[252] Administration of parathyroid extract results in small increases in serum Mg concentrations.[386] In primary hypoparathyroidism, a negative Mg balance without hypermagnesemia has been reported.[345] Hypermagnesemia also has been reported in patients with hypothyroidism[387] possibly associated with a reduced exchangeable Mg pool.[109]

During pregnancy, Mg sulfate administration for preeclampsia may result rarely in serum Mg concentrations as high as 14 mg/dl in the mother and 11.5 mg/dl in umbilical cord serum.[388,389] Prematurity and perinatal asphyxia may aggravate hypermagnesemia, presumably because of decreased renal Mg excretion.[390] The use of Mg-containing enemas in animals,[391] human adults,[392] and neonates[393] can cause hypermagnesemia even in the absence of renal failure.

Clinical and Laboratory Features

Hypermagnesemia results in toxic manifestations.[394] Symptoms of toxicity may occur when serum Mg concentrations are above 3.5 mg/dl.[381] The major effects of hypermagnesemia are on neuromuscular and cardiovascular functions. In adults with hypermagnesemia, hypotension and urinary retention occur at serum Mg concentrations >3.5 mg/dl; electrocardiographic changes (increased atrioventricular and ventricular conduction time) occur at >6.0 mg/dl; respiratory depression and cardiac arrest may occur at >12 mg/dl.[381] In healthy volunteers in whom serum Mg concentrations were elevated to >17.5 mg/dl, there was profound skeletal muscle paralysis, but the subjects remained conscious and had normal pain sensation.[394,395] In women with normal pregnancy and with eclampsia, high-dose parenteral Mg sulfate results in elevated maternal and fetal plasma Mg concentrations, but spinal fluid Mg concentrations remained similar to normal pregnant and nonpregnant women who did not receive Mg treatment.[396] Presumably, the central nervous system depression of Mg may result from either very prolonged hypermagnesemia or circumstances in which a defect in the blood-brain barrier accompanies hypermagnesemia, where direct contact of Mg with nervous tissue occurs. The effect of hypermagnesemia on reducing intestinal motility is controversial.[314]

Acute hypermagnesemia may induce hypocalcemia in animals[397] and humans, particularly in the Mg-treated eclamptic mother.[201] This effect may be mediated in part by a decreased PTH secretion[200,201] and possibly by altering end organ responsiveness to PTH,[398] and renal Ca loss due to competition of Mg ions with Ca for reabsorption primarily in the loop of Henle. In the offspring of mothers treated with Mg, serum Ca may be decreased[399] or increased compared to controls.[390] It is possible that bone exchange of Mg for Ca may have been the reason for increased serum Ca concentrations in these infants. Other endocrine effects of hypermagnesemia such as stimulation of CT or renin secretion are not well defined.[394]

Chronic Mg excess probably occurs only in the patient with advanced renal failure and its clinical importance has not been clearly defined. Bone Mg is considerably increased in

Table 3
CAUSES OF HYPOMAGNESEMIA

Decreased intake
- Mg-free parenteral nutrition
- Prolonged starvation
- Specific intestinal Mg malabsorption
- Malabsorption states including intestinal resection

Increased loss from:
- Intestinal fistula/diarrhea
- Hepatobiliary disorders
- Decreased renal tubular reabsorption
  - Congenital
  - Drugs—loop diuretics, thiazide diuretic, aminoglycoside, *cis*-DDP
  - Osmotic diuresis
  - Phosphate depletion
  - Extracellular fluid expansion

Specific for neonates
- Maternal diabetes
- Maternal hyperparathyroidism
- Small for gestational age infants
- "Exchange" (citrated) blood transfusion
- Uremic infants

Miscellaneous
- Chronic alcoholism
- Hyperthyroidism
- Severe burns

uremia. However, only 30% of the increased Mg is in the surface-exchangeable pool of bone, and it is only this fraction that normalizes after renal transplantation.[400] In contrast, Mg content of other tissue stores such as muscle are normal.[9] Excess Mg may produce abnormal bone in animals[401] Mg stabilizes amorphous Ca carbonates and phosphates and may disturb apatite crystallization,[7,402] but its role in renal osteodystrophy remains unclear. Mg is also present in visceral Ca-phosphate deposits.[403]

An adequate history and physical examination frequently point to the cause of hypermagnesemia. Laboratory investigations in confirmed cases of hypermagnesemia are predominantly oriented to the possible complications listed earlier.

Therapy

Therapy of hypermagnesemia includes cessation of Mg administration; immediate but transient reversal of toxicity may be effected with i.v. Ca gluconate (10%), 0.1 to 0.3 ml/kg; symptomatic support of cardiorespiratory system should be instituted if needed.[389]

Peritoneal or hemodialysis with Mg-free dialysate may be needed. "Exchange" blood transfusions have been used successfully in the treatment of hypermagnesemia in neonates. This mode of therapy may not be successful in all cases,[389] probably because the volume of distribution of Mg is somewhat larger than the volume of extracellular space.

If the patient has adequate renal function, urinary Mg excretion may be augmented by furosemide and saline-induced diuresis, as described in the section "Therapy" on therapy for hypercalcemia. In addition, Ca gluconate (10%), 1 ml/kg over a 4-h period, may be used to augment urinary Mg excretion.

Hypomagnesemia

Causes (Table 3)

Mg depletion is a common clinical disorder. It is frequently overlooked because of the

Table 4
SERUM CHANGES

	Ca	Phosphorus	PTH	25OHD	1,25(OH)$_2$D
In Hypocalcemia					
Hypoparathyroidism	↓	↑	↓	—	↓
Pseudohypoparathyroidism	↓	↑	↑	—	↓
Vitamin D deficiency (rickets)	↓	↓	↑	↓ ↓	Variable
Vitamin D dependency (rickets)	↓	↓	↑	—	↓
End organ resistance to 1,25(OH)$_2$D	↓	↓	↑	—	↑
Hyperphosphatemic (phosphate load)	↓	↑	↑	—	Variable
Hypomagnesemia	↓	± ↑	↓	—	—
In Hypercalcemia					
Hyperparathyroidism	↑	↓	↑	—	↑
Familial hypocalcuric hypercalcemia	↑	—			
Malignancy	↑	—	↓	—	↓
Immobilization	↑	—	↓	—	↓
Hypervitaminosis D	↑	—	↓	↑ ↑	± ↑
Infantile hypercalcemia	↑	—	↓	—	—

typically complex clinical setting and the frequently associated electrolyte disorders seen in Mg-deficient patients.[110,404] Since >99% of total body Mg is found in intracellular fluid or in complexes in skeleton, the precise clinical determination of Mg deficiency remains a problem. Variable turnover periods of Mg from exchangeable Mg pools by compartmental analysis would suggest that certain tissues such as erythrocyte Mg content are reduced only after prolonged Mg deprivation.[405] The determination of Mg concentration in a tissue sample may only relate to the Mg status for that tissue (pool).[406] Substantial intracellular Mg depletion can occur when serum Mg concentrations are normal or even elevated and symptomatic Mg deficiency may occur when serum values are only moderately below the normal range.[407,408] However, to date, all adverse clinical effects that are unequivocally due to Mg deficiency are accompanied by severe hypomagnesemia and are reversed by restoration of the serum Mg concentrations to normal.[110,235,404,408]

Mg deficiency and hypomagnesemia (serum Mg <1.5 mg/dl) may occur as a result of reduced intake or intestinal absorption, and/or increased losses from the gastrointestinal tract or renal route.

Mg deficiency secondary to reduced intake has been reported in patients receiving long-term Mg-free i.v. therapy.[404,409,410] It is difficult to develop Mg deficiency from dietary habits alone unless they are extreme, since Mg is present in virtually all food sources.[92] Prolonged starvation may result in hypomagnesemia; there is a continuing urinary and fecal Mg loss even with total starvation,[411] probably as a result of ketoacidosis and muscle catabolism.

Selective malabsorption of Mg can result in Mg deficiency. It may occur as an isolated defect[116,117] or may be familial, although the exact mode of inheritance is unclear.[116] More commonly, Mg deficiency may occur as part of a condition with malabsorption of multiple nutrients, for example, a steatorrheic state.[412] Surgical resections of jejunum and ileum will particularly affect Mg balance, because the major portion of intestinal Mg absorption appears to occur at these sites.[412] There is a small but significant correlation between serum Mg concentrations and the degree of steatorrhea, and the stool Mg content is directly proportional to the stool fat content.[412]

As a group, Mg losses, whether from the gastrointestinal tract or kidney, are the most common cause of Mg depletion and hypomagnesemia. The Mg content in gastric fluid, bile, and pancreatic secretion varies from 0.5 to 1.3 mg/dl, while diarrheal Mg content may be as high as 17 mg/dl. In severe diarrheal stools, fecal Mg excretion, as for potassium, is related to the total stool water content.[413] Since the typical deficit required to produce symptomatic hypomagnesemia is approximately 12 to 24 mg/kg of body weight, fluid losses from diarrhea and chronic intestinal or biliary fistula may become significant sources of Mg loss.[110]

Renal Mg losses may be increased because of an intrinsic defect in Mg reabsorption or some extrarenal or intrarenal stimulus. Intrinsic renal Mg wasting may be sporadic[182] or familial.[181] It may occur as an isolated defect or in association with a potassium losing nephropathy with or without an accompanying distal renal tubular acidosis. In Bartter's syndrome,[414] the renal Mg leak is mild, and severe hypomagnesemia does not occur in this condition.

Drugs such as aminoglycosides[180] and *cis*-dichlorodiammine platinum (*cis*-DDP),[179] when used in high doses and for prolonged periods, may induce renal tubular damage or dysfunction resulting in renal Mg wasting. The defect in renal Mg handling may persist for months after the *cis*-DDP has been discontinued. Loop diuretics such as furosemide can increase urinary excretion of Mg as well as Ca.[139] Thiazide diuretics increase urinary Mg excretion in contrast to its action in decreasing urinary Ca losses.[140]

A variety of systemic and local factors can influence renal Mg handling (see ''Factors Influencing Renal Regulation of Magnesium''), since urinary Mg excretion is interlinked with a number of other cations. Acute infusion of sodium, particularly in the form of saline, and Ca salts can markedly increase urinary Mg excretion. Excesses of other readily metabolizable substrates such as glucose, galactose, protein, and alcohol may result in increased urinary Mg loss and possibly hypomagnesemia. Phosphate depletion also may result in hypomagnesemia.[134] Chronic hypercalciuric states associated with high-dose vitamin D administration also may produce Mg wasting.

Hypomagnesemia has been reported with primary[415] and secondary[416] hyperaldosteronism, but it is not certain whether aldosterone directly increases urine Mg. Volume expansion and the resultant diuresis and magnesiuria account for the hypomagnesemia that occasionally occurs in the syndrome of inappropriate secretion of antidiuretic hormone.[417] Hypomagnesemia frequently occurs in patients with diabetic ketoacidosis.[384,418,419] This probably results from a combination of factors including complexing of Mg to acetoacetate, the osmotic effect of glycosuria, and a redistribution of Mg during treatment of the diabetic ketoacidosis. Other factors such as superimposed acute pancreatitis[420] or potassium phosphate therapy[418,419] may contribute to hypomagnesemia.

Although hypermagnesemia occurs in acute renal failure and acidosis, it is important to remember that there is a net loss of Mg upon reestablishment of urinary flow, which may result in hypomagnesemia.[384,421]

Conditions with multiple contributory factors leading to Mg depletion and hypomagnesemia include alcoholism, certain endocrine disorders such as hyperthyroidism, and severe burns. The mechanism of alcohol-induced Mg depletion may, in part, be due to inadequate Mg intake, ketoacidosis, or ethanol per se.[175,404,409] Other factors which may contribute to hypomagnesemia in alcoholics include phosphate depletion[134] and low exchangeable Mg.[422] The degree of hypomagnesemia has been reported to correlate directly with the severity of the hyperthyroid state.[423] The mechanisms involved are not clear. The exchangeable Mg pool is increased,[109] and Mg balance is normal in hyperthyroid humans[424] and animals given exogenous thyroxine.[425]

Diminished Mg intake and loss of Mg through the denuded skin may contribute to the hypomagnesemia seen in patients with severe burns.[426] Hypomagnesemia can occur as early as 4 d to as late as 72 d following the burn.

Most of the previously described situations can occur in infants and result in Mg deficiency and hypomagnesemia.[228] In addition, there are a number of causes of hypomagnesemia which are unique for the neonate. The normally present feto-maternal circulating Mg concentration gradient in favor of the fetus may not be sufficient to protect the fetus if the mother is Mg depleted.[427] Infants of diabetic mothers typically have hypomagnesemia, hypocalcemia, and decreased parathryoid function.[428,429] The severity and prevalence of hypomagnesemia is directly related to the severity of maternal diabetes. Hypomagnesemia is also related to maternal hypomagnesemia and young mothers. Hypomagnesemia occurs more frequently in infants with intrauterine growth retardation (IUGR) than in appropriate-for-gestational-age infants. The hypomagnesemia in IUGR occurs particularly in young primiparous mothers, especially those who have toxemia of pregnancy.[430]

The serum Mg concentration is usually normal in hyperparathyroidism.[431] Hypomagnesemia and hypocalcemia have been reported in an infant of a mother with hyperparathyroidism.[432] The mechanism of hypomagnesemia is not clear, but negative Mg balance has been reported in patients with hyperparathyroidism;[431] the maternal negative Mg balance may, in turn, contribute to the neonatal hypomagnesemia. Alternatively, neonatal hypoparathyroidism secondary to maternal hyperparathyroidism may lead to hypomagnesemia because of lack of PTH-related mobilization of bone Mg.

Other causes of hypomagnesemia peculiar to the neonate and infants include repeated ''exchange'' blood transfusions utilizing citrate as anticoagulant, which results in complexing of citrate with Mg.[433] Infants with uremia may have hypomagnesemia, possibly related to the higher dietary phosphate intake.[434] Infants with congenital biliary atresia and neonatal hepatitis may have low serum Mg concentrations;[435] the exact mechanism involved is unclear.

Clinical and Laboratory Features

Severe hypomagnesemia may be asymptomatic. Adaptive mechanisms appear to protect intracellular function from alterations of serum Mg concentration, both by preventing wide fluctuations in intracellular levels and by reducing the concentration needed to activate certain key enzymes.[436] When symptoms do occur, the serum Mg concentration is almost always <1.2 mg/dl. Typically, these patients have generalized weakness, anorexia, nausea, and apathy. They may develop muscle fasiculations, tremors, positive Chvostek and Trousseau signs, and occasionally tetany. Tetany may occur in the absence of hypocalcemia or alkalosis. Biochemical abnormalities include hypomagnesemia, hypocalcemia, and hypokalemia. In animals, the Mg depletion syndrome can occur spontaneously or be induced experimentally.[437] In humans, Mg depletion syndrome has been reported in alcoholics[404,409] and can be induced experimentally.[110]

As with other disturbances of circulating Ca and Mg concentration, an adequate history and physical examination and limited numbers of laboratory investigations are needed to determine the majority of the causes of Mg depletion and hypomagnesemia. The renal Mg waster can easily be separated from the individual with Mg deficiency of a gastrointestinal cause by measuring urinary Mg prior to Mg replacement. Normally, the kidney has a great capability for conserving Mg when the serum Mg concentration is reduced.[110] However, in the renal Mg wasting syndromes, high fractional clearances of Mg are present whether serum Mg concentrations are normal or subnormal.[408]

It has been proposed that high Mg retention after a Mg load may reflect Mg deficiency.[438] However, infants generally retain large amounts of infused Mg and there are large variations in response; the clinical utility of this test in infants thus appears limited.

Therapy

Mild asymptomatic hypomagnesemia requires no treatment if the patient is eating normally and provided the cause of hypomagnesemia can be eliminated. In view of the obligatory urinary Mg losses, it would seem reasonable to supplement i.v. fluids with Mg to prevent Mg depletion if the patient requires maintenance i.v. fluid for periods in excess of 3 to 4 d. The daily requirement of Mg ranges from 1 to 2 mg/kg for an adult to approximately 6 mg/kg for an infant.[4,26]

For patients with acute symptomatic hypomagnesemia, Mg can be given intramuscularly or as i.v. infusion after diluting with 10 to 100 ml (depending on size of patient) of standard i.v. fluids (5% dextrose or dextrose-saline solution) over 10 min. The dose used for adults is 2 to 4 ml of 50% Mg sulfate ($MgSO_47H_2O$) (49 mg elemental Mg per milliliter) and for infants is 0.1 to 0.2 ml of 50% $MgSO_4$.[228,413,438,439] The dose may be repeated every 4 h pending serum Mg measurements.

Concomitant oral Mg supplements can be started if oral fluids are tolerated. Fifty percent $MgSO_4$ can be given at a dose of 0.2 ml/kg/d. In patients with specific Mg malabsorption or chronic Mg wasting from various gastrointestinal and renal causes, daily oral doses of 1 ml/kg/d may be required. Daily serum Mg concentrations should be measured initially until values are stable, in order to evaluate efficacy and safety. Oral Mg salts are not well absorbed and large doses may cause diarrhea. The maintenance Mg supplement should be diluted five- to sixfold to allow for more frequent administration to maximize gut absorption and minimize side effects. Chloride and oxide of Mg such as $MgCl_2$ $6H_2O$ or MgO also may be used instead of $MgSO_4$. Large doses of Ca or vitamin D appear to be detrimental to Mg status, because increased intestinal Ca absorption may decrease intestinal Mg absorption and increased Ca excretion would also increase Mg losses.

Depending on the underlying cause of hypomagnesemia, the overall deficit of Mg may be in excess of the amount recommeded above. The intracellular deficit may take several weeks to be replenished.[110,413,440] It has been estimated that patients with diabetic ketoacidosis may retain 480 to 960 mg of Mg over 2 to 6 d following recovery.[384] In alcoholic subjects, deficits of up to 12 mg/kg have been found.[440]

In patients with compromised renal function, Mg should be given cautiously, with close monitoring of serum Mg concentrations and the status of the cardiovascular system. Systemic hypotension, prolonged atrioventricular conduction time, and sinoauricular or atrioventricular block are possible complications of rapid i.v. injection and diminished ability to excrete any excess Mg.

ACKNOWLEDGMENTS

This work was supported by grants from NIH 1RO1 HD 18505-01A1, NIH RR 00123, and NIH RR 00068 (CLINFO). W.K. is a recipient of NIH Clinical Associate Physician Award 3MO1 RR 00123-21S1.

REFERENCES

1. **Widdowson, E. M., and McCance, R. A.,** The metabolism of calcium, phosphorus, magnesium and strontium, *Pediatr. Clin. North Am.*, 12, 595, 1965.
2. **Koo, W. W. K. and Tsang, R.,** Bone mineralization in infants, *Prog. Food Nutr. Sci.*, 8, 229, 1984.
3. **Koo, W. W. K. and Tsang, R. C.,** Calcium, magnesium and phosphorus, in *Nutrition in Infancy*, Tsang, R. C. and Nichols, B., Eds., Hanley and Belfus, Philadelphia, in press.
4. Committee on Dietary Allowances, *Minerals, Food and Nutrition Board. Recommended Dietary Allowances*, National Academy of Sciences, Washington, D. C., 1980, 125.

5. **Shoback, D. M., Thatcher, J., Leombruno, R., and Brown, E. M.,** Relationship between parathyroid hormone secretion and cytosolic calcium concentration in dispersed bovine parathyroid cells, *Proc. Natl. Acad. Sci. U.S.A.*, 81, 3113, 1984.
6. **Taylor, A. and Weindhager, E. E.,** Cytosolic calcium and its role in the regulation of transepithelial ion and water transport, in *The Kidney, Physiology and Pathophysiology,* Vol. 2, Seldon, D. W. and Giebisch, G., Eds., Raven Press, New York, 1985, 1297.
7. **Posner, A. S.,** Crystal chemistry of bone mineral, *Physiol. Rev.*, 49, 760, 1969.
8. **Garty, J., Nito, H., and Massry, S. G.,** The role of the kidney in the control of serum calcium, *Clin. Res.*, 21, 688, 1973.
9. **Alfrey, A. C., Miller, N. L., and Butkus, D.,** Evaluation of body magnesium stores, *J. Lab. Clin. Med.*, 84, 153, 1974.
10. **Walser, M.,** Magnesium metabolism, *Ergeb. Physiol. Biol. Chem. Exp. Pharmakol.*, 59, 185, 1967.
11. **Levine, B. S. and Coburn, J. W.,** Magnesium, the mimic/antagonist of calcium, *N. Engl. J. Med.*, 310, 1253, 1984.
12. **MacManus, J. and Heaton, F. W.,** The influence of magnesium on calcium release from bone in vitro, *Biochim. Biophys. Acta,* 215, 360, 1970.
13. **Elin, R. J., Armstrong, W. D., and Singer, L.,** Body fluid electrolyte composition of chronically magnesium-deficient and control rats, *Am. J. Physiol.*, 220, 543, 1971.
14. **Natelson, S., Richelson, M. R., Sheid, B., and Bender, S. L.,** X-ray spectroscopy in the clinical laboratory. I. Calcium and potassium, *Clin. Chem.*, 5, 519, 1959.
15. **Moore, E. W.,** Ionized calcium in normal serum, ultrafiltrates, and whole blood determined by ion exchange electrodes, *J. Clin. Invest.*, 49, 318, 1970.
16. **Carruthers, B. M., Copp, D. H., and McIntosh, H. W.,** Diurnal variation in urinary excretion of calcium and phosphate and its relation to blood levels, *J. Lab. Clin. Med.*, 63, 959, 1964.
17. **McLean, F. C. and Hastings, A. B.,** The state of calcium in the fluids of the body. I. The conditions affecting the ionization of calcium, *J. Biol. Chem.*, 108, 285, 1935.
18. **Loken, H. F., Havel, R. J., Gordon, G. S., and Whittington, S. L.,** Ultracentrifugal analysis of protein bound and free calcium in human serum, *J. Biol. Chem.*, 235, 3654, 1960.
19. **Walser, M.,** Divalent cations: physicochemical state in glomerular filtrate and urine and renal excretion, in *Handbook of Physiology. Renal Physiology,* Orloff, J. and Berliner, R. W., Eds., Waverly Press, Baltimore, 1973, 555.
20. **Harris, C. A., Baer, P. G., Chirito, E., and Dirks, J. H.,** Composition of mammalian glomerular filtrate, *Am. J. Physiol.*, 227, 972, 1974.
21. **Markowitz, M., Rotkin, L., and Rosen, J. F.,** Circadian rhythms of blood minerals in humans, *Science,* 213, 672, 1981.
22. **Young, D. S.,** Normal laboratory values in SI units, *N. Engl. J. Med.*, 292, 795, 1975.
23. **Lowenstein, F. W. and Stanton, M. F.,** Serum magnesium levels by age, sex and two racial groups in the United States, First National Health and Nutrition Examination Survey (NHANES I), 1971—1974.
24. **Silverman, S. H. and Gardner, L. I.,** Ultrafiltration studies on serum magnesium, *N. Engl. J. Med.*, 250, 938, 1954.
25. **Massry, S. G.,** Pharmacology of magnesium, *Annu. Rev. Pharmacol. Toxicol.*, 17, 67, 1977.
26. **Greer, F. R. and Tsang, R. C.,** Calacium, phosphorus, magnesium, and vitamin D requirements for the preterm infant, in *Vitamin and Mineral Requirements in Preterm Infants,* Tsang, R. C., Ed., Marcel Dekker, New York, 1985, 99.
27. **Avioli, L. V.,** Calcium and osteoporosis, *Annu. Rev. Nutr.*, 4, 471, 1984.
28. **Benjamin, H. R., Gordon, H. H., and Marples, E.,** Calcium and phosphorus requirements of premature infants, *Am. J. Dis. Child.*, 65, 412, 1943.
29. **Barltrop, D., Mole, R. H., and Sutton, A.,** Absorption and endogenous fecal excretion of calcium by low birthweight infants on feeds with varying contents of calcium and phosphate, *Arch. Dis. Child.*, 52, 41, 1977.
30. **Moya, M. and Domenech, E.,** Role of calcium-phosphate ratio of milk formulae on calcium balance in birth weight infants during the first three days of life, *Pediatr. Res.* 16, 675, 1982.
31. **Ireland, P. and Fordtran, J. S.,** Effect of dietary calcium and age on jejunal calcium absorption in humans studied by intestinal perfusion, *J. Clin. Invest.*, 52, 2672, 1973.
32. **Heaney, R. P., Saville, P. D., and Recker, R. R.,** Calcium absorption as a function of calcium intake, *J. Lab. Clin. Med.*, 85, 881, 1975.
33. **Tellez, M., Reeve, J., Royston, J. P., Veall, N., and Wooton, R.,** The reproducibility of double-isotope deconvolution measurements of intestinal calcium absorption, *Clin. Sci.*, 59, 169, 1980.
34. **Pansu, D., Bellaton, C., and Bronner, F.,** Effect of Ca intake on saturable and nonsaturable components of duodenal Ca transport, *Am. J. Physiol.*, 240, G32, 1981.
35. **Wasserman, R. H.,** Intestinal absorption of calcium and phosphorus, *Fed. Proc.*, 40, 68, 1981.

36. **Hylander, E., Ladefoged, K., and Jarnum, S.,** The importance of the colon in calcium absorption following small-intestinal resection, *Scand. J. Gastroenterol.*, 15, 55, 1980.
37. **DeVizia, B., Fomon, S. J., Nelson, S. E., Edwards, B. E., and Ziegler, E. E.,** Effect of dietary Ca on metabolic balance of normal infants, *Pediatr. Res.*, 19, 800, 1985.
38. **Haussler, M. R., Nagode, L. A., and Rasmussen, H.,** Induction of intestinal brush border alkaline phosphatase by vitamin D and identity with Ca-ATPase, *Nature*, 228, 1199, 1970.
39. **Franceschi, R. T. and DeLuca, H. F.,** The effect of inhibitors of protein and RNA synthesis on 1,25-dihydroxyvitamin D_3-dependent calcium uptake in cultured embryonic chick duodenum, *J. Biol. Chem.*, 256, 3848, 1981.
40. **Wasserman, R. H. and Taylor, A. N.,** Vitamin D_3-induced calcium binding protein in chick intestinal mucosa, *Science*, 152, 791, 1966.
41. **Christakos, S., Friedlander, E. J., Frandsen, B. R., and Norman, A. W.,** Studies on the mode of action of calciferol. XIII. Development of a radioimmunoassay for vitamin D-dependent chick intestinal calcium binding protein and tissue distribution, *Endocrinology*, 104, 1495, 1979.
42. **Feher, J. J. and Wasserman, R. H.,** Calcium absorption and intestinal calcium binding protein: quantitative relationship, *Am. J. Physiol.*, 236, E556, 1979.
43. **Kowarski, S. and Schachter, D.,** Vitamin D and adenosine triphosphatase dependent on divalent cations in rat intestinal mucosa, *J. Clin. Invest.*, 52, 2765, 1973.
44. **Armbrecht, H. J., Zenser, T. V., and Davis, B. B.,** Effect of vitamin D metabolites on intestinal calcium absorption and calcium-binding protein in young and adult rats, *Endocrinology*, 106, 469, 1980.
45. **Spencer, R., Charman, M., Wilson, P. W., and Lawson, D. E. M.,** The relationship between vitamin D-stimulated calcium transport and intestinal calcium-binding protein in chicken, *Biochem. J.*, 170, 93, 1978.
46. **Morrissey, R. L., Zolock, D. T., Bikle, D. D., Bucci, T. J., and Empson, R. N., Jr.,** Intestinal response to 1,25-dihydroxycholecalciferol. I. RNA polymerase, alkaline phosphatase, calcium and phosphorus uptake in vitro, and in vivo calcium transport and accumulation, *Biochim. Biophys. Acta*, 538, 23, 1978.
47. **Wilson, P. W. and Lawson, D. E. M.,** 1,25-Dihydroxyvitamin D stimulation of specific membrane proteins in chick intestine, *Biochem. Biophys. Acta*, 497, 805, 1977.
48. **Weringer, E. J., Oldham, S. B., and Bethune, J. E.,** A proposed cellular mechanism for calcium transport in intestinal epithelial cells, *Calcif. Tissue Res.*, 26, 71, 1978.
49. **Hay, A. W. M., Mawer, E. B., Hassam, A. G., Crawford, M. A., and Stevens, P. A.,** The role of essential fatty acids in vitamin D dependent calcium absorption in the intestine, in *Vitamin D, Chemical, Biochemical and Clinical Endocrinology of Calcium Metabolism*, Norman, A. W., Schaefer, K., Herrath, D. V., and Grigoleit, H. G., Eds., Walter de Gruyter, Berlin, 1982, 309.
50. **Halloran, B. P. and DeLuca, H. F.,** Calcium transport in small intestine during early development: role of vitamin D, *Am. J. Physiol.*, 239, G473, 1980.
51. **Massaro, E. R., Simpson, R. U., and DeLuca, H. F.,** Glucocorticoids and appearance of 1,25 dihydroxyvitamin D_3 receptor in rat intestine, *Am. J. Physiol.*, 244, E230, 1983.
52. **Day, G. M., Chance, G. W., Radde, I. C., Reilly, B. J., Park, E., and Sheepers, J.,** Growth and mineral metabolism in very low birth weight infants, II. Effects of calcium supplementation on growth and divalent cations, *Pediatr. Res.*, 9, 568, 1975.
53. **Senterre, J., Putet, G., Salle, B., and Rigo, J.,** Effects of vitamin D and phosphorus supplementation on calcium retention in preterm infants fed banked human milk, *J. Pediatr.*, 103, 305, 1983.
54. **Senterre, J.,** Endogenous faecal calcium, total digestive juice calcium, net and true calcium absorption in premature infants, in Stemberg, Z., Polacek, K., and Sabata, V., Eds., *Perinatal Medicine: 4th European Congr. of Perinatal Medicine*, Georg Thieme Verlag, Stuttgart, 1975, 287.
55. **Moore, L. J., Machlan, L. A., Lim, M. O., Yergey, A. L., and Hansen, J. W.,** Dynamics of calcium metabolism in infancy and childhood. I. Methodology and quantification in the infant, *Pediatr. Res.*, 19, 329, 1985.
56. **Ehrenkranz, R. A., Ackerman, B. A., Nelli, C. M., and Janghorbani, M.,** Absorption of calcium in premature infants measured with a stable isotope ^{46}Ca extrinsic tag, *Pediatr. Res.*, 19, 178, 1985.
57. **Gallagher, J. C., Riggs, B. L., Eisman, J., Hamstra, A., Arnaud, S. B., and DeLuca, H. F.,** Intestinal calcium absorption and serum vitamin D metabolites in normal subjects and osteoporotic patients: effect of age and dietary calcium, *J. Clin. Invest.*, 64, 729, 1979.
58. **Ziegler, E. E. and Fomon, S. J.,** Lactose enhances mineral absorption in infancy, *J. Pediatr. Gastroenterol. Nutr.*, 2, 288, 1983.
59. **Cochet, B., Jung, A., Griessen, M., Bartholdi, P., Schaller, P., and Donath, A.,** Effects of lactose on intestinal calcium absorption in normal and lactase-deficient subjects, *Gastroenterology*, 84, 935, 1983.
60. **Pansu, D., Bellaton, C., and Bronner, F.,** Effect of lactose on duodenal calcium-binding protein and calcium absorption, *J. Nutr.*, 109, 508, 1979.
61. **Williams, M. L., Rose, C. S., Morrow, G., III, Sloan, S. E., and Barness, L. A.,** Calcium and fat absorption in neonatal period, *Am. J. Clin. Nutr.*, 23, 1322, 1970.

62. **Bach, A. C. and Babayan, V. K.,** Medium chain triglyceride: an update, *Am. J. Clin. Nutr.*, 36, 950, 1982.
63. **Tantibhedhyangkul, P. and Hashim, S.,** Medium-chain triglyceride feeding in premature infants: effects on calcium and magnesium absorption, *Pediatrics*, 61, 537, 1978.
64. **Okamoto, E., Muttart, C. R., Zucker, C. L., and Heird, W. C.,** Use of medium-chain triglycerides in feeding the low-birth-weight infant, *Am. J. Dis. Child.*, 136, 428, 1982.
65. **Huston, R. K., Reynolds, J. W., Jensen, C., and Buist, N. R. M.,** Nutrient and mineral retention and vitamin D absorption in low-birth-weight infants: effect of medium-chain triglycerides, *Pediatrics*, 72, 44, 1983.
66. **Barltrop, D. and Oppe, T.,** Absorption of fat and calcium by low birthweight infants from milks containing butterfat and olive oil, *Arch. Dis. Child.*, 48, 496, 1973.
67. **Southgate, D. A. T., Widdowson, E. M., Smits, B. J., Cooke, W. T., Walker, C. H. M., and Mathers, N. P.,** Absorption and excretion of calcium and fat by young infants, *Lancet*, 1, 487, 1969.
68. **Drenick, E. J.,** The influence of ingestion of calcium and other soap-forming substances on fecal fat, *Gastroenterology*, 41, 242, 1961.
69. **Lutwak, L., Laster, L., Gitelman, H. J., Fox, M., Whedon, G. D., Wolfe, D. E., and Woodson, M. L.,** Effects of high dietary calcium and phosphorus on calcium, phosphorus, nitrogen and fat metabolism in children, *Am. J. Clin. Nutr.*, 14, 76, 1964.
70. **MacLaurin, J. C., Watson, J., Murphy, W., Stewart, M. E., Griffin, J. F., and Samuel, P. D.,** Fat, calcium, and nitrogen balance in full-term infants, *Postgrad. Med. J.*, 51(Suppl. 3), 45, 1975.
71. **Katz, L. and Hamilton, J.,** Fat absorption in infants of birth weight less than 1300 gm, *J. Pediatr.*, 85, 608, 1974.
72. **Chappell, J. E., Clandinin, M. T., Kearney-Wolpe, C., Reichman, B., and Swyer, P. W.,** Fatty acid balance studies in premature infants fed human milk or formula: effect of calcium supplementation, *J. Pediatr.*, 108, 439, 1986.
73. **Barltrop, D. and Oppe, T. E.,** Calcium and fat absorption by low birth weight infants from a calcium supplemented milk formula, *Arch. Dis. Child.*, 48, 580, 1973.
74. **Lindeman, R. D., Adler, S., Yiengst, M. J., and Beard, E. S.,** Influence of various nutrients on urinary divalent cation excretion, *J. Lab. Clin. Med.*, 70, 236, 1967.
75. **Margen, S., Chu, J. Y., Kaufmann, N. A., and Calloway, D. H.,** Studies in calcium metabolism. I. The calciuretic effects of dietary protein, *Am. J. Clin. Nutr.*, 27, 584, 1974.
76. **Chu, J. Y., Margen, S., and Costa, F. M.,** Studies in calcium metabolism. II. Effects of low calcium and variable protein intake on human calcium metabolism, *Am. J. Clin. Nutr.*, 28, 1028, 1975.
77. **Spencer, H., Kramer, L., Osis, D., and Norris, C.,** Effect of a high protein (meat) intake on calcium metabolism in man, *Am. J. Clin. Nutr.*, 31, 2167, 1978.
78. **Spencer, H., Menczel, J., Lewin, I., and Samachson, J.,** Effect of high phosphorus intake on calcium and phosphorus metabolism in man, *J. Nutr.*, 86, 125, 1965.
79. **Malm, O. J.,** On phosphates and phosphoric acid as dietary factors in the calcium balance of man, *Scand. J. Clin. Lab. Invest.*, 5, 75, 1952.
80. **Bell, R. R., Draper, H. H., Tzeng, D. Y. M., Shin, H. K., and Schmidt, G. R.,** Physiological responses of human adults to foods containing phosphate additives, *J. Nutr.*, 107, 42, 1977.
81. **Widdowson, E. M., McCance, R. A., Harrison, G. E., and Sutton, A.,** Effect of giving phosphate supplements to breast fed babies on absorption and excretion of calcium, strontium, magnesium and phosphorus, *Lancet*, 2, 1250, 1963.
82. **Clark, I.,** Importance of dietary Ca:PO_4 ratios on skeletal, Ca, Mg, and PO_4 metabolism, *Am. J. Physiol.*, 217, 865, 1969.
83. **Hurwitz, S., Dubrov, D., Eisner, U., Risenfeld, G., and Bar, A.,** Phosphate absorption and excretion in the young turkey, as influenced by calcium intake, *J. Nutr.*, 108, 1329, 1978.
84. **Leichsenring, J. M., Norris, L. M., Lamison, S. A., Wilson, E. D., and Patton, M. B.,** The effect of level of intake on calcium and phosphorus metabolism in college women, *J. Nutr.*, 45, 407, 1951.
85. **Hsu, F. S., Krook, L., Pond, W. G., and Duncan, J. R.,** Interactions of dietary calcium with toxic levels of lead and zinc in pigs, *J. Nutr.*, 105, 112, 1975.
86. **Forbes, R. M., Erdman, J. W., Jr., Parker, H. M., Kondo, H., and Ketelsen, S. M.,** Bioavailability of zinc in coagulated soy protein (tofu) to rats and effect of dietary calcium at a constant phytate:zinc ratio, *J. Nutr.*, 113, 205, 1983.
87. **Spencer, H., Asmussen, C. R., Holtzman, R. B., and Kramer, L.,** Metabolic balances of cadmium, copper, manganese, and zinc in man, *Am. J. Clin. Nutr.*, 32, 1867, 1979.
88. **Steinhardt Bour, N. J., Soullier, B. A., and Zemel, M. B.,** Effect of level and form of phosphorus and level of calcium intake on zinc, iron and copper bioavailability in man, *Nutr. Res.*, 4, 371, 1984.
89. **Spencer, H., Osis, D., Kramer, L., and Rgan, K. J.,** *Nutrients in Foods*, The Nutrition Guild, Massachusetts, 1983, 2.

90. **Jowsey, J. and Riggs, B. L.,** Effect of concurrent calcium ingestion on intestinal absorption of fluoride, *Metabolism,* 27, 971, 1978.
91. **Barton, J. C., Conrad, M. E., and Parmley, R. T.,** Calcium inhibition of inorganic iron absorption in rats, *Gastroenterology,* 84, 90, 1983.
92. **Leveille, G. A., Zabick, M. E., and Morgan, K. J.,** *Nutrients in Foods,* The Nutrition Guild, Massachusetts, 1983, 2.
93. **Lemons, J. A., Moye, L., Hall, D., and Simmons, M.,** Differences in the composition of preterm and term human milk during early lactation, *Pediatr. Res.,* 16, 113, 1982.
94. **Committee on Nutrition from American Academy of Pediatrics,** Nutritional needs of low birth weight infants, *Pediatrics,* 60, 519, 1977.
95. **Graham, L. A., Caesar, J. J., and Burger, A. S. V.,** Gastrointestinal absorption and excretion of Mg^{28} in man, *Metabolism,* 9, 646, 1960.
96. **Brannon, P. G., Vergne-Marini, P., Pak, C. Y. C., Hull, A. R., and Fordtran, J. S.,** Magnesium absorption in human small intestine. Results in normal subjects, patients with chronic renal disease, and patients with absorptive hypercalciuria, *J. Clin. Invest.,* 57, 1412, 1976.
97. **Jones, J. E., Manalo, R., and Flink, E. B.,** Magnesium requirements in adults, *Am. J. Clin. Nutr.,* 20, 632, 1967.
98. **Miller, D. L. and Schedl, H. P.,** Effects of diabetes on intestinal magnesium absorption in the rat, *Am. J. Physiol.,* 231, 1039, 1976.
99. **Wilz, D. R., Gray, R. W., Dominguez, J. H., and Lemann, J., Jr.,** Plasma 1,25-$(OH)_2$-vitamin D concentrations and net intestinal calcium, phosphate, and magnesium absorption in humans, *Am. J. Clin. Nutr.,* 32, 2052, 1979.
100. **Danielson, B., Johansson, G., Jung, B., Ljunghall, S., Lundquist, H., and Malmborg, P.,** Gastrointestinal magnesium absorption, *Miner. Electrolyte Metab.,* 2, 116, 1979.
101. **Slater, J. E.,** Retentions of nitrogen and minerals by babies one week old, *Br. J. Nutr.,* 15, 83, 1961.
102. **Meintzer, R. B. and Steenbock, H.,** Vitamin D and magnesium absorption, *J. Nutr.,* 56, 285, 1955.
103. **Miller, E. R., Ullrey, D. E., Zutaut, C. L., Hoefer, J. A., and Luecke, R. W.,** Mineral balance studies with the baby pig: effects of dietary vitamin D_2 level upon calcium, phosphorus and magnesium balance, *J. Nutr.,* 85, 255, 1965.
104. **Hodgkinson, A., Marshall, D. H., and Nordin, B. E. C.,** Vitamin D and magnesium absorption in man, *Clin. Sci.,* 57, 121, 1979.
105. **Schmulen, A. C., Lerman, M., Pak, C. Y. C., Zerwekh, J., Morawski, S., Fordtran, J. S., and Vergne-Marini, P.,** Effect of 1,25-$(OH)_2D_3$ on jejunal absorption of magnesium in patients with chronic renal disease, *Am. J. Physiol.,* 238, G349, 1980.
106. **Gray, T., Walton, J., Williams, M., and Pool, D.,** 1,25-Dihydroxyvitamin D and jejunal absorption of calcium and phosphate in normal subject, in *Vitamin D: Biochemical, Chemical, and Clinical Aspects Related to Calcium Metabolism,* Norman, A., Schaefer, K., Coburn, J., et al., Eds., Walter de Gruyter, Berlin, 1977, 623.
107. **McCance, R. A. and Widdowson, E. M.,** The fate of calcium and magnesium after intravenous administration to normal persons, *Biochem. J.,* 33, 523, 1939.
108. **Silver, L., Robertson, J. S., Dahl, L. K., Heine, M., and Tassinari, L.,** Magnesium turnover in the human studied with Mg^{28}, *J. Clin. Invest.,* 39, 420, 1960.
109. **Avioli, L. V., Lynch, T. N., and Berman, M.,** Digital computer compartmental analysis of Mg^{28} kinetics in normal subjects, Paget's disease and thyroid disease, *J. Clin. Invest.,* 42, 915, 1963.
110. **Shils, M. E.,** Experimental human magnesium depletion, *Medicine,* 48, 61, 1969.
111. **McCance, R. A., Widdowson, E. M., and Lehmann, H.,** Effect of protein intake on absorption of calcium and magnesium, *Biochem. J.,* 36, 686, 1942.
112. **Greene, L. W., Fontenot, J. P., and Webb, K. E., Jr.,** Effect of dietary potassium on absorption of magnesium and other macroelements in sheep fed different levels of magnesium, *J. Anim. Sci.,* 56, 1208, 1983.
113. **Lotz, M., Zisman, E., and Bartter, F. C.,** Evidence for a phosphorus-depletion syndrome in man, *N. Engl. J. Med.,* 278, 409, 1968.
114. **Heaton, F. W. and Parsons, F. M.,** The metabolic effect of high magnesium intake, *Clin. Sci.,* 21, 273, 1961.
115. **Urakabe, S., Nakata, K., Ando, A., Orita, Y., and Abe, H.,** Hypokalemia and metabolic alkalosis resulting from overuse of magnesium oxide, *Jpn. Circ. J.,* 39, 1135, 1975.
116. **Stromme, J. H., Nesbakken, R., Normann, T., Skyrten, F., Skyberg, D., and Johannessen, B.,** Familial hypomagnesemia: biochemical, histological and hereditary aspects studied in two brothers, *Acta Paediatr. Scand.,* 58, 433, 1969.
117. **Woodard, J. C., Webster, P. D., and Carr, A. A.,** Primary hypomagnesemia with secondary hypocalcemia, diarrhea and insensitivity to parathyroid hormone, *Am. J. Dig. Dis.,* 17, 612, 1972.

118. **Strauss, E. W.,** Effects of calcium and magnesium ions upon fat absorption by sacs of everted hamster intestine, *Gastroenterology,* 73, 421, 1977.
119. **Saunders, D. R., Sillery, J., Huibregtse, K., and Tytgat, G. N.,** Effects of Mg^{++} and Ca^{++} on water absorption from rodent and human intestine, *Digestion,* 22, 80, 1981.
120. **Szymanski, A. and Klos, A.,** Certain aspects of the mechanism of intestinal transport of sodium, potassium, calcium and magnesium in rats, *Acta Physiol. Pol.,* 31, 317, 1980.
121. **Rouse, D., Ng, R. C. K., and Suki, W. N.,** Calcium transport in the pars recta and thin descending limb of Henle of the rabbit, perfused in vitro, *J. Clin. Invest.,* 65, 37, 1980.
122. **Suki, W. N., Rouse, D., Ng, R. C. K., and Kokko, J. P.,** Calcium transport in the thick ascending limb of Henle. Heterogeneity of function in the medullary and cortical segments, *J. Clin. Invest.,* 66, 1004, 1980.
123. **Ng, R. C. K., Peraino, R. A., and Suki, W. N.,** Divalent cation transport in isolated tubules, *Kidney Int.,* 22, 492, 1982.
124. **Shareghi, G. R. and Agus, Z. S.,** Magnesium transport in the cortical thick ascending limb of Henle's loop of the rabbit, *J. Clin. Invest.,* 69, 759, 1982.
125. **Bourdeau, J. E. and Burg, M. B.,** Voltage dependence of calcium transport in the thick ascending limb of Henle's loop, *Am. J. Physiol.,* 242, F357, 1979.
126. **Bourdeau, J. E. and Hellstrom-Stein, R. J.,** Voltage-dependent calcium movement across the cortical collecting duct, *Am. J. Physiol.,* 242, F285, 1982.
127. **Massry, S. G. and Kleeman, C. R.,** Calcium and magnesium excretion during acute rise in glomerular filtration rate, *J. Lab. Clin. Med.,* 80, 654, 1972.
128. **Beck, L. H. and Goldberg, M.,** Effects of acetazolamide and parathyroidectomy on renal transport of sodium, calcium and phosphate, *Am. J. Physiol.,* 224, 1136, 1973.
129. **Gonda, A., Wong, N., Seely, J. F., and Dirks, J. H.,** The role of hemodynamic factors on urinary calcium and magnesium excretion, *Can. J. Physiol. Pharmacol.,* 47, 619, 1969.
130. **Agus, Z. S., Chiu, P. J. S., and Goldberg, M.,** Regulation of urinary calcium excretion in the rat, *Am. J. Physiol.,* 232, F545, 1977.
131. **Kleeman, C. R., Bohannan, J., Bernstein, D., Ling, S., and Maxwell, M. H.,** Effect of variations in sodium intake on calcium excretion in normal humans, *Proc. Soc. Exp. Biol. Med.,* 115, 29, 1964.
132. **Lennon, E. J. and Piering, W. F.,** A comparison of the effects of glucose ingestion and NH_4Cl acidosis on urinary calcium and magnesium excretion in man, *J. Clin. Invest.,* 49, 1458, 1970.
133. **Coburn, J. W. and Massry, S. G.,** Changes in serum and urinary calcium during phosphate depletion: studies on mechanisms, *J. Clin. Invest.,* 49, 1073, 1970.
134. **Dominguez, J. H., Gray, R. W., and Lemann, J., Jr.,** Dietary phosphate deprivation in women and men: effects on mineral and acid balances, parathyroid hormone and the metabolism of 25-OH-vitamin D, *J. Clin. Endocrinol. Metab.,* 43, 1056, 1976.
135. **Buckalew, V. M., Jr., McCurdy, D. K., Ludwig, G. D., Chaykin, L. B., and Elkinton, J. R.,** Incomplete renal tubular acidosis. Physiologic studies in three patients with a defect in lowering urine pH, *Am. J. Med.,* 45, 32, 1968.
136. **Massry, S. G., Coburn, J. W., Chapman, L. W., and Kleeman, C. R.,** Role of serum Ca, parathyroid hormone, and NaCl infusion on renal Ca and Na clearances, *Am. J. Physiol.,* 214, 1403, 1968.
137. **Edwards, B. R., Sutton, R. A. L., and Dirks, J. H.,** Effect of calcium infusion on renal tubular reabsorption in the dog, *Am. J. Physiol.,* 227, 13, 1974.
138. **Koo, W. W. K., Tsang, R. C., Steichen, J. J., et al.,** Parenteral nutrition for infants: effect of high vs low calcium and phosphorus content, *J. Pediatr. Gastroenterol. Nutr.,* in press.
139. **Koo, W. W. K., Guan, Z. P., Tsang, R. C., Laskarzewski, P., and Neumann, V.,** Growth failure and decreased bone mineral of newborn rats with chronic furosemide therapy, *Pediatr. Res.,* 20, 74, 1986.
140. **Brickman, A. S., Massry, S. G., and Coburn, J. W.,** Changes in serum and urinary calcium during treatment with hydrochlorothiazide: studies on mechanisms, *J. Clin. Invest.,* 51, 945, 1972.
141. **Sutton, R. A. L., Wong, N. L. M., and Dirks, J. H.,** Effects of metabolic acidosis and alkalosis on sodium and calcium transport in the dog kidney, *Kidney Int.,* 15, 520, 1979.
142. **Jones, J. H., Peters, D. K., Morgan, D. B., Coles, G. A., and Mallick, N. P.,** Observations on calcium metabolism in the nephrotic syndrome, *Q. J. Med.,* 36, 301, 1967.
143. **DeFronzo, R. A., Cooke, C. R., Andres, R., Faloona, G. R., and Davis, P. J.,** The effect of insulin on renal handling of sodium potassium, calcium, and phosphate in man, *J. Clin. Invest.,* 55, 845, 1975.
144. **Agus, Z. S., Gardner, L. B., Beck, L. H., and Goldberg, M.,** Effects of parathyroid hormone on renal tubular reabsorption of calcium, sodium and phosphate, *Am. J. Physiol.,* 224, 1143, 1973.
145. **Bourdeau, J. E. and Burg, M. B.,** Effect of PTH on calcium transport across the cortical thick ascending limb of Henle's loop, *Am. J. Physiol.,* 239, F121, 1980.
146. **Shareghi, G. R. and Stoner, L. C.,** Calcium transport across segments of the rabbit distal nephron in vitro, *Am. J. Physiol.,* 235, F367, 1978.

147. **Harris, C. A., Burnatowska, M. A., Seely, J. F., Sutton, R. A. L., Quamme, G. A., and Dirks, J. H.,** Effects of parathyroid hormone on electrolyte transport in the hamster nephron, *Am. J. Physiol.*, 236, F342, 1979.
148. **Quamme, G. A., Carney, S. L., Wong, N. L. M., and Dirks, J. H.,** Effect of parathyroid hormone on renal calcium and magnesium reabsorption in magnesium deficient rats, *Pfluegers Arch.*, 386, 59, 1980.
149. **Paillard, F., Ardaillou, R., Malendin, H., Fillastre, J. P., and Prier, S.,** Renal effects of salmon calcitonin in man, *J. Lab. Clin. Med.*, 80, 200, 1972.
150. **Quamme, G. A.,** Effect of calcitonin on calcium and magnesium transport in rat nephron, *Am. J. Physiol.*, 238, E573, 1980.
151. **Suki, W. N. and Rouse, D.,** Hormonal regulation of calcium transport in thick ascending limb renal tubules, *Am. J. Physiol.*, 241, F171, 1981.
152. **Kawashima, H. and Kurokawa, K.,** Localization of 1,25 dihydroxyvitamin D_3 receptors along the rat nephron: evidence for the presence of specific binding in both proximal and distal nephron, in *Vitamin D: Chemical, Biochemical and Clinical Endocrinology of Calcium Metabolism,* Norman, A. W., Schaefer, K., Herrath, D. V., and Grigoleit, H. G., Eds., Walter de Gruyter, Berlin, 1982, 95.
153. **Puschett, J. B., Moranz, J., and Kurnick, W. S.,** Evidence for a direct action of cholecalciferol and 25 hydroxycholecalciferol on the renal transport of phosphate, sodium and calcium, *J. Clin. Invest.*, 51, 373, 1972.
154. **Puschett, J. B., Fernandez, P. C., Boyle, I. T., Gray, R. W., Omdahl, J. L., and DeLuca, H. F.,** The acute renal tubular effects of 1,25-dihydroxycholecalciferol, *Proc. Soc. Exp. Biol. Med.*, 141, 379, 1972.
155. **Sutton, R. A. L., Harris, C. A., Wong, N. L. M., and Dirks, J. H.,** Effects of vitamin D on renal tubular calcium transport, in *Vitamin D: Biochemical, Chemical and Clinical Aspects Related to Calcium Metabolism,* Norman, A. W., Schaefer, K., Coburn, J. W., et al., Eds., Walter de Gruyter, Berlin, 1977, 451.
156. **Chen, T. C. and DeLuca, H. F.,** Stimulation of [^{3}H] uridine incorporation into nuclear RNA of rat kidney by vitamin D metabolites, *Arch. Biochem. Biophys.*, 156, 321, 1973.
157. **Roth, J., Brown, D., Norman, A. W., and Orci, L.,** Localization of the vitamin D-dependent calcium-binding protein in mammalian kidney, *Am. J. Physiol.*, 243, F243, 1982.
158. **Morrissey, R. L. and Rath, D. F.,** Purification of human renal calcium binding protein from necropsy specimens, *Proc. Soc. Exp. Biol. Med.*, 145, 699, 1974.
159. **Santos, F., Vernon Smith, M. J., and Chan, J. C. M.,** Hypercalciuria associated with long term administration of calcitriol, *Am. J. Dis. Child.*, 140, 139, 1986.
160. **Quamme, G. A. and Dirks, J. H.,** Magnesium transport in the nephron, *Am. J. Physiol.*, 239, F393, 1980.
161. **Poujeol, P., Chabardes, D., Roinel, N., and DeRouffignac, C.,** Influence of extracellular fluid volume expansion on magnesium, calcium and phosphate handling along the rat nephron, *Pfluegers Arch.*, 365, 203, 1976.
162. **Wong, N. L. M., Quamme, G. A., Sutton, R. A. L., and Dirks, J. H.,** Effects of mannitol on water and electrolyte transport in the dog kidney, *J. Lab. Clin. Med.*, 94, 683, 1979.
163. **Le Grimellec, C., Roinel, N., and Morel, F.,** Simultaneous Mg, Ca, P, K, Na and Cl analysis in rat tubular fluid. II. During acute Mg plasma loading, *Pfluegers Arch.*, 340, 197, 1973.
164. **Wong, N. L. M., Dirks, J. H., and Quamme, G. A.,** Tubular reabsorptive capacity for magnesium in the dog kidney, *Am. J. Physiol.*, 244, F78, 1983.
165. **Shareghi, G. R. and Agus, Z. S.,** Magnesium transport in the cortical thick ascending limb of Henle's loop of the rabbit, *J. Clin. Invest.*, 69, 759, 1982.
166. **Barker, E. S., Elkinton, J. R., and Clark, J. K.,** Studies of the renal excretion of magnesium in man, *J. Clin. Invest.*, 38, 1733, 1959.
167. **Kelly, H. G., Gross, H. C., Turton, M. R., and Hatcher, J. D.,** Renal and cardiovascular effects induced by intravenous infusion of magnesium sulphate, *Can. Med. Assoc. J.*, 82, 866, 1960.
168. **Massry, S. G. and Seelig, M. S.,** Hypomagnesemia and hypermagnesemia, *Clin. Nephrol.*, 7, 147, 1977.
169. **Aikawa, J. K., Rhoades, E. L., and Gordon, G. S.,** Urinary and fecal excretion of orally administered Mg^{28}, *Proc. Soc. Exp. Biol. Med.*, 98, 29, 1958.
170. **Avioli, L. V. and Berman, M.,** Mg^{28} kinetics in man, *J. Appl. Physiol.*, 21, 1688, 1966.
171. **Wesson, L. G., Jr.,** Magnesium, calcium, and phosphate excretion during osmotic diuresis in the dog, *J. Lab. Clin. Med.*, 60, 422, 1962.
172. **Brunette, M., Wen, S. F., Evanson, R. L., and Dirks, J. H.,** Micropuncture study of magnesium reabsorption in the proximal tubule of the dog, *Am. J. Physiol.*, 216, 1510, 1969.
173. **Kreusser, W. J., Kurokawa, K., Aznar, E., Sachtjen, E., and Massry, S. G.,** Effect of phosphate depletion on magnesium homeostasis in rats, *J. Clin. Invest.*, 61, 573, 1978.

174. **Wong, N. L. M., Quamme, G. A., O'Callaghan, T. J., Sutton, R. A. L., and Dirks, J. H.,** Renal tubular transport in phosphate depletion: a micropuncture study, *Can. J. Physiol. Pharmacol.*, 58, 1063, 1980.
175. **Kalbfleish, J. M., Lindeman, R. D., Ginn, H. E., and Smith, W. O.,** Effects of ethanol administration on urinary excretion of magnesium and other electrolytes in alcoholic and normal subjects, *J. Clin. Invest.*, 42, 1471, 1963.
176. **Wacker, W. E. C. and Parisi, A. F.,** Magnesium metabolism, *N. Engl. J. Med.*, 278, 772, 1968.
177. **Wong, N. L. M., Quamme, G. A., Sutton, R. A. L., and Dirks, J. H.,** Effects of mannitol on water and electrolyte transport in the dog kidney, *J. Lab. Clin. Med.*, 94, 683, 1979.
178. **Wong, N. L. M., Quamme, G. A., and Dirks, J. H.,** Effects of urea on electrolyte transport in the dog kidney, *J. Lab. Clin. Med.*, 98, 741, 1981.
179. **Lyman, N. W., Hemalatha, C., Viscuso, R. L., and Jacobs, M. G.,** Cisplatin-induced hypocalcemia and hypomagnesemia, *Arch. Intern. Med.*, 140, 1513, 1980.
180. **Keating, M. J., Sethi, M. R., Bodey, G. P., and Samaan, N. A.,** Hypocalcemia with hypoparathyroidism and renal tubular dysfunction associated with aminoglycoside therapy, *Cancer,* 39, 1410, 1977.
181. **Booth, B. E. and Johanson, A.,** Hypomagnesemia due to renal tubular defect in reabsorption of magnesium, *J. Pediatr.*, 85, 350, 1974.
182. **Michelis, M. F., Drash, A. L., Linarelli, L. G., De Rubertis, F. R., and Davis, B. B.,** Decreased bicarbonate threshold and renal magnesium wasting in a sibship with distal renal tubular acidosis (evaluation of the pathophysiologic role of parathyroid hormone), *Metabolism,* 21, 905, 1972.
183. **Butler, A. M.,** Diabetic coma, *N. Engl. J. Med.*, 243, 648, 1950.
184. **Whang, R. and Welt, L. G.,** Observations in experimental magnesium depletion, *J. Clin. Invest.*, 42, 305, 1963.
185. **Baldwin, D., Robinson, P. K., Zierler, K. L., and Lilenthal, J. L., Jr.,** Interrelations of magnesium, potassium, phosphorus, and creatine in skeletal muscle of man, *J. Clin. Invest.*, 31, 850, 1952.
186. **House, W. A. and Bird, R. J.,** Magnesium tolerance in goats fed two levels of potassium, *J. Anim. Sci.*, 41, 1134, 1975.
187. **Quamme, G. A. and Dirks, J. H.,** Magnesium: cellular and renal exchanges, in *The Kidney: Physiology and Pathophysiology,* Seldin, D. W. and Giebisch, G., Eds., Raven Press, New York, 1985, 1269.
188. **LeGrimellec, C., Roinel, N., and Morel, F.,** Simultaneous Mg, Ca, P, K, Na and Cl analysis in rat tubular fluid. III. During acute Ca plasma loading, *Pfluegers Arch.*, 346, 171, 1974.
189. **Quamme, G. A.,** Effect of hypercalcemia on renal tubular handling of calcium and magnesium, *Can. J. Physiol. Pharmacol.*, 60, 1275, 1982.
190. **Massry, S. G., Ahumada, J. J., Coburn, J. W., and Kleeman, C. R.,** Effect of Mg Cl_2 infusion in urinary Ca and Na during reduction in their filtered loads, *Am. J. Physiol.*, 219, 881, 1970.
191. **Quamme, G. A. and Dirks, J. H.,** Intraluminal and contraluminal magnesium on magnesium and calcium transfer in the rat nephron, *Am. J. Physiol.*, 238, F187, 1980.
192. **Poujeol, P., Touvay, C., Roinel, N., and de Rouffignac, C.,** Stimulation of renal magnesium reabsorption by calcitonin in the rat, *Am. J. Physiol.*, 239, F524, 1980.
193. **Blum, J. W., Fischer, J. A., Schwoerer, D., Hunziker, W., and Binswanger, U.,** Acute parathyroid hormone response: sensitivity, relationship to hypocalcemia, and rapidity, *Endocrinology,* 95, 753, 1974.
194. **Koo, W. W. K. and Tsang, R. C.,** Neonatal hypocalcemia, in *Perspectives in Pediatrics I. Pediatric Endocrinology,* Lifshitz, F., Ed., Marcel Dekker, New York, 1985, 387.
195. **Mayer, G. P. and Hurst, J. G.,** Sigmoidal relationship between parathyroid hormone secretion rate and plasma calcium concentration in calves, *Endocrinology,* 102, 1036, 1978.
196. **Brown, E. M., Gardner, D. G., Windeck, R. A., and Aurbach, G. D.,** Relationship of intracellular 3′,5′-adenosine monophosphate accumulation to parathyroid hormone release from dispersed bovine parathyroid cells, *Endocrinology,* 103, 2323, 1978.
197. **Habener, J. F. and Potts, J. T., Jr.,** Relative effectiveness of magnesium and calcium on the secretion and biosynthesis of parathyroid hormone in vitro, *Endocrinology,* 98, 197, 1976.
198. **Chase, L. R. and Slatopolsky, E.,** Secretion and metabolic efficacy of parathyroid hormone in patients with severe hypomagnesemia, *J. Clin. Endocrinol. Metab.*, 38, 363, 1974.
199. **Rude, R. K., Oldham, S. B., and Singer, F. R.,** Functional hypoparathyroidism and parathyroid hormone end-organ resistance in human magnesium deficiency, *Clin. Endocrinol.*, 5, 209, 1976.
200. **Targovnik, J. H., Rodman, J. S., and Sherwood, L. M.,** Regulation of parathyroid hormone secretion in vitro: quantitiative aspects of calcium and magnesium ion control, *Endocrinology,* 88, 1477, 1971.
201. **Cholst, I. N., Steinberg, S. F., Tropper, P. J., Fox, H. E., Segre, G. V., and Bilezikian, J. P.,** The influence of hypermagnesemia on serum calcium and parathyroid hormone levels in human subjects, *N. Engl. J. Med.*, 310, 1221, 1984.
202. **MacManus, J., Heaton, F. W., and Lucas, P. W.,** A decreased response to parathyroid hormone in magnesium deficiency, *J. Endocrinol.*, 49, 2353, 1971.

203. **Austin, L. A. and Heath, H.,** III, Calcitonin. Physiology and pathophysiology, *N. Engl. J. Med.*, 304, 269, 1981.
204. **Sugimoto, T., Fukase, M., Tsutsumi, M., Imai, Y., Hishikawa, R., Yoshimoto, Y., and Fujita, T.,** Additive effects of parathyroid hormone and calcitonin on adenosine 3',5'-monophosphate release in newly established perfusion system of rat femur, *Endocrinology*, 117, 1901, 1985.
205. **Kennedy, J. W., III, Tanzer, F. S., and Talmage, R. V.,** Plasma phosphate and the hypocalcemic response of intact, parathyroidectomized and nephrectomized rats to thyrocalcitonin, *Endocrinology*, 85, 657, 1969.
206. **Raisz, L. G. and Niemann, I.,** Effect of phosphate, calcium and magnesium on bone resorption and hormonal responses in tissue culture, *Endocrinology*, 85, 446, 1969.
207. **Gray, T. K., Bieberdorf, F. A., and Fordtran, J. S.,** Thyrocalcitonin and the jejunal absorption of calcium, water, and electrolytes in normal subjects, *J. Clin. Invest.*, 52, 3084, 1973.
208. **Juan, D., Liptak, P., and Gray, T. K.,** Absorption of inorganic phosphate in the human jejunum and its inhibition by salmon calcitonin, *J. Clin. Endocrinol. Metab.*, 43, 517, 1976.
209. **Halloran, B. P., Portale, A. A., Castro, M., Morris, R. C., Jr., and Goldsmith, R. S.,** Serum concentration of 1,25 dihydroxyvitamin D in the human: diurnal variation, *J. Clin. Endocrinol. Metab.*, 60, 1104, 1985.
210. **Markstad, T.,** Plasma concentrations of 1,25 dihydroxyvitamin D, 24,25 dihydroxyvitamin D and 25,26 dihydroxyvitamin D in the first year of life, *J. Clin. Endocrinol. Metab.*, 57, 755, 1983.
211. **Stern, P. H., Taylor, A. B., Bell, N. H., and Epstein, S.,** Demonstration that circulating 1,25-dihydroxyvitamin D is loosely regulated in normal children, *J. Clin. Invest.*, 68, 1374, 1981.
212. **Mawer, E. B., Berry, J. L., Sommer-Tsilenis, E., Beykirch, W., Kuhlwein, A., and Rohde, B. T.,** Ultraviolet irradiation increases serum 1,25 dihydroxyvitamin D in vitamin D replete adults, *Miner. Electrolyte Metab.*, 10, 117, 1984.
213. **Lo Cascio, C., Adami, S., Galvanini, G., Ferrari, M., Cominicini, L., and Tartarotti, D.,** Substrate product relation of 1 hydroxylase activity in primary hyperparathyroidism, *N. Engl. J. Med.*, 313, 1123, 1985.
214. **Kadowaki, S. and Norman, A. W.,** Time course study of insulin secretion after 1,25 dihydroxyvitamin D_3 administration, *Endocrinology*, 117, 1765, 1985.
215. **Manolagas, S. C. and Deftos, L. J.,** The vitamin D endocrine system and the hematolymphopoietic tissue, *Ann. Intern. Med.*, 100, 144, 1984.
216. **Garabedian, M., Tanaka, Y., Holick, M. F., and DeLuca, H. F.,** Response of intestinal calcium transport and bone calcium mobilization to 1,25-dihydroxyvitamin D_3 in thyroparathyroidectomized rats, *Endocrinology*, 94, 1022, 1974.
217. **Bilezikian, J. P., Canfield, R. E., Jacobs, T. P., Polay, J. S., D'Adamo, A. P., Eisman, J. A., and DeLuca, H. F.,** Response of 1,25-hydroxyvitamin D_3 to hypocalcemia in human subjects, *N. Engl. J. Med.*, 299, 437, 1978.
218. **Trechsel, U., Eisman, J. A., Fischer, J. A., Bonjour, J. P., and Fleisch, H.,** Calcium-dependent, parathyroid hormone-independent regulation of 1,25-dihydroxyvitamin D, *Am. J. Physiol.*, 239, E119, 1980.
219. **Portale, A. A., Halloran, B. P., Murphy, M. M., and Morris, R. C., Jr.,** Oral intake of phosphorus can determine the serum concentration of 1,25 dihydroxyvitamin D by determining its production rate in humans, *J. Clin. Invest.*, 77, 7, 1986.
220. **Cantley, L. K., Russell, J., Lettieri, D., and Sherwood, L. M.,** 1,25 Dihydroxyvitamin D_3 suppresses parathyroid hormone secretion from bovine parathyroid cells in tissue culture, *Endocrinology*, 117, 2114, 1985.
221. **Slatopolsky, E., Weerts, C., Thielan, J., Horst, R., Harter, H., and Martin, K. J.,** Marked suppression of secondary hyperparathyroidism by intravenous administration of 1,25 dihydroxycholecalciferol in uremic patients, *J. Clin. Invest.*, 74, 2136, 1984.
222. **Adams, N. D., Gray, R. W., and Lemann, J., Jr.,** The effects of oral Ca CO_3 loading and dietary calcium deprivation on plasma 1,25-dihydroxyvitamin D concentrations in healthy adults, *J. Clin. Endocrinol. Metab.*, 48, 1008, 1979.
223. **Eisman, J. A., Prince, R. L., Wark, J. D., and Moseley, J. M.,** Modulation of plasma 1,25 dihydroxyvitamin D in man by stimulation and suppression tests, *Lancet*, 2, 931, 1979.
224. **Lund, B., Sorensen, O. H., Lund, B., Bishop, J. E., and Norman, A. W.,** Stimulation of 1,25-dihydroxyvitamin D production by parathyroid hormone and hypocalcemia in man, *J. Clin. Endocrinol. Metab.*, 50, 480, 1980.
225. **Blum, J. W., Trechsel, U., Born, W., Tobler, P. H., Taylor, C. M., Binswanger, U., and Fischer, J. A.,** Rapidity of plasma 1,25-dihydroxyvitamin D responses to hypo- and hypercalcemia in steers, *Endocrinology*, 113, 523, 1983.
226. **Horsting, M. and DeLuca, H. F.,** In vitro production of 25-hydroxycholecalceferol, *Biochem. Biophys. Res. Commun.*, 36, 251, 1969.

227. **Bhattacharyya, M. H. and DeLuca, H. F.,** Subcellular location of rat liver calciferol-25-hydroxylase, *Arch. Biochem. Biophys.*, 160, 58, 1974.
228. **Koo, W. W. K. and Tsang, R. C.,** Calcium and magnesium homeostatis in the newborn, in *Neonatology: Pathophysiology and Management of the Newborn,* Avery, G. B., Ed., Lippincott, Philadelphia, in press.
229. **Boonstra, C. E. and Jackson, C. E.,** Serum calcium survey for hyperparathyroidism: results in 50,000 clinic patients, *Am. J. Clin. Pathol.*, 55, 523, 1971.
230. **Walser, M., Robinson, B. H., and Duckett, J. W., Jr.,** The hypercalcemia of adrenal insufficiency, *J. Clin. Invest.*, 42, 456, 1963.
231. **Annesley, T. M., Burritt, M. F., and Kyle, R. A.,** Artifactual hypercalcemia in multiple myeloma, *Mayo Clin. Proc.*, 57, 572, 1982.
232. **Ladenson, J. H., Lewis, J. W., and Boyd, J. C.,** Failure of total calcium corrected for protein, albumin and pH to correctly assess free calcium status, *J. Clin. Endocrinol. Metab.*, 46, 986, 1978.
233. **Ladenson, J. H., Lewis, J. W., McDonald, J. M., Slatopolsky, E., and Boyd, J. C.,** Relationship of free and total calcium in hypercalcemic conditions, *J. Clin. Endocrinol. Metab.*, 48, 393, 1979.
234. **DeRemme, R. A. and Banks, P. M.,** Non Hodgkin's lymphoma associated with hypercalcemia and increased activity of serum angiotensin-converting enzyme, *Mayo Clin. Proc.*, 61, 714, 1986.
235. **Agus, Z. S., Wasserstein, A., and Goldfarb, S.,** Disorders of calcium and magnesium homeostasis, *Am. J. Med.*, 72, 473, 1982.
236. **Jamieson, M. J.,** Hypercalcemia, *Br. Med. J.*, 290, 378, 1985.
237. **Massry, S. G. and Kaptein, E. M.,** Hypercalcemic and hypocalcemic states, in *The Kidney: Pathophysiology and Pathology,* Seldin, D. W. and Giebisch, G., Eds., Raven Press, New York, 1985, 1337.
238. **Root, A. W.,** Calcium, phosphate and magnesium disorders, in *Pediatric Pathophysiology,* Colon, A. R. and Ziai, M., Eds., Little, Brown, Boston, 1985, 97.
239. **Arnaud, C. D., Nissenson, R. A., and Strewler, G. J.,** Hypercalcemia of malignancy in humans, in *Bone Resorption, Metastasis, and Diphosphonates,* Garattini, S., Ed., Raven Press, New York, 1984, 27.
240. **Goltzman, D., Stewart, A. F., and Broadus, A. E.,** Malignancy-associated hypercalcemia: evaluation with a cytochemical bioassay for parathyroid hormone, *J. Clin. Endocrinol. Metab.*, 53, 899, 1981.
241. **Yoneda, T. and Mundy, G. R.,** Prostaglandins are necessary for osteoclast-activating factor production by activated peripheral blood leukocytes, *J. Exp. Med.*, 149, 279, 1979.
242. **Seyberth, H. W., Raisz, L. G., and Oates, J. A.,** Prostaglandins and hypercalcemic states, *Annu. Rev. Med.*, 29, 23, 1978.
243. **Heath, H., III, Hodgson, S. F., and Kennedy, M. A.,** Primary hyperparathyroidism: incidence, morbidity and potential economic impact in a community, *N. Engl. J. Med.*, 302, 189, 1980.
244. **Cope, O.,** The story of hyperparathyroidism at the Massachusetts General Hospital, *N. Engl. J. Med.*, 274, 1174, 1966.
245. **Shane, E. and Bilezikian, J. P.,** Parathyroid carcinoma: a review of 62 patients, *Endocrinol. Rev.*, 3, 218, 1982.
246. **Goldsmith, R. E., Sizemore, G. W., Chen, I. W., Zalme, E., and Altemeier, W. A.,** Familial hyperparathyroidism: description of a large kindred with physiologic observations and a review of the literature, *Ann. Intern. Med.*, 84, 36, 1976.
247. **Marx, S. J., Spiegel, A. M., Brown, E. M., and Aurbach, G. D.,** Family studies in patients with primary parathyroid hyperplasia, *Am. J. Med.*, 62, 698, 1977.
248. **Tsang, R. C., Noguchi, A., and Steichen, J. J.,** Pediatric parathyroid disorders, *Pediatr. Clin. North Am.*, 26, 223, 1979.
249. **Stuart, C., Aceto, T., Jr., Kuhn, J. P., and Terplan, K.,** Intrauterine hyperparathyroidism, *Am. J. Dis. Child.*, 133, 67, 1979.
250. **Spiegel, A. M., Harrison, H. E., Marx, S. J., Brown, E. M., and Aurbach, G. D.,** Neonatal primary hyperparathyroidism with autosomal dominant inheritance, *J. Pediatr.*, 90, 269, 1977.
251. **Goldbloom, R. B., Gillis, D. A., and Prasad, M.,** Hereditary parathyroid hyperplasia: a surgical emergency of early infancy, *Pediatrics,* 49, 514, 1972.
252. **Marx, S. J., Spiegel, A. M., Levine, M. A., et al.,** Familial hypocalciuric hypercalcemia. The relation to primary parathyroid hyperplasia, *N. Engl. J. Med.*, 307, 416, 1982.
253. **Burman, K. D., Monchik, J. M., Earll, J. M., and Wartofsky, L.,** Ionized and total serum calcium and parathyroid hormone in hyperthyroidism, *Ann. Intern. Med.*, 84, 668, 1976.
254. **Eskildsen, P. C., Lund, Bj., Sorensen, O. H., Lund, Bi., Bishop, J. E., and Norman, A. W.,** Acromegaly and vitamin D metabolism: effect of bromocriptine treatment, *J. Clin. Endocrinol. Metab.*, 49, 484, 1979.
255. **Heath, H., III. and Edis, A. J.,** Pheochromocytoma associated with hypercalcemia and ectopic secretion of calcitonin, *Ann. Intern. Med.*, 91, 208, 1979.
256. **Jorgensen, H.,** Hypercalcemia in adrenocortical insufficiency, *Acta Med. Scand.*, 193, 175, 1973.

257. **Barbour, G. L., Coburn, J. W., Slatopolsky, E., Norman, A. W., and Horst, R. L.,** Hypercalcemia in an anephric patient with sarcoidosis: evidence for extrarenal generation of 1,25-dihydroxyvitamin D, *N. Engl. J. Med.*, 305, 440, 1981.
258. **Reiner, M., Sigurdsson, G., Nunziata, V., Malik, M. A., Poole, G. W., and Joplin, G. F.,** Abnormal calcium metabolism in normocalcaemic sarcoidosis, *Br. Med. J.*, 2, 1473, 1976.
259. **Bell, N. H. and Bartter, F. C.,** Studies of ^{47}Ca metabolism in sarcoidosis: evidence for increased sensitivity of bone to vitamin D, *Acta Endocrinol.*, 54, 173, 1967.
260. **Abbasi, A. A., Chemplavil, J. K., Farah, S., Muller, B. F., and Arnstein, A. R.,** Hypercalcemia in active pulmonary tuberculosis, *Ann. Intern. Med.*, 90, 324, 1979.
261. **Stoeckle, J. D., Hardy, H. L., and Weber, A. L.,** Chronic beryllium disease: long-term follow-up of sixty cases and selective review of the literature, *Am. J. Med.*, 46, 545, 1969.
262. **Lee, J. C., Catanzaro, A., Parthemore, J. G., Roach, B., and Deftos, L. J.,** Hypercalcemia in disseminated coccidioidomycosis, *N. Engl. J. Med.*, 297, 431, 1977.
263. **Walker, J. V., Baran, D., Yakub, Y. N., and Freeman, R. B.,** Histoplasmosis with hypercalcemia, renal failure, and papillary necrosis. Confusion with sarcoidosis, *J.A.M.A.*, 237, 1350, 1977.
264. **Feinstein, E. I., Akmal, M., Goldstein, D. A., Telfer, N., and Massry, S. G.,** Hypercalcemia and acute widespread calcificiations during the oliguric phase of acute renal failure due to rhabdomyolysis, *Miner. Electrolyte Metab.*, 2, 193, 1979.
265. **Feinstein, E. I., Akmal, M., Telfer, N., and Massry, S. G.,** Delayed hypercalcemia with acute renal failure associated with nontraumatic rhabdomyolysis, *Arch. Intern. Med.*, 141, 753, 1981.
266. **Thomas, M.A.B. and Ibels, L.S.,** Rhabdomyolysis and acute renal failure, *Aust. N.Z. J. Med.*, 15, 623, 1985.
267. **Akmal, M., Goldstein, D. A., Telfer, N., Wilkinson, E., and Massry, S. G.,** Resolution of muscle calcification in rhabdomyolysis and acute renal failure, *Ann. Intern. Med.*, 89, 928, 1978.
268. **Llach, F., Felsenfeld, A. J., and Haussler, M. R.,** The pathophysiology of altered calcium metabolism in rhabdomyolysis-induced acute renal failure: interactions of parathyroid hormone, 25-hydroxycholecalciferol, and 1,25-dihydroxycholecalciferol, *N. Engl. J. Med.*, 305, 117, 1981.
269. **Massry, S. G., Coburn, J. W., Popovtzer, M. M., Shinaberger, J. H., Maxwell, M. H., and Kleeman, C. R.,** Secondary hyperparathyroidism in chronic renal failure: the clinical spectrum in uremia, during hemodialysis, and after renal transplantation, *Arch. Intern. Med.*, 124, 431, 1969.
270. **Freeman, R. M., Lawton, R. L., and Chamberlain, M. A.,** Hard-water syndrome, *N. Engl. J. Med.*, 276, 1113, 1967.
271. **Meyrier, A., Marsac, J., and Richet, G.,** The influence of a high calcium carbonate intake on bone disease in patients undergoing hemodialysis, *Kidney Int.*, 4, 146, 1973.
272. **Papadimitriou, M., Gingell, J. C., and Chisholm, G. D.,** Hypercalcemia from calcium ion-exchange resin in patients on regular hemodialysis, *Lancet*, 2, 948, 1968.
273. **Koppel, M. H., Massry, S. G., Shinaberger, J. H., Hartenbower, D. L., and Coburn, J. W.,** Thiazide-induced rise in serum calcium and magnesium in patients on maintenance hemodialysis, *Ann. Intern. Med.*, 72, 895, 1970.
274. **Brickman, A. S., Coburn, J. W., Massry, S. G., and Norman, A. W.,** 1,25-Dihydroxy-vitamin D_3 in normal man and patients with renal failure, *Ann. Intern. Med.*, 80, 161, 1974.
275. **Balsan, S., Garabedian, M., Sorgniard, R., Holick, M. F., and DeLuca, H. F.,** 1,25-Dihydroxyvitamin D_3 and 1,alpha-hydroxyvitamin D_3 in children: biologic and therapeutic effects in nutritional rickets and different types of vitamin D resistance, *Pediatr. Res.*, 9, 586, 1975.
276. **Parfitt, A. M.,** Hypercalcemic hyperparathyroidism following renal transplantation: differential diagnosis, management and implications for cell population control in the parathyroid gland, *Miner. Electrolyte Metab.*, 8, 92, 1982.
277. **Penn, I.,** Malignancies associated with immunosuppressive or cytotoxic therapy, *Surgery*, 83, 492, 1978.
278. **Davies, M. and Adams, P. H.,** The continuing risk of vitamin-D intoxication, *Lancet*, 2, 621, 1978.
279. **Lee, D. B. N., Brautbar, N., and Massry, S. G.,** Renal production and biologic actions of vitamin D metabolites, *Semin. Nephrol.*, 1, 335, 1981.
280. **Belchetz, P. E., Cohen, R. D., O'Riordan, J. L. H., and Tomlinson, S.,** Factitious hypercalcemia, *Br. Med. J.*, 1, 690, 1976.
281. **Frame, B., Jackson, C. E., Reynolds, W. A., and Umphrey, J. E.,** Hypercalcemia and skeletal effects in chronic hypervitaminosis A, *Ann. Intern. Med.*, 80, 44, 1974.
282. **Parfitt, A. M.,** The interactions of thiazide diuretics with parathyroid hormone and vitamin D: studies in patients with hypoparathyroidism, *J. Clin. Invest.*, 51, 1879, 1972.
283. **Parfitt, A. M.,** Chlorothiazide-induced hypercalcemia in juvenile osteoporosis and primary hyperparathyroidism, *N. Engl. J. Med.*, 281, 55, 1969.
284. **Meyer-Sabellek, W., Brautbar, N., Moser, S., Montcalm, A., and Massry, S. G.,** Effect of thiazides on the hypercalciuria of phosphate depletion, *J. Lab. Clin. Med.*, 96, 830, 1980.

285. **Malluche, H. H., Meyer-Sabellek, W., Singer, F. R., and Massry, S. G.,** Evidence for a direct effect of thiazides on bone, *Miner. Electrolyte Metab.*, 4, 89, 1980.
286. **Popovtzer, M. M., Subryan, V. L., Alfrey, A. C., Reeve, E. B., and Schrier, R. W.,** The acute effect of chlorothiazide on serum-ionized calcium: evidence for a parathyroid hormone-dependent mechanism, *J. Clin. Invest.*, 55, 1295, 1975.
287. **Christiansen, C., Baastrup, P. C., Lindgreen, P., and Transbol, I.,** Endocrine effects of lithium. II. ''Primary'' hyperparathyroidism, *Acta Endocrinol.*, 88, 528, 1978.
288. **Orwoll, E. S.,** The milk-alkali syndrome: current concepts, *Ann. Intern. Med.*, 97, 242, 1982.
289. **Shike, M., Harrison, J. E., Sturtridge, W. C., Tam, C. S., Bobechko, P. E., Jones, G., Murray, T. M., and Jeejeebhoy, K. N.,** Metabolic bone disease in patients receiving long-term total parenteral nutrition, *Ann. Intern. Med.*, 92, 343, 1980.
290. **Frame, B., Jackson, G. M., Kleerekoper, M., Rao, D. S., DeLorenzo, A. J. D., and Garcia, M.,** Acute severe hypercalcemia a al Munchausen, *Am. J. Med.*, 70, 316, 1981.
291. **Lyon, A. J., McIntosh, N., Wheeler, K., and Brooke, O. G.,** Hypercalcemia in extremely low birth weight infants, *Arch. Dis. Child.*, 59, 1141, 1984.
292. **Bruin, W. J., Baylink, D. J., and Wergedal, J. E.,** Acute inhibition of mineralization and stimulation of bone resorption mediated by hypophosphatemia, *Endocrinology*, 96, 394, 1975.
293. **Koo, W. W. K. and Tsang, R. C.,** Rickets in infants, in *Current Therapy in Neonatal-Perinatal Medicine*, Nelson, N. M., Ed., Decker BC, Philadelphia, 1984, 299.
294. **Bergstrom, W. H.,** Hypercalciuria and hypercalcemia complicating immobilization, *Am. J. Dis. Child.*, 132, 553, 1978.
295. **Stewart, A. F., Adler, M., Byers, C. M., Segre, G. V., and Broadus, A. E.,** Calcium homeostasis in immobilization: an example of resorptive hypercalciuria, *N. Engl. J. Med.*, 306, 1136, 1982.
296. **Goodyer, P. R., Frank, A., and Kaplan, B. S.,** Observations on the evolution and treatment of idiopathic infantile hypercalcemia, *J. Pediatr.*, 105, 771, 1984.
297. **Fraser, D.,** Hypophosphatasia, *Am. J. Med.*, 22, 730, 1957.
298. **Barltrop, D.,** Hypercalcemia associated with neonatal subcutaneous fat necrosis, *Arch. Dis. Child.*, 38, 516, 1963.
299. **Taylor, A. B., Stern, P. H., and Bell, N. H.,** Abnormal regulation of circulating 25 hydroxyvitamin D in the Williams syndrome, *N. Engl. J. Med.*, 306, 972, 1982.
300. **Culler, F. L., Jones, K. L., and Deftos, L. J.,** Impaired calcitonin secretion in patients with Williams syndrome, *J. Pediatr.*, 107, 720, 1985.
301. **Llach, F., Brickman, A. S., Ben-Isaacs, C., Coburn, J. W., and Massry, S. G.,** Phosphate loading as a test for primary hyperparathyroidism, in *Phosphate Metabolism. Kidney and Bone*, Avioli, L., Bordier, P., Fleisch, H., Massry, S. G., and Slatopolsky, E., Eds., Nouvelle Imprimerie Fourniew, Toulouse, France, 1975, 171.
302. **Brown, E. M., Broadus, A. E., Brennan, M. F., et al.,** Direct comparison in vivo and in vitro of suppressibility of parathyroid function by calcium in primary hyperparathyroidism, *J. Clin. Endocrinol. Metab.*, 48, 604, 1979.
303. **Insogna, K. L., Mitnick, M. A. E., Stewart, A. F., Burtis, W. J., Mallette, L. E., and Broadus, A. E.,** Sensitivity of the parathyroid hormone-1,25 dihydroxyvitamin D axis to variations in calcium intake in patients with primary hyperparathyroidism, *N. Engl. J. Med.*, 313, 1126, 1985.
304. **Broadus, A. E., Horst, R. L., Lang, R., Littledike, E. T., and Rasmussen, H.,** The importance of circulating 1,25-dihydroxyvitamin D in the pathogenesis of hypercalciuria and renal-stone formation in primary hyperparathyroidism, *N. Engl. J. Med.*, 302, 421, 1980.
305. **Suki, W. N., Yium, J. J., Von Minden, M., Saller-Herbert, C., Eknoyan, G., and Martinez-Maldonado, M.,** Acute treatment of hypercalcemia with furosemide, *N. Engl. J. Med.*, 283, 836, 1970.
306. **Massry, S. G., Mueller, E., Silverman, A. G., and Kleeman, C. R.,** Inorganic phosphate treatment of hypercalcemia, *Arch. Intern. Med.*, 121, 307, 1968.
307. **Fulmer, D. H., Dimich, A. B., Rothschild, E. Q., and Myers, W. P. L.,** Treatment of hypercalcemia: comparison of intravenously administered phosphate, sulfate and hydrocortisone, *Arch. Intern. Med.*, 129, 923, 1972.
308. **Breuer, R. I. and LeBauer, J.,** Caution in the use of phosphates in the treatment of severe hypercalcemia, *J. Clin. Endocrinol.*, 27, 695, 1967.
309. **Laflamme, G. H. and Jowsey, J.,** Bone and soft tissue changes with oral phosphate supplements, *J. Clin. Invest.*, 51, 2834, 1972.
310. **Kammerman, S. and Canfield, R. E.,** Effect of porcine calcitonin on hypercalcemia in man, *J. Clin. Endocrinol.*, 31, 70, 1970.
311. **West, T. E. T., Joffe, M., Sinclair, L., and O'Riordan, J. L. H.,** Treatment of hypercalcemia with calcitonin, *Lancet*, 1, 675, 1971.
312. **Foster, G. V., Joplin, G. F., MacIntyre, I., and Melvin, K. E. W.,** Effect of thyrocalcitonin in man, *Lancet*, 1, 107, 1966.

313. **Smith, I. E. and Powles, T. J.,** Mithramycin for hypercalcemia associated with myeloma and other malignancies, *Br. Med. J.*, 1, 268, 1975.
314. **Koo, W. W. K. and Tsang, R. C.,** Neonatal hypocalcemia, in *Perspectives in Pediatrics I. Pediatric Endocrinology*, Lifshitz, F., Ed., Marcel Dekker, New York, 1985, 387.
315. **Specker, B. L., Lichtenstein, M. D., Mimouni, F., Gormley, C., and Tsang, R. C.,** Calcium regulating hormones and animals from birth to 18 months: across sectional study. II. Effects of sex, race, age, season, and diet on serum minerals, parathyroid hormone, and calcitonin, *Pediatrics*, 77, 891, 1986.
316. **Peden, V. H.,** True idiopathic hypoparathyroidism as a sex-linked recessive trait, *Am. J. Hum. Genet.*, 12, 323, 1960.
317. **Richter, P. L. and Chutorian, A. M.,** Familial hypoparathyroidism: case reports and a review of the literature, *Neurology (Minneapolis)*, 18, 75, 1968.
318. **DiGeorge, A. M.,** Congenital absence of the thymus and its immunologic consequence: concurrence with congenital hypoparathyroidism, *Birth Defects*, 4, 116, 1968.
319. **Harvey, J. C., Dungan, W. T., Elders, M. J., and Hughes, E. R.,** Third and fourth pharyngeal pouch syndrome, associated vascular anomalies and hypocalcemia seizures, *Clin. Pediatr.*, 9, 496, 1970.
320. **Eipe, J., Johnson, S. A., Kiamko, R. T., and Bronsky, D.,** Hypoparathyroidism following ^{131}I therapy for hyperthyroidism, *Arch. Intern. Med.*, 121, 270, 1968.
321. **Salander, H. and Tisell, L. E.,** Incidence of hypoparathyroidism after radical surgery for thyroid carcinoma and autotransplantation of parathyroid glands, *Am. J. Surg.*, 134, 358, 1977.
322. **Brezis, M., Shalev, O., Leibel, B., Bernheim, J., and Ben-Ishay, D.,** Phosphorus retention and hypoparathyroidism associated with transfusional iron overload in thalassaemia, *Miner. Electrolyte Metab.*, 4, 57, 1980.
323. **Neufeld, M., MacLaren, N. K., and Blizzard, R. M.,** Two types of autoimmune Addison's disease associated with different polyglandular autoimmune (PGA) syndromes, *Medicine*, 60, 355, 1981.
324. **Levine, M. A., Downs, R. W., Jr., Moses, A. M., et al.,** Resistance to multiple hormones in patients with pseudohypoparathyroidism: association with deficient activity of guanine nucleotide regulatory protein, *Am. J. Med.*, 74, 545, 1983.
325. **Nusynowitz, M. L., Frame, B., and Kolb, F. O.,** The spectrum of the hypoparathyroid states: a classification based on physiologic principles, *Medicine*, 55, 105, 1976.
326. **Mann, J. B., Alterman, S., and Hills, A. G.,** Albright's hereditary osteodystrophy comprising pseudohypoparathyroidism and pseudo-pseudohypoparathyroidism, *Ann. Intern. Med.*, 56, 315, 1962.
327. **Albright, F., Forbes, A. P., and Henneman, P. H.,** Pseudo-pseudohypoparathyroidism, *Trans. Assoc. Am. Physicians*, 65, 337, 1952.
328. **Daneman, D., Kooh, S. W., and Fraser, D.,** Hypoparathyroidism and pseudohypoparathyroidism in childhood, *Clin. Endocrinol. Metab.*, 11, 211, 1982.
329. **Farfel, Z., Brickman, A. S., Kaslow, H. R., Brothers, V. M., and Bourne, H. R.,** Defect of receptor-cyclase coupling protein in pseudohypoparathyroidism, *N. Engl. J. Med.*, 303, 237, 1980.
330. **Farfel, Z. and Bourne, H. R.,** Deficient activity of receptor-cyclase coupling protein in platelets of patients with pseudohypoparathyroidism, *J. Clin. Endocrinol. Metab.*, 51, 1202, 1980.
331. **Bourne, H. R., Kaslow, H. R., Brickman, A. S., and Farfel, Z.,** Fibroblast defect in pseudohypoparathyroidism, type I: reduced activity of receptor-cyclase coupling protein, *J. Clin. Endocrinol. Metab.*, 53, 636, 1981.
332. **Downs, R. W., Jr., Levine, M. A., Drezner, M. K., Burch, W. M., Jr., and Spiegel, A. M.,** Deficient adenylate cyclase regulatory protein in renal membranes from a patient with pseudohypoparathyroidism, *J. Clin. Invest.*, 71, 231, 1983.
333. **Downs, R. W., Sekura, R. D., Levine, M. A., and Spiegel, A. M.,** The inhibitory adenylate cyclase coupling protein in pseudohypoparathyroidism, *J. Clin. Endocrinol. Metab.*, 61, 351, 1985.
334. **Drezner, M., Neelon, F. A., and Lebovitz, H. E.,** Pseudohypoparathyroidism type II: a possible defect in the reception of the cyclic AMP signal, *N. Engl. J. Med.*, 289, 1056, 1973.
335. **Rodriguez, H. J., Villarreal, H., Jr., Klahr, S., and Slatopolsky, E.,** Pseudohypoparathyroidism type II: restoration of normal renal responsiveness to parathyroid hormone by calcium administration, *J. Clin. Endocrinol. Metab.*, 39, 693, 1974.
336. **Bronsky, D., Kushner, D. S., Dubin, A., and Snapper, I.,** Idiopathic hypoparathyroidism and pseudohypoparathyroidism: case reports and review of the literature, *Medicine*, 37, 317, 1958.
337. **Massry, S. G., Arieff, A. I., Coburn, J. W., Palmieri, G., and Kleeman, C. R.,** Divalent ion metabolism in patients with acute renal failure: studies on the mechanism of hypocalcemia, *Kidney Int.*, 5, 437, 1974.
338. **Llach, F., Massry, S. G., Singer, F. R., Kurokawa, K., Kaye, J. H., and Coburn, J. W.,** Skeletal resistance to endogenous parathyroid hormone in patients with early renal failure. A possible cause for secondary hyperparathyroidism, *J. Clin. Endocrinol. Metab.*, 41, 339, 1975.
339. **Haldimann, B., Goldstein, D. A., Akmal, M., and Massry, S. G.,** Renal function and blood levels of divalent ions in acute pancreatitis. A prospective study in 99 patients, *Miner. Electrolyte Metab.*, 3, 190, 1980.

340. **Bernstein, D., Kleeman, C. R., Dowling, J. T., and Maxwell, M. H.,** Steatorrhea, functional hypoparathyroidism, and metabolic bone defect, *Arch. Int. Med. Exp.* in press.
341. **Goldstein, D. A., Haldimann, B., Sherman, D., Norman, A. W., and Massry, S. G.,** Vitamin D metabolites and calcium metabolism in patients with nephrotic syndrome and normal renal function, *J. Clin. Endocrinol. Metab.*, 152, 116, 1981.
342. **Liberman, U. A., Eil, C., and Marx, S. J.,** Resistance to 1,25-dihydroxyvitamin D. Association with heterogeneous defects in cultured skin fibroblasts, *J. Clin. Invest.*, 71, 192, 1983.
343. **Chen, T. L., Hirst, M. A., Cone, C. M., Hochberg, Z., Tietze, H. U., and Feldman, D.,** 1,25-Dihydroxyvitamin D resistance, rickets and alopecia: analysis of receptors and bioresponse in clutured fibroblasts from patients and parents, *J. Clin. Endocrinol. Metab.*, 59, 383, 1984.
344. **Hirst, M. A., Hochman, H. I., and Feldman, D.,** Vitamin D resistance and alopecia: a kindred with normal 1,25 dihydroxyvitamin D binding, but decreased receptor affinity for deoxyribonucleic acid, *J. Clin. Endocrinol. Metab.*, 60, 490, 1985.
345. **King, R. G. and Stanbury, S. W.,** Magnesium metabolism in primary hyperparathyroidism, *Clin. Sci.*, 39, 281, 1970.
346. **Engel, F. L., Martin, S. P., and Taylor, H.,** On the relation of potassium to the neurological manifestations of hypocalcemic tetany, *Johns Hopkins Hosp. Bull.*, 84, 285, 1949.
347. **Venkataraman, P. S., Tsang, R. C., Greer, F. R., et al.,** Late infantile tetany and secondary hyperparathyroidism in infants fed humanized cow milk formula, *Am. J. Dis. Child.*, 139, 664, 1985.
348. **Davis, R. F., Eichner, J. M., Bleyer, W. A., and Okamoto, G.,** Hypocalcemia hyperphosphatemia, and dehydration following a single hypertonic phosphate enema, *J. Pediatr.*, 90, 484, 1977.
349. **Zusman, J., Brown, D. M., and Nesbit, M. E.,** Hyperphosphatemia, hyperphosphaturia and hypocalcemia in acute lymphoblastic leukemia, *N. Engl. J. Med.*, 289, 1335, 1973.
350. **Chaudhuri, T. K. and Chaudhuri, T. K.,** Mechanism of hypocalcemic effect of inorganic phosphate, *N. Engl. J. Med.*, 285, 691, 1971.
351. **Hill, G. L., Guinn, E. J., and Dudrick, S. J.,** Phosphorus distribution in hyperalimentation induced hypophosphatemia, *J. Surg. Res.*, 20, 527, 1976.
352. **Goldsmith, R. S.,** Multiple effects of phosphate therapy, *N. Engl. J. Med.*, 282, 927, 1970.
353. **Cooper, L. J. and Anast, C. S.,** Circulating immunoreactive parathyroid hormone levels in premature infants and the response to calcium therapy, *Acta Paediatr. Scand.*, 74, 669, 1985.
354. **Hillman, L. S., Rojanasatkit, S., Slatopolsky, E., and Haddad, J. G.,** Serial measurements of serum calcium, magnesium, parathyroid hormone, calcitonin, and 25 hydroxyvitamin D in premature and term infants during the first week of life, *Pediatr. Res.*, 11, 739, 1977.
355. **Koo, W. W. K., Tsang, R. C., Poser, J. W., Laskarzewski, P., Buckley, D., Johnson, R., and Steichen, J. J.,** Elevated serum osteocalcin and calcium from 1,25-dihydroxyvitamin D_3 in preterm infants: a prospective randomized study, *Am. J. Dis. Child,* in press.
356. **Seamonds, B., Towfighi, J., and Arvan, D. A.,** Determination of ionized calcium in serum by use of an ion-selective electrode. I. Determination of normal values under physiologic conditions, with comments on the effects of food ingestion and hyperventilation, *Clin. Chem.*, 18, 155, 1972.
357. **Schneider, A. B. and Sherwood, L. M.,** Pathogenesis and management of hypoparathyroidism and other hypocalcemic disorders, *Metabolism,* 24, 871, 1975.
358. **Scott, S. M., Ladenson, J. H., Aguanna, J. J., Walgate, J., and Hillman, L. S.,** Effect of calcium therapy in the sick premature infant with early neonatal hypocalcemia, *J. Pediatr.*, 104, 747, 1984.
359. **Shanes, A. M.,** Electrochemical aspects of physiological and pharmacological action in excitable cells. II. The action potential and excitation, *Pharmacol. Rev.*, 10, 165, 1958.
360. **Walsh, F. B. and Murray, R. G.,** Ocular manifestations of disturbances in calcium metabolism, *Am. J. Ophthalmol.*, 36, 1657, 1953.
361. **Levin, P.,** Intracranial calcification associated with hypoparathyroidism, *Bull. N.Y. Acad. Med.*, 38, 632, 1962.
362. **Ireland, A. W., Hornbrook, J. W., Neale, F. C., and Posen, S.,** The crystalline lens in chronic surgical hypoparathyroidism, *Arch. Intern. Med.*, 122, 408, 1968.
363. **Gass, J. D. M.,** The syndrome of keratoconjunctivitis, superficial moniliasis, idiopathic hypoparathyroidism and Addison's disease, *Am. J. Ophthalmol.*, 54, 660, 1962.
364. **Nikiforuk, G. and Fraser, D.,** The etiology of enamel hypoplasia: a unifying concept, *J. Pediatr.*, 98, 888, 1981.
365. **Noren, J. G.,** Enamel structure in deciduous teeth from low birth weight infants, *Acta Odontol. Scand.*, 41, 355, 1983.
366. **Scharer, K. and Komposch, G.,** Etiology of enamel hypoplasia, *J. Pediatr.*, 100, 673, 1982.
367. **Troughton, O. and Singh, S. P.,** Heart failure and neonatal hypocalcaemia, *Br. Med. J.*, 4, 76, 1972.
368. **Chopra, D., Janson, P., and Sawin, C. T.,** Insensitivity to digoxin associated with hypocalcemia, *N. Engl. J. Med.*, 296, 917, 1977.

369. **Connor, T. B., Rosen, B. L., Blaustein, M. P., Applefeld, M. M., and Doyle, L. A.,** Hypocalcemia precipitating congestive heart failure, *N. Engl. J. Med.*, 307, 869, 1982.

369a. **Watanabe, K., Hara, K., Miyazaki, S., and Hakamada, S.,** Neurophysiological study of newborns with hypocalcemia, *Neuropediatrics,* 13, 34, 1982.

370. **Basser, L. S., Neale, F. C., Ireland, A. W., and Posen, S.,** Epilepsy and electroencephalographic abnormalities in chronic surgical hypoparathyroidism, *Ann. Intern. Med.*, 71, 507, 1969.

371. **Halbert, K. and Tsang, R. C.,** Treatment of parathyroid disease: hypoparathyroidism and pseudohypoparathyroidism, in *Current Pediatric Therapy,* 12th Ed., Gellis, S. S. and Kagan, B. M., Eds., W. B. Saunders, Philadelphia 1986, 306.

372. **Ireland, A. W., Clubb, J. S., Neale, F. C., Posen, S., and Reeve, T. S.,** The calciferol requirements of patients with surgical hypoparathyroidism, *Ann. Intern. Med.*, 69, 81, 1968.

373. **Kanis, J. A.,** Vitamin D metabolism and its clinical application, *J. Bone Jt. Surg.*, 64B, 542, 1982.

374. **Porter, R. H., Cox, B. C., Heaney, D., Hostetter, T. H., Stinebaugh, B. J., and Suki, W. N.,** Treatment of hypoparathyroid patients with chlorthalidone, *N. Engl. J. Med.*, 298, 577, 1978.

375. **Riedesel, M. L. and Folk, G. E.,** Serum magnesium and hibernation, *Fed. Proc.*, 15, 151, 1956.

376. **Hannon, J. P., Larson, A. M., and Young, D. W.,** Effect of cold acclimatization on plasma electrolyte levels, *J. Appl. Physiol.*, 13, 239, 1958.

377. **Hamburger, J.,** Electrolyte disturbances in acute uremia, *Clin. Chem.*, 3, 332, 1957.

378. **Mordes, J. P., Swartz, R., and Arky, R. A.,** Extreme hypermagnesemia as a cause of refractory hypotension, *Ann. Intern. Med.,* 83, 657, 1975.

379. **Coburn, J. W., Popovtzer, M. M., Massry, S. G., and Kleeman, C. R.,** The physicochemical state and renal handling of divalent ions in chronic renal failure, *Arch. Intern. Med.*, 124, 302, 1969.

380. **Alfrey, A. C., Terman, D. S., Brettschneider, L., Simpson, K. M., and Ogden, D. A.,** Hypermagnesemia after renal homotransplantation, *Ann. Intern. Med.*, 73, 367, 1970.

381. **Randall, R. E., Jr., Cohen, M. D., Spray, C. C., Jr., and Rossmeisl, E. C.,** Hypermagnesemia in renal failure. Etiology and toxic manifestations, *Ann. Intern. Med.*, 61, 73, 1964.

382. **Cruikshank, D. P., Pitkin, R. M., Donnelly, E., and Reynolds, W. A.,** Urinary magnesium, calcium and phosphate excretion during the magnesium sulfate infusion, *Obstet. Gynecol.*, 58, 430, 1981.

383. **Grantham, J. J., Tu, W. H., and Schloerb, P. R.,** Acute magnesium depletion and excess induced by hemodialysis, *Am. J. Physiol.*, 198, 1211, 1960.

384. **Nabarro, J. D. N., Spencer, A. G., and Stowers, J. M.,** Metabolic studies in severe diabetic ketosis, *Q. J. Med.*, 21, 225, 1952.

385. **Madajewicz, S., Szymendera, J., and Zulawski, M.,** Hypermagnesemia in malignant neoplasms, *Pol. Med. J.*, 10, 599, 1971.

386. **Palmieri, G. M. A., Thompson, J. S., and Eliel, L. P.,** Modifications of plasma magnesium by thyrocalcitonin, parathyroid extract and cortisone, *Endocrinology,* 84, 1509, 1969.

387. **Wallach, S., Cahill, L. N., Rogan, F. H., and Jones, H. L.,** Plasma and erythrocyte magnesium in health and disease, *J. Lab. Clin. Med.*, 59, 195, 1962.

388. **Stone, S. R. and Pritchard, J. A.,** Effect of maternally administered magnesium sulfate on the neonate, *Obstet. Gynecol.*, 35, 574, 1970.

389. **Lipsitz, P. J.,** The clinical and biochemical effects of excess magnesium in the newborn, *Pediatrics,* 47, 501, 1971.

390. **Donovan, E. F., Tsang, R. C., Steichen, J. J., Strub, R. J., Chen, I. W., and Chen, M.,** Neonatal hypermagnesemia: effect on parathyroid hormone and calcium homeostasis, *J. Pediatr.*, 96, 305, 1980.

391. **Andrews, B. F., Campbell, D. R., and Thomas, P.,** Effects of hypertonic magnesium sulphate enemas on newborn and young lambs, *Lancet,* 2, 64, 1965.

392. **Stevens, A. R., Jr. and Wolff, H. G.,** Magnesium intoxication. Absorption from the intact gastrointestinal tract, *Arch. Neurol. Psychiatry,* 63, 749, 1950.

393. **Outerbridge, E. W., Papageorgiou, A., and Stern, L.,** Magnesium sulfate enema in a newborn, *J.A.M.A.*, 224, 1392, 1973.

394. **Mordes, J. P. and Wacker, W. E. C.,** Excess magnesium, *Pharmacol. Rev.*, 29, 273, 1978.

395. **Somjen, G. and Kato, G.,** Effects of magnesium and calcium on neurones in the central nervous system, *Brain Res.*, 9, 161, 1968.

396. **Pritchard, J. A.,** The use of the magnesium ion in the management of eclamptogenic toxemias, *Surg. Gynecol. Obstet.*, 100, 131, 1955.

397. **Nielsen, S. P. and Jorgensen, F. S.,** Mechanism of the hypocalcemic action of parenterally administered magnesium, *Acta Endocrinol.*, 70, 476, 1972.

398. **Slatopolsky, E., Mercado, A., Morrison, A., Yates, J., and Klahr, S.,** Inhibitory effects of hypermagnesemia on the renal action of parathyroid hormone, *J. Clin. Invest.*, 58, 1273, 1976.

399. **Savory, J. and Monif, G. R. G.,** Serum calcium levels in cord sera of the progeny of mothers treated with magnesium sulfate for toxemia of pregnancy, *Am. J. Obstet. Gynecol.*, 110, 556, 1971.

400. **Alfrey, A. C. and Miller, N. L.,** Bone magnesium pools in uremia, *J. Clin. Invest.,* 52, 3019, 1973.
401. **Cunningham, I. J.,** Magnesium in animal diets, *N.Z. J. Sci. Technol.,* 15, 191, 1933.
402. **Bachra, B. N., Trautz, O. R., and Simon, S. L.,** Precipitation of calcium carbonates and phosphates. III. The effect of magnesium and fluoride ions on the spontaneous precipitation of calcium carbonates and phosphates, *Arch. Oral Biol.,* 10, 371, 1965.
403. **Alfrey, A. C., Solomons, C. C., Ciricillo, J., and Miller, N. L.,** Extraosseous calcification: evidence for abnormal pyrophosphate metabolism in uremia, *J. Clin. Invest.,* 57, 692, 1976.
404. **Vallee, B., Wacker, W. E. C., and Ulmer, D. D.,** The magnesium-deficiency tetany syndrome in man, *N. Engl. J. Med.,* 262, 155, 1960.
405. **Watson, W. S., Hilditch, T. E., Horton, P. W., Davies, D. L., and Lindsay, R.,** Magnesium metabolism in blood and the whole body in man using ^{28}magnesium, *Metabolism,* 28, 90, 1979.
406. **Elin, R. J.,** Overview of problems in diagnosis of magnesium status, *J. Am. Coll. Nutr.,* 4, 330, 1985.
407. **Seelig, M. S., Berger, A. R., and Spielholz, N.,** Latent tetany and anxiety, marginal magnesium deficit, and normocalcemia, *Dis. Nerv. Syst.,* 36, 461, 1975.
408. **Alfrey, A. C.,** Disorders of magnesium metabolism, in *The Kidney: Physiology and Pathophysiology,* Seldin, D. W. and Giebisch, G., Eds., Raven Press, New York, 1985, 1281.
409. **Flink, E. B., Stutzman, F. L., Anderson, A. R., Konig, T., and Fraser, R.,** Magnesium deficiency after prolonged parenteral fluid administration and after chronic alcoholism complicated by delirium tremens, *J. Lab. Clin. Med.,* 43, 169, 1954.
410. **Baron, D. N.,** Magnesium deficiency after gastro-intestinal surgery and loss of secretions, *Br. J. Surg.,* 48, 344, 1960.
411. **Jones, J. E., Albrink, M. J., Davidson, P. C., and Flink, E. B.,** Fasting and refeeding of various suboptimal isocaloric diets, *Am. J. Clin. Nutr.,* 19, 320, 1966.
412. **Booth, C. C., Babouris, N., Hanna, S., and MacIntyre, I.,** Incidence of hypomagnesaemia in intestinal malabsorption, *Br. Med. J.,* 2, 141, 1963.
413. **Thoren, L.,** Magnesium deficiency in gastrointestinal fluid loss, *Acta Chir. Scand.,* [Suppl] 306 (Suppl.), 1, 1963.
414. **Brouhard, B. H.,** Prostaglandins and hypokalemia, *J. Pediatr.,* 107, 738, 1985.
415. **Horton, R. and Biglieri, E. G.,** Effect of aldosterone on the metabolism of magnesium, *J. Clin. Endocrinol. Metab.,* 22, 1187, 1962.
416. **Cohen, M. I., McNamara, H., and Finberg, L.,** Serum magnesium in children with cirrhosis, *J. Pediatr.,* 76, 453, 1970.
417. **Hellman, E. S., Tschudy, D. P., and Bartter, F. C.,** Abnormal electrolyte and water metabolism in acute intermittent porphyria: the transient inappropriate secretion of antidiuretic hormone, *Am. J. Med.,* 32, 734, 1962.
418. **Ziff, W. B., Bacon, G. E., and Spencer, M. L.,** Hypocalcemia, hypomagnesemia and transient hypoparathyroidism during therapy with potassium phosphate in diabetic ketoacidosis, *Diabetes Care,* 2, 265, 1979.
419. **Winter, R. J., Harris, C. J., Phillips, L. S., and Green, O. C.,** Diabetic ketoacidosis. Induction of hypocalcemia and hypomagnesemia by phosphate therapy, *Am. J. Med.,* 67, 897, 1979.
420. **Edmondson, H. A., Berne, C. J., Homann, R. E., Jr., and Wertman, M.,** Calcium, potassium, magnesium and amylase disturbances in acute pancreatitis, *Am. J. Med.,* 12, 34, 1952.
421. **McMullen, J. K.,** Asystole and hypomagnesaemia during recovery from diabetic ketoacidosis, *Br. Med. J.,* 1, 690, 1977.
422. **Mendelson, J. H., Barnes, B., Mayman, C., and Victor, M.,** The determination of exchangeable magnesium in alcoholic patients, *Metabolism,* 14, 88, 1965.
423. **Marks, P. and Ashraf, H.,** Apathetic hyperthyroidism with hypomagnesaemia and raised alkaline phosphatase concentration, *Br. Med. J.,* 1, 821, 1978.
424. **Tibbetts, D. M. and Aub, J. C.,** Magnesium metabolism in health and disease. III. In exophthalmic goiter, basophilic adenoma, Addison's disease and steatorrhea, *J. Clin. Invest.,* 16, 511, 1937.
425. **Aikawa, J. K.,** Effect of thyroxine and propylthiouracil on magnesium metabolism in the rabbit. Study with Mg^{28}, *Proc. Soc. Exp. Biol. Med.,* 104, 594, 1960.
426. **Broughton, A., Anderson, I. R. M., and Bowden, C. H.,** Magnesium-deficiency syndrome in burns, *Lancet,* 2, 1156, 1968.
427. **Dancis, J., Springer, D., and Cohlan, S. Q.,** Fetal homeostasis in maternal malnutrition. II. Magnesium deprivation, *Pediatr. Res.,* 5, 131, 1971.
428. **Tsang, R. C., Strub, R., Brown, D. R., Steichen, J., Hartman, C., and Chen, I. W.,** Hypomagnesemia in infants of diabetic mothers: perinatal studies, *J. Pediatr.,* 89, 115, 1976.
429. **Tsang, R. C., Brown, D. R., and Steichen, J. J.,** Diabetes and calcium: calcium disturbances in infants of diabetic mother, in *The Diabetic Pregnancy: A Perinatal Perspective,* Merkatz, E. and Adam, P., Eds., Grune & Stratton, Orlando, 1979, 207.

430. **Tsang, R. C. and Oh, W.,** Serum magnesium levels in low birth weight infants, *Am. J. Dis. Child.,* 120, 44, 1970.
431. **Hanna, S., North, K. A. K., MacIntyre, I., and Fraser, R.,** Magnesium metabolism in parathyroid disease, *Br. Med. J.,* 2, 1253, 1961.
432. **Monteleone, J. A., Lee, J. B., Tashjian, A. H., Jr., and Cantor, H. E.,** Transient neonatal hypocalcemia, hypomagnesemia, and high serum parathyroid hormone with maternal hyperparathyroidism, *Ann. Intern. Med.,* 82, 670, 1975.
433. **Bajpai, P. C., Sugden, D., Stern, L., and Denton, R. L.,** Serum ionic magnesium in exchange transfusion, *J. Pediatr.,* 70, 193, 1967.
434. **Ghazali, S., Hallett, R. J., and Barratt, T. M.,** Hypomagnesemia in uremic infants, *J. Pediatr.,* 81, 747, 1972.
435. **Kobayashi, A. and Shiraki, K.,** Serum magnesium level in infants and children with hepatic diseases, *Arch. Dis. Child.,* 42, 615, 1967.
436. **Rude, R. K., Baker, A. J., and Watts, R. B.,** Renal cortical adenylate cyclase: characterization of magnesium activation, *Endocrinology,* 113, 1348, 1983.
437. **Wacker, W. E. C. and Parisi, A. F.,** Magnesium metabolism, *N. Engl. J. Med.,* 278, 658, 1968.
438. **Byrne, P. A. and Caddell, J. L.,** The magnesium load test. II. Correlation of clinical and laboratory data in neonates, *Clin. Pediatr.,* 14, 460, 1975.
439. **Parfitt, A. M. and Kleerekoper, M.,** Clinical disorders of calcium, phosphorus and magnesium metabolism, in *Clinical Disorders of Fluid and Electrolyte Metabolism,* Maxwell, M. H. and Kleeman, C. R., Eds., McGraw-Hill, New York, 1980, 947.
440. **Jones, J. E., Shane, S. R., Jacobs, W. H., and Flink, E. B.,** Magnesium balance studies in chronic alcoholism, *Ann. N.Y. Acad. Sci.,* 162, 934, 1969.

PHOSPHATE METABOLISM

Myron M. Warshaw

INTRODUCTION

It is interesting that while phosphate is clearly an essential component of biological structure and function, its measurement in body fluids plays a relatively minor role in medical diagnoses; and yet, one cannot underestimate the importance of understanding the full role of phosphate metabolism in health and disease. The mechanisms for phosphate homeostasis must necessarily be strong and effective. Indeed, in most disease states large disruption of phosphate levels are rare and smaller disruptions are usually well tolerated.[67] Although plasma phosphate concentration is not as tightly controlled as is the calcium concentration (± 50% of mean vs. ± 10% of mean for Ca), the control mechanisms for both are essential and related. However, significant disturbances in phosphate levels do occur in malnutrition, parathyroid disease, and renal failure, and in conditions involving specific defects of phosphate control such as pseudohypoparathyroidism and familial hypophasphatemic rickets.

Phosphorus, strongly bound to oxygen as phosphate, is present in all body tissues. Phosphate is either in the form of "inorganic phosphate" associated or complexed with cations (Table 1) or in the form of organic phosphate covalently bound within various smaller molecules and especially within various macromolecules such as nucleic acids (Table 2). At the normal physiologic pH of 7.4 the $HPO_4^-/H_2PO_4^-$ ratio is 4:1, with a pK_A for dihydrogen phosphate of 6.8. Consequently, there are about 1.8 equivalents per mole of phosphate.

Phosphate is the major anion found within the cells. Intracellular inorganic phosphate concentration is about 40 times that in the plasma and perhaps as much as 100 times that found in interstitial fluid.[12] The typical intracellular concentration of inorganic phosphate can range from 30 to 50 mmol/l and the total intracellular phosphate content can be as high as 100 mmol/l (Table 2).

The average 70-kg man has some 500 to 700 g of phosphorus, 80 to 85% of which is localized in the skeleton as hydroxyapatite and intermediate forms of calcium phosphate.[11,14] Approximately 10% of phosphorus is found in muscle. As diagrammed in Figure 1, 80% of the approximately 800 to 1200 mg ingested by the average, well-nourished individual is absorbed in the small intestine, primarily in the jejunum. Since 30 to 40% of this dietary phosphate appears in the feces, it follows that 10 to 20% of the intake is secreted back into the intestine. Under conditions of phosphate balance the remainder (60 to 70% of the intake) appears in the urine. Because 85 to 90% is reabsorbed in the proximal renal tubules, only about 10 to 15% of the filtered phosphate load is excreted.

Phosphate has a number of important biological functions, some of which are noted in Table 3.

Serum phosphate concentrations are significantly higher in newborns and children than in adults. The reference ranges for serum phosphate levels at various ages are summarized in Table 4. The maximum tubular reabsorption of phosphorus is highest in the 9- to 11-year cohort (up to 95% of the filtered load is reabsorbed).[20] This coincides with the period of maximum growth hormone activity.[15,21] Growth hormone increases tubular reabsorption of phosphate which could explain the higher phosphate levels seen in children and adolescents.[15,22] The difference between older men and post-menopausal women has not been explained.[23]

REGULATION OF PLASMA PHOSPHATE

Plasma phosphate homeostasis can be viewed as a net flux of phosphate from the intestine

Table 1
DISTRIBUTION OF INORGANIC PHOSPHATE SPECIES IN NORMAL PLASMA[a]

	Conc. (mmol/l)	Percent of total
Free $HPO_4^=$	0.51	44
Free $H_2PO_4^-$	0.13	11
Protein bound	0.13	11
$NaHPO_4^-$	0.31	27
$CaHPO_4$	0.04	3
$MgHPO_4$	0.03	3
	1.15[b]	100

[a] Adapted and calculated using data from various sources.[8-10]
[b] 1.15 mmol/l is approximately 36 mg/l as phosphorus.

Table 2
CONCENTRATION OF THE MAJOR SPECIES OF PHOSPHATE IN SERUM AND THE ERYTHROCYTE[a]

	Concentration of phosphorus (mg/l)			
	Serum		Erythrocyte	
Phosphate species	Mean	Range	Mean	Range
Inorganic	35	27—45	24	10—33
Organic	84	72—92	696	630—890
Pyrophosphate (ATP, etc.)	2.0	1.0—7.0	106	40—150
2,3-Diphosphoglycerate	0.3	0.1—4.0	292	190—400
Hexose phosphates	0.4	0.1—2.0	75	30—110
Phospholipids	81	49—110	119	70—130
Nucleic acids	—	—	62	50—70
Total	119	100—140	720	610—920

[a] Adapted from Natelson and Natelson[12] and modified with data from various sources.[11,13]

(dietary absorption), bone (net of resorption and deposition), soft tissue (intracellular), and kidney (the net of glomerular filtration rate [GFR] and the rate of tubular reabsorption of phosphate) into the plasma pool (Figure 1). The primary regulatory process controlling the steady-state balance is tubular reabsorption.[11,15] Bone and soft tissue are either a source or depository for phosphate and no phosphate-specific regulatory process has been demonstrated. The active vitamin D metabolite, 1,25-OH_2D_3, is thought to have some regulatory control via intestinal phosphate absorption and tubular reabsorption.[24-27] No major evidence has been presented for intestinal secretion being a source for control. All four major components can be and are factors in conditions of phosphate imbalance.

Role of the Kidney

The renal handling of phosphate which has been extensively reviewed[1,24,28] is the major component in the regulation of plasma phosphate levels. In a healthy individual 85% or more of the filtered load is reabsorbed by the tubules. The reabsorptive process takes place throughout the tubules, but primarily in the proximal tubule via a sodium-dependent active transport across the brush-border cell membrane.[24] This process functions effectively as a

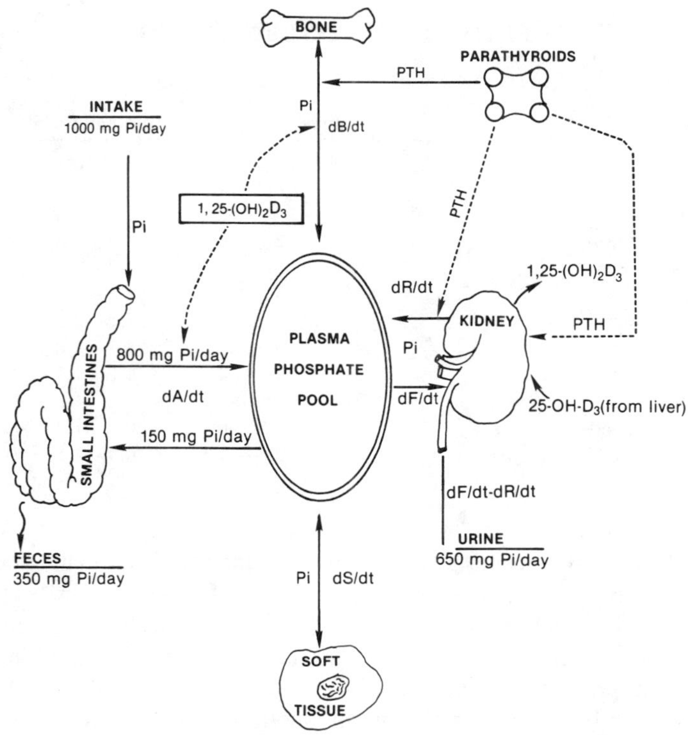

FIGURE 1. Generalized schematic of phosphate homeostasis. According to Robertson,[15] one can consider dA/dt, dB/dt, and dS/dt as the net steady-state flux into the plasma pool from the intestines, bone, and soft tissues, respectively. The rate of tubular reabsorption of phosphate is dR/dt and dF/dt in the renal phosphate filtration rate, such that the rate of urinary phosphate excretion is dF/dt minus dR/dt. In the normal nonfasting steady state, intestinal absorption is balanced by urinary excretion. In the normal fasting state the phosphate excretion is low and primarily balanced by flux from the soft tissue.[11,15] (Adapted from Natelson and Natelson[12] and Robertson.[15])

Table 3
SOME ROLES AND ACTIONS OF PHOSPHATE

Energy storage: high-energy nucleotide pyrophosphates, creatine phosphate
Transport processes: phosphorylation across membranes
Second messenger: ATP > cyclic AMP + pyrophosphate, protein phosphorylation, and dephosphorylation via kinases and diesterases
Intermediary metabolism: glycolysis, glycogenolysis, fructose metabolism, etc.
Cell membrane structure and function: as phospholipids
Intracellular calcium regulation: via mitrochondrial calcium uptake
Hemoglobin-oxygen binding: influences synthesis of 2,3-diphosphoglycerate
Bone cell function: influences activity of osteoblasts, enhances hypocalcemic effect of calcitonin
Phosphate homeostasis: through its feedback, influences renal tubular absorption of phosphate
Bone structure and composition: as calcium hydroxyapatite
Urine buffering: primary buffer at about pH 6
Plasma ionized calcium: via formation of complexes
Intestinal absorption of calcium and phosphate: influences synthesis of active vitamin D by influencing renal 25-OH-D_3-1-hydroxylase

Table 4
SERUM PHOSPHATE CONCENTRATIONS[a]

	Concentration		
	In mg/l of P		In mmol/l of P
	Mean	Range	Range
Cord blood	55	48—62	1.5—2.0
Premature birth	68	56—80	1.8—2.6
Full-term birth	64	50—78	1.7—2.5
Newborn	66	42—92	1.4—2.9
Child, 1 year	50	38—62	1.2—2.0
Child, 10 years	46	36—56	1.2—1.8
Child, 15—16 years	39	24—54	0.8—1.7
Adult	35	25—45	0.8—1.5
Adult male (60—75 years)	30	20—40	0.6—1.3
Adult female (70—75 years)	39	29—49	0.9—1.6

[a] Prepared using data from various sources.[16-19]

threshold.[11,15] When the serum phosphate is above the threshold (normally about 30 mg/l) net urinary excretion occurs; when below almost all of the filtered phosphate is reabsorbed. As the serum phosphate concentration increases, the tubular reabsorption rate reaches a maximum and any further increases in serum phosphate concentration result in an excretion rate that increases in proportion to the GFR. From this point, all of the additional filtered phosphate is not absorbed. The serum concentration at which saturation of the tubular reabsorption is reached is called the Threshold_{max} and marks the point where tubular reabsorption reaches its maximum, T_mP.[11,29]

Parathyroid hormone (PTH) acts as the proximal tubules to effectively lower the threshold, resulting in decreased tubular reabsorption of phosphate. The PTH binds to specific receptors at the basolateral membranes of the brush-border cells of the proximal tubules. This binding turns on adenylcyclase to generate the cyclic AMP messenger cascade. Thus, one can observe increased urinary phosphate and cyclic AMP under conditions of elevated PTH. Conversely, the T_mP is increased in the absence of PTH.[30] The net result is a generally higher or lower steady-state serum phosphate level in the hypo- or hyperparathyroid individual, respectively. For these individuals, phosphate intake is still balanced by phosphate excretion just as in a normal individual.[11,15,29,30]

A simple diagram of the factors which influence renal phosphate transport is shown in Figure 2.[28] The major factors are the availability of phosphate, ionized calcium, and hydrogen ions. The role of each will be discussed further.

Phosphate availability can rapidly modulate tubular reabsorption of phosphate either directly or indirectly via its effect on ionized calcium. Animal and *in vitro* experiments demonstrate that low phosphate diets directly block the phosphaturic response to increased PTH, but not the cyclic AMP or anticalciuric responses. The mechanism, independent of PTH, has not been elucidated, but may involve more than one process.[31,33]

Man responds in a few hours to changes in phosphate intake (dietary or i.v.), even in the hyper- or hypoparathyroid state.[33] Initially, dietary restriction will increase the fractional phosphate reabsorption (TRP), whereas increased phosphate intake or i.v. infusion will result in a reduced TRP and phosphaturia.[34,35] Over a more prolonged period a corresponding change will occur in the maximum threshold, the mechanism for which has not been yet identified.[25,54]

Tubular reabsorption of phosphate is also affected indirectly through the ionized calcium-PTH mechanism. Because circulating plasma phosphate complexes with calcium, effectively lowering the ionized calcium level, a phosphate infusion would result in increased PTH

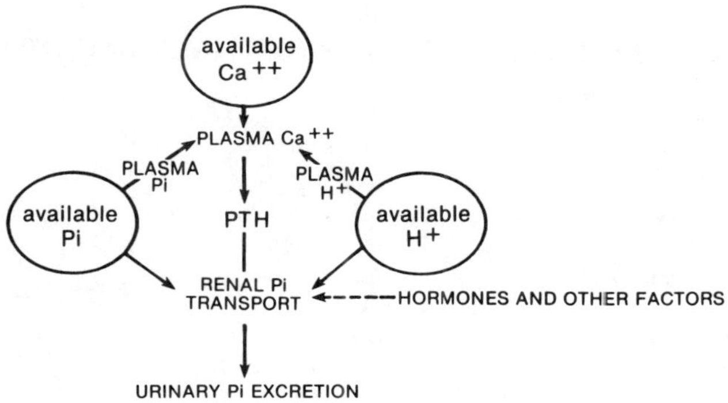

FIGURE 2. Generalized overview for the interaction of the major factors affecting renal phosphate control. The term "available" denotes the total pool, whether of exogenous or endogenous origin. (Adapted from Mizgala and Quamme.[28])

secretion.[36] The subsequent phosphaturic effect of PTH counters the increased phosphate load. The parathyroids do not respond directly to changes in serum phosphate concentration.[37]

A unified explanation on the effects of acidosis and alkalosis upon plasma phosphate and tubular reabsorption mechanisms is lacking. Elucidation of the mechanism is complicated by the many factors that must be experimentally controlled such as phosphate availability and load, bicarbonate, PTH, and ionized calcium. Conclusions also depend on whether the pH imbalance is acute or chronic, extracellular or intracellular, and tubular vs. systemic, animal vs. human, and *in vitro* vs. *in vivo*. A number of reviews have attempted to explain the current research.[28,38-41]

Although acute experimental maneuvers do not usually extrapolate to clinical situations, a few observations are accepted. The hypercapnia of respiratory acidosis increases urinary phosphate excretions in animals in the presence[42] and absence[43] of PTH. The metabolic alkalosis produced by bicarbonate infusions is also phosphaturic.[40,44] Respiratory alkalosis has been shown to be antiphosphaturic[44-46] in man, but hypophosphatemic because of redistribution into the intracellular pool (see Table 7).

A number of other factors and hormones which can affect serum phosphate levels and/or phosphate excretion (see Table 5) have been recently reviewed.[23,28] The mechanisms of action can be extrarenal, renal, or both and are still the subject of much ongoing research.[1,2] A few mechanisms have been discussed above; a few will be mentioned later (see Tables 7 and 8).

There may be conflicting observations on the effect of vitamin D and its metabolites on phosphate excretion. Lee et al.[11,26] suggest that increased vitamin D tends to be plasma phosphate elevating through an antiphosphaturic effect in states of negative phosphate balance, phosphaturia, or hypophosphatemia; whereas vitamin D has a phosphaturic action and a plasma phosphate lowering influence in hyperphosphatemic or phosphate-repleted states.

The measurement of renal phosphate excretion and its relation to tubular reabsorption have been well described.[11,29] Traditionally, the TRP was determined as the ratio of the tubular reabsorption rate to the filtered phosphate load (GFR times $[PO_4]$). The calculation only requires determination of inorganic phosphate and creatinine on simulaneously collected urine and serum samples. The volumes cancel in the ratio and need not be measured.

$$\text{TRP} = 1 - \frac{(\text{phosphate clearance})}{(\text{creatinine clearance or GFR})}$$

$$= 1 - \left(\frac{[PO_4]}{[\text{Creat}]}\right)_{\text{ur}} \left(\frac{[\text{Creat}]}{[PO_4]}\right)_{\text{ser}}$$

Table 5
FACTORS AFFECTING RENAL PHOSPHATE EXCRETION[a]

Increase[b]	Decrease[b]
High P_i diet	Low P_i diet
PTH	Thyroparathyroidectomy
Calcitonin[c]	Intracellular Ca^{2+}
Chronic vitamin D	Acute vitamin D
Glucagon[c]	Insulin[c]
Glucocorticoids[c]	Growth hormone
Volume expansion	Volume contraction
Increased pCO_2	Decreased pCO_2
Acidosis[c]	Alkalosis[c]
Diuretics	
ADH at low P_i load[c]	
Thyroid hormones[6]	

[a] Adapted from several sources.[4,11,23,28,29]
[b] Many of the factors do not necessarily influence the serum phosphate levels significantly and the mechanism may be very complex. The effect may be additionally extrarenal. For example, thyroid hormones increase the GFR, T_mP/GFR, and the flux from bone and probably soft tissue and intestine. Hyperthyroid individuals tend to have higher serum phosphate with increased excretion.
[c] Data are conflicting.

The TRP depends on the filtered load, fasting state, GFR, and threshold (T_mP/GFR). It is, therefore, too variable to be used successfully as a diagnostic tool and does not always represent the true tubular reabsorption rate.[29]

The mean threshold is defined as the maximum tubular reabsorption rate divided by the GFR and is usually moderately lower than the maximum threshold mentioned above. T_mP/GFR can be calculated from the TRP when it is less than 0.8 as during phosphate infusion:

$$T_mP/GFR = (TRP)[PO_4]_{ser}$$

However, nomograms have been developed to enable one to calculate T_mP/GFR from TRP and serum phosphate without the need for phosphate loading. Measurements can then be made in the fasting state. The normal range for T_mP/GFR is 25 to 42 mg/l. Thyrotoxic and hypoparathyroid individuals usually have values greater than 42, and hyperparathyroid individuals are usually less than 25. There clearly is a better correlation to disease than the TRP alone.[29]

Intestinal Phosphate

Dietary phosphate is absorbed in the small intestine by passive diffusion and by sodium-dependent active transport.[11,47] The active transport occurs primarily in the jejunum and to lesser degrees in the ileum and duodenum (the primary site for active transport of Ca).[26] Direct administration of 1,25-$(OH)_2D_3$ increases the rate of phosphate passage from the intestinal lumen independently of its enhancement of calcium absorption. 1,25-$(OH)_2D_3$ action is on the active transport and may trigger the phosphate carrier protein.[48] Because PTH and phosphate influence synthesis of 1,25-$(OH)_2D_3$, the active phosphate transport across the intestinal lumen must have some regulatory role.[3,11,26]

CELLULAR PHOSPHATE TRANSPORT

Transport of phosphate across the cell can be viewed as occurring in three distinct steps:

(1) phosphate enters the cell across the luminal brush-border membrane (2) crosses from the mucosal side within the cell to the serosal side, and (3) exits the cell across the basolateral membrane into the extracellular space.

There is some general commonality of phosphate transport across the brush-border cells of intestinal and renal tubular epithelium, but renal tubular processes have been more extensively studied.[47,49] For example, both exhibit a sodium-dependent, carrier-mediated phosphate uptake across the brush-border cell membrane. In the kidney this process is dramatically reversed by PTH and in the intestine enhanced by 1,25-$(OH)_2D_3$. In the renal tubule $HPO_4^=$ is the preferentially absorbed species, whereas $H_2PO_4^-$ is more predominantly absorbed in the intestine (for discussion, see references). More detailed discussions appear in several reviews.[11,28,38]

The driving force for active phosphate transport across the brush-border membrane is maintained by the ATP-ATPase sodium pump located at the basolateral membrane. On the luminal side two sodiums then couple electroneutrally with $HPO_4^=$, via a transport protein, and proceed down the Na gradient across the brush-border cell membrane of the tubule.[11,48] This transmembrane process supports the net transfer of phosphate and appears to be rate limiting. The same transporter functions for $H_2PO_4^-$ to a much lesser degree, but the mechanism is not yet clearly elucidated. The role of the abundant alkaline phosphatase present within the brush-border cell membrane is not yet clear.[28]

Factors controlling transport of phosphate through the cell cytosol to the basolateral side are not yet well established. As it crosses it becomes incorporated into large and small organelles (for example, to form ATP in the mitochondria). The resulting chemical concentration gradient is believed to provide the necessary driving force for diffusion of phosphate across the basolateral membrane into the intracellular space, perhaps facilitated by a transport protein.[24,28,46]

DISTURBANCES IN SERUM PHOSPHATE

Most discussions of phosphate regulation center on the serum phosphate concentration changing to below normal (hypophosphatemia) or above normal (hyperphosphatemia) or not changing. With some oversimplification, the potential mechanisms which cause serum phosphate disturbances involve three main locations: kidney, tissue, and intestine. Generally, under- or overrenal retention, redistribution into or out of the cellular compartments, and reduced or increased intestinal intake are potentially hypophosphatemic or hyperphosphatemic, respectively. These mechanisms are detailed in Table 6 and are used to explain the effect of the specific clinical conditions referenced in Tables 7 and 8. The clinical conditions affecting serum phosphate levels or urinary phosphate excretion have been extensively reviewed.[7,11,49-51]

Severe hypophosphatemia is usually more clinically symptomatic and less tolerated than is hyperphosphatemia. Severe hypophosphatemia (less than 10 mg/l) produces malaise, weakness, anorexia, hypercalciuria, bone pain, hemolytic anemia, myopathy, and paresthesias. These are rapidly reversed by supplemental phosphate therapy.[49] Most of the symptoms can be related to depletion of intracellular ATP and red cell 2,3-DPG.

Hypophosphatemia most commonly occurs transiently during glucose infusion and in the chronic withdrawing alcoholic (see Table 7). Dialysis patients on excessive antacids and suffering from vitamin D deficiency secondary to malabsorption sometimes require phosphate therapy. The hypophosphatemia during respiratory alkalosis does not require treatment.[49,51]

Combined treatment with phosphate and vitamin D will reverse the debilitating osteomalacia and renal phosphate wasting seen in vitamin D-resistant rickets (VDRR, also known as X-linked hypophosphatemic rickets). VDRR is characterized by an unresponsiveness to the factors which affect tubular phosphate reabsorption and by poor intestinal absorption of calcium and phosphorus.

Table 6
GENERAL MECHANISMS FOR SERUM PHOSPHATE DISTURBANCES

Hypophosphatemia	Hyperphosphatemia
Renal Control	
(1) Increased renal excretion of phosphate because tubular reabsorption is reduced (A) Elevated serum PTH levels: primary (1°) or secondary (2°) (B) Renal tubular disorders (C) Osmotic diuresis (D) Other	(4) Decreased renal excretion of phosphate because GFR is reduced (<20 ml/min) Tubular reabsorption is usually also reduced (5) Decreased renal excretion of phosphate because tubular reabsorption of phosphate is increased (A) Lower to absent serum PTH levels (B) Tubular uptake unresponsive to PTH level (C) Lowered phosphate stores (D) Other, also possible effects on tubular reabsorption of phosphate from other hormones such as insulin, thyroxine, growth hormone
Redistribution to Cells and Bone	
(2) Increased cellular uptake from extracellular fluid (A) Rapid formation of intermediary phosphometabolites (B) Rapid bone deposition	(6) Increased intracellular spillage into extracellular fluid (A) Hemolysis, severe necrosis (B) Increased bone resorption
Intake	
(3) Reduced phosphate absorption across intestine (A) Dietary deprivation (very rare) or phosphate binders (B) Malabsorption (diarrhea, vomiting, vitamin D deficiency)	(7) Increased phosphate absorption across intestine (A) Increased dietary intake (not common as a single cause) (B) Increased intestinal absorption (excess vitamin D) (C) Excessive parenteral delivery

Table 7
SOME CONDITIONS EXHIBITING OR EFFECTORS OF HYPOPHOSPHATEMIA[a]

Conditions	Cause(s) of disturbance[b]	Ref.
Severe hypophosphatemia ([PO_4] <10 mg/l)		
Chronic alcoholism, especially during withdrawal	3A,B,C 1A—2°	11,56
Recovery (treatment) phase of diabetic ketoacidosis	2A	57
Excessive use of antacids	3A	6
Respiratory alkalosis also secondary to severe heat stroke	2A,5C	44,45,58
Hyperalimentation with insufficient phosphate or high caloric infusions, especially fructose	2A,3A	59,60
Nutritional recovery syndrome	2A	61
Recovery stage from severe extensive third degree burns	1C,2A	11
Post fructose load, hereditary fructose intolerance (transient)	2A	62
Recovery and treatment of hypothermia	2A ?	63
Hyperthermia	1B	64

Table 7 (continued)

Moderate hypophosphatemia ([PO_4] between 10—25 mg/l)		
Hyperparathyroidism	1A	32
Fanconi syndrome—renal tubular acidosis, heavy metal poisoning, Wilson's disease, salicylate poisoning	1A,1B	65
Hypomagnesemia	1A	66
Vitamin D deficiency, malabsorption, osteomalacia	3B,1A	24,25
Vitamin D-resistant rickets (familial X-linked hypophosphatemia)	1B,3B	52
Vitamin D-dependent rickets	3B,1A	52
Acute untreated gout	1B,2A	67
Oncogenic hypophosphatemic osteomalacia	?	68
Carbohydrate administration, ingestion, or enhanced carbohydrate metabolism (transient)	2A	50,59
Hemodialysis and post transplantation	1A,1B,1C	69
Various malabsorption states	3B,1A	11

[a] General source was Agus et al.,[49] Stoff,[51] and Knochel.[55]
[b] The possible mechanism(s) refer to the categories tabulated in Table 6.

Table 8
SOME CONDITIONS EXHIBITING OR EFFECTORS OF HYPERPHOSPHATEMIA[a]

Conditions	Cause(s) of disturbance[b]	Ref.
Severe hyperphosphatemia ([PO_4] >60 mg/l)		
Chronic renal failure (GFR <20 ml/min)	4	70
Hypoparathyroidism	5a	71
Pseudohypoparathyroidism (sensitive to phosphate intake)	5b (7a)	72
Chemotherapy of malignancy: leukemia, lymphomas	6a	73,74
Acute phosphate poisoning: hypertonic phosphate enemas, laxatives, hyperalimentation	7a	75
Cell destruction, rhabdomyolysis, malignant hyperpyrexia	6a	76
Tumoral calcinosis	5	49
Moderate elevation ([PO_4] between 40—60 mg/l)		
Hyperthyroidism	5d, 6b	77
Acromegaly, increased growth hormone levels as in childhood	5d	22
Vitamin D intoxication (depends on PO_4 stores)	7b, 6b	78
Diphosphonates (use in Paget's)	4, 5, 6B	79

[a] General source was Lee et al.[11] and Agus et al.[49]
[b] The possible mechanism(s) refer to the categories tabulated in Table 6.

The most important consequences of hyperphosphatemia are hypocalcemia and extraosseous deposition of calcium phosphate salts when the calcium-phosphate product is greater than 75.[53] The most common occurrence is in chronic renal failure. Phosphate levels above 120 mg/l imply additional disturbances such as increased load or severe secondary hyperparathyroidism. Therapy including reduced intake and diuresis is indicated when the phosphate levels approach 90 mg/l.[49]

Some abnormal conditions exhibit normal serum phosphate levels due to the adequate balancing of renal control. An essentially normal serum phosphate is observed in early stages of renal failure, uncontrolled diabetes, and experimental starvation.[49-51]

There is an important distinction between serum phosphate levels, on the one hand, and phosphate stores and demand, on the other.[11,28,38,54] It has been demonstrated that renal tubular phosphate reabsorption is enhanced in situations of increased phosphate requirement such as rapid growth, dietary restriction, pregnancy, and lactation. Conversely, decreased reabsorption occurs under conditions of low phosphate demand such as slow growth, dietary excess, or chronic renal failure.[28] Research is being directed toward the possibility of measuring the status of total body phosphorus.

REFERENCES

BOOKS

1. **Massry, S. G. and Fleisch, H., Eds.,** *Renal Handling of Phosphate,* Plenum Press, NewYork, 1980.
2. **Massry, S. G., Maschio, G., and Ritz, E., Eds.,** *Phosphate and Mineral Metabolism,* Plenum Press, New York, 1984.
3. **Bronner, F. and Peterlik, M., Eds.,** *Epithelial Calcium and Phosphate Transport: Molecular and Cellular Aspects,* Alan R. Liss, New York, 1984.
4. **Brenner, B. M. and Rector, F. C., Jr., Eds.,** *The Kidney,* 3rd ed., W. B. Saunders, Philadelphia, 1986.
5. **Maxwell, M. H. and Kellman, C. R., Eds.,** *Clinical Disorders of Fluid and Electrolyte Metabolism,* McGraw-Hill, New York, 1979.

ARTICLES

6. **Lotz, M., Zisman, E., and Bartter, F. C.,** Evidence for phosphorus depletion syndrome in man, *N. Engl. J. Med.,* 278, 409, 1968.
7. **Knochel, J. P.,** Models of hypophosphatemia and phosphate depletion, in *Regulation of Phosphate and Mineral Metabolism,* Massry, S. G., Letteri, J. M., and Ritz, E., Eds., Plenum Press, New York, 1982, 191.
8. **Walser, M.,** Ion association. VI. Interactions between calcium, magnesium, inorganic phosphate, citrate and protein in normal human plasma, *J. Clin. Invest.,* 40, 723, 1961.
9. **Marshall, R. W.,** Plasma fractions, in *Calcium, Phosphate, and Magnesium Metabolism,* Nordin, B. E. C., Ed., Churchill Livingstone, Edinburgh, 1976, 162.
10. **Parfitt, A. M. and Keerekoper, M.,** The divalent ion homeostatic system: physiology and metabolism of calcium, phosphorus, magnesium and bone, in *Clinical Disorders of Fluid and Electrolyte Metabolism,* Maxwell, M. H. and Kleeman, C. R., Eds., McGraw-Hill, New York, 1979, 269.
11. **Lee, D. B. N., Brautbar, N., and Kleeman, C. R.,** Disorders of phosphorus metabolism,in *Disorders of Mineral Metabolism. III,* Bronner, F. and Coburn, J. W., Eds., Academic Press, New York, 1981, 283.
12. **Natelson, S. and Natelson, E. A.,** *Principles of Applied Clinical Chemistry. I,* Plenum Press, New York, 1975, 169.
13. **Smith, C. C., Wolfe, G. F., and Cartwright, G. E.,** Blood electrolytes: primates, in *Biology Data Handbook. III,* 2nd ed., Altman, P. L. and Dittmer, D. S., Eds., Fed. American Society Experimental Biology, Bethesda, 1974, 1752.
14. **Widdowson, E. M. and Dickerson, J. W. T.,** Chemical composition of the body, in *Mineral Metabolism. 2A,* Bronner, F. and Comar, C. L., Eds., Academic Press, New York, 1964, 1.
15. **Robertson, W. G.,** Plasma phosphate homeostasis, in *Calcium, Phosphate and Magnesium Metabolism: Clinical Physiology and Diagnostic Procedures,* Nordin, B. E. C., Ed., Churchill Livingstone, Edinburgh, 1976, 217.
16. **Meites, S., Ed.,** *Pediatric Clinical Chemistry,* 2nd ed., American Association of Clinical Chemistry, Washington, D.C., 1981, 258.
17. **Morgan, B.,** *Osteomalacia, Renal Osteodystrophy and Osteoporosis,* Charles C Thomas, Springfield, IL, 1973.
18. **Keating, F. R., Jones, J. D., Elveback, L. R., and Randall, R. V.,** The relation of age and sex to distribution of values in healthy adults of serum calcium, inorganic phosphorus, magnesium, alkaline phosphatase, total protein, albumin, blood urea, *J. Lab. Clin. Med.,* 73, 825, 1969.
19. **Cheng, M. H., Lipsey, A. I., Blanco, V., et al.,** Microchemical analysis for 13 constituents of plasma from healthy children, *Clin. Chem.,* 25, 692, 1979.
20. **Thalassinos, N. C., Leese, B., Latham, S. S., and Joplin, G. F.,** Urinary excretion of phosphate in normal children, *Arch. Dis. Child.,* 45, 269, 1970.

21. **Hunter, W. M. and Greenwood, F. C.,** A radio-immunoelectrophoretic assay for human growth hormone, *Biochem. J.*, 91, 43, 1964.
22. **Corvilain, J. and Abramow, M.,** Effect of growth hormone on tubular transport of phosphates in normal and parathyroidectomized dogs, *J. Clin. Invest.*, 48, 1608, 1964.
23. **Ritz, E., Kreusser, W. and Bommer, J.,** Effects of hormones other than parathyroid hormone on the renal handling of phosphate, in *Renal Handling of Phosphate,* Massry, S. G. and Fleisch, H., Eds., Plenum Press, New York, 1980, 137.
24. **Bonjour, J. P. and Caverzasio, J.,** Phosphate transport in the kidney, *Rev. Physiol. Biochem. Pharmacol.*, 100, 162, 1984.
25. **Bonjour, J. P., Preston, C., and Fleisch, H.,** Effect of 1,25-dihydroxyvitamin D3 on the renal handling of Pi in thyroparathyroidectomized rats, *J. Clin. Invest.*, 60, 1419, 1977.
26. **Lee, D. B. N., Levine, B. S., Walling, M. W., et al.,** Regulation of serum phosphorus by 1,25-$(OH)_2$D3, in *Vitamin D: Basic Research and Clinical Applications,* Norman, A. W., Schaefer, K., von Herrath, D., et al., Eds., de Gruyter, Berlin, 1979, 939.
27. **Gloor, H. J., Bonjour, J. P., Caverzasio, J., and Fleisch, H.,** Resistance to the phosphaturic and calcemic actions of parathyroid hormone during phosphate depletion and prevention by 1,25-dihydroxy-vitamin D_3, *J. Clin. Invest.*, 63, 371, 1979.
28. **Mizgala, C. L. and Quamme, G. A.,** Renal handling of phosphate, *Physiol. Rev.*, 65, 431, 1985.
29. **Bijovet, O. L. M.,** Indices for measurement of renal handling of phosphate, in *Renal Handling of Phosphate,* Massry, S. G. and Fleisch, H., Eds., Plenum Press, New York, 1980, 1.
30. **Bijovet, O. L. M., Morgan, D. B., and Fourman, P.,** The assessment of phosphate reabsorption, *Clin. Chim. Acta,* 26, 15, 1969.
31. **Troehler, U., Bonjour, J. P., and Fleisch, H.,** Inorganic phosphate homeostasis: renal adaptation to dietary intake in intact and thyroparathyroidectomized rats, *J. Clin. Invest.*, 57, 264, 1976.
32. **Steele, T. H.,** Renal resistance to parathyroid hormone during phosphorus deprivation, *J. Clin. Invest.*, 58, 1461, 1976.
33. **Caverzasio, J. and Bonjour, J. P.,** Parathyroid hormone (PTH) status and phosphate (Pi) transport in proximal brush border membrane vesicles (BBMV) and whole kidney, in *Phosphate and Mineral Metabolism,* Massry, S. G., Maschio, G., and Ritz, E., Eds., Plenum Press, New York, 1984, 33.
34. **Chambers, E. L., Godon, G. S., Goldman, L., and Reifenstein, E. C.,** Tests for hyperparathyroidism: tubular reabsorption of phosphate, phosphate deprivation, and calcium infusion, *J. Clin. Endocrinol.*, 16, 1507, 1956.
35. **Goldman, R. and Bassett, S.,** Renal regulation of phosphorus excretion, *J. Clin. Endocrinol.*, 18, 981, 1958.
36. **Peraino, R. A. and Wadi, N. S.,** Influence of calcium on renal handling of phosphate, in *Renal Handling of Phosphate,* Massry, S. G. and Fleisch, H., Eds., Plenum Press, New York, 1980, 287.
37. **Sherwood, L. M., Mayer, G. P., Ramberg, C. F., Jr., et al.,** Regulation of parathyroid hormone secretion: proportional control by Ca, lack of effect of phosphorus, *Endocrinology,* 83, 1043, 1968.
38. **Knox, F. G., Hoppe, A., Kempson, S. A., Shah, S. V., and Dousa, T. P.,** Cellular mechanisms of phosphate transport, in *Renal Handling of Phosphate,* Massry, S. G. and Fleisch, H., Eds., Plenum Press, New York, 1980, 79.
39. **Knox, F. G. and Haramati, A.,** Renal regulation of phosphate excretion, in *The Kidney: Physiology and Pathophysiology,* Seedin, D. W. and Geibisch, G., Eds., Raven Press, New York, 1985, 1382.
40. **Bank, N. and Malnic, G.,** Effect of urinary alkalinization on the renal phosphate reabsorption, in *Renal Handling of Phosphate,* Massry, S. G. and Fleisch, H., Eds., Plenum Press, New York, 1980, 209.
41. **Murer, H. and Burckardt, G.,** Membrane transport of anions across epithelia of mammalian small intestine and kidney proximal tubule, *Rev. Physiol. Biochem. Pharmacol.*, 96, 1, 1983.
42. **Gray, S. P., Morris, J. E. W., and Brooks, C. J.,** Renal handling of calcium, magnesium, inorganic phosphate, hydrogen ions during prolonged exposure to elevated carbon dioxide concentrations, *Clin. Sci. Mol. Med.*, 45, 75l, 1973.
43. **Hoppe, A. and Knox, F. G.,** Effect of acute acid-base changes on fractional excretion in TPTX hamsters, *Clin. Res.*, 26, 542, 1978.
44. **Mostellar, M. E. and Tuttle, E. P.,** Effects of alkalosis on plasma concentration and urinary excretion of inorganic phosphate in man, *J. Clin. Invest.*, 43, 138, 1964.
45. **Hoppe, A., Metler, M., Berndt, T. J., Knox, F. G., and Angielski, S.,** Effect of respiratory alkalosis on renal phosphate excretion, *Am. J. Physiol.*, 239, F471, 1982; *Renal Fluid Electrolyte Physiol.*, 12.
46. **Hoffman, N., Thees, M., and Kinne, R.,** Phosphate transport by isolated renal brush border vesicles, *Pfluegers Arch.*, 362, 147, 1976.
47. **Danisi, G., Bonjour, J. P., and Staub, R. N.,** Regulation of Na-dependent phosphate influx across the mucosal border of the duodenum by 1,25-dihyroxycholecalciferol, *Pfluegers Arch.*, 388, 227, 1980.
48. **DeLuca, H. F.,** Vitamin D metabolism and function, *Arch. Intern. Med.*, 138, 936, 1978.

49. **Agus, Z. S., Goldfarb, S., and Wasserstein, A.,** Disorders of calcium and phosphate balance, in *The Kidney,* 2nd ed., Brenner, B. M. and Rector, F. C., Jr., Eds., W. B. Saunders, Philadelphia, 1981.
50. **Knochel, J. P. and Jacobson, H. R.,** Renal handling of phosphorus, clinical hypophosphatemia and phosphorus deficiency, in *The Kidney,* 3rd ed., Brenner, B. M. and Rector, F. C., Jr., Eds., W. B. Saunders, Philadelphia, 1986, 619.
51. **Stoff, J. S.,** Phosphate homeostasis and hypophosphatemia, *Am. J. Med.,* 72, 489, 1982.
52. **DeFronzo, R. A. and Thier, S. O.,** Inherited disorders of renal tubule function, in *The Kidney,* 3rd ed., Brenner, B. M. and Rector, F. C., Jr., Eds., W. B. Saunders, Philadelphia, 1986, 1297.
53. **Parfitt, A. M.,** Soft tissue calcification in uremia, *Arch. Intern. Med.,* 124, 544, 1969.
54. **Bonjour, J. P. and Fleisch, H.,** Tubular adaptation to the supply and requirement of phosphate, in *Renal Handling of Phosphate,* Massry, S. G. and Fleisch, H., Eds., Plenum Press, New York, 1980, 243.
55. **Knochel, J. P.,** The pathophysiology and clinical characteristics of severe hypophosphatemia, *Arch. Intern. Med.,* 137, 203, 1977.
56. **Territo, M. C. and Tanaka, K. R.,** Hypophosphatemia in chronic alcoholism, *Arch. Intern. Med.,* 134, 445, 1974.
57. **Keller, U. and Berger, W.,** Prevention of hypophosphatemia by phosphate infusion during treatment of diabetic ketoacidosis and hyperosmolar coma, *Diabetes,* 29, 87, 1980.
58. **Relman, A. S.,** Metabolic consequences of acid-base disorders, *Kidney Int.,* 1, 347, 1972.
59. **Sheldon, G. F. and Grzyle, S.,** Phosphate depletion and repletion. Relation to parenteral nutrition and oxygen transport, *Ann. Surg.,* 182, 683, 1975.
60. **Miller, R. R., Menke, J. A., and Mentser, M. I.,** Hypercalcemia associated with phosphate depletion in the neonate, *J. Pediatr.,* 105, 814, 1984.
61. **Schnitker, M. A., Mattman, P. E., and Bliss, T. L.,** A clinical study of malnutrition in Japanese prisoners of war, *Ann. Int. Med.,* 35, 69, 1951.
62. **Froesch, E. R.,** Hereditary fructose intolerance, in *The Metabolic Basis of Inherited Disease,* 4th ed., Stanbury, J. B., Wyngaarden, J. B., and Fredrickson, D. S., Eds., McGraw-Hill, New York, 1978, 121.
63. **Levy, L. A.,** Severe hypophosphatemia is a complication of treatment of hypothermia, *Arch. Intern. Med.,* 140, 128, 1980.
64. **Guntapalli, K. K., Sladen, A., Selker, R. G., et al.,** Effects of induced total-body hyperthermia on phosphorus metabolism in humans, *Am. J. Med.,* 77, 250, 1984.
65. **Seldin, D. W. and Wilson, J. D.,** Renal tubular acidosis in *The Metabolic Basis of Inherited Disease,* 4th ed., Stanbury, J. B., Wyngaarden, J. B., and Fredrickson, D. S., Eds., McGraw-Hill, New York, 1978, 1618.
66. **Rude, R. K., Oldham, S. B., Sharp, C. F., Jr. et al.,** Parathyroid hormone secretion in magnesium deficiency, *J. Clin. Endocrinol. Metab.,* 47, 800, 1978.
67. **DeOliveira, H. D. and Lans-Filho, J. A.,** Letter: hypophosphatemia in gout, *Lancet,* 2, 215, 1961.
68. **Lau, K., Stom, M. C., Goldberg, M., et al.,** Evidence for a humoral phosphaturic factor in oncogenic hypophosphatemic osteomalacia, *Clin. Trd.,* 27, 421A, 1979.
69. **Ahmed, K. Y., Willis, M. R., Skinner, R. K., et al.,** Persistent hypophosphatemia and osteomalacia in dialysis patients not on oral phosphate binders: response to DHT, *Lancet,* 2, 439, 1976.
70. **Kopple, J. D. and Coburn, J. W.,** Metabolic studies of low protein diets in uremia. II. Calcium, phosphorus and magnesium, *Medicine,* 52, 597, 1973.
71. **Parfitt, A. M. and Frame, B.,** Phosphate loading and depletion in vitamin D treated hypoparathyroidism, in *Phosphate Metabolism Kidney and Bone,* Alvioli, L., Bordier, P. H., Fleisch, H., Massry, S. G., and Slatopolsky, E., Eds., Armour Montagu, Paris, 1976, 179.
72. **Potts, J. T., Jr.,** Pseudohypoparathyroidism, in *The Metabolic Basis of Inherited Disease,* 4th ed., Stanbury, J. B., Wyngaarden, J. B., and Fredrickson, D. S., Eds., McGraw-Hill, New York, 1978, 1350.
73. **Ettinger, D. S., Harker, W. G., Gerry, H. W., Sanders, R. C., and Saral, R.,** Hyperphosphatemia. hypercalcemia and transient renal failure. Results of cytotoxic treatment of acute lymphoblastic leukemia, *JAMA,* 239, 2472, 1978.
74. **Betro, M. G. and Pain, R. W.,** Hypophosphatemia and hyperphosphatemia in a hospital population, *Br. Med. J.,* 1, 273, 1972.
75. **Sotos, J. F., Cutler, E. A., Finkel, M. A., and Doody, D.,** Hypocalcemic coma following two pediatric phosphate enemas, *Pediatrics,* 60, 305, 1977.
76. **Pollok, P. A. and Watson, R. L.,** Malignant hyperthermia associated with hypocalcemia, *Anesthesiology,* 34, 188, 1971.
77. **Bommer, J., Bonjour, J. P., Ritz, E., and Fleisch, H.,** Parathyroid-independent changes in renal handling of phosphate in hyperthyroid rats, *Kidney Int.,* 15, 325, 1979.
78. **Bikle, D. D.,** Osteomalacia and rickets, in *Cecil Textbook of Medicine,* Wyngaarden, J. B. and Smith, L. H., Jr. Eds., W.B. Saunders, Philadelphia, 1985, 1423.
79. **Recker, R. R., Hassing, G. S., Lau, J. R., and Saville, P. D.,** The hyperphosphatemic effect of EHDP: renal handling of phosphorus and renal response to parathyroid hormone, *J. Lab. Clin. Med.,* 81, 258, 1973.

NORMAL AND ABNORMAL DISTRIBUTION OF FLUID IN THE BODY (EDEMA)

Norman C. Kramer

INTRODUCTION

The control of body fluid and electrolyte volume and concentration is one of the fundamental functions in the maintenance of the "internal milieu". Bernard pointed out that the cells of most higher organisms do not come in contact with the external environment, but exist in their own internal sea — the extracellular fluid which surrounds them and through which all metabolic exchanges take place. Bernard deduced from a very few facts the need for constancy of this internal environment as a requirement for free and independent existence.[1] Strauss, paraphrasing a commentary by Starling (1909), stated: "Man is provided with distinct mechanisms for the regulation of the amount as well as the composition of body fluids. Thirst and appetite under the regulation of the central nervous system and the kidneys under the influence of circulatory, hormonal, and possible neurogenic factors subserve this office."[2] In this chapter, we will examine both the normal composition and regulation of body fluids and the alterations in regulation which result in abnormal retention and distribution of water in man.

ANATOMY OF BODY FLUIDS: BODY FLUID COMPARTMENTS

The body is composed of both solids and water. Water comprises from 50 to 70% of the body mass. This figure varies depending on the percentage of body mass derived from fat. Since adipose tissue is relatively low in water content, the higher the fat content the lower the body water. Total body water (TBW) is distributed between two major compartments, intracellular fluid (ICF) and extracellular fluid (ECF). The ICF is the larger of the two, comprising 30 to 40% of the TBW. The ECF represents 20% of the TBW and is composed of the intravascular (IV) and the interstitial compartments (IC). These contain 4 and 16% of the body mass, respectively. Lymph, which constitutes 2 to 3% of body mass, is included in the IC. A third subdivision of the ECF, the transcellular compartment, is composed of a number of small volumes. The transcellular compartment represents 1 to 3% of body mass and includes cerebrospinal, intraocular, pleural, peritoneal, and synovial fluids plus the digestive secretions. It is different from the IC in that it is separated from the blood by an epithelium in addition to the capillary endothelium. Because of this characteristic, movement of fluid involving these compartments are slower to occur in response to factors causing change in fluid volumes.[3]

Composition of the fluid in the various compartments differs significantly from one another. The ICF has a high concentration of potassium and a low sodium concentration, while the ECF exhibits a high sodium concentration and a low potassium concentration. These differences in concentration are maintained by metabolic pumps operating at the cell membrane which keep the concentrations at their present levels against the gradient present at the cell surface. The metabolic activity is part of the means of control of the concentration and volume of the compartments outlined above and will be discussed later.

NORMAL REGULATION OF VOLUME IN THE BODY

Body fluid volume is controlled to very precise limits by a complex series of mechanisms that monitor the distribution of TBW between the various compartments and adjust these

Table 1
ELECTROLYTE CONCENTRATIONS IN EXTRACELLULAR AND INTRACELLULAR FLUIDS

Ion	Plasma (meq/l)	Interstitial fluid (meq/l)	Plasma water (meq/l)	Muscle cell water (meq/l)
Sodium	140.0	145.5	151.0	12.0
Potassium	4.5	4.8	5.0	160.0
Calcium	5.0	2.8	5.5	—
Magnesium	1.5	1.0	1.6	34.0
Chloride	104.0	116.6	112.0	2.0
Bicarbonate	24.0	27.4	26.0	10.0
Sulfate	1.0	1.2	1.1	—
Phosphate	2.0	2.3	2.2	140.0
Protein	15.0	2.0	16.0	54.0
Other anions	5.0	5.6	5.3	—

volumes by water excretion or reabsorption, depending on the presence of an excess or deficiency of water in the ECF.[4] Water is added to the body by ingestion via the gastrointestinal tract and by production within the body as the result of oxidative metabolic activity. Excess is excreted via the urinary system, the respiratory system, and through the skin by perspiration. As long as the body receives adequate amounts of water from external sources, internal regulation of volume is automatic and precise, with the excess being rapidly excreted. In the presence of small deficits, decreased urine output will maintain the integrity of the body fluid volume. Since water moves freely from one compartment to another, even across most cell membranes, it is the quantity of solute that determines the size of each compartment. Each fluid compartment has a single solute which is limited largely to that space and, therefore, determines the size of that compartment. Serum proteins serve that purpose for the IV compartment, sodium for the IC, and potassium for the intracellular compartment.

THE INTRAVASCULAR VOLUME

The intravascular volume is maintained relatively constant due to the oncotic effect of the serum proteins within the vascular space.[5] The serum proteins are a diverse group of proteins totaling 7 to 8 g/100 ml plasma. Serum albumin is the primary single species and is present in a concentration of about 5 g/100 ml, the rest being composed of globulins. The albumin is a homogeneous solute having a molecular weight of 69,000 Da. The globulins are diverse and are much larger in size. The albumin is present in about twice the amount of the other proteins and contributes about 80% of the oncotic pressure within the vascular space. The total osmotic pressure contribution of the serum proteins is about 0.7 mOsm, but since they are the only components restricted to the vascular space their contribution represents the effective osmotic pressure difference between the IV and interstitial spaces. This pressure of about 15 mmhg counteracts the outward directed hydrostatic pressure generated by the pumping action of the heart and thereby maintains the IV volume while allowing continuous circulation and mixing of the entire interstitial fluid.[5]

The Extracellular Fluid Volume

The regulation of the ECF volume has been studied extensively by both clinicians and physiologists. Many facts have been established about the factors controlling the ECF volume, but there is still much controversy about other factors involved.[6] As the serum proteins function to maintain in IV volume, so sodium concentration functions to determine the volume of the ECF compartment. Increasing the concentration of sodium in the ECF produces expansion of the volume and return of sodium concentration to its predetermined level. Excess sodium in the body is excreted under normal circumstances by action of the nephrons. Sodium is filtered freely through the glomerulus and reaches the proximal tubule in a filtrate isotonic with plasma. Approximately 80% of filtered sodium is reabsorbed by the proximal tubule, again, as an isotonic solution. The remainder is reabsorbed by action of the distal tubule and collecting duct in exchange for potassium and/or hydrogen ion under the influence of aldosterone.[7] Prostaglandins, and arginine vasopressin also exercise some control over the final excretion of sodium and water by the kidney.[8] Renal blood flow (RBF) and glomerular filtration rate (GFR) contribute to the amount of proximal reabsorption that takes place. In the presence of decreased RBF and GFR more contact time between the filtrate and the proximal tubule is permitted and increased isotonic reabsorption can occur. When RBF and GFR are increased because of increased cardiac output or expansion of IV volume, a larger amount of sodium and water is presented to the distal tubule and collecting duct and diuresis may occur.[9] Arginine vasopressin serves as the final regulator of body water in response to osmoreceptors situated in the carotid body and other locations. If body water is increased or tonicity decreased, vasopressin release is inhibited and a larger volume of dilute urine low in sodium concentration is produced. At the same time, the sensation of thirst is inhibited unless serum sodium is markedly reduced. If body water is reduced or tonicity increased, vasopressin release is increased resulting in decreased water excretion and a more concentrated urine in an effort to conserve water. In this case, the thirst mechanism is stimulated and the individual increases his water intake as an additional compensatory act.[10] In the healthy individual these actions produce a steady state and provide excellent control of the temporary deficiencies and excesses of both sodium and water as they occur in response to the environment.

The Intracellular Fluid Volume

Potassium is the primary cation of ICF and serves with sodium to regulate intracellular volume.[11] Since cell membranes are freely permeable to water the tonicity of the intracellular space is the same as that of the extracellular space. The major differences between the intracellular and extracellular spaces are in the electrolyte composition of the two compartments. The concentration gradients for sodium and potassium, high to low from the intracellular side for potassium and the reverse for sodium, are maintained by the action of ''sodium-potassium pumps'' operating at the cell membrane.[12] Thus, the integrity of these concentration gradients and, therefore, the regulation of the tonicity and volume of the cells require active mtabolic work. This concept becomes important when we examine pathologic states resulting in intracellular volume changes.

ETIOLOGY AND PATHOPHYSIOLOGY OF ABNORMAL RETENTION OF BODY FLUID

Hypervolemia is the localized or generalized accumulation of water in one or more of the body fluid compartments and can be associated with either hypertonicity or hypotonicity. Edema is a manifestation of either extracellular or intracellular hypervolemia. At all sites of edema formation, regardless of cause, there is increased ultrafiltration of sodium and water from the vascular to the IC. This movement is produced by physical forces operating across the capillary wall. Increased capillary hydraulic pressure, reduced colloid osmotic pressure, or an increased permeability of the capillary membrane favor edema formation. The etiologies of the edematous states are as diverse as the mechanisms for control of body water and will be discussed as we describe the various edematous states.

Localized Edema

The occurrence of swelling of a limited area of the body is one of the most common manifestations of the body's response to chemical, physical, or immunological injury. The localized edema that occurs in response to these types of injury may be the result of: (1) obstruction to venous or lymphatic flow; (2) inflammatory response to vasoactive amines, cytotoxins, complement activation; or (3) cell injury. In some circumstances, when the insult is large, the result may appear to be more generalized and have systemic manifestations. A single bee sting, for example, may cause localized edema, while multiple bee stings can produce a systemic reaction and generalized edema approaching anasarca. Edema due to venous or lymphatic obstruction may involve a small area or an entire extremity depending on the site of the obstruction and its severity. Localized edema is usually self-limiting in duration, resolving as the cause diminishes or the obstruction is corrected.

Generalized Edema

The more serious causes of expansion of body water such as cardiac, renal, or liver disease, by virtue of their systemic manifestations, usually result in generalized edema and can produce debilitating illness. The specific pathophysiology of edema in each of these categories is different and therefore will be discussed individually.

Cardiac Edema

Edema is one of the major manifestations of cardiac disease and is the result of decrease in cardiac output. A cascade of alterations follow which result in expansion of the ECF compartment and eventually the signs and symptoms of edema.[13] When edema occurs suddenly, as in acute pulmonary edema due to acute left ventricular decompensation, there is a rapid redistribution of the ECF and interstitial fluid volume expands in the lungs at the expense of plasma volume. In more chronic generalized cardiac edema, there is an expansion of the interstitial fluid volume in the lungs and/or peripheral tissues with no change or an increase in plasma volume and an increase in ECF volume. Chronic cardiac edema is always accompanied by retention of salt and water by the kidney. The kidneys, therefore, participate significantly in the production of cardiac edema. The heart fails as a pump when it can no longer satisfy the requirements of the peripheral tissues for oxygen and other nutrients while operating at normal filling pressures. The circulatory system, in its attempt to compensate for the reduced cardiac output, alters pressures and tone within the vascular system so that while cardiac output and blood pressure may be restored to normal, organs other than the brain and the heart may receive decreased blood flow (skin, muscle, kidneys, and the splanchnic bed). The compensatory mechanisms brought into play include: (1) activation of the sympathetic nervous system and increasing the secretion of norepinephrine, which causes increased arteriolar vasoconstriction, increased heart rate, and increased force of cardiac contraction; (2) activation of the renin-angiotensin system, with increased secretion of angiotensin II, which further increases peripheral and splanchnic vasoconstriction and acts intrarenally to increase renal vascular resistance; and (3) increased aldosterone secretion from angiotensin II stimulation which induces sodium retention.[14] The initial response to these adjustments is the production of a state of "compensation" which maintains cardiac output at the expense of an expanded ECF volume and increases in venous or atrial filling pressures. Eventually, this results in the continued expansion of the interstitial space and eventually the production of edema.

Edema Due to Liver Disease

The end result of chronic disease of the liver is the development of abnormal salt and water retention with the production of edema and ascites.[15] Similar problems occur with severe acute disease as well. Oliguria and renal insufficiency are frequently associated with hepatic failure and are the cause of much of the morbidity associated with chronic liver disease. This renal component of the "hepatorenal syndrome" has been demonstrated to be a functional impairment of the kidney, since, when the kidney is removed and placed in a better environment, it functions well.[16] In fact, kidneys from patients with hepatorenal disease have been used as kidney donors for transplantation with good results. The series of events leading to salt and water retention in the patient with cirrhosis of the liver can be used as a model for the pathogenesis of abnormal water distribution in patients with liver disease. Loss of parenchymal cells of the liver results in the development of hypoalbuminemia secondary to inadequate hepatic synthesis of serum albumin. The hypoalbuminemia is responsible for a reduction in effective blood volume, which, as we have seen in congestive heart failure, results in sodium and water retention. Because of the increase in portal vein pressure and the hypoalbuminemia, ascites develops which further increases the sodium and water retention. The liver may also be a source of a "third factor" or natriuretic factor, the absence of which may lead to early sodium reabsorption by the kidney and contribute to the edematous state.[17,18]

Nephrotic Syndrome

Edema in the nephrotic syndrome also results from the combined effect of a number of contributing factors.[19] The prime factor is the development of hypoalbuminemia which is an early concomitant of the heavy excretion, increased catabolism, and decreased synthesis of albumin that has been demonstrated in nephrotics.[20] The heavy proteinuria alone cannot be the cause of the hypoproteinemia, even though 24 h of excretion of protein may exceed 20 g. Evidence for an increased rate of protein catabolism has been presented by a number of investigators, but the site of the increased catabolism has not been fully ascertained. Some evidence suggests that the kidney is the major site of this increased catabolic activity.[21] Edema results from the decrease in plasma colloid osmotic pressure and the movement of water into the ECF. The kidneys respond to the reduction in plasma volume by retaining sodium and water, which further enhances the development of edema by the mechanisms discussed in relation to cardiac and hepatic edema.

Other Causes of Generalized Edema

In addition to the major causes of edema listed above there are a variety of edematous states which do not fall into the three categories described above. These include: (1) idiopathic recurrent (cyclical) edema; (2) premenstrual syndrome; (3) protein-losing enteropathies; (4) capillary leak syndrome; and (5) iatrogenic edematous states.

Idiopathic recurrent (cyclical) edema is a syndrome, occurring almost exclusively in women, characterized by bouts of edema, headaches, nervous irritability, and abdominal swelling in the absence of cardiac, hepatic, or renal disease.[22] The pathogenesis of this syndrome is unclear. Sodium retention due to the effects of estrogen, aldosterone, potassium depletion, prolactin, and increased capillary permeability to protein have all been suggested as possible etiologies. A major consideration in some cases is the long-term use of combined sodium restriction and diuretics which can result in edema secondary to hyponatremia when water intake is high.

Premenstrual syndrome occurs in some women of child-bearing age and is characterized by sodium retention with edema of the ankles, fingers, and periorbital areas during the premenstrual period. It is frequently associated with headaches, irritability, and emotional lability and can represent a cause of recurring temporary disability. The etiology of the sodium retention is unclear and endocrine disturbances such as abnormal estrogen:progesterone ratio or elevated prolactin levels have been suggested as causes. The administration of bromocriptine has been successful in controlling the edema and lends some credence to the relationship of prolactin elevation to the syndrome.

Protein-losing enteropathies such as non-tropical and tropical sprue can present with generalized edema secondary to marked hypoalbuminemia.[23] Diuresis occurs when the protein loss is corrected.

Capillary leak syndrome produces edema and ascites due to the movement of protein through the capillary wall, with reduction of the colloid osmotic pressure within the vascular compartment. One form has been noted in association with a persistent monoclonal IgG paraprotein. Another possible cause is myxedema due to the accumulation of anionic mucopolysaccharides which bind sodium and water.

Among the most common iatrogenic causes of edema is the overvigorous use of diuretics by the patient or physician. This results in the development of hyponatremia and continued stimulation of thirst.[10] The effect is intensified by the rigorous restriction of sodium during diuretic administration. The continued drinking of water under these circumstances further aggravates the dilutional syndrome and increases the production of angiotensin which is a potent stimulator of the thirst response. Some non-steroidal anti-inflammatory drugs (phenylbutazone) appear to bind to mineralocorticoid receptors in the distal tubule and exert an aldosterone-like effect, while others seem to exert their sodium-retaining effect by inhibiting prostaglandin synthetase. Vasodilator drugs such as minoxidil, hydralazine, and diazoxide, when given alone for reduction of blood pressure, cause sodium retention by reducing peripheral vascular resistance and require the concomitant use of diuretics to prevent the development of edema.[24]

MODERN METHODS FOR THE CONTROL OF EDEMATOUS STATES

Control of edema requires application of the knowledge of the pathophysiology of the cause of edema in order to be effective. Since it is beyond the scope of this chapter to present a complete discourse on management of edema, I will briefly review only the underlying principles. The first effort should be directed to control of the primary disease or underlying causes, including medications that may be contributing to the retention of salt and water. In chronic disease this may not always be possible, but should at least be explored. Dietary therapy of fluid-retaining states requires an estimate of salt losses and adjustment of salt intake to a level that creates a negative balance. The use of a 40-meq/d sodium intake in an individual that is excreting only 25 meq/d will not correct the underlying sodium excess. Initial assessment of 24-hr urinary sodium excretion and subsequent follow-up determinations 24 to 48 h after an adjustment in therapy are very useful in monitoring the severely ill patient. Similarly, water intake must be adjusted to provide for a negative water balance. Since many patients with severe edema are oliguric, simple restriction to 1000 ml/d is also inadequate. The best estimate of water balance is the use of daily weight, preferably at the same time of day on the same scale. The use of fractional urine sodium excretion (FvNa) can also be used to evaluate effects of therapeutic intervention. Bed rest as a part of the management of edema is firmly grounded in physiology. Cardiac output is increased by at least 400 ml by as little as 4 h of bed rest and mobilization of fluid from the interstitial compartment is increased. Circulating arginine-vasopressin level is reduced with bed rest as is the level of circulating aldosterone. In addition, bed rest frequently will render an individual more responsive to diuretics. In the presence of edema of the lower extremities the use of elastic support stockings will assist in the mobilization of the trapped fluid.

Diuretics have become the mainstay of therapy for edema as the number and type of diuretic agents have increased. Acetazolamide acts in the proximal tubule by inhibiting carbonic anhydrase. Its effect, when used alone, is short-lived and it is used only in association with other agents acting more distal in the nephron. Mannitol also acts in the proximal tubule, although its effect can be seen more distally as well. Mannitol, because it is nonreabsorbable in the tubule, acts as an osmotic agent preventing the isotonic absorption of water, usually taking sodium with it. Thiazide diuretics act predominantly in the distal cortical-diluting segment, blocking reabsorption of sodium and decreasing free water clearance, thereby providing increased delivery of sodium and water to the distal segment and collecting duct. Plasma volume and venous pressure are reduced, resulting in increased mobilization of fluid from the interstitial space. The "loop" diuretics, furosamide and ethnacrynic acid, are more potent and are frequently effective when patients are refractory to the thiazides. Care must be taken in the use of these agents to prevent excessively rapid diuresis and hypotension from rapid contraction of the plasma volume. Spironolactone is a clinically useful inhibitor of aldosterone and it can be used as the primary agent when dealing with high aldosterone states. It is especially useful when added to thiazides after secondary hyperaldosteronism has developed.

In refractory patients there are some useful points which should be remembered to increase the effectiveness and reduce the morbidity from the use of diuretics. These are:

1. Because of extensive edema of the intestinal mucosa, oral medication may not be adequately absorbed, and the i.v. route should be used.
2. The renal tubule may become refractory to a given diuretic when used for prolonged periods of time. A switch to another agent may be effective.
3. Use diuretics sequentially, using that agent which has its effect in the area in which maximal reabsorption is taking place rather than one which acts more distal to that site. Follow it, if necessary, with one that acts more distally.
4. Pre-treat patients who are hypoalbuminemic with albumin or other hyperoncotic colloids prior to the use of a diuretic to permit the beneficial effect of the diuretic to occur.
5. Reduction of afterload in patients with congestive heart failure using a peripheral vasodilator such as nitroprusside can increase the effectiveness of the diuretic.
6. Avoid using thiazides in patients with impaired renal function, as these agents frequently further reduce renal function.

Table 2
CAUSES FOR EDEMA FORMATION IN MAN

Type of edema	Primary pathophysiological alteration	Clinical condition
	Localized edema	
Hydrostatic	Increased venous hydrostatic pressure Increased lymphatic hydrostatic pressure	Venous obstruction or occlusion Lymphatic obstruction Thoracic duct obstruction
Inflammatory	Inflammatory response to vasoactive amines, cytotoxins, complement activation, or cell injury	Multiple insect stings, bee stings Lupus erythematosis Angioneurotic edema
	Generalized edema	
Hydrostatic	Increased venous hydrostatic pressure	Hepatic vein occlusion Portal vein occlusion or hypertension Vena cava occlusion
Osmotic	Hypoalbuminemia resulting in increased diffusion from arterial capillary and decreased reabsorption in the venous capillary	Nephrotic syndrome Malnutrition Protein-losing enteropathy Hepatic cirrhosis
Primary sodium retention	Increased total body sodium due to increased sodium retention by the kidney due to reduction in renal blood flow	Hypertension and/or treatment Congestive heart failure Nephrotic syndrome

Table 2 (continued)
CAUSES FOR EDEMA FORMATION IN MAN

Type of edema	Primary pathophysiological alteration	Clinical condition
		Hepatic cirrhosis
	Increased sodium retention by the kidney due to increased secretion or decreased inactivation of renin-angiotensin	Hypertension and/or treatment Congestive heart failure Nephrotic syndrome Hepatic cirrhosis Chronic diuretic use
	Increased sodium retention by the kidney due to primary increased aldosterone secretion	Primary hyperaldosteronism Non-steroidal anti-inflammatory drugs (aldosterone-like effect)
Lymphatic obstruction	Increased hydrostatic pressure within the lympatics resulting in decreased lymph drainage	Lymphatic obstruction due to congenital disease Tumor invasion Surgical disruption Parasite infestation
Capillary leak	Increased permeability of capillary wall to normal or abnormal proteins	Monoclonal IgG paraproteinemia
	Increased anionic mucopolysaccharides in tissues causing increased binding of sodium and water	Myxedema
Hormonal	Increased secretion of estrogens and progesterone resulting in sodium and water retention via aldosterone	Premenstrual syndrome
Neurogenic	Damage to osmoreceptors and stretch receptors of the brain and vascular tree	Syndrome of inappropriate secretion of ADH Generalized edema due to damage to hypothalamus
Idiopathic	Undetermined origin (see text)	Idiopathic recurrent (cyclical) edema

REFERENCES

1. **Bernard, C.,** *Lecons sur les Propriétés Physiologiques et les Altérations Pathologiques des Liquides de l'Organisme,* Vol., 1, J.-B. Bailliere, Paris, 1859.
2. **Strauss, M. B.,** *Body Water in Man,* Little, Brown, Boston, 1957.
3. **Edelman, I. S. and Leibman, J.,** Anatomy of body water and electrolytes, *Am. J. Med.,* 27, 256, 1959.
4. **Skorecki, K. L. and Brenner, B. M.,** Body fluid homeostasis in man. A contemporary overview, *Am. J. Med.,* 70, 77, 1981.
5. **Wiederhielm, C. A.,** Dynamics of transcapillary fluid exchange, *J. Gen. Physiol.,* 52, 29, 1968.
6. **Gauer, O. H., Henry, J. P., and Behn, C.,** The regulation of extracellular volume, *Annu. Rev. Physiol.,* 32, 547, 1970.
7. **Burg, M. B. and Orloff, J.,** Control of fluid reabsorption in the renal proximal tubule, *J. Clin. Invest.,* 47, 2016, 1968.
8. **Lee, J. B.,** Cardiovascular-renal effects of prostaglandins, *Arch. Int. Med.,* 133, 56, 1974.
9. **Wright, F. S. and Briggs, J. P.,** Feedback control of glomerular bloodflow, pressure, and filtration rate, *Physiol. Rev.,* 59, 958, 1979.
10. **Epstein, A. N.,** The physiology of thirst, *Can. J. Physiol. Pharmacol.,* 54, 639, 1976.
11. **Leaf, A.,** Regulation of intracellular fluid volume and disease, *Am. J. Med.,* 49, 291, 1970.
12. **Schmidt, U., Schmid, J., Schmid, H., et al.,** Sodium and potassium activated ATPase. A possible target of aldosterone, *J. Clin. Invest.,* 55, 655, 1975.
13. **Warren, J. V. and Stead, E. A., Jr.,** Fluid dynamics in chronic congestive heart failure, *Arch. Int. Med.,* 73, 138, 1944.
14. **Oparil, S. and Haber, E.,** The renin-angiotensin system, *N. Engl. J. Med.,* 291, 389, 1974.
15. **Epstein, M.,** Renal sodium handling in cirrhosis: a reappraisal, *Nephron,* 23, 211, 1979.
16. **Epstein, M.,** Increased sympathetic activity in cirrhosis, *N. Engl. J. Med.,* 308, 1030, 1983.
17. **deWardener, H. E.,** Natriuretic hormone, *Clin. Sci. Mol. Med.,* 53. 1, 1977.

18. **Kramer, H. J.,** Natriuretic hormone-its possible role in fluid and electrolyte disturbances in chronic liver disease, *Postgrad. Med.,* 51, 532, 1975.
19. **Paller, M. S. and Schrier, R. W.,** Pathogenesis of sodium and water retention in edematous disorders, *Am. J. Kidney Dis.,* 2, 241, 1982.
20. **Johansson, S. V., Odar-Cederlof, I., Plantin, L. O., et al.,** Albumin metabolism and gastrointestinal loss of protein in chronic renal failure, *Acta Med. Scand.,* 201, 353, 1977.
21. **Reineck, H. J.,** Mechanisms of edema formation in nephrotic syndrome, in *Contemporary Issues in Nephrology,* Brenner, B. M. and Stein, J. H., Eds., Churchill Livingstone, New York, 1982, 31.
22. **Gill, J. R., Jr.,** Idiopathic edema, *Semin. Nephrol.,* 3, 205, 1983.
23. **Weinbren, I., Taggart, P. I., and Glass, H. I.,** Edema due to abnormal distribution of protein, *Lancet,* 3, 512, 1965.
24. **Davis, J. O. and Freeman, R. H.,** Mechanisms regulating renal renin release, *Physiol. Rev.,* 56, 1, 1976.

BOOKS AND REVIEW ARTICLES

Brenner, B. M. and Stein, J. H., *Sodium and Water Homeostasis,* Vol. 1, Churchill Livingstone, New York, 1978.

Carroll, H. J. and Oh, M. S., *Water, Electrolyte and Acid Base Metabolism: Diagnosis and Management,* Lippincott, Philadelphia, 1978.

Martinez-Maldonado, M., Ed., *Edematous Diseases,* Vol. 3 (No. 3), Grune & Stratton, New York, 1983.

Maxwell, M. H. and Kleeman, C. R., *Clinical Disorders of Fluid and Electrolyte Metabolism,* 3rd ed., McGraw-Hill, New York, 1980.

Valtin, H., *Renal Dysfunction: Mechanisms Involved in Fluid and Solute Imbalance,* Little, Brown, Boston, 1979.

MAINTENANCE OF WATER, pH, AND ELECTROLYTE BALANCE BY THE KIDNEY

A. D. Baines

INTRODUCTION

A human kidney is made up of 1 million nephrons (Figure 1). Each nephron consists of a capillary tangle — the glomerulus — and a long tubule, which descends from cortex to papilla where it forms a hairpin bend and ascends back to rejoin its glomerulus in the cortex. At this point of contact the ascending tubule lies between the afferent and efferent arterioles creating a specialized structure called the juxtaglomerular apparatus. The apparatus contains specialized tubular cells (the macula densa), specialized arteriolar smooth muscle cells that produce renin and mesangila cells that are insinuated into the glomerulus to form a contractile scaffolding for its capillaries. After leaving the juxtaglomerular apparatus, the tubule proceeds back through the cortex to descend again into the papilla, joining other tubules en route to form the collecting duct system.

Every hour the kidney produces a volume of glomerular filtrate greater than the total plasma volume and reabsorbs more than 99% of the filtered water along with selected solutes. Glomerular filtration rate (GFR) in a healthy adult is closely related to dietary protein and can vary by more than 50% due to acute or chronic alterations of protein intake.[2] Glomerular filtration and tubular reabsorption increase and decrease in parallel. The mechanisms responsible for maintaining this glomerular-tubular balance are incompletely understood, but they are crucial for homeostatic control of body fluid volume, osmolality, and pH. Almost all the glomerular filtrate is reabsorbed, whatever the GFR may be. Subtle changes in the composition and volume of the reabsorbed fluid regulate urine flow and composition. For example if, while GFR remains constant, reabsorption of water decreases from 99.5 to 99%, then urine flow doubles. The rapidity and wide range of the renal response to alterations of body fluid composition depend upon control of the small difference between the large GFR and tubular reabsorption rate.

The activities of the nephron are segregated in discrete anatomic segments. Each segment has its own unique complement of enzymes and hormonal receptors.[12] The *proximal tubule* is responsible for reabsorbing sodium, chloride, bicarbonate, and essential organic nutrients, secreting toxic substances, and producing ammonia. In this segment 60 to 80% of the filtered water, and most of the glucose, amino acids, and bicarbonate are reabsorbed in an isotonic solution. The *thick ascending limb* produces a hypotonic fluid by reabsorbing sodium chloride without water. The *distal* and *cortical-collecting tubules* secrete potassium and hydrogen ions. Finally, in the *collecting duct* system urine may be either further diluted by solute removal or concentrated by water removal. Concentrating urine requires the concerted action of all the nephrons in a counter-current system.

REGULATION OF RENAL FUNCTION

Total kidney failure (uremia) rapidly produces life-threatening edema, hypertension, hyperkalemia, and acidosis. These drastic consequences of kidney dysfunction indicate the kidney's importance in regulation of extracellular fluid homeostasis. To carry out its homeostatic role the kidney must continuously receive information from sensors that reflect the volume and composition of the extracellular fluid.[6,24]

Stretch receptors in the cardiac atria and possibly also the liver monitor blood volume in the low-pressure side of the cardiovascular system. Arterial stretch receptors in the carotid

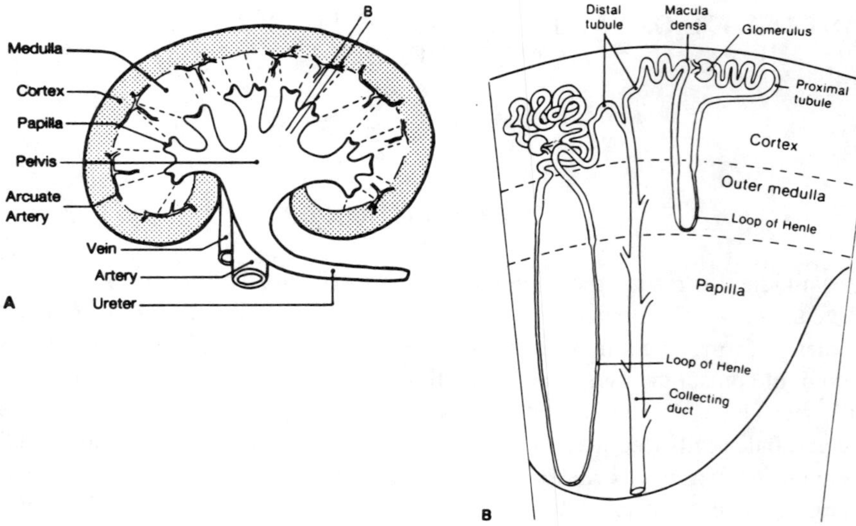

FIGURE 1. (A) Cross section of the human kidney; (B) enlargement of a section through the area indicated in (A). The loop of Henle includes the straight part of the proximal tubule, the descending thin limb, ascending thin limb, medullary TAL, and cortical TAL. Note that loops of Henle are longer in juxtamedullary than in superficial nephrons. Short-looped nephrons do not have an ascending thin limb. (From ABCD, Gornall, A. G., Ed. With permission.)

sinus, and the kidney itself monitor blood pressure in the high-pressure side of the cardiovascular system. Osmoreceptors in the hypothalamus monitor extracellular fluid osmolality.[21] The adrenal gland responds to changes in the ratio of K^+/Na^+ in adrenal blood. Chemosensors in the carotid body and the brain monitor pH.

The kidney also responds directly to changes in arterial pressure, pH, osmolality, and composition of glomerular filtrate and peritubular blood. In this way the kidney acts as both detector and effector in correction of homeostatic imbalances.

Some of the signals from these various sensors are modified and integrated in the brain before being sent to the kidney. The hypothalamic region integrates messages from stretch receptors, chemoreceptors, and osmoreceptors in the atria, blood vessels, liver, and brain to regulate thirst and vasopressin (antidiuretic hormone, ADH) release. The brain stem integrates similar information to regulate activity of the sympathetic nervous system. The kidney participates in this flow of information through afferent renal nerves which are sensitive to chemical and osmotic changes in the papilla and elsewhere within the kidney. Damage to the integrating centers in the hypothalamic region or brain stem can disrupt regulation of extracellular fluid volume, blood pressure, or body fluid osmolality.

Signals from the brainstem and hypothalamus are delivered to the kidney through sympathetic nerves and vasopressin, respectively. Sympathetic nervous activity regulates blood pressure, GFR renin release, and electrolyte reabsorption. Vasopressin not only modifies water reabsorption, but also acts synergistically with catecholamines and the renin-angiotensin system to regulate vascular tone and, therefore, arterial blood pressure.

Vasopressin

The combination of thirst, which regulates water intake, and vasopressin, which regulates water excretion, sets the level of body fluid osmolality, which is crucial for maintenance of cell volume.[21] Osmolality is inextricably linked to sodium balance, since sodium salts constitute the bulk of osmotically active extracellular solutes. Any change in body sodium content produces a change in osmolality which normally is rapidly corrected by appropriate changes in water intake and excretion. Glucose and urea do not stimulate vasopressin release.

The excretory response to altered extracellular fluid volume is slower in onset and less closely regulated than the response to altered osmolality. Consequently, extracellular fluid volume increases and decreases in parallel with sodium content. This can be seen when salt intake is abruptly increased; several days elapse before urinary salt output equals intake and in the interim the subject retains water, gains several kilograms in weight, and his blood pressure increases by several millimeters of mercury. The increased weight indicates increased extracellular fluid volume. The increased fluid volume raises cardiac output and blood pressure.

Blood volume as reflected by pressure in the cardiac atria also regulates vasopressin release. Stretching the atrial wall stimulates vagal afferent nerves that inhibit the hypothalamic stimuli for vasopressin release. The response to changes in blood volume is much less sensitive than the response to osmolality. Variations of 1 to 2% in osmolality are sufficient to alter vasopressin release, but blood volume must change by approximately 10% before an effect is detectable. Volume status influences the response to osmolality; thus, a volume-depleted patient may release vasopressin even when plasma osmolality is low.

Blood Pressure

The increase in blood pressure produced by an increased salt intake is an important long-term regulator of salt excretion.[13] A minimum mean pressure of 60 to 70 mmHg is required for glomerular filtration to occur. Above this threshold urine flow and salt excretion are directly related to arterial pressure even though the intrinsic autoregulatory capacity of the kidney may maintain relatively constant blood flow and GFR. The physical mechanisms responsible for pressure natriuresis are not fully understood.

The setting of the natriuretic response to blood pressure can be altered by hormonal or neural stimuli or renal disease processes acting on blood vessels or tubules. Renal disease shifts the pressure-natriuresis curve to the right. This means that blood pressure must rise further to maintain salt excretion in balance with intake. This form of hypertension should be ameliorated by reduced salt intake and diuretic therapy.

Angiotensin-Aldosterone

Aldosterone has a major role in regulating renal excretion of sodium, potassium, and hydrogen ion. Aldosterone production and secretion by the adrenal cortex requires ACTH and is stimulated by an increased ratio of K^+/Na^+ in adrenal blood and by angiotensin II and inhibited by dopamine. The production of angiotensin is controlled primarily by the kidney and autonomic nervous system. Angiotensin II not only regulates aldosterone secretion, but also constricts systemic and renal blood vessels and stimulates salt reabsorption in the kidney.

Renal Nerves and the Autonomic Nervous System

The kidney is richly innervated by sympathetic nerves with most of the terminals located on arteries, arterioles, and veins, but some are adjacent to proximal tubules, ascending thick limbs and distal tubules. There are α 1, α 2, β 1, β 2, and dopamine receptors located in blood vessels and tubules. The autonomic nervous system acts primarily to stimulate renin-angiotensin release and to reduce renal blood flow and urinary salt and water excretion.

Atrial Natriuretic Hormone(s) (ANH)

The cardiac atria produce a peptide or peptides with potent vasodilatory and natriuretic effects.[16] These peptides are released in response to atrial stretch and perhaps to stimulation by the autonomic nervous system. Plasma levels of ANH antigen rise with extracellular fluid volume expansion. The peptides may operate as a counterbalance to the renin-angiotensin-aldosterone system. ANH antagonizes the vasoconstrictor action of angiotensin II and reduces aldosterone secretion by the adrenal. ANH also increases GFR without increasing renal blood flow (see below).

Metabolic Control

Dietary protein has two effects on the kidney: it increases the load of non-volatile acid that must be excreted and it increases GFR.[2] The mechanisms responsible for metabolic effects on glomerular filtration are not understood, but they appear to be related to changes in a few amino acids. Hyperglycemia vasodilates the kidney and raises GFR, which may explain why the onset of insulin-dependent diabetes mellitus is associated with renal vasodilation and increased GFR.

RENAL HORMONES

Endocrine Hormones

The kidney releases two important systemic hormones, erythropoietin and renin, to the blood. Erythropoietin, produced in the renal cortex, stimulates red cell production in bone marrow. Renin, produced in the juxtaglomerular apparatus, initiates the production of angiotensin II from angiotensinogen, a precursor polypeptide produced in the liver. Angiotensin II is a strong vasoconstrictor and stimulates the release of aldosterone from the adrenal cortex.

Renin release from specialized, richly innervated, afferent arteriolar cells adjacent to the glomerulus is triggered by neural stimulation of β-adrenoceptors and by a reduced transmural pressure gradient in juxtaglomerular arterioles (i.e., arterial hypotension or increased tissue pressure). Renal nerve stimulation and reduced arteriolar pressure, which act synergistically, are the major regulators of renin release.[16] Na^+ or Cl^- reabsorption by the macula densa cells also influences renin release. This so-called "natriostat" function of specialized cells in the thick ascending limb is inhibited by furosemide, K^+ deficiency, and aldosterone deficiency. The significance of this mechanism in renin release is under debate. Some experiments indicate that increased NaCl delivery to the macular densa increases renin release; other experiments contradict this conclusion.

Reflexes originating in the cardiopulmonary and arterial baroreceptors regulate renin release. Renin acts upon angiotensinogen to release angiotensin I, a decapeptide. This inactive peptide is converted to the active octapeptide, angiotensin II, by angiotensin-converting enzyme. Converting enzyme is found in the kidney and, systemically, especially in the lungs. Circulating angiotensin II stimulates aldosterone production by the adrenal cortex and vasoconstricts systemic arteries. Angiotensinases in capillaries destroy circulating angiotensin II.

An inactive form of renin, prorenin, is found in plasma and kidney. Plasma prorenin levels are two to ten times greater than plasma renin. Prorenin can be activated *in vitro* by acidification, proteolysis, or freeze-thawing. Evidence of *in vivo* activation has not been obtained. Active renin comes almost exclusively from the kidney and disappears from the plasma of nephrectomized patients, but prorenin continues to circulate in these patients; therefore, prorenin is produced outside the kidney. It has been suggested that kallikrein (see later) converts to renin in the kidney.

Paracrine Hormones

Paracrine hormones have short half-lives which limits their actions to the region where they are produced. Renal tissue produces a number of paracrine hormones that are not unique to the kidney, for example, prostaglandins, leukotrienes, kinins, and adenosine. Prostaglandins modulate several signals from extra-renal sources. They are released by neural stimulation and antagonize the vasoconstrictor action of norepinephrine. They are also released by vasopressin and antagonize its action on water permeability of the collecting ducts. Prostaglandins produced in the kidney also stimulate renin release and are important in the sustained release of renin during chronic volume depletion. Prostaglandins do not mediate the acute response to reduced renal artery pressure or nerve stimulation. Leukotrienes vasoconstrict the kidney and may participate in the inflammatory response. Adenosine, which escapes from renal cells when ATP consumption exceeds the rate of ATP production, is a renal vasoconstrictor and also sensitizes renin release to neural stimuli. Kallikrein is a proteolytic enzyme produced by granular cells of the distal tubule and cortical-collecting tubules. Kallikrein releases vasodilatory kinins from circulating kininogen procursors. These kinins are destroyed within the kidney by angiotensin-converting enzyme. The significance of these substances is not known, but they may act to limit the vasoconstrictor actions of angiotensin.

Within the kidney the renin-angiotensin system may have a paracrine function.[15,20] All the synthetic components of this system are available in the juxtaglomerular apparatus. Angiotensin constricts the efferent arteriole more than the afferent arteriole. Angiotensin also contracts mesangial cells. Extraglomerular mesangial cells, contiguous with those in the glomerulus, form a cushion between the afferent arteriole and the macula densa. These cells probably play an important role in tubulo-glomerular feedback (see below).

REGULATION OF GLOMERULAR FILTRATION

The nature of the filtration process has been explored using micropuncture with small glass pipettes (tips less than 10 μm) to sample fluid and pressures in superficial glomerular capillaries and tubules of rat kidneys.

The glomerular filter permits free passage of crystalloids and substances with effective molecular radii less than 20 Å and completely excludes substances with radii greater than 42 Å. This means that inulin (2000 Da) concentration in the filtrate is the same as in plasma water, but that albumin (68,000 Da) concentration is only 1% of that in plasma.

Filtration is driven by a hydrostatic gradient from capillary lumen to proximal tubular fluid. Unlike other capillary beds, flow through the glomerulus is regulated by muscular arterioles at both inflow (afferent) and outflow (efferent) ends; thus, glomerular capillary pressure depends upon muscle tone of afferent and efferent arterioles as well as systemic arterial pressure. The pressure in proximal tubules is normally constant (10 to 14 mmHg), but may be increased by reduced fluid reabsorption along the tubule or obstruction to outflow anywhere along the urinary tract. Oncotic pressure due to the presence of albumin in the capillary fluid, but not in the filtrate, opposes the hydrostatic pressure gradient from capillary to tubule. This oncotic pressure is equivalent to 20 mmHg in blood entering the glomerular capillaries. The oncotic pressure rises as water is filtered out of the blood during its passage through the glomerular capillaries. The increase of oncotic pressure reduces net filtration pressure and may bring filtration to a halt before blood reaches the end of the capillary network. Under these conditions the rate of blood flow through the glomerulus becomes one of the factors limiting GFR (Table 1).

Hydraulic (or water) permeability and surface area of the glomerular filter determine the rate of filtration in response to the net filtration pressure. (In micropuncture experiments the product of hydraulic permeabilty times surface area [called K_f] is measured). The filtering surface consists of basement membrane sandwiched between a fenestrated endothelial layer on the capillary side and the mesh of epithelial foot processes on the tubular side. The basement membrane, which probably contributes most of the resistance to water flow, is a fibrinous gel containing many negatively charged maromolecules rich in scialic acid and heparan sulfate.

Table 1
REGULATION OF GLOMERULAR FILTRATION RATE

Factor	Modifiers
Arterial pressure	Cardiac output, peripheral resistance
Arteriolar resistance	Constricted by α agonists, nerves, angiotensin, adenosine, prostaglandins, leukotrienes, vasopressin, thromboxanes
	Dilated by histamine, acetylcholine, glucocorticoids, dopamine, prostaglandins (PGE_2)
	Efferent constriction and increased GFR: atrial natriuretic hormone
Capillary surface	Reduced by angiotensin, vasopressin, α agonists
Glomerular permeability	Reduced by immunological attack, microthrombi
Tubular pressure	Water reabsorption, tubular obstruction
Glomerular tubular feedback	Delivery of Cl^- to macula densa

Glomerular diseases that alter the chemical composition and negative charge in the basement membrane may decrease hydraulic permeability and simultaneously increase permeability to proteins.

The core or backbone of the glomerular capillary mesh contains mesangial cells. These cells probably contribute to the production and turnover of glomerular basement membrane. They also contain contractile fibers that respond to angiotensin II, norepinenphrine, and vasopressin. Contraction of mesangial cells reduces the filtering surface area.

A number of hormones regulate glomerular filtration (Table 1). Angiotensin II and norepinephrine constrict efferent and to a lesser extent afferent arterioles; they also reduce K_f. The net result is a reduction in glomerular blood flow with little change in GFR. A number of vasodilators relax both efferent and afferent arterioles, but also stimulate renin release. These hormones, for example, acetylcholine, histamine, prostaglandin E_1, and bradykinin, increase glomerular blood flow, but by releasing renin to produce angiotensin II they also reduce K_f and leave GFR unchanged or decreased. Glucocorticoids vasodilate without changing K_f and thereby raise GFR. ANH increase GFR by a combination of afferent arteriolar dilation, efferent constriction, and increased K_f.

The renal vasculature, especially in the afferent arteriole, appears to have an inherent myogenic autoregulatory response. This myogenic response causes afferent arteriolar resistance to increase as arterial pressure rises above 80 mmHg and is partly responsible for the autoregulation of blood flow relative to arterial pressure.

Tubulo-glomerular feedback[21] is also involved in the autoregulatory phenomenon. This mechanism maintains constant delivery of fluid to the distal tubule in the following way. GFR and reabsorption in the proximal tubule and Henle's loop determine the flow of water and NaCl past the macula densa. Increased chloride delivery to the macula densa reduces filtration in the adjacent glomerulus by triggering constriction of the afferent arteriole and mesangial cells. Activation of the tubuloglomerular feedback mechanisms may be responsible for the functional reduction of GFR associated with toxic or ischemic damage in the proximal tubule and thick ascending limb. In this situation reduced proximal reabsorption leads to increased delivery of salt to the macula densa cells. The mechanism may also contribute to the effect of protein intake on GFR. A high protein diet stimulates reabsorption in the proximal tubule or loop; this decreases distal delivery of salt and thereby activates an increase in glomerular filtration.

Glomerular filtration is increased in the first two trimesters of pregnancy as a result of afferent arteriolar dilation.

Removal of the kidney tissue prompts the remaining glomeruli to increase their filtration rate.[3] Part of this response is functional due to reduced arteriolar resistance and part is structural due to hypertrophy and hyperplasia. The response is detectable within hours after the kidney mass is reduced, but takes weeks to reach its maximum. The magnitude of the response is proportional to the extent of kidney tissue loss and dietary protein intake and inversely proportional to the patient's age.

The immune events responsible for some glomerular diseases may reduce surface area by activating adenylcyclase and releasing angiotensin II within the glomerulus. In addition, immune-mediated glomerulonephritis may reduce GFR by plugging capillary lumens with leukocytes, proliferating endothelial cells, and fibrin clots. Ultimately, the disease process causes sclerosis or part of all of a glomerulus.

SODIUM, CHLORIDE, AND BICARBONATE REABSORPTION

Na-K ATPase in the basolateral membranes of tubular cells drives most of the Na^+ reabsorption throughout the nephron. Na^+ transport out of the cell makes the cell interior electronegative to both the luminal and peritubular fluid and reduces intracellular Na^+ to 10 to 20 mmol/l. This creates a large electrochemical gradient for Na^+ entry into the cell. Passive Na^+ transport systems in the luminal membranes regulate Na^+ access to the cell interior and control the rate of Na^+ reabsorption. Most of these transport systems link Na^+ entry to movement of another solute. In this way the energy generated by active Na^+ transport across the basolateral membrane is used to drive selective reabsorption of many different solutes.

Proximal Tubule

Reabsorption in the proximal tubule is isotonic; that is to say, water follows solute reabsorption so that osmotic pressure of tubular fluid is never more than 1 or 2 mOsm below the osmotic pressure of peritubular fluid.[19] Na^+ enters proximal tubular cells from the lumen through coupled carriers, either by exchanging with a proton or co-transport with another solute. The co-transport is usually electrogenic, because one Na^+ accompanies a neutral molecule and more than one Na^+ accompanies an organic anion. Na^+ moves down the electrochemical gradient from lumen to cell and simultaneously drives uphill transport of glucose, amino acids, organic acids, and phosphate. Some coupled NaCl reabsorption also occurs. The availability of co-transported substances can limit Na entry into the cell.

A Na^+-proton antiporter in the brush-border membranes passively exchanges one luminal Na^+ for one cellular proton.[17] Secreted H^+ combines with bicarbonate in the tubule lumen to form carbonic acid which breaks down to CO_2 and water; this process is accelerated by carbonic anhydrase in the brush border. Within the cell the OH^- ion created by H^+ secretion reacts with CO_2 to form HCO_3^-, which exits by passive carrier-mediated diffusion through the basolateral membrane. The net effect is to exchange a luminal bicarbonate for an intracellular bicarbonate. A maximum tenfold gradient for protons can be maintained between tubular lumen (pH 6.5) and blood. At this pH most of the luminal bicarbonate is converted to CO_2 and water. Reduced intracellular pH allosterically enhances proton secretion.

Reabsorption of Na^+ with HCO_3^-, phosphate, and organic anions leaves an excess of Cl^- in the tubular lumen. The tight junction between cells in downstream parts of the proximal tubule is more permeable to Cl^- than to HCO_3^-; therefore, Cl^- diffuses out of the lumen between the cells and Na^+ follows passively. This may account for $^1/_3$ of the total solute reabsorption in the proximal tubule.[19]

Na^+-proton exchange also promotes reabsorption of weak acids that form lipid-soluble undissociated acids when they combine with H^+ at neutral pH. In their undissociated form, weak acids such as acetate and propionate diffuse passively into the cell. Stronger acids, such as urate, may be reabsorbed by a carrier-mediated exchange of luminal anion for intracellular hydroxyl ion.

Na^+-proton exchange is stimulated by increased pCO_2, insulin, glucocorticoids, and possibly by thyroid hormones and is inhibited by parathyroid hormone.

Phosphate reabsorption is coupled to sodium.[18] The activity of the phosphate carrier is enhanced by a low phosphate diet and reduced by parathyroid hormone stimulation of adenylate cyclase. Increased H^+ concentration allosterically reduces the transporter affinity for sodium and indirectly for phosphate; thus, increased luminal H^+ decreases Na^+-phosphate reabsorption and increased intracellular H^+ concentration stimulates reabsorption by facilitating the release of sodium and phosphate from the transporter on the cytoplasmic surface of the brush border. Turnover of the Na^+-phosphate transporter in the brush border is rate limiting for phosphate reabsorption. Reabsorbed phosphate enters the cellular metabolic pool and may be incorporated into organic molecules during glycolysis. Gluconeogenesis releases phosphate from the organic pool and increased gluconeogenesis is usually associated with decreased phosphate reabsorption.[9a]

All evidence points to active Na^+ reabsorption coupled to passive cotransport systems as the driving force for solute and water transport. The interspaces between cells created by the complex folding of the basolateral membranes probably trap solutes after they have been transported out of the cells. Current opinion favors the idea that water passes through the cell drawn by small solute gradients built up in the peritubular spaces.

Peritubular protein stimulates the reabsorption of NaCl and water in two ways. First, protein directly stimulates active NaCl transport across the basolateral membrane of the cell. Second, after solutes and water have moved through the cells they must be taken up into peritubular capillaries. This process depends upon the balance between hydrostatic and oncotic pressures in capillaries and peritubular fluid. Fluid that is not reabsorbed either diffuses back to the lumen through the tight junctions between cells or enters the renal lymphatics.

Peritubular oncotic and hydrostatic pressure, or physical factors as they are called, are strongly influenced by glomerular arteriolar resistance and renal blood flow. Efferent arteriolar constriction increases GFR relative to blood flow and raises peritubular oncotic pressure as a consequence of increased fluid filtration in the glomerulus. Peritubular hydrostatic pressure is lowered at the same time. These changes, favoring reabsorption from the proximal tubule, are seen when the sympathetic nervous system is activated by reduced effective circulating blood volume and blood pressure in patients with heart failure or hepatic failure. The increased proximal tubular reabsorption contributes to fluid retention and reduces distal delivery of fluid for excretion of dilute urine etc. (see later).

Ammonia production occurs primarily in the proximal tubule and will be discussed under the heading of acid-base regulation.

Genetic defects in proximal tubular transport are manifested by increased urinary excretion of individual substances such as bicarbonate (proximal renal tubular acidosis), cystine, glucose, phosphate, or calcium, or by increased excretion of several substances such as bicarbonate, phosphate, glucose, and amino acids as in the Fanconi syndrome.

Acquired defects in proximal tubular reabsorption can be produced by toxins, certain nephrotoxic drugs, and by severe ischemia. The proximal tubular damage may be manifested by bicarbonaturia, glucosuria, aminoaciduria, phosphaturia, and mild proteinuria.

Glomerlar-Tubular Balance

Under most conditions reabsorption from the proximal tubule matches GFR so that the fraction of filtered water and sodium is roughly constant. This balance is achieved by action of peritubular physical forces just described and by the load-dependent nature of reabsorption for substances, such as glucose and amino acids, that are co-transported with sodium. Reabsorption of these substances is virtually complete at normal plasma concentrations and the reabsorptive mechanisms are not saturated; therefore, reabsorption increases as filtered load increases. Bicarbonate and phosphate reabsorption are also load dependent.

Most factors (e.g., glucocorticoids, dopamine, protein feeding, or i.v. glycine and alanine) that increase GFR (Table 1) also increase proximal or distal tubular reabsorption so that fractional excretion of water and sodium is unchanged. Atrial natriuretic hormone appears to be the major exception to this rule; it increases GFR and depresses fractional water and salt reabsorption.

Thick Ascending Limb (TAL)

This portion of the nephron (TAL) is impermeable to water and has more mitrochondria per cell than any other part of the nephron. It is here that the major work required to dilute and concentrate the urine is carried out. The energy is again provided by Na-K-ATPase in the basolateral surface of the tubular cells; but in the TAL each Na^+ enters through the luminal membrane along with one K^+ and two Cl^- transported down a composite electrochemical gradient.[11] The "loop" diuretic furosemide specifically inhibits this passive carrier-mediated transport. Most of the K^+ leaks back into the lumen to be recycled into the cell. Back diffusion of K^+ down its concentration gradient creates a lumen-positive potential difference across the epithelial cell. The positive potential drives Na^+ passively through the tight junction between cells. Cl^- movement from cell to peritubular fluid is driven by the strong intracellular negative electrical potential. The net result of this complex series of events is reabsorption of two Na^+ and two Cl^- by the active transport of one Na^+ from cell to peritubular fluid.

The TAL is water impermeable even though it is highly permeable to Na^+. The medullary segment of the TAL participates in the counter-current concentrating mechanism by depositing NaCl into the medullary interstitium. NaCl reabsorption in this segment is stimulated by renal nerves acting on β adrenergic receptors and possibly also by vasopressin.

At the end of the ascending limb, where it rejoins its glomerulus, the luminal concentrations of Na^+, Cl^-, and K^+ are low and osmolality is $^1/_3$ of that in proximal tubular fluid.

Distal Tubules, Collecting Tubules, and Collecting Ducts

In the late distal tubule, collecting tubules, and collecting ducts Na^+ reabsorption, controlled by aldosterone, can reduce urine Na^+ to less than 50 m*M*. With severe salt depletion urine Na^+ concentration is less than 10 mmol/l. Aldosterone acts primarily on collecting tubules where it increases Na^+ reabsorption, K^+ secretion, and H^+ secretion. These changes are produced by or associated with increased luminal permeability to Na^+, H^+, and K^+, increased H^+-ATPase activity, Na-K-ATPase activity, and citrate synthase activity. The enhancement of Na^+ permeability and citrate synthase activity requires *de novo* protein synthesis, but the effects on K^+ and H^+ permeability do not. Increased Na-K-ATPase activity is a prominent response, but appears to be secondary to increased Na^+ entry into the cell.

The kidney escapes from the Na^+ retaining action of sustained excess aldosterone after several days. Escape occurs after salt and water retention has increased extracellular fluid volume and blood pressure. Decreased reabsorption in Henle's loop, due perhaps to altered peritubular pressure and blood flow in the outer medulla, restores the balance between intake and output of salt and water. K^+ secretion continues to be stimulated even though the Na^+ balance has been restored; thus, the patient with hyperaldosteronism becomes progressively K^+ depleted.

Cl^- reabsorption in the late distal tubule and collecting tubules is stimulated by β-adrenergic agonists and may be controlled prostaglandins.

HCO_3^- is both secreted and reabsorbed as an anion in this part of the nephron, but secretion probably plays a minor role in the regulation of HCO_3^- excretion. In patients eating a North American acid-ash diet, very little HCO_3^- reaches the distal tubule. Protons secreted in the distal tubule combine with HCO_3^- to form carbonic acid, but since there is no carbonic anhydrase in the distal tubular lumen, carbonic acid is not dehydrated to CO_2 and water as it is in the proximal tubule. The carbonic acid is excreted in urine where it gradually breaks down to release CO_2, thereby raising urine pCO_2. This delayed hydration of carbonic acid makes it possible to estimate distal proton secretion from the difference between urine and blood pCO_2 (urine-blood pCO_2) when the patient is excreting a relatively alkaline urine-containing unreabsorbed HCO_3^-. The validity of this estimate is not accepted by all renal physiologists.

POTASSIUM

K^+ reabsorption in the proximal tubule keeps the luminal concentration equal to peritubular fluid as water is reabsorbed. Extensive reabsorption in the TAL lowers the K^+ concentration to less than 1 mmol/l. In the distal tubule and collecting tubules K^+ can be secreted by principal cells to create seven- to tenfold higher concentration in tubular fluid than in blood. Due to distal secretion K^+ excretion can be more than twice the quantity of K^+ filtered by the glomerulus.[8] The urinary K^+ concentration increases as water is reabsorbed in the collecting ducts. A useful estimate of K^+ secretion in the cortical collecting tubules can be obtained by correcting for urine osmolality as follows: urine/plasma (K^+/osmolality). Values greater than 7 indicate significant aldosterone-stimulated K^+ secretion. This correction only applies to hypertonic urine.

Distal secretion depends upon diffusion of K^+ from the cell where concentration is high (>100 mmol/l) to the tubular lumen. The rate of diffusion varies with luminal membrane permeability and the electrochemical gradient for K^+. Na^+ reabsorption creates a negative potential in distal tubular fluid. This potential is dissipated by entry of either K^+ or H^+ into the lumen or Cl^- exit from the lumen. Thus, K^+ secretion depends upon Na^+ reabsorption and competes indirectly with H^+ secretion and Cl^- reabsorption. Aldosterone increases the negative electrical potential; it may also be enhanced by the presence of nonreabsorbed anions such as sulfate or certain drugs (e.g., penacillamine). Excretion of these non-reabsorbed anions may increase K^+ secretion.

Tubular fluid flow rate limits the quantity of K^+ that can be secreted before the electrochemical diffusion gradient is dissipated. Flow of tubular fluid through the distal nephron is reduced by NaCl depletion or extracellular volume depletion or any disease state that reduces GFR and stimulates reabsorption in the proximal tubule. All of these conditions limit K^+ secretory capacity. Aldosterone is the major hormonal stimulus for K^+ secretion, but it is ineffective in the absence of adequate fluid delivery to the distal tubule; thus, extracellular fluid depletion may stimulate aldosterone release without increasing K^+ excretion. K^+ retention and hyperkalemia may develop as a result of reduced distal sodium and water delivery. Hyperkalemia may also develop when aldosterone release is impaired as in Addison's disease.

Aldosterone stimulates K^+ secretion in several ways. The primary action appears to be by increasing luminal K^+ and Na^+ permeability. This facilitates K^+ exit and Na^+ entry across the luminal membrane. Increased intracellular Na^+ secondarily increases Na^+,K^+,ATPase activity in the basolateral membranes of distal tubule and collecting tubule cells. Consequently, intracellular K^+ tends to rise, lumen-negative potential is increased, and permeation of K^+ from cell to lumen is enhanced.

Chronic ingestion of a high K^+ diet increases the capacity of the kidney to secrete K^+.[9] Adaptation is associated with increased infolding of the basolateral membranes and increased Na^+,K^+,ATPase activity in the chief cells in distal tubules and collecting tubules. Aldosterone is required for this adaptation to occur.

Glucocorticoids do not directly stimulate K^+ secretion, but since they increase GFR they act permissively to maintain distal fluid delivery.

Vasopressin increases K^+ secretion in the collecting tubules and improves K^+ tolerance. During a solute diuresis, due, for example, to glucosuria, systemic levels of AVP are high and K^+ secretion is promoted. During a water diuresis AVP levels are low and urinary K^+ excretion is lower.

β-1 Adrenergic agonists stimulate distal Cl^- reabsorption and inhibit K^+ excretion. Circulating epinephrine is most likely to produce this effect, but renal nerves may also modify K^+ excretion.

ACID-BASE EXCRETION

The kidney excretes acid buffered by NH_3 and weak bases, primarily phosphate.[22] Very little free H^+ is excreted because the minimal urine pH is 4.5. In the proximal tubule Na^+-H^+ countertransport and some ATP-dependent H^+ secretion reduce tubular fluid pH to approximately 6.5. At this pH most of the filtered bicarbonate is converted to CO_2 and reabsorbed. This reaction is accelerated by carbonic anhydrase in the brush border of proximal tubular cells. Some weak acids are also titrated to their un-ionized form at this pH. Increased intracellular pCO_2 or decreased intracellular pH promotes the secretion of H^+ and reabsorption of HCO_3^-.

In the intercalated cells of distal tubules and collecting tubules H^+ secretion is ATP dependent. These cells are filled with carbonic anhydrase and are structurally distinct from the principal cells that carry out K^+ secretion. Most of the secreted H^+ is buffered either by phosphate and other weak acids with appropriate pKa's such as β hydroxybutyric acid or by ammonia. The quantity of acid buffered by phosphate etc. can be measured by back titration to pH 7.4 and it is called titratable acid for this reason. Excretion of titratable acid depends not only on H^+ secretion, but also on availability of the buffers delivered from the proximal tubule.

Normally, little HCO_3^- escapes from the proximal tubule to enter the distal nephron. Ingestion of large quantities of HCO_3^- can increase distal HCO_3^- delivery to the point that significant amounts of HCO_3^- appear in the urine. According to the Henderson-Haselbach equation and the pK for HCO_3^-/H_2CO_3 of 6.1 there will be significant bicarbonaturia when urine pH exceeds pH 6.5.

The kidney appears to have an unlimited capacity to excrete bicarbonate if GFR is adequate. Excretion of HCO_3^- is achieved by reduced reabsorption in the proximal tubule. The threshold plasma HCO_3 concentration which saturates proximal tubular reabsorption is influenced by intracellular pH and by pCO_2. Above this threshold any increase in filtered HCO_3^- escapes to enter the distal tubule. Since H^+ secretion by the distal nephron is limited it is easily overwhelmed by increased delivery of HCO_3^- from the proximal tubule. It is almost impossible for a normal person to overwhelm the renal bicarbonate excretory capacity by ingestion of HCO_3^-.

Ammonia

Ammonia is produced by deamination and deamidation of glutamine in the proximal tubule.[10] The 2-oxoglutarate produced in this process must be disposed of either by oxidation or by conversion to glucose. ATP produced by oxidation of 2-oxoglutarate is used almost exclusively for transport of Na^+; there is, therefore, a correlation between Na^+ reabsorption and ammonia production.[14] Gluconeogenesis provides another means of removing 2-oxoglutarate without generating excess ATP. Acidosis stimulates both ammoniagenesis and gluconeogenesis in the proximal tubule. With chronic acidosis the ammonia-generated capacity of the proximal tubule increases tenfold. Ammonia production is dependent upon an adequate supply of glutamine from muscle.

Almost all the ammonia excreted in urine is present in tubular fluid at the end of the proximal tubule.[10] Ammonia enters the tubular lumen largely by diffusion, but its selective accumulation indicates some specific secretion of either NH_3 or NH_4^+ into the proximal tubular lumen. The pKa of NH_4^+/NH_3 is approximately 9; therefore, at pH 7 the ratio is 100:1 and at pH 6 it is 1000;1. Assuming diffusion equilibrium for the non-ionized NH_3 molecule there will be tenfold more NH_4^+ in the tubule lumen at pH 6 than in the cell at pH 7. This diffusion equilibrium accounts for most of the NH_4 secretion in the proximal tubule, but there is also evidence of some preferential selective secretion.

Some NH_4^+ is reabsorbed in the thick ascending limb. This transport probably depends upon NH_4^+ substituting for K^+ in reactions with Na^+,K^+,ATPase and perhaps on the Na^+,K^+,$2Cl^-$ carrier of the thick ascending limb. The reabsorbed NH_4^+ diffuses into the descending thin limbs and is trapped by counter-current flow to create a high NH_3 concentration in the medulla. NH_3 diffuses from the medulla into the collecting ducts when it is trapped and excreted as NH_4^+.

The amount of NH_4^+ excreted depends upon production in the proximal tubule, countercurrent trapping in the medulla, and luminal flow and pH in the collecting ducts.

Chronic renal disease or loss of nephrons reduces NH_4^+ excretion by limiting production in the proximal tubule and by impairing NH_4^+ transfer from medulla to collecting ducts.

Renal Tubular Acidoses (RTA)

Impaired renal H^+ excretion results from several different defects which can occur alone as acquired or inherited defects or as part of more generalized renal abnormalities as in the Fanconi syndrome.[4]

The most common cause of defective H^+ excretion is advanced renal failure. In this condition, distal H^+ secretion may still be able to lower urine pH, but ammonia production is reduced by the loss of proximal tubular cells, and bicarbonate reabsorption in the proximal tubule is reduced.

Other less common forms of RTA are related to defective distal H^+ secretion (type 1), defective proximal bicarbonate reabsorption (type 2), a mixture of proximal and distal defects (type 3), or inability to respond to aldosterone (type 4). Type I to III are associated with urinary K^+ loss and hypokalemia; type IV is associated with hyperkalemia.

In patients with distal RTA, Na^+ and K^+ replace H^+ and NH_4^+ and are excreted along with unreabsorbed anions such as phosphate.

In patients with proximal RTA large quantities of HCO_3^- are not reabsorbed and the unreabsorbed HCO_3 overwhelms the H^+-secretory capacity of the distal nephron. Consequently, the unreabsorbed HCO_3 is excreted with Na^+ and K^+. The secondary hyperaldosteronism which develops as a result of volume depletion increases distal K^+ secretion and exacerbates the hypokalemia.

Type IV RTA occurs as part of reduced aldosterone secretion in Addison's disease or due to defective renal responsiveness to aldosterone.[7] Impaired distal responsiveness to aldosterone may be an inherited or acquired defect.

URINE CONCENTRATION

The driving force for urine concentration is provided by the medullary (TALs) which reabsorb salt without water to raise the salt content of outer medullary interstitial fluid. When concentrated urine is being produced the interstitial solute concentration increases to the tip of the papilla. When dilute urine is being produced osmolality of the papilla falls below that in the outer medulla (see the next section).

The essential structural feature of renal-concentrating mechanisms is counter-current tubular and capillary flow in the medulla and papilla (Figure 2). Tubular fluid flows down the descending limb of Henle's loop, which penetrates into the medulla and then reverses sharply to flow up the ascending thin and thick limbs. After passing through the cortical distal tubules and collecting tubules, tubular fluid re-enters the medulla in the collecting

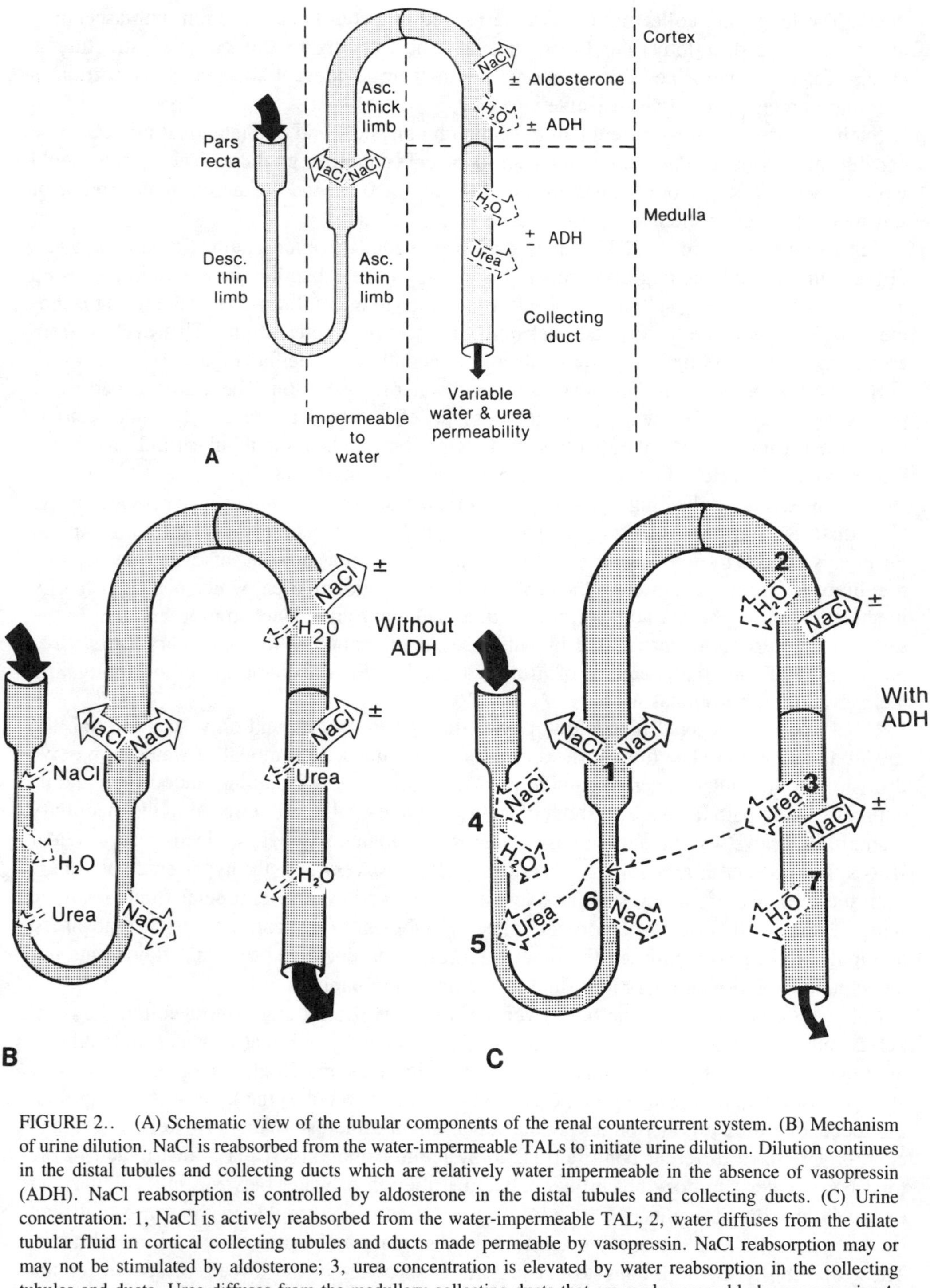

FIGURE 2.. (A) Schematic view of the tubular components of the renal countercurrent system. (B) Mechanism of urine dilution. NaCl is reabsorbed from the water-impermeable TALs to initiate urine dilution. Dilution continues in the distal tubules and collecting ducts which are relatively water impermeable in the absence of vasopressin (ADH). NaCl reabsorption is controlled by aldosterone in the distal tubules and collecting ducts. (C) Urine concentration: 1, NaCl is actively reabsorbed from the water-impermeable TAL; 2, water diffuses from the dilate tubular fluid in cortical collecting tubules and ducts made permeable by vasopressin. NaCl reabsorption may or may not be stimulated by aldosterone; 3, urea concentration is elevated by water reabsorption in the collecting tubules and ducts. Urea diffuses from the medullary collecting ducts that are made permeable by vasopressin; 4, NaCl diffuses into the descending thin limbs from the medullary interstitium; 5, urea diffuses into the descending thin limbs and is recycled especially in short-looped nephrons. The high concentration of NaCl and urea in the medullary interstitium draws water from descending thin limbs. Entry of NaCl and urea plus exit of water increases osmolality of the descending limb fluid; 6, NaCl diffuses out of the ascending thin limb as it returns to the cortex; 7, water diffuses from the medullary collecting ducts into the hypertonic medullary interstitium. (From ABCD, Gornall, A. G., Ed. With Permission.)

ducts. The loops and collecting ducts are arranged in a coordinated fashion with ascending and descending postglomerular blood vessels. The integrity of this complex structure is crucial for effective urine concentration. Maximal urine-concentrating capacity correlates with the number and length of Henle's loops.[1]

Such a counter-current system can be shown by physical and mathematical models to be capable of producing and maintaining an axial concentration gradient. The system could work in several ways, but all require counter-current flow and the active reabsorption of salt without water in the ascending limb.

Salt reabsorbed from the TAL enters the outer medullary interstitium. Some of this salt diffuses into the descending thin limb and is carried into the papilla; then as fluid passes up the thin ascending limb, salt diffuses back from lumen to interstitium. When the fluid reaches the TAL, salt is actively secreted back into the medullary interstitium. This cycling from ascending to descending limbs traps salt in the medulla and papilla (Figure 2).

Removal of salt from the TAL occurs without water reabsorption. The unreabsorbed water passes on to the cortex where, in the presence of vasopressin, water (but not urea) is reabsorbed from the cortical-collecting ducts to produce an isotonic tubular fluid that is rich in urea. This urea-rich fluid enters the medullary collecting ducts.

In the medullary collecting ducts vasopressin increases urea as well as water permeability. Urea diffuses from the collecting ducts to the medullary and papillary interstitium. Some of the urea is trapped by counter-current blood flow in the papilla and some enters the descending thin limbs to be recycled through the nephron. This recycling of urea, which occurs primarily in short-looped nephrons, adds a large cortico-medullary urea concentration gradient to the salt concentration gradient created by active salt reabsorption in the medullary TAL. Urea recycling facilitates the excretion of urine with high urea concentration without obligating the excretion of additional water.

Ninety percent of renal blood flow passes through the cortex and only 10% through the medulla and papilla. The high cortical blood flow is suited for removal of water by passive diffusion from tubular lumen to capillaries. The relatively low medullary blood flow and its counter-current path limit the washout of salt from the medulla. Salt trapped in the medullary interstitium draws water from the descending thin limb and papillary collecting duct. Water leaves the medulla in ascending venous capillaries as a very slightly hypertonic solution.

In the presence of vasopressin passive water reabsorption renders tubular fluid isosmotic with plasma by the end of the cortical collecting tubules. Water continues to be reabsorbed as tubular fluid passes through the medullary-collecting ducts and the final fluid urine has the same osmolality as interstitial fluid at the tip of the papilla.

Reflux of concentrated urine back over the tip of the papilla that protrudes into the renal pelvis, probably contributes to the concentrating process by removal of water and addition of solutes to the upstream portions of the collecting ducts and Henle's loop.

Urine osmolality depends upon the amount of urea delivered to the kidney. A low protein diet decreases urea production and reduces maximum urine osmolality. However, urea and electrolytes make different contributions to systemic osmolality. Urea is distributed in the total body water and does not influence the distribution of water between intracellular and extracellular compartments. Na^+ and Cl^- are "effective" osmoles in the sense that they remain extracellular and thereby determine the distribution of water across cell membranes. Thus, the ability of the kidney to concentrate electrolytes in the urine is essential for the maintenance of effective systemic osmolality. The ability to concentrate urea is advantageous because it permits excretion of a major waste product in a small volume of water.

Maximal urine concentration of salt is reduce by a low protein diet. It appears that urea also amplifies the effectiveness of the counter current system for concentrating salt in the papilla. The exact mechanism is not known, but may involve an osmotic effect on water reabsorption from either the descending thin limb or the collecting ducts.

In summary, urine concentration depends upon (1) active reabsorption of salt from medullary TAL, (2) counter-current trapping of salt and urea in the Henle's loops and capillaries of the medulla, (3) vasopressin to increase water permeability and reabsorption in the cortical-collecting ducts, and (4) vasopressin to increase water and urea reabsorption in the medullary-collecting ducts. It is important to note that most of the water reabsorption occurs in the cortex. Production of concentrated urine requires low urine flow, usually less than 1% of the GFR. In humans the maximum urine concentration is three to four times the plasma osmolality. The solutes required to produce this concentration are contained in an isotonic fluid volume three to four times that of the final urine; therefore, isotonic tubular fluid flow from the cortical collecting ducts to the medullary collecting ducts needs to be only 4% of the GFR. In other words, 96% of the fluid filtered in the glomerulus is reabsorbed before it enters the medullary collecting ducts. At least 60% of this reabsorption occurs in the proximal tubule and 20% in Henle's loop. The remainder occurs in the distal tubules and cortical collecting ducts.

Osmotic Diuresis

The excretion of urea can be contrasted with that of non-reabsorbable solutes such as mannitol. Mannitol, which is sometimes given to increase urine flows, is filtered by the glomerulus but not reabsorbed. Because the proximal tubule is very water permeable, water remains in the lumen with the unreabsorbed mannitol and tubular fluid is isosmotic with peritubular plasma. In the descending part of Henle's loop luminal mannitol hinders water reabsorption usually driven by the high interstitial NaCl concentration. The resulting increase in flow through the tubule washes solutes out of the medulla and papilla and impairs the counter-current exchange system. There is no passive recycling of mannitol through the counter-current system; consequently, as mannitol enters the renal pelvis from the papillary-collecting ducts it obligates the excretion of water. Other solutes are drawn along with the flow.

The most common cause of an osmotic diuresis is unreabsorbed glucose due to hyperglycemia. An i.v. infusion of urea augments urine flow, but because of recyling in the papilla the osmotic diuretic effect is much less than that of mannitol or unreabsorbed glucose. During an osmotic diuresis the volume of water entering medullary-collecting ducts increases and washes out the solutes from the medulla.

Vasopressin

When vasopressin binds to its receptor on the basolateral surface of collecting tubule cells, adenylate cyclase is activated to produce cAMP. cAMP stimulates protein kinases that modify water permeability in both cortical and medullary-collecting tubules and ducts. Urea permeability of medullary-collecting ducts is also increased in some species. Increased water permeability results from insertion of water channels into the luminal membrane; this process depends upon cytoskeletal alterations.

The actions of vasopressin are antagonized by prostaglandins (PGE_2) released by receptor activation. PGE_2 inhibits adenylate cyclase. Hypercalcemia interferes with vasopressin binding to its receptor and activates a phosphodiesterase that destroys cAMP (Table 2).

Any form of moderate to severe renal failure decreases the ability of the kidney to respond to vasopressin. This is most evident in tubulo-interstitial or obstructive forms of renal disease. The loss of responsiveness is partly due to structural changes, most important being the loss of nephrons, that impair the counter-current system.

Normally, vasopressin secretion is associated with increased plasma osmolality with or without volume depletion.[21] Under these circumstances less than 5% of the filtered water and solute enters the medullary-collecting ducts. However, excess vasopressin secretion by some tumors, most commonly oat cell carcinomas of the lung, causes water retention and

Table 2
CAUSES OF ENHANCED AND IMPAIRED RENAL REPONSE TO VASOPRESSIN

	Impaired
Congenital	Nephrogenic diabetes insipidus
Acquired	
Structural damage	Advanced renal failure, postobstructive diuresis
Osmotic diuresis	Hyperglycemia, mannitol, urea
Electrolyte disorders	Hypercalcemia, hypokalemia
Dietary	Excess water intake, low salt intake, low protein intake
Drugs	Lithium, demeclocycline, furosemide, α-2 adrenergic agonists
Other	Sickle cell anemia
	Enhanced
Low blood pressure	
Prostaglandin inhibition	Chlorpropamide and non-steroidal anti-inflammatory agents

hyponatremia (syndrome of inappropriate ADH, SIADH). Hypervolemia and hyponatremia may increase GFR and reduce reabsorption in the proximal tubule; this increases delivery of fluid to the medullary-collecting ducts. Increased delivery of fluid to the medullary ducts washes out the medullary solute concentration gradient and reduces urine concentration. In this way patients may escape from the effects of chronic excess vasopressin.[5]

A number of drugs antagonize the actions of vasopressin either directly at the cellular level or by changing fluid delivery to the collecting ducts (Table 2). For example, lithium interferes with the vasopressin effect on water permeability of collecting-duct cells and may also reduce water reabsorption in the proximal tubule. A number of other drugs may reduce the response to vasopressin by their effects on reabsorption from the proximal tubule (*cis*-platin) or loop (furosemide). Inherited insensitivity to vasopressin is called nephrogenic diabetes insipidus.

The response to vasopressin may be modestly increased by preventing the release of PGE_2. Chlorpropamide and some nonsteroidal, antiinflammatory agents act in this way. Increased protein intake by augmenting GFR and urea production may also improve the response to vasopressin. Blood pressure by its effect on GFR and distal tubular fluid delivery is a significant modulator as well (see later).

Urine Dilution

Urine dilution begins in the TAL where Na^+ and Cl^- are extensively reabsorbed to create a hypotonic fluid (100 mOsm/kg) that enters the distal tubule. In the absence of vasopressin in the distal tubules, collecting tubules, and collecting ducts are relatively water impermeable and little water is reabsorbed. However, it must be remembered that collecting tubules and ducts are not completely water impermeable; therefore, during water diuresis some of the large volume of dilute fluid entering the medullary collecting ducts is reabsorbed. This reabsorbed water dilutes NaCl deposited by the TAL in the medullary interstitium. Urine dilution proceeds despite the reabsorption of water from medullary collecting ducts, because NaCl reabsorption exceeds the rate of water reabsorption. The final urine Na^+ may be less than 10 mmol/l.

Hypotonic fluid left within the tubular lumen will eventually reach osmotic equilibrium with peritubular fluid even in the total absence of vasopressin. This means that the ability to produce a dilute urine depends upon an adequate flow of fluid through the distal nephron. Flow dependence of urine dilution can be seen in patients with total diabetes insipidus (no vasopressin). When these patients are volume depleted, blood pressure and GFR fall, proximal tubular reabsorption increases, and distal fluid delivery is severely curtailed. Urine osmolality in such patients rises slightly above plasma levels.

The capacity to excrete a dilute urine (free water) is compromised by reduced distal delivery of water and salt. This occurs commonly in patients with heart failure in whom GFR falls and proximal reabsorption increases. Similar changes can occur with hepatic failure or any condition reduced GFR and increases proximal reabsorption such as hypothyroidism.

EXTRACELLULAR FLUID REGULATION BY THE KIDNEY

The major factors regulating normal renal response to fluid imbalances are blood pressure, renin-angiotensin-aldosterone, vasopressin, and the autonomic nervous system. The pattern of normal responses to changes in salt intake suggest that a sensory organ(s) monitors the sodium content of the body and sends a signal(s) to a separate organ that compares the signal with a set point for sodium homeostasis. The sodium error signal resulting from this comparison is sent to a control organ which transmits a controlling signal to the kidney to regulate sodium excretion. Surprisingly little is known about the sensors, the comparitors, and their relation to the controlling signals. Some models have been made using aldosterone as the sole control for maintenance of body sodium stores, but it is evident that vasopressin, renin-angiotensin, the sympathetic nervous system, and atrial natriuretic factor are also involved.[6,24] Table 3 lists other hormones that may influence renal function.

Altered sensory input to a comparitor due to redistribution of circulating blood flow appears to explain renal fluid retention in patients with heart failure or hypoalbuminemia. These disease states disrupt fluid distribution across capillary beds because either hydrostatic pressure is increased or oncotic pressure is reduced. They lead to reduced effective circulating blood volume. Reduction of blood volume and pressure is detected by volume receptors and baroreceptors. These receptors activate the renin-angiotensin-aldosterone system, vasopressin release, and sympathetic nervous activity. When a new, but edematous, steady state is achieved these regulatory systems are not excessively active and can respond to daily fluctuations in fluid and solute balance.

Defects can also develop as a result of altered comparator function. For example, damage to the centers regulating vasopressin release can lead to hypernatremic states due to the loss of osmotic sensory control.

Of course, derangements of renal function can impair the response as well. The normal half-time for adjusting urinary salt excretion to an abrupt change in salt intake is 1 to 1.5 d. This time may be prolonged in hypertensive patients and by renal disease.

The mechanisms regulating extracellular fluid volume, osmolality, and pH cannot be separated. In the proximal tubule, where most of the filtrate is reabsorbed, Na^+ and HCO_3^- reabsorption and H^+ excretion are linked by a common transport system; ammonia production depends upon Na^+ reabsorption to consume the 2-oxoglutarate produced by deamination and deamidation of glutamine. In the distal nephron Na^+, H^+, Cl^-, and K^+ reabsorption and secretion are intertwined and modified by aldosterone. Excretion of a dilute urine requires Na^+ reabsorption in the TAL and collecting tubules. Conservation of water by excretion of a concentrated urine also requires Na^+ reabsorption. Vasopressin, the hormone that increases water permeability of the collecting tubules and ducts, is also a potent vasoconstrictor and may have a role in regulation of blood pressure and GFR. In some animals vasopressin also stimulates Na^+ reabsorption and K^+ secretion. The counter-current system that concentrates

Table 3
HORMONAL REGULATION OF SALT AND WATER EXCRETION BY THE KIDNEY

Angiotensin II	Increases proximal Na^+ reabsorption
α-1 agonists	Increase proximal Na^+ reabsorption
β-agonists	Increase distal Cl^- reabsorption Increase distal K reabsorption
Aldosterone	Increases distal Na^+ reabsorption and K^+ secretion
Vasopressin	Increases H_2O permeability in collecting ducts Increases urea permeability in medullary collecting ducts Increases K^+ secretion
Atrial natriuretic hormone	Increases GFR, decreases fractional Na^+ reabsorption Antagonizes renin and aldosterone release
Progesterone	Antagonizes aldosterone
Dopamine	Increases GFR
Insulin	Increases PO_4^- and Na^+ reabsorption
Glucagon	Decreases Na^+ and PO_4^- reabsorption
Somatostatin	Antagonizes the hydroosmotic effect of vasopressin
Glucocorticoids	Increase GFR, permissive role in Na reabsorption and urine dilution
Parathyroid hormone	Decreases proximal PO_4^-, Na^+, $HCO3^-$ reabsorption Increases distal Ca^{++} reabsorption
Thyroid hormones	Increase renal growth, GFR, renal Na^+-K^+-ATPase, free water excretion
Calcitonin	Decreases distal Ca^+, Na^+, and PO_4^- reabsorption

salt and urea also concentrates ammonia for urinary excretion. Thus, a disturbance in one function ramifies through the whole regulatory system.

Without adequate blood pressure and GFR none of these functions can occur. Without adequate delivery of fluid to the distal nephron all the subtle regulatory mechanisms are compromised. Heart failure can show the consequences of inadequate distal delivery. As the heart fails blood pressure falls and increased sympathetic nervous activity constricts the efferent arterioles to reduce renal blood flow and raise filtration fraction. Glomerulo-tubular balance is reset by altered peritubular physical conditions favoring reabsorption in the proximal tubule. Distal delivery of fluid and the ability to produce dilute urine decrease; consequently, hyponatremia may develop. Decreased tubular fluid flow reduces the ability to secrete K^+ despite high levels of aldosterone. Thus, heart failure impairs the capacity of the structurally intact kidney to regulate salt, potassium, and water balance. Hypovolemic shock, hepatic failure, and hypoalbuminemia are other examples of diseases that impair the kidney's ability to regulate homeostasis.

REFERENCES

1. **Bankir, L. and de Rouffignac, C.,** Urinary concentrating ability: insights from comparative anatomy, *Am. J. Med.*, 249, R643, 1985.
2. **Bergstrom, J., Ahlberg, M., and Alvestrand, A.,** Influence of protein intake on renal hemodynamics and plasma hormone concentrations in normal subjects, *Acta Med. Scand.*, 217, 189, 1985.

3. **Brenner, B. M.,** Nephron adaptation to renal injury or ablation, *Am. J. Physiol.,* 249, F324, 1985.
4. **Chan, J. C. M. and Alon, U.,** Tubular disorders of acid-base and phosphate metabolism, *Nephron,* 40, 257, 1985.
5. **Cooke, C. R., Turin, M. D., and Walker, W. G.,** The syndrome of inappropriate antidiuretic hormone secretion (SUADH): pathophysiologic mechanisms in solute and volume regulation, *Medicine,* 58, 240, 1979.
6. **Daniels, F. H., Cortell, S., and Leonard, E. F.,** Formulating and evaluating quantitative models of control of sodium stores, *Am. J. Physiol.,* 249, R624, 1985.
7. **DeFronzo, R. A.,** Hyperkalemia and hyporeninemic hypoaldosteronism, *Kidney Int.,* 17, 118, 1980.
8. **Field, M. J. and Giebisch, G. J.,** Hormonal control of renal potassium excretion, *Kidney Int.,* 27, 379, 1985.
9. **Garg, L. C. and Narang, N.,** Renal adaptation to potassium in the adrenalectomized rabbit, *J. Clin. Invest.,* 76, 1065, 1985.

9a. **Gmaj, P. and Murer, H.,** Cellular mechanisms of inorganic phosphate transport in kidney, *Physiol. Rev.,* 66, 36, 1986.

10. **Good, D. W. and Knepper, M. A.,** Ammonia transport in the mammalian kidney, *Am. J. Physiol.,* 17, F459, 1985.
11. **Greger, R.,** Ion transport mechanisms in thick ascending limb of Henle's loop, *Physiol. Rev.,* 65, 658, 1985.
12. **Guder, W. G. and Ross, B. D.,** Enzyme distribution along the nephron, *Kidney Int.,* 26, 101, 1984.
13. **Guyton, A. C., Coleman, T. G., Cowley, A. W., Jr., Manning, R. D., Jr., Norman, R. A., Jr., and Ferguson, J. D.,** A systems analysis approach to understanding long-range arterial blood pressure control and hypertension, *Circ. Res.,* 35, 159, 1974.
14. **Halperin, M. L. and Jungas, R. L.,** Metabolic production and renal disposal of hydrogen ion, *Kidney Int.,* 24, 709, 1983.
15. **Harris, P. J. and Navar, L. G.,** Tubular transport responses to angiotensin, *Am. J. Physiol.,* 248, F621, 1985.
16. **Laragh, J. H.,** Atrial natriuretic hormone, the renin-aldosterone axis, and blood pressure-electrolyte homeostasis, *N. Engl. J. Med.,* 313, 1330, 1985.
17. **Mahnensmith, R. L. and Aronson, P. S.,** The plasma membrane sodium-hydrogen exchanger and its role in physiological and pathophysiological processes, *Circ. Res.,* 57, 773, 1985.
18. **Mizagala, C. L. and Quamme, G. L.,** Renal handling of phosphate, *Physiol. Rev.,* 65, 431, 1985.
19. **Rector, F. C., Jr.,** Sodium, bicarbonate and chloride absorption by the proximal tubule, *Am. J. Physiol.,* 244, F461, 1983.
20. **Schnermann, J. and Briggs, J.,** Concentration-dependent sodium chloride transport as the signal in feedback control of glomerular filtration rate, *Kidney Int.,* 22, S82, 1982.
21. **Schrier, R. W., Berl, T., and Anderson, R. J.,** Osmotic and nonosmotic control of vasopressin release, *Am. J. Physiol.,* 236, F321, 1979.
22. **Schwart, W. B. and Cohen, J. J.,** The nature of the renal response to chronic disorders of acid-base equilibrium, *Am. J. Med.,* 64, 417, 1978.
23. **Tannen, R. L.,** Relationship of renal ammonia production and potassium homeostasis, *Kidney Int.,* 11, 453, 1977.
24. **Walser, M.,** Phenomenological analysis of renal regulation of sodium and potassium balance, *Kidney Int.,* 27, 837, 1985.

GENERAL REFERENCES

1. **Seldin, D. W. and Giebisch, G.,** *The Kidney Physiology and Pathophysiology,* Vols. I and II, Raven Press, New York, 1985.
2. **Brenner, B. M. and Rector, F. C., Jr.,** *The Kidney,* Vols.I and II, 2nd ed., W.B. Saunders, Philadelphia, 1981.

THE SWEAT GLANDS IN MAINTAINING pH AND ELECTROLYTE BALANCE

Paul Wolf

INTRODUCTION

The sweat glands are important in the homeostatic maintenance of water and electrolyte balance and the maintenance of body temperature (Figure 1). The primary clinical purpose for performing biochemical analysis of sweat is in the diagnosis of cystic fibrosis. The sweat gland abnormality with increased sodium and chloride in the sweat was discovered in the early 1950s. Prior to the identification of the sweat chloride test, the diagnosis of cystic fibrosis was made on clinical data such as pancreatic insufficiency, malabsorption, and recurrent pulmonary infection.

FACTORS CONTROLLING THE VOLUME AND COMPOSITION OF SWEAT

The normal baseline sodium and chloride concentration in sweat is up to 40 meq/l. Various factors determine the volume and composition of sweat. Sweating is under control of the nervous system. Psychological stimuli such as apprehension and excitement are associated with increased sweating.[1] In addition, increased physical activity, especially in a warm environment, will result in increased perspiration.[2] However, with continuous exercise, if dehydration ensues, the concentration of sodium and chloride in sweat will decrease. Aldosterone has a direct effect on the sweat gland and influences the concentration of sodium and chloride in sweat.[3]

Diseases associated with increased serum aldosterone cause sweat sodium and chloride concentrations to decrease, in contrast to the increased concentration of sodium and chloride in sweat in Addison's disease, which is associated with decreased serum aldosterone levels.[4]

SUBSTANCES OTHER THAN Na, K, AND Cl IN SWEAT IN HEALTH AND DISEASE

Various other substances are present in sweat. These include nitrogenous compounds such as urea and creatinine.[5,6] In chronic renal failure there is a deposition of these substances in increased amounts in the skin which causes a uremic frost appearance. Other cations in sweat are potassium, calcium, iron, and zinc.[7,8] Continuous and persistent sweating has resulted in zinc deficiency and iron deficiency, especially when the patient is in a warm environment. Amino acids are also present in small amounts in sweat, while ammonia also appears in sweat.[9] Hydrogen ion combines with ammonia with the excretion of ammonium in sweat (Table 1).

Immunoglobulins are present in sweat and may be partially responsible for the antibacterial activity found there.[10-12] A number of enzymes are also present.[10] Both killakrein and amylase have also been identified in sweat.[10-13,31-37]

ABNORMALITIES IN SWEAT AND OTHER BODY FLUIDS IN CYSTIC FIBROSIS (FIGURE 2)

Mucoviscidosis (cystic fibrosis) is the most common genetic disorder of Caucasians, occurring in 1 in every 200 births. Ten million people in the U.S. carry the gene; 33,000 people have the disease. The genetic defect has not been identified. The disease appears

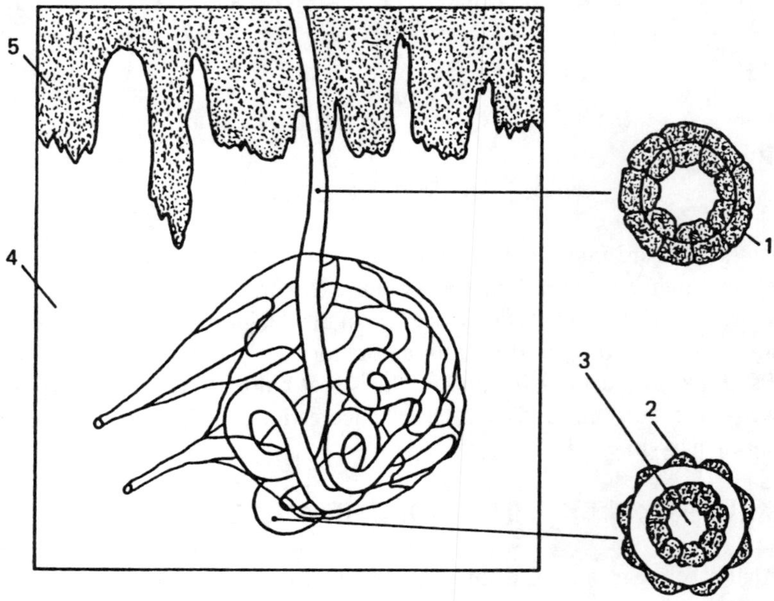

FIGURE 1. The eccrine sweat gland, composed of the secretory coil, which is surrounded by contractile myoepithelial cells, and a capillary network, is shown draining via the excretory duct. (1) Excretory duct, (2) myoepithelial cell, (3) secretory coil, (4) subcutaneous tissue, (5) epidermis. (Reproduced from Natelson, S. and Natelson, E., *Principles of Applied Clinical Chemistry,* Vol. I. Plenum Press, New York, 1975, 358. With permission.)

Table 1
COMPOSITION OF HUMAN SWEAT[31-36]

Component	Normal range	Component	Normal range
Volume/24 h/1.73 m²	600—700 ml	Amino acid N	1.1—10.2 mg/100 ml
pH	4.0—6.5	Nonprotein N	17—64 mg/100 ml
Specific gravity	1.001—1.006	Ammonia N	5—9 mg/100 ml
Sodium	15—52 meq/l	Urea N	8—30 mg/100 ml
Potassium	7—37 meq/l	Creatinine	8.4—1.3 mg/100 ml
Chloride	10—45 meq/l	Uric acid	0.5—2 mg/100 ml
Calcium	2—6 mg/100 ml	Nitrogen (total)	20—90 mg/100 ml
Magnesium	0.15—4.0 mg/100 ml	Lactic acid	30—80 mg/100 ml
Phosphorus	1—3 mg/100 ml	Thiamin	0.2—1.0 μg/100 ml
Sulfur	7—7.4 μg/100 ml	Riboflavin	0.1—0.7 μg/100 ml
Iron	10—20 μg/100 ml	Folic acid	0.5—0.9 μg/100 ml
Iodine	0.15—0.6 μg/100 ml	Ascorbic acid	40—100 μg/100 ml
Copper	2—12 μg/100 ml	Inositol	15—36 μg/100 ml
Glucose	3—23 mg/100 ml	Phenols	1.5—4 mg/100 ml
Manganese	3—8 μg/100 ml	Corticoids	4—8 μg/100 ml

Reproduced from Natelson, S. and Natelson, E. A., *Principles of Applied Clinical Chemistry,* Plenum Press, New York, 1975, 365. With permission.

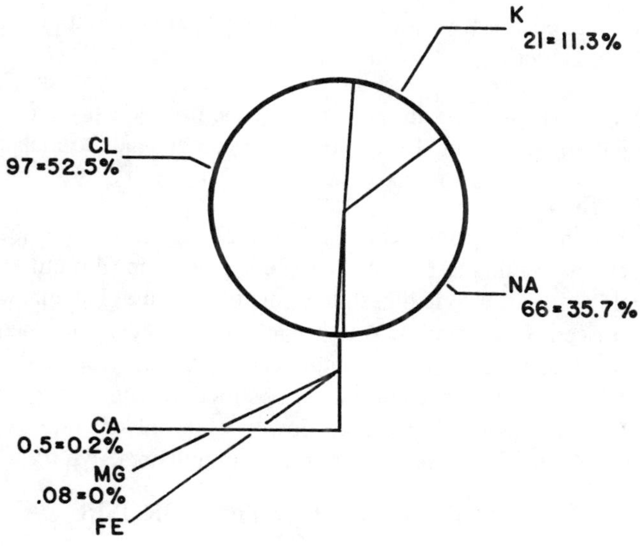

FIGURE 2. Chemical makeup of cystic fibrosis sputum in millimole per liter units. Presented in Poster Session at the Annual Meeting of the American Association of Clinical Chemists, Washington, D.C., July 31, 1984.

clinically only in the homozygous state. The major manifestations of the disease are related to chronic lung disease and malabsorption due to chronic pancreatic insufficiency.[14] An abnormality is present in the secretions of the exocrine glands.[15] In addition to the chronic pulmonary disease and pancreatic insufficiency, many of the major organs also develop serious abnormalities. The acute and chronic pulmonary infection leads to pulmonary abscesses and bronchiectasis.[16] Pulmonary fibrosis occurs with pulmonary hypertension, right ventricular hypertrophy, congestive heart failure, and death. Chronic hypercapnia leads to cerebro-vascular dilatation and cerebral edema. Hyperviscosity of bile leads to cholestasis, biliary cirrhosis, portal hypertension, and bleeding esophageal varices. Pancreatic insufficiency results in malabsorption. The pancreatic pathology leads to glucose intolerance. The early gastrointestinal manifestations include meconium ileus, intussusception, and meconium peritonitis. Other minor manifestations include nasal polyps and abnormalities in the reproductive system. Aspermia and decreased fertility are present. Obliteration of the vas deferens occurs. In young girls there is thick vaginal mucus secretion.

Hematological and Biochemical Abnormalities in Cystic Fibrosis

1. Hypoxia
2. Hypercapnia
3. Increased serum bicarbonate
4. Hyperglycemia
5. Cholestasis causing increased hepatic alkaline phosphatase
6. Decreased serum amylase
7. Decreased serum immunoreactive trypsin
8. Hypoalbuminemia
9. Osteomalacia with increased bone alkaline phosphatase
10. Amyloidosis of kidney secondary to chronic pulmonary infection; nephrotic syndrome with proteinuria and elevation of serum creatinine.[17]
11. Anemia of chronic disease

12. Neurophilic leukocytosis with left shift secondary to bacterial pneumonitis
13. Elevated sweat chloride and sodium

The cystic fibrosis patient early in the course of the disease shows a failure to thrive. A genetic familial pattern is present. Pancreatic insufficiency leads to malabsorption.

Tests for Cystic Fibrosis

A test which was popular many years ago for the diagnosis of cystic fibrosis has been shown to be unreliable; in this test a fecal sample is placed on film utilized in radiographs. If the feces contains trypsin, tryptic digestion of the film occurs. Patients with cystic fibrosis do not possess sufficent fecal trypsin. Another test which is currently being evaluated as a diagnostic test for cystic fibrosis is the immunoreactive trypsin test. In the neonate who has cystic fibrosis, there is early pancreatic ductal obstruction with regurgitation of immunorreactive trypsin into the neonate's plasma. However, the older child eventually develops atrophy and fibrosis of the pancreas and the serum immunoreactive trypsin is decreased.

DIAGNOSIS OF CYSTIC FIBROSIS

The gold standard for the diagnosis of cystic fibrosis is still the sweat test.[38,39] For the sweat chloride test to be reliable at least 100 mg of sweat must be collected. Currently, the sweat is obtained by a process termed iontophoresis. Pilocarpine is driven into the skin by an applied electrical voltage.[18] The pilocarpine stimulates the sweat glands. The sweat is collected on a cotton gauze. The sodium chloride is eluted from the gauze with deionized water. The chloride is measured quantitatively utilizing an ion-selective chloride electrode or by the Cotlove chloride titrator where all chloride ions are combined with silver ions with the formation of silver chloride as the end point.[19] In direct-reading electrodes attached to the skin, the procedure may at times be inaccurate because the sweat production rate may be too low.[20] Also, evaporation of sweat may occur causing an increased concentration in the direct-reading electrode test. In the more-complicated procedures, quantitative sodium chloride in fingernails or hair is determined.[21]

One must be cognizant that cystic fibrosis results in chloride values in sweat above 40 meq/l, which is the value which is consistent with the disease. The gray zone where the possibility of the disease exists is 40 to 60 meq/l. A value above 60 meq/l is highly suggestive of the disease. It is not diagnostic because several other conditions cause an elevated sweat chloride concentration. The condition which especially causes an increased sweat chloride other than cystic fibrosis is Addison's disease with decreased serum aldosterone. As previously discussed, aldosterone directly affects sweat sodium chloride concentration in hypoaldosterone disease states, the urine concentration of sodium and chloride is elevated, and potassium is decreased. Other conditions which cause an elevated sweat chloride are skin ectodermal dysplasia, cretin hypothyroid infants, diabetes insipidus of the nephrogenic type, and glycogen storage disease type 1. All of the above are clinically different from cystic fibrosis. A false-negative test occurs if insufficient sweat is obtained or in edematous hypoalbuminemic states.[22] Neonates do not have a high sweat rate; thus, the sweat chloride test is difficult to utilize in neonates.[23] Thus, one test which may substitute for the sweat chloride test in newborns is the meconium albumin test.[24] Since a cystic fibrosis neonate may lack proteases in the stool such as trypsin, albumin appears in the meconium. The albumin is derived from swallowed amniotic fluid which contains albumin.[25,26]

Thus, the sweat chloride test is an important test and should be requested in children with recurrent pneumonia, asthma, failure to thrive, or malabsorption.

GENETIC ABNORMALITY IN CYSTIC FIBROSIS

At the 1985 annual meeting of the American Society of Human Genetics, it was reported that a genetic abnormality or marker that appears to accompany the genetic defect has been identified. A general probe has been tested on the genes of 50 Canadian families who had had two or more children with cystic fibrosis. The probe binds to a genetic abnormality that appears to occur only in carriers of the cystic fibrosis defect on a strand of genetic material. To make the test accurate, a second abnormality lying in the opposite side of the cystic fibrosis defect must be found. By finding the two abnormalities or markers the cystic fibrosis gene may be identified; it has now been found that the genetic defect is located on chromosome 7.

SIGNIFICANCE OF THE CHEMICAL ABNORMALITIES IN THE SALIVA OF CYSTIC FIBROSIS PATIENTS

Recently, sputum specimens from cystic fibrosis patients were collected and they demonstrated increased concentrations of sodium, potassium, chloride, magnesium, calcium, iron, and peptide calcitonin and calmodulin.[27,28] A summary of the ionic results is reproduced in Figure 2. The clinical significance of increased concentrations of some of these ions in the pathophysiology of chronic lung disease has been established. The increased iron is necessary for promotion of bacterial growth. The increased magnesium may correlate with the presence of mucoid *Pseudomonas aeroginosa.* This mucoid coat enables this organism to resist phagocytosis and digestion by neutrophilic leukocytes.

The increased sodium, potassium, and chloride may be related to a defect in chloride transport in the cystic fibrosis fibroid lung.[29] The defect also may be due to a decreased sodium reabsorption. The elevated calmodulin and calcitonin in the sputum may be an expression of the genetic defect found in cystic fibrosis. The increased concentrations of these important calcium-regulating substances may thus lead to increased calcium in sputum, which combined with sputum dehydration causes an increased level of mucus.[40]

MANAGEMENT OF THE PATIENT WITH CYSTIC FIBROSIS

Calcium promotes the binding of mucus which promotes the proliferation of mucoid-coated bacterial species.[30] Calcium channel blockers such as Verapamil or Nifedipine could lower intracellular calcium levels and prevent mucus thickening and dehydration of sputum. By introducing these agents prior to the onset of lung infection, it may be possible to inhibit infection before it occurs. The calcium channel blockers have been successfully used to treat pulmonary hypertension which may complicate cystic fibrosis with the development of cor pulmonale.[31] Thus, the ionic composition of the cystic fibrosis sputum appears to be a selective environment for the colonization of the cystic fibrosis sputum by only a few bacterial species. The genetic defect appears to be a chloride ion channel defect affecting many tissues such as sweat glands, lungs, pancreas, and gastrointestinal and urogenital tracts.

REFERENCES

1. **Kennard, D. W.,** The nervous regulation of the sweating apparatus of the human skin and emotive sweating in thermal sweating area, *J. Physiol. (London),* 165, 457, 1963.
2. **van Beaumont, W. and Bullard, R. W.,** Sweating: its rapid response to muscular work, *Science,* 141, 643, 1963.
3. **Collins, K. J.,** The action of exogenous aldosterone on the secretion and composition of drug-induced sweat, *Clin. Sci.,* 30, 207, 1966.
4. **Shuster, S.,** Adrenal control of eccrine sweat secretion, *Proc. R. Soc. Med.,* 55, 719, 1962.

5. **Brusilow, S. W. and Gordes, E. H.,** Ammonia secretion in sweat, *Am. J. Physiol.*, 214, 513, 1968.
6. **Coltman, C. A., Jr., Row, N. J., and Atwell, R. J.,** The amino acid content of sweat in normal adults, *Am. J. Clin. Nutr.*, 18, 373, 1966.
7. **Prasad, A. S., Schulert, A. R., Sandstead, H. H., Miale, A., Jr., and Farid, Z.,** Zinc, iron, and nitrogen content of sweat in normal and deficient subjects, *J. Lab. Clin. Med.*, 62, 84, 1963.
8. **Calloway, D. H., Odell, A. C. F., and Margen, S.,** Sweat and miscellaneous nitrogen losses in human balance studies, *J. Nutr.*, 101, 775, 1971.
9. **Cier, J. F., Manuel, U., and Lacour, J. R.,** Electrophoretic studies of the proteins of human sweat with paper, gel and immunoelectrophoresis, *R. Soc. Biol.*, 157, 1623, 1963.
10. **Fraki, J. E., Jansen, C. T. and Hopsu Havu, V. K.,** Human sweat kallikrein. Biochemical demonstration and chromatographic separation from several other esteropeptidases in the sweat, *Acta Dermatol. Venereol.*, 50, 321, 1970.
11. **Page, C. O., Jr. and Remington, J. S.,** Immunologic studies in normal human sweat, *J. Lab. Clin. Med.*, 69, 634, 1967.
12. **Jirka, M. and Masoput, J.,** Immunochemical behavior of proteins in human sweat, *Biochem. Biophys. Acta*, 71, 217, 1963.
13. **Astrand, I.,** Lactate content in sweat, *Acta Physiol. Scand.*, 58, 359, 1963.
14. **Gurwitz, D., Corey, M., Francis, P. W. J., Crozier, D., and Levision, H.,** Perspectives in cystic fibrosis, *Pediatr. Clin. North Am.*, 26, 603, 1979.
15. **diSant Agnese, P. A. and Talamo, R. C.,** Pathogenesis and physiopathology of cystic fibrosis of the pancreas mucoviscidosis, *N. Engl. J. Med.*, 277, 1287, 1967.
16. **Lloyd-Still, J. D., Khaw, K. T., and Shwachman, H.,** Severe respiratory disease in infants with cystic fibrosis, *Pediatrics*, 53, 678, 1974.
17. **Biberstein, M., Wolf, P., Pettross, B., Fanestil, D., and Vazquez, M.,** Amyloidosis complicating cystic fibrosis, *Am. J. Clin. Pathol.*, 5, 752, 1983.
18. **Gibson, L. E. and Cooke, R. E.,** A test for concentration of electrolytes in sweat in cystic fibrosis of the pancreas utilizing pilocarpine by iontophoresis, *Pediatrics*, 23, 545, 1959.
19. **Kopito, L. and Shwachman, J.,** Studies in cystic fibrosis: determination sweat electrolytes in situ with direct reading electrodes, *Pediatrics*, 43, 794, 1969.
20. **Szabo, L., Kenny, M. A., and Lee, W.,** Direct measurement of chloride in sweat with an ion-selective electrode, *Clin. Chem.*, 19, 727, 1973.
21. **Johnson, G. F., Thompson, M. F., Fetteroff, S., and Fasano, A. N.,** Neutron activation analysis technique for nail sodium concentration in cystic fibrosis patients, *Pediatrics*, 47, 88, 1971.
22. **McLean, W. C., Jr. and Tripp, R. W.,** Cystic fibrosis with edema and falsely negative sweat test, *J. Pediatr.*, 83, 86, 1973.
23. **Hardy, J. D., Davison, S. H. H., Higgins, M. U., and Polycarpou, P. N.,** Sweat tests in the newborn period, *Arch Dis. Child.*, 48, 316, 1973.
24. **Green, M. N. and Shwachman, H.,** Presumptive tests for cystic fibrosis based on serum protein in meconium, *Pediatrics*, 41, 989, 1968.
25. **Prosser, R., Owen, H., Bull, F., Parry, B., Smerkinich, J., Goodwin, H. A., and Dathan, J.,** Screening for cystic fibrosis by examination of meconium, *Arch. Dis. Child.*, 49, 597, 1974.
26. **Stephan, U., Busch, E. W., Kollberg, H., and Hellsing, K.,** Cystic fibrosis detection by means of a test-strip, *Pediatrics*, 55, 35, 1975.
27. **Kilbourn, J. P.,** Composition of sputum from patients with cystic fibrosis, *Curr. Microbiol.*, 11, 19, 1984.
28. **Kilbourn, J. P. and Wolf, P. L.,** Biochemical abnormalities in sputum from cystic fibrosis patients possibly enhancing lung infection, *Clin. Chem.*, 30, 946, 1984.
29. **Knowles, M. R., Stutts, M. J., Spock, A., Fischer, N., Gatzy, J. T., and Boulder, R. L.,** Abnormal ion permeation through cystic fibrosis respiratory epithelium, *Science*, 221, 1067, 1983.
30. **Lam, J., Chan, R., Lam, K., and Costerton, J. W.,** Production of mucoid microcolonies by Pseudomonas aeruginosa within infected lungs in cystic fibrosis, *Infect. Immun.*, 28, 546, 1980.
31. **Rubin, L. J., Nicod, P., Hillis, L. D., and Firth, B. G.,** Treatment of primary pulmonary hypertension with nifepidine, *Ann. Intern. Med.*, 99, 433, 1983.
32. **Brown, G. and Dobson, R. L.,** Sweat sodium excretion in normal women, *J. Appl. Physiol.*, 23, 97, 1967.
33. **Consolazio, C. F., Matoush, L. O., Nelson, R. A., Isaac, G. J., and Canham, J. E.,** Comparisons of nitrogen, calcium and iodine excretion in arm and total body sweat, *Am. J. Clin. Nutr.*, 18, 443, 1966.
34. **Ahlmann, K. L., Eranko, O., Karvonen, M. J., and Leppanen, V.,** Mineral composition of thermal sweat in healthy persons, *J. Clin. Endocrinol.*, 13, 773, 1953.
35. **Levin, S., Shapira, E., Garin, T., Seligson, S., and Strassberg, D.,** Effect of age, ethnic background and disease on sweat chlorides, *Isr. J. Med. Sci.*, 2, 333, 1966.
36. **Danton, R. A. and Nyhan, W. L.,** Concentrations of uric acid in the sweat of control and mongoloid children, *Proc. Soc. Exp. Biol. Med.*, 121, 270, 1966.

37. **Coltman, C. A., Jr. and Rowe, N. J.,** The iron content of sweat in normal adults, *Am. J. Clin. Nutr.,* 18, 270, 1966.
38. **Wilcken, W. and Chalmers, G.,** Reduced morbidity in patients with cystic fibrosis detected by neonatal screening, *Lancet,* December 14, 1985.
39. Ad Hoc Committee Task Force on Neonatal Screening, Cystic Fibrosis Foundation, Neonatal screening for cystic fibrosis: position paper, *Pediatrics,* 72, 741, 1983.
40. **Wolf, P., Hall, C., and Kilbourn, J. P.,** Demonstration of calcitonin and calmodulin by immunoperoxide in the cystic fibrosis lung, *Chest,* 89, 327, 1986.

CURRENT ASSAY METHODS FOR ESTIMATING ELECTROLYTE CONCENTRATIONS

Samuel Natelson

Tables 1 to 6 outline the methods in current use in the clinical laboratories for assaying for the electrolytes, Na, K, Cl, Ca, Mg, and phosphate, and total CO_2, as an estimate of the bicarbonate level. Tables 7 to 11 describe the procedures for pCO_2, pO_2, oxygen saturation, and osmolarity, which are associated commonly with the electrolyte balance system in the body.

In these tables, it can be seen that pH, pCO_2, Na, K, Cl, Ca, and pO_2 can now be performed potentiometrically. This development was stimulated by the need for simple procedures which lend themselves to ready automation. It was aided considerably by the development of precise methods for the measurement of voltage and current, using the high technology techniques developed for the computer.

ION-SELECTIVE ELECTRODES FOR ELECTROLYTE ASSAY

The ion-selective electrode is a device based on the phenomenon, discovered by Volta, that a chemical reaction can generate an electromotive force (EMF). For example, if a tin rod is dipped into a dilute acid solution, some of the tin will dissolve, and the reaction $Sn^0 \rightleftharpoons Sn^{2+} + 2e$ will take place. A voltage will develop. If the solution contains tin ions, the reverse reaction will be stimulated, modifying the voltage developed. Ions other than tin will not have this property, and, thus, this is an "ion-selective electrode". A reference is required to define the voltage. For this the voltage of the hydrogen electrode ($H_2^0 \rightleftharpoons 2H^+$) was taken as zero, provided the conditions were standard; that is, the solution had an activity which was molar with respect to hydrogen ion and the temperature was 298 K (25°C). The EMF values developed uder these standard conditions are represented by the symbol E^0. These standard voltages vary with the chemical reaction under study, and Table 12 lists the E^0 values for the reactions of various elements. E^0 values for various chemical reactions are listed in the *Handbook of Chemistry and Physics.*

Blood pH with the Hydrogen Electrode

This selective electrode exploits the reaction $H_2^0 \rightleftharpoons 2H^+ + 2e$. Hydrogen gas, containing 5.26% CO_2 (pCO_2 = 40 mm), is blown against a platinum black electrode which adsorbs the hydrogen. Palladium black has also been used (Figure 1).

The presence of CO_2 in the gas prevents the removal of the CO_2 of the plasma, but the pCO_2 of the plasma is now 40 mm. It should be pointed out that this is not necessarily the pCO_2 of the blood originally. Thus, pH values obtained on blood or plasma with the hydrogen electrode are not acceptable for use in the management of the patient. However, it is the reference method for measuring the pH of buffer solutions. In this case pure hydrogen gas is blown against the electrode.

A salt bridge and reference electrode, such as a calomel or Ag-AgCl electrode, permit the measurement of the voltage developed. The hydrogen ion concentration and, thus, the pH are calculated from the Nernst Equation for the hydrogen electrode*

$$E = \frac{RT}{F} \ln a_{H^+}$$

* For a complete explanation of the Nernst Equation see the Appendix on page 155.

Table 1
CURRENT METHODS FOR THE ASSAY OF SODIUM

	Table ref.
By Emission Spectroscopy (Flame Photometry)	
The sample is diluted and sprayed into a propane-air flame; the Na atoms are stimulated to an excited state; as they return to the ground state they emit the characteristic yellow sodium doublet at about 589 nm; this line is used to evaluate the Na content of the sample; a lithium (677-nm emission) or cesium (852-nm emission) internal standard is used to correct for variations in the flame or aspirator	1, 2
Potentiometrically	
A sodium glass electrode (sodium, aluminum-silicate) is used to form a half-cell with the sodium ions in the sample and, thus, develop an EMF; the sodium concentration in the sample is calculated from the Nernst Equation; the serum may be read directly or diluted with a pH 8.0 buffer containing no K, NH_4, or Na ions; usually magnesium is used as the cation in the buffer	3, 4
Spectrophotometrically	
Sodium forms colored chelate compounds with certain synthetic macrocyclic compounds as mentioned previously. These have been used for Na^+ assays, colorimetrically;[5] potassium interference is very low	5

Table References:

1. **Natelson, S.,** *Am. J. Clin. Pathol.,* 20, 463, 1950; *Am. J. Clin. Pathol.,* 21, 1153, 1951.
2. **Schaffer, R. et al.,** *Clin. Chem.,* 27, 1824, 1981.
3. **Ladenson, J. H.,** *Clin. Chem.,* 25, 757, 1979.
4. **Harff, G. A.,** *Clin. Chem.,* 28, 1232, 1982.
5. **Kumar, A. et al.,** *Clin. Chem.,* 34, 1709, 1988.

Table 2
CURRENT METHODS FOR THE ASSAY OF POTASSIUM

	Table ref.
By Emission Spectroscopy (Flame Photometry)	
The reference method for the assay of potassium in serum is by the flame photometer, as it is for sodium; the same internal standard is used (lithium or cesium); potassium concentration is estimated by emission at 766 nm	1, 2
Potentiometrically	
Potassium, in contrast to sodium, readily forms chelate compounds; the potassium complex with valinomycin is used most commonly to form an ion-selective electrode; the serum is used directly or diluted with a buffer which contains no K, NH_4, or Na ions	3, 4
Spectrophotometrically	
Like Na, macrocyclic compounds have been prepared which produce internal colors with potassium and little with sodium. These are useful in the colorimetric assay of K in serum and plasma	5

Table 2 (continued)
CURRENT METHODS FOR THE ASSAY OF POTASSIUM

Table ref.

Table References:

1. **Natelson, S.,** *Am. J. Clin. Pathol.*, 20, 463, 1950; *Am. J. Clin. Pathol.*, 21, 1153, 1951.
2. **Velapoldi, R. A. et al.,** NBS special publ. 260-63, U.S. Department of Commerce, Washington, D.C., 1978.
3. **Ladenson, J. H.,** *Clin. Chem.*, 25, 757, 1979.
4. **Durst, R. A.,** *Clin. Chem. Acta*, 80, 225, 1977.
5. **Kumar, A. et al.,** *Clin. Chem.*, 34, 1709, 1988.

Table 3
CURRENT METHODS FOR CALCIUM ASSAY

	Table ref.
Atomic Absorption	
The calcium salts are vaporized by the acetylene-air flame; a hollow cathode calcium tube generates monochromatic light at 424 nm; this traverses the flame; absorption of Ca emission from the source by calcium in the ground state is recorded as a measure of the calcium concentration	1, 2
Spectrophotometric	
With *O*-cresolphthalein (Aldrich No. 8-577-8); calcium forms a red-colored complex with an absorption maximum at 520 nm; alternatively, *O*-cresolphthalein complexone[4] (Aldrich No. 18,819-0) can be used, when the absorption maximum is at 575 nm; in both cases 8-hydroxy quinoline is added to complex magnesium ion; the reaction is carried out at strongly alkaline pH, usually by adding an organic amine such as diethylamine along with KCN	3, 4
Titrimetrically	
Calcein blue (4-methyl-umbelliferone-6-methylene imino-diacetic acid, Aldrich No. 22-909-1) forms a fluorescent complex with calcium ion; titration with EGTA at alkaline pH quenches the fluorescence to the baseline value; excitation is at 490 nm and emission is at 520 nm[5]	5
Potentiometrically for Calcium Ion	
In one version (Orion) calcium is complexed to an insoluble matrix which is immersed in a reference solution and serves as one electrode;[6] the assembly is separated from the sample (usually in a flow-through system) by a membrane which permits calcium ion exchange; the reference source of EMF is a calomel electrode[7]	6, 7

Table References:

1. **Trudeau, D. L. and Freir, E. F.,** *Clin. Chem.*, 13, 101, 1967.
2. **Gimblet, E. G., Marney, A. F., and Bonsnes, R. W.,** *Clin. Chem.*, 13, 204, 1967.
3. **Kessler, G. and Wolfman, M.,** *Clin. Chem.*, 10, 686, 1964.
4. **Baginski, E. S., Alawa, S. M., and Zak, B.,** *Selected Methods for the Small Clinical Laboratory*, Vol. 9, Faulkner, W. R. and Meites, S., Eds., American Association for Clinical Chemical Publishers, Washington, D.C., 1982, 125.
5. **Jackson, J. E., Breem, M., and Cheng, C.,** *J. Lab. Clin. Med.*, 60, 700, 1962.
6. **Li, T.-K. and Piechocki, J. T.,** *Clin. Chem.*, 17, 411, 1971.
7. **Rehfeld, S. J., Barkley, J., and Loken, H. F.,** *Clin. Chem.*, 30, 304, 1984.

Table 4
CURRENT ASSAY METHODS FOR MAGNESIUM

	Table ref.
By Atomic Absorption	
The magnesium salts are vaporized into an acetylene-air flame and monochromatic light from a hollow cathode lamp (285 nm) is passed through the flame; the absorbance by the Mg atoms in the ground state is measured; strontium or lanthanum salts are added to complex phosphate, which interferes in the test	1, 2
Spectrophotometrically	
With calmagite [3-hydroxy-4(2 hydroxy-5 methyl) phenylazo-1-napthalene sulfonic acid, Aldrich Chem. Co. No. C20-4]; a violet color with maximum absorbance at 532 nm is formed; the color is stabilized with polyvinyl pyrrolidone solution; EGTA and KCN are added to prevent Ca and heavy metal interference	3, 4
With Titan Yellow (Thiazole Yellow, Clayton Yellow, Sigma No. T-0513); magnesium forms a color with maximum absorbance at 540 nm; the lake formed is stabilized with polyvinyl alcohol; commercial samples of the dye need to be purified	5, 6, 7
With Methyl Thymol Blue [3,3′-bis (*N,N*-dicarboxymethyl amino-methyl) thymolsulfonphthalein sodium salt, Aldrich, Compound No. B 4, 200-4]; absorbance maxima are at 510 and 600 nm; chelators (e.g., EGTA) are added to prevent calcium interference	8, 9

Table References:

1. **Gimblet, E. G., Marney, H. F., and Bonsnes, R. W.,** *Clin. Chem.,* 13, 204, 1967.
2. **Weatherburn, M. W. and Trotman, R. B. B.,** *Clin. Chem.,* 26, 1022, 1980.
3. **Gindler, E. M. and Heth, D. A.,** *Clin. Chem.,* 17, 663, 1971.
4. **Khayam-Bashi, H., Liv, T., II, and Walters, V.,** *Clin. Chem.,* 23, 289, 1977.
5. **Orange, M. and Rhein, H. C.,** *J. Biol. Chem.,* 189, 379, 1951.
6. **Spencer, W. W. and Gentile, S. J.,** *Selected Methods for the Small Clinical Laboratory,* Faulkner, W. R. and Meites, S., Eds., American Association for Clinical Chemical Publishers, Washington, D.C., 1982, 283.
7. **King, H. G. C. and Pruden, G.,** *Analyst,* 92, 83, 1967.
8. **Sadek, F. S. and Reilly, C. N.,** *J. Lab. Clin. Med.,* 54, 621, 1959.
9. DuPont, ACA Analyzer method.

Table 5
CURRENT METHODS FOR CHLORIDE ESTIMATION

	Table ref.
Amperometric (Coulometric)	
A constant current is passed through a sample (e.g., serum, urine plasma) diluted with a solution of 0.1 *M* nitric acid, 1.5 *M* with respect to acetic acid and containing a small amount of thymol and thymol blue	1, 2
Silver dissolved from the anode reacts and precipitates with chloride ion; hydrogen deposits at the platinum cathode and polarizes it	
When all of the chloride ion has reacted, the excess silver begins to deposit on the cathode, thus depolarizing it; a sudden increase in current indicates the end point of the titration	
Potentiometric (Ion-Selective Electrode)	
A silver, silver chloride electrode permits the direct measurement of the chloride concentration, according to the Nernst equation	3—5
This electrode has also been used for measuring the end point in the titration of the chloride ion with silver or mercuric nitrate	

Table 5 (continued)
CURRENT METHODS FOR CHLORIDE ESTIMATION

	Table ref.
Spectrophotometrically	
A solution is prepared to contain, per liter, 50 m*M* Fe $(NO_3)_2$, 2 m*M* $Hg(SCN)_2$ 0.3 m*M* $Hg(NO_2)_2$, and adjusted to a pH of 1.5 to 2.0 with HNO_3; 1 ml is added at 37 °C to a small sample (e.g., 10 μl) of the biological fluid to be assayed and the concentration of ferric thiocyanate is measured at 525 nm	6, 7
Titrimetrically	
The sample is diluted about 20—100 times with 0.07 *M* HNO_3; a few drops of methanolic solution of diphenylcarbazone (phenylazoformic acid-2-phenyl hydrazide, Aldrich, No. 12, 453-2) is added as the indicator; the solution is titrated with M/100 mercuric nitrate solution in 0.07 *M* HNO_3; the chloride binds the Hg ion; excess Hg reacts with the indicator to form a blue color at the end point	8—10

Table References:

1. **Cotlove, E. and Nishi, H. H.,** *Clin. Chem.,* 7, 285, 1961.
2. **Siggard-Anderson, O.,** *Am. J. Clin. Pathol.,* 48, 444, 1967.
3. **Lustgarten, J. A. et al.,** *Clin. Chem.,* 20, 1217, 1974.
4. **Warwick, W. L. and Hansen, L.,** *Clin. Chem.,* 24, 2050, 1978.
5. **Seligson, D., McCormick, G. J., and Sleeman, K.,** *Clin. Chem.,* 4, 159, 1958.
6. **Levinson, S. S.,** Chloride colorimetrically, in *Selected Methods of Clinical Chemistry,* Vol. 9, Faulkner, W. R. and Meites, S., Eds., American Association for Clinical Chemical Publishers, Washington, D. C., 1982, 143.
7. **Hamilton, R. H.,** *Clin. Chem.,* 12, 1, 1966.
8. **Schales, O. and Schales, S. S.,** *J. Biol. Chem.,* 140, 879, 1941.
9. **Dubsky, J. V. and Trtilek, J.,** *Z. Anim. Chem.,* 93, 345, 1933.
10. **Natelson, S.,** *Am. J. Clin. Pathol.,* 27, 1153, 1951.

Table 6
CURRENT ASSAY METHODS FOR PHOSPHATE

	Table ref.
Direct Spectrophotometric	
By direct reading of the phosphomolybdate complex absorbance at 340 nm	1, 2
Spectrophotometric (Indirect)	
By reduction of the phosphomolybdate complex to a blue color, reading at 660 to 820 nm; reducing agents are stannous chloride, with and without hydrazine, 1-amino-2-naphthol 4-sulfonic acid, or ferric ammonium sulfate, as examples	3—5

Table References:

1. **Munoz, M. A., Balon, M., and Fernandez, C.,** *Clin. Chem.,* 29, 372, 1983.
2. **Daly, J. A. and Ertingshausen, G.,** *Clin. Chem.,* 18, 263, 1972.
3. **Fiske, C. H. and Subbarrow, Y.,** *J. Biol. Chem.,* 66, 375, 1925.
4. **Baginski, E. S., Marie, S. S., and Zak, B.,** *Microchem. J.,* 19, 285, 1974.
5. **Goldenberg, H. and Fernandez, A.,** *Clin. Chem.,* 12, 871, 1966.

Table 7
CURRENT METHODS FOR CO_2 CONTENT MEASUREMENT (TOTAL CO_2)

	Table ref.
Gasometric	
The CO_2 is liberated from the serum sample (30 μl) and its volume is brought to a preset volume (0.2 ml); the pressure of the liberated gas, measured on a Hg manometer, is a measure of the CO_2 content of the serum	1
With the pH Glass Electrode	
The CO_2 is liberated by the addition of acid, and the CO_2 gas is allowed to diffuse into a dilute sodium bicarbonate solution through a plastic membrane; the change in pH of the bicarbonate solution is a measure of the pCO_2 of the gaseous phase and, thus, a measure of the total CO_2 of the serum	1, 2
Spectrophotometrically	
The serum is diluted and acidified; the liberated CO_2 diffuses through a silicone rubber membrane to a bicarbonate solution containing Cresol Red indicator; the change in color at 430 nm is read and serves as a measure of the CO_2 content of the serum	3
Enzymatically	
Alkali is added to the serum to bring it to a pH of about 8.5; at this pH practically all of the CO_2 is in the form of bicarbonate ion; phosphoenolpyruvate (PEP, Sigma sodium salt, No. P-7002) is present in the buffer; phosphoenolpyruvate carboxylase (Sigma No. P-2023 or P-7150) is added and the PEP is converted to oxalacetate and inorganic phosphate; the oxalacetate is promptly reduced to malic acid by NADPH; this reaction is mediated by malic dehydrogenase (Sigma No. M-9004); the decrease in absorbance at 340 nm or decrease in fluorescence is a measure of the NADPH oxidized to NADP, and thus a measure of the CO_2 originally presented in the serum (used in the Duport ACA Analyzer®)	4
By Calculation	
pH and pCO_2 of the whole blood are measured, and these values are substituted in the Henderson-Hasselbalch equation pK′ is assumed to be constant and at 6.1; the sum of the numerator (HCO_3^-) of the fraction and the denominator is the total CO_2 value	

Table References:

1. **Natelson, S.,** *Am. J. Clin. Pathol.*, 21, 1153, 1951.
2. **Hicks, J. M., Aldrich, F. T., and Josephsohn, M.,** *Clin. Chem.*, 22, 1868, 1976.
3. **Gambino, R. S. and Schreiber, H.,** *Am. J. Clin. Pathol.*, 45, 406, 1966.
4. **Astrup, P.,** *Scand. J. Clin. Lab. Invest.*, 8, 33, 1956.

Table 8
CURRENT METHODS FOR pCO_2 MEASUREMENT

	Table ref.
With the pH Electrode[1]	
The CO_2 of whole heparinized blood is allowed to diffuse through a thin plastic membrane (e.g., Teflon® or silicone rubber) so as to change the pH of a dilute bicarbonate buffer solution; the change in pH of the buffer is a measure of the pCO_2 level (see Figure 3)	1

Table 8 (continued)
CURRENT METHODS FOR pCO_2 MEASUREMENT

	Table ref.
By Calculation[2]	
pCO_2 can be calculated from the pH and total CO_2 values with the Henderson-Hasselbalch equation, assuming a constant pK′ value of 6.1 (see discussion in the introductory Chapter to this Section *et seq.*)	2
With Tonometered Blood[3]	
Blood is tonometered with a gas at pCO_2 = 20 m*M* Hg, and the pH is measured; this is repeated with a gas mixture at pCO_2 = 60 m*M*; these two values permit a line to be drawn plotting pH vs. pCO_2; the measured pH of the blood initially permits the interpolation of the line to estimate the blood pCO_2; this also tacitly assumes a constant pK_1' value	3

Table References:

1. **Severinghaus, J. W.,** *J. Appl. Physiol.*, 21, 1108, 1966.
2. **Natelson, S. and Barbour, J. H.,** *Am. J. Clin. Pathol.*, 22, 426, 1952.
3. **Astrup, P.,** *Clin. Chem.*, 7, 1, 1961.

Table 9
CURRENT METHODS FOR pO_2 MEASUREMENT

	Table ref.
By the Clark Electrode[1-3]	
This technique is an application of the science of Polarography. As can be seen from Figure 3 a current is passed through a dilute KCl solution, 0.01 M with respect to KOH, with a voltage of about 0.67 V at the Pt cathode. The solution is retained by a polypropylene membrane which permits the oxygen to pass but not the water or salts.	
When the voltage applied (abscissa) is plotted against the current generated (ordinate), the current increases until a plateau is reached at the potential necessary for oxygen reduction. Further increases in potential does not result in a current increase until the potential is reached where hydrogen is liberated. The constant current with increasing EMF. is due to the current being limited by the rate of diffusion of the oxygen to the electrode. This rate is a function of the oxygen concentration. The location of this plateau is then due to the oxygen concentration, which in turn is due to the pO_2 of the blood.	
At the cathode, reduction of peroxide formed from O_2 and alkali is taking place, mainly. However, some direct reduction of O_2 also takes place. In either case the product is water.	
Transcutaneously[4-6]	
A Clark-type electrode makes close contact with the skin; the oxygen diffuses through the skin and the plastic membrane to depolarize the platinum cathode and yield an estimate of blood pO_2	3—5

Table References:

1. **Clark, L. C.,** *Trans. Am. Soc. Artif. Intern. Organs*, 2, 41, 1956.
2. **Huch, R., Lubbers, D. W., and Huch, A.,** *Arch. Dis. Child.*, 49, 213, 1974.
3. **Linek, V. et al.,** *Measurement of Oxygen by Membrane Covered Probes*, John Wiley & Sons, New York, 1988.
4. **Oda, Y.,** *Birth Defects*, 15, 585, 1979.
5. **Graham, G. and Kenny, M. A.,** *Clin. Chem.*, 26, 1860, 1980.
6. **Marshall, T. A., Kathwinkel, J., Berry, F. A., and Shaw, A.,** *J. Pediatr. Surg.*, 16, 794, 1980.

Table 10
CURRENT METHODS FOR THE ESTIMATION OF OXYGEN SATURATION OF BLOOD

	Table ref.
Gasometrically[1,2]	
Whole heparinized blood is sampled and mixed with alkaline cyanide reagent; the hemoglobin is oxidized to ferrihemoglobin (Methemoglobin); this releases the oxygen, and its pressure is measured at constant volume; the oxygen is reabsorbed into sodium hydrosulfite and the pressure is measured again at the same constant volume; the difference between these two values (the second is the blank) yields the oxygen content of the sample of whole blood; the measurement is repeated after saturating a sample of whole blood with oxygen; this value is the *oxygen capacity*	1, 2
If the oxygen content is divided by the oxygen capacity and then multiplied by 100, the *% of oxygen saturation* is obtained	
Spectrophotometrically[3,4]	
Reduced hemoglobin has a single absorption peak at about 556 nm; its Soret band is at 430 nm	3, 4
Oxyhemoglobin has two peaks: one at 577 nm and the other at 541 nm; from the absorbance values at their peaks, as measured with a thin film of whole blood, the hemoglobin-to-oxyhemoglobin ratio can be computed, and from the measurement of the total hemoglobin concentration, colorimetrically, the % of the oxygen saturation can be determined	
Transcutaneously[5]	
A source of infrared light and a detector are arranged on an ear clip, so that the infrared light traverses the ear lobe and can be used to estimate the % of oxygen saturation	
This is useful for monitoring during surgery, or in newborns in the respirator	

Table References:

1. **Natelson, S. and Menning, C.,** *Clin. Chem.*, 1, 167, 1955.
2. **Van Slyke, D. D. and Neil, J. J.,** *Biol. Chem.*, 61, 523, 1924.
3. **Molyneau, L. and Pask, E. A.,** *Br. Heart J.*, 17, 169, 1955.
4. **Drabkin, D. L. and Schmidt, C. F.,** *J. Biol. Chem.*, 157, 69, 1945.
5. **Graham, G. and Kenny, M. A.,** *Clin. Chem.*, 26, 1860, 1980.

Table 11
SERUM AND URINE OSMOLARITY

	Table ref.
Freezing Point Depression[1,2,3]	
The sample is cooled gradually until ice crystals form; the initiation of the process of freezing releases heat, resulting in a sudden rise in temperature; the temperature thus remains constant, instead of dropping steadily	1, 2, 4
This phenomenon is sensed by heat-sensitive transistor and the freezing point is recorded; the difference between 0°C and this point is the freezing point depression; this value divided by −1.858°C (the freezing point of a 1000-mOsm solution) multiplied by 538.2, (1000/1.858 = 538.2), yields the osmolarity of the solution	

Table 11 (continued)
SERUM AND URINE OSMOLARITY

	Table ref.
Vapor Pressure Depression[3,4]	
The cooling effect of a drop of serum deposited on a temperature-sensitive thermistor is a function of its vapor pressure and heat of vaporization; for aqueous solutions, the vapor pressure depression is directly related to the difference between the cooling effect of pure water and the aqueous solution	3, 4
This principle is used for the vapor pressure depression measurement of serum or urine samples and their osmolarity, as a micro method	

Table References:

1. Osmometer Manual, Advanced Instruments Inc., Needham Heights, MA.
2. **Johnston, R. B. and Hoch, H.,** *Osmolality of Serum and Urine in Standard Methods of Clinical Chemistry,* Vol. 5, Academic Press, New York, 1965, 159.
3. Osmometer Manual, Wescor Inc., Logan, UT.
4. **Dorvart, W. V. and Chalmers, L.,** *Clin. Chem.,* 21, 190, 1975.

Table 12
STANDARD OXIDATION POTENTIALS AT 0°C IN ACID SOLUTION

Half-cell	Volts	Half-cell	Volts
$Li^0 \rightarrow Li^+ + e$	+3.02	$Cd^0 \rightarrow Cd^{2+} + 2e$	0.0402
$Cs^0 \rightarrow Cs^+ + e$	3.02	$Co^0 \rightarrow Co^{2+} + 2e$	0.277
$Rb^0 \rightarrow Rb^+ + e$	2.99	$Ni^0 \rightarrow Ni^{2+} + 2e$	0.250
$K^0 \rightarrow K^+$	2.92	$Su^0 \rightarrow Sn^{2+} + 2e$	0.136
$Ba^0 \rightarrow Ba^{2+} + 2e$	2.90	$H^0 \rightarrow H^+ + e$	0.000
$Sr^0 \rightarrow Sr^{2+} + 2e$	2.89	$Cu^0 \rightarrow Cu^{2+} + 2e$	−0.3448
$Ca^0 \rightarrow Ca^{2+} + 2e$	2.87	$I^- \rightarrow {}^1/_2 I_2^0 + e$	−0.5345
$Na^0 \rightarrow Na^{2+} + 2e$	2.71	$H_2O_2 \rightarrow O_2 + 2\,H^+ + 2e$	−0.682
$Mg^0 \rightarrow Mg^{2+} + 2e$	2.34	$Hg^0 \rightarrow Hg^+ + e$	−0.7986
$Al^0 \rightarrow Al^{3+} + 3e$	1.67	$Ag^0 \rightarrow Ag^+ + e$	−0.7995
$Mn^0 \rightarrow Mn^{2+} + 2e$	1.05	$H_2O \rightarrow {}^1/_2 O_2 + 2H^+ + 2e$	−0.815
$Zn^0 \rightarrow Zn^{2+} + 2e$	0.762	$Pt \rightarrow Pt^{2+} + 2e$	−1.20
$Fe^0 \rightarrow Fe^{2+} + 2e$	0.440	$Au^0 \rightarrow Au^+ + e$	−1.68

Note: In giving off an electron, the electrode is actually negative. Thus, the Zn-Cu couple is 0.762 + 0.3448 or a total of 1.10 V, with Cu being the positive role, according to present convention, at 0°C and in contact with a solution of molar hydrogen ion activity.

where R = 8.3147 J, F = 96,500 C, and T = 298 K. Thus, $E = 0.02568 \ln a_{H^+}$. For an HCl solution with an activity of 0.1 *M* and a pH of 1.0, E = 0.05913 V or 59.13 mV.

The primary standard for the estimation of pH with this electrode was 0.1 *M* HCl. If the HCl were fully dissociated, 0.1 *M* HCl would have a pH of 1.0. From conductivity measurements, it was established that the activity coefficient of 0.1 *M* HCl was 0.796. From these data it was estimated that the pH of 0.1 *M* HCl was actually 1.0796, or 1.08 as used by Cullen[1] and Hastings.[2]

With 0.1 *M* HCl as a primary standard, more convenient standard buffers could be prepared. The British Bureau of Standards selected 0.05 *M* KHphthalate as a primary

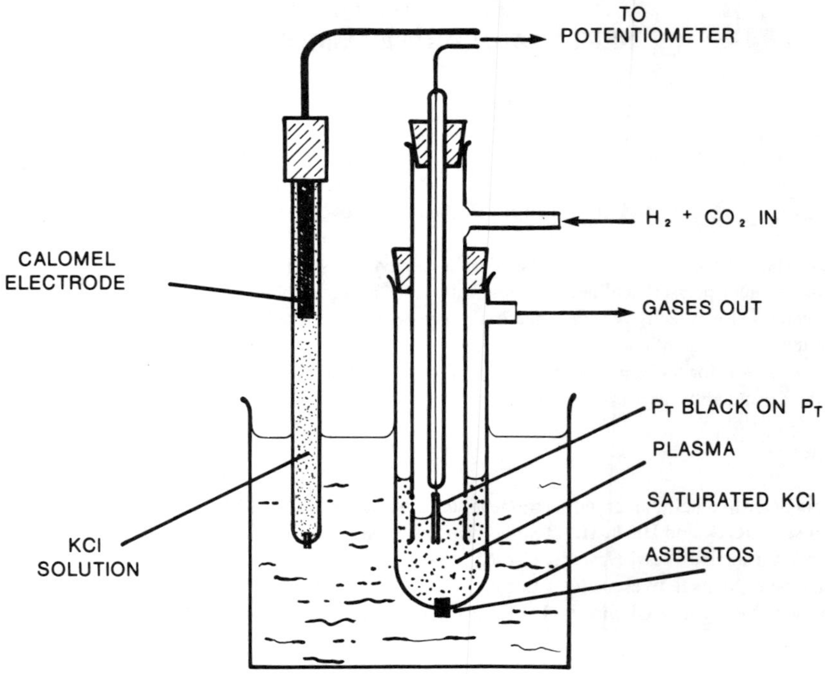

FIGURE 1. The hydrogen electrode. A gas composed of H_2 (720 mmHg) and CO_2 (40 mmHg) is blown against a deposit of platinum black on platinum. The plasma is contained in a test tube into which the hydrogen electrode dips. A reference electrode dips into a saturated KCl solution to make contact with the plasma through an asbestos wick. The voltage is interpreted in terms of pH by a potentiometer.

standard, designating its pH as 4.01 at 25°C. The U.S. National Bureau of Standards selected several standards for the calibration of pH meters, and these are listed in Table 13. These buffers are preserved readily with 0.1% sodium azide.

Blood pH and Electrolytes with Ion Selective Electrode

The glass electrode is an ion-selective electrode which was developed at the turn of the century. However, it was not used for measuring blood pH until the 1930s.

The glass electrode is one of a group of ion-selective electrodes, all of which are formed into the shape of a membrane containing a compound from which the ion to be tested is poorly dissociated. In the case of the glass pH electrode, hydrogen is complexed with silicic acid. Contact is made with a calomel or Ag-AgCl electrode by a buffer solution containing hydrogen ion in the form of HCl and acetic acid, behind the membrane, to complete the half-cell. A reference electrode completes the circuit. This is shown schematically in Figure 2.

In the reaction silicic acid $\rightleftarrows$ silicate + H^+, the amount dissociated is negligible, and the silicic acid can be considered as a solid source of hydrogen ions (Equation 1 applies). It was assumed that the glass electrode was permeable to the hydrogen ion. This was shown not to be the case when tritium was shown not to pass the glass membrane.[4]

If SiO_2 is mixed with Al_2O_3 and the oxide of an alkali metal, the glass formed can serve as an ion-selective electrode for that electrolyte. For example, in mole %, 11% Na_2O, 18% Al_2O_3, plus 71% SiO_2 will make a sodium electrode, and 15% LiO, 25% Al_2O_3, plus 60% SiO_2 form a lithium electrode.[5] In each case the thin film of glass is backed by a conducting solution containing the ion to be tested.

Table 13
REFERENCE BUFFER SOLUTIONS

Composition	Temp (°C)	pH
HCl, 0.1080 *M*	25	1.000
	37	1.000
KH tartrate, saturated at 25°C	25	3.557
	37	3.548
KH_2 citrate, 0.05 *M*	25	3.776
	37	3.759
KH phthalate, 0.05 *M*	25	4.008
	37	4.030
Solution, 0.025 *M* KH_2PO_4 and 0.025 *M* Na_2HPO_4	25	6.895
	37	6.840
Solution, 0.0087 *M* KH_2PO_4 and 0.0302 *M* Na_2HPO_4	25	7.413
	37	7.384
$Na_2B_4O_7$, 0.01 *M*	25	9.180
	37	9.081
$NaHCO_3$, 0.025 *M* and Na_2CO_3, 0.025 *M*	25	10.012
	37	9.925
$Ca(OH)_2$, saturated at 25°C	25	12.454
	37	12.043

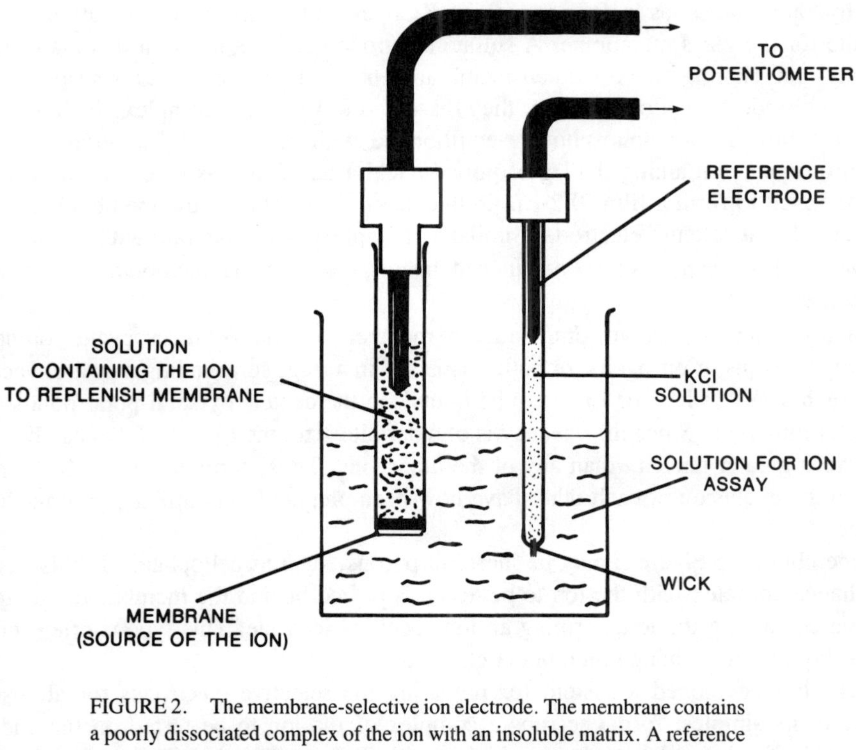

FIGURE 2. The membrane-selective ion electrode. The membrane contains a poorly dissociated complex of the ion with an insoluble matrix. A reference electrode (e.g., calomel) serves to complete the circuit. Some porous membranes (not glass) permit the maintenance of the ion at constant concentration at the membrane by a buffer behind the membrane.

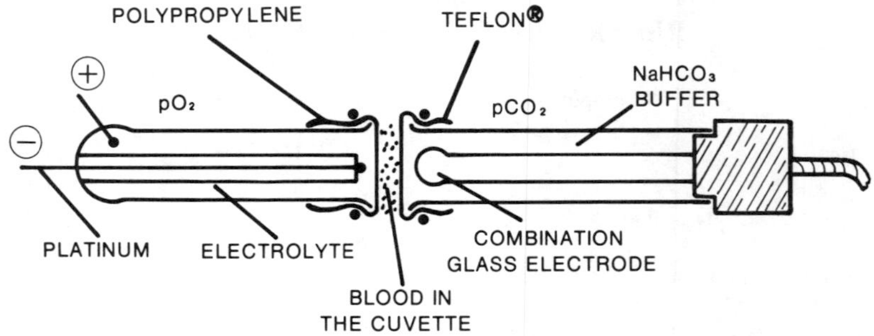

FIGURE 3. Combined pO_2 and pCO_2 electrode. CO_2 diffuses to a buffer whose change in pH is monitored by a pH glass electrode. The pO_2 concentration is measured by the change in current due to a change in diffused oxygen concentration. (Adapted from Natelson, S. and Natelson, E., *Techniques of Clinical Chemistry,* 3rd ed., Charles C Thomas, Springfield, IL, 1971. With permission.)

There is some cross-reaction between the monovalent cations K, Na, Li, and NH_4, as they compete for sites on the glass. However, with blood, both Na and K are assayed simultaneously, and corrections can be made for the interference.

For some ions it is practicable to compress a poorly dissociated complex into a thin disk with a hydraulic press, as is done in making KCl pellets for infrared examination. This can substitute for the glass membrane. A sulfate electrode has been made in this manner.[6]

Moody and Thomas[7] have studied membranes using the ion complexes suspended in a polyvinylchloride film. For example, they dissolved 0.4 g of the complex, such as didecyl phosphoric-dioctyl phenylphosphonate equilibrated with molar $CaCl_2$ solution in 6 ml of tetrahydrofurane, containing 0.17 g of polyvinylchloride. This was spread on a glass plate and evaporated to form a film, 0.2 μm in thickness. This acted as the membrane in Figure 2 and served as a calcium electrode. Similarly, complexes of potassium with valinomycin,[8] dicyclohexyl-18-crown-6, or monactin can be dispersed in the membrane to act as the potassium electrode.

Valinomycin and the Crown compounds are members of a class of macrocyclic compounds containing oxygens or nitrogens, or both, dispersed in a ring, surrounding a central vacancy. An ion such as K^+, Na^+, or Li^+ may be bound, in the center, by semi-polar bonds to the oxygens or nitrogens. Since the vacancy is of a definite size, the ion which fits can be bound with a much greater tenacity than any of the other ions. Thus, compounds specific for Na^+ or K^+, to the exclusion of each other have now been prepared, and utilized in ion-selective electrodes.[9]

The membrane in Figure 2 may be inert but porous, such as cellophane. In this case, an ion exchange complex with the ion to be tested is placed behind the membrane, along with a solution containing the ion, forming an ion-specific electrode. This type is often referred to as the liquid ion exchange membrane electrode.

Ruzicka has developed a system for preparing ion-selective electrodes for almost any element. In its simplest form, he rubs a complex of the ion to be tested on the end of a graphite rod. He then "hydrophobizes" this with Teflon. The thin Teflon film acts as a porous film and the poorly dissociated complex provides the ion to be used to measure that ion's concentration in the solution being tested. The carbon rod then acts as the conductor, which is then connected to the potentiometer. These electrodes are called "Selectrodes".[8]

BLOOD pH SPECTROPHOTOMETRICALLY

The pH of a bicarbonate solution was followed spectrophotometrically, in the orginal

Table 14
SOME COMMONLY USED INDICATORS

Indicator	Chemical name	pH range	pK [a]	Max (nm)	Color change[b]
Cresol red (acid range)	*O*-Cresolsulfonephathalein	0.2—1.8	—	—, —	R-Y
Thymol blue (acid range)	Thymolsulfonephthalein	1.2—2.8	1.65	544, 430	R-Y
Methyl orange	Dimethylaminoazo-benzene sulfonate, Na salt	3.1—4.4	3.40	522, 464	R-O
Bromophenol blue	Tetrabromophenolsulfonephthalein	3.0—4.6	3.85	436—592	Y-BV
Bromocresol green	Tetrabromo-*m*-cresolsulphonephthalein	4.0—5.6	4.68	444—617	Y-B
Phenol red	Phenolsulfonephthalein	6.8—7.4	7.9	—, 553	C-R
Cresol red (alkaline range)	*O*-Cresolsulfonephthalein	7.2—8.8	8.2	434, 572	Y-R
Thymol blue (alkaline range)	Thymolsulfonephthalein	8.0—9.6	8.9	430—596	Y-B
Phenolphthalein	Phenolphthalein	8.0—10.0	9.4	—, 558	C-R
Thymolphthalein	Thymolphthalein	9.4—10.6	10.0	—, 598	C-B

[a] pH at which equal amounts of the colors indicated are formed.
[b] B = blue, C = colorless, BV = blue-violet, O = orange, R = red, Y = yellow.

Autoanalyzer, as a measure of the bicarbonate content, using phenol or cresol red as the indicator. The pH of the plasma has also been measured using indicators.[10]

The Sørenson equation applies to indicators, since they are acids or conjugate acids of bases. Thus,

$$pH = pK' + \log \frac{[\text{Indicator}^-]}{[\text{H Indicator}]}$$

For phenolsulfonephthalein, the dissociated form (in alkali) is red and indicates a high pH (Table 14), since most of the indicator has lost a proton (numerator of the fraction). With acid the pH is low, since most of the indicator is undissociated (the denominator). Between these extremes, different shades of pink will be observed. From Table 14 it can be seen that the red color begins to appear at pH = 6.8 and becomes darker as it approaches 8.4. Thus, it encompasses the human blood physiological range. If the plasma is separated, free of hemoglobin, then the intensity of the color with phenol red will indicate its pH.

Bromothymol blue has been used to estimate blood pH, since its range is 6.2 to 7.6 when it goes from a yellow to a blue color, through various shades of green. Comparison with pH standards, treated as for the plasma, permits estimation of the pH.

Fluorescent compounds may be used for measuring blood pH. They have the advantage that low concentrations of indicator are required, and less interference from visual colors are experienced. Umbelliferone and orcinaurine do not fluoresce below pH 6.5 and reach their maximum fluorescence at pH 8.0. Umbelliferone gives a blue fluorescence and orcinaurine yields a green, when irradiated with a Woods UV lamp. Both are soluble in water at a concentration of 0.03%, and 10 μl of this solution suffices for 1 ml of serum. This technique of measuring pH may be of interest in automated systems.

APPENDIX — NERNST EQUATION

The Nernst equation derives from Hemlholtz's concept of *Maximum Work* (Arbeit [A]).

Imagine a piston retaining 2 l of gas in a 1-l volume, in a cylinder under pressure at 2 atm pressure at 25°C. The piston is released and the gas is allowed to expand against the atmosphere to 2 l. Heat is applied to return the temperature to 25°C at the 2-l volume. The work done is 1 l-atm which is equivalent to 24.2 cal.

Helmholtz raised the question as to whether this machine (the steam engine) could be made to do more work. Of course, more coal would be consumed. He proposed to do this by relieving the pressure gradually. Thus, the pressure on the piston would be, successively, 2, then 1.9, then 1.8, etc. . . . atm, until the pressure was 1 atm. During the process the gas would be heated so as to remain at 25°C. It is apparent that more work would be done this way, since there was greater resistance to the expansion. More heat would have to be applied to compensate for this extra work. In this case p is a variable and the work done would be pdV, since the volume is changing by minute increments. From the gas laws, p must always be equal to nRT/V. Thus, the maximum work would be given by the equation A = nRT(dV/V). Integrating this expression from $V = V_1$, to $V = V_2$, we obtain $A = nRT\ln(V_2/V_1)$. Substituting in this formula, for the gas constant, R, 1.99 cal, and 298 K (25°C) for T and 2/22.4 for n, the number of moles of gas, we get 36.7 cal. Compare this with the 24.2 cal obtained by releasing the piston suddenly. This process is called a reversible process since to bring it back to the compressed condition one would be encountering an increasing pressure as one compressed the gas to 2 atm.

Gibbs pointed out that volume and pressure were functions of each other, and that pressure was equivalent to concentration. Thus, chemical reactions, where concentrations of the components change without necessarily a change in volume or pressure, should also obey the same law laid down by Helmholtz, namely, that processes in nature change so that the Maximum Work is given off, if they are reversible processes. Gibbs laid down the principle that in chemical reactions the maximum work differential between reactants and products is what drives a reaction. He therefore proposed a more general term. *Free Energy* ($-\Delta F$), for the maximum work. The minus sign indicates that energy is given off. The delta signifies the change in free energy, which he visualized as a property of the system. At present we prefer to use A for maximum work, when pressure and volume change, and $-\Delta F$ for chemical reactions. Currently, $-\Delta G$ is used commonly for $-\Delta F$ in honor of Gibbs. We can thus write:

$$\text{Free energy change} = -\Delta G = RT \ln \frac{[c]\ [d]}{[a]\ [b]} = RT \ln K_{eq}$$

Nernst adapted this equation to predict the voltage of a chemical reaction set up as a half-cell. He pointed out that the work a cell does is the EMF(E) multiplied by the number of charges moved (coulombs). He used the number of coulombs required to transport one equivalent weight, the Faraday (F), multiplied by its valence, n, to denote the total number of charges present in 1 mol of the ion,

$$E \times nF = RT \ln K_{eq} \text{ and } E = (RT/nF) \ln K_{eq}$$

To adjust for the standard voltages for the different ions, this must be subtracted from the standard voltages listed in Table 12. Thus, we have

$$E = E^{o} - (RT/nF) \ln K_{eq} \text{ (Nernst Equation)}$$

For oxidation reduction reactions, values for E^0 are given in the handbooks of chemistry.

Let us apply this to the hydrogen electrode (H_2^0 gas $\leftrightarrows 2H^+ + e$). In the case of hydrogen, E^0 is zero. We also need to replace concentration by activity, since only the ionic concentration affects the voltage. Thus,

$$E = -(RT/nF) \ln [a_{H+} / a_{Ho}]$$

However, the activity of H_2 gas is 1, as is that for all the solid metals in Table 12. The equation then reduces to $E = -(RT/F) \ln (a_{H+})$ for the hydrogen electrode, since $n = 1$.

REFERENCES

1. **Cullen, G. E., Keller, H. R., and Robinson, H. W.** The pK′ of the Henderson-Hasselbalch equation for hydrion concentration of serums, *J. Biol. Chem.*, 66, 301, 1925.
2. **Hastings, A. B. and Sendroy, J. Jr.**, The effect of variation in ionic strength on the apparent first and second dissociation constants of carbonic acid, *J. Biol. Chem.* 65, 445, 1925.
3. **Dill, D. B., Daly, C., and Forbes, W. H.** The pK′ of serum and red cells, *J. Biol. Chem.* 117, 569, 1937.
4. **Schwabe, K. and Dahms, H.**, Comparative studies of the electromotive characteristics and chemical behavior of glass electrodes with the aid of radioactive indicators, *Z. Elektrochemie*, 518, 65, 1961.
5. **Carey, H. H. and Baxter, W. P.**, Glass Electrodes U.S. Patent No. 2,462,843, 1945.
6. **Gardner, G. and Nancollas, G. H.**, Complex formation in lead sulfate solutions, *Anal. Chem.* 42, 794, 1970.
7. **Moody, G. J., Oke, R. B., and Thomas, J. D. R.**, A calcium-sensitive electrode based on a liquid ion exchanger in a polyvinyl chloride matrix, *Analyst*, 95, 910, 1970.
8. **Ruzicka, J., Hansen, E. A., and Kjell, J.**, Selectrode — The universal ion-selective electrode employing a new ion exchanger in a non-porous membrane and a solid state reference electrode, *Anal. Chim. Acta*, 67, 155, 1973.
9. **Kumar, A., Chapateau, E., Czech, B. P., Gebauer, C. R., Chimenti, M. Z., and Raimondo, O.**, Chromogenic ionophore-based methods for spectrophotometric assay of sodium and potassium in serum and plasma, *Clin. Chem.*, 34, 1709, 1988.
10. **Singer, R. B., Shohl, J., and Bluemle, D. B.**, Simultaneous determination of pH, CO_2 content and cell volume in 0.1 ml aliquots of cutaneous blood, *Clin. Chem.* 1, 287, 1955.
11. **Morf, W. E.**, *The Principles of Ion Selective Electrodes and of Membrane Transport*, Elsevier, New York, 1981, 264.

Trace Elements

J. Thomas Hindmarsh, M.D.

This section is dedicated to the fond memory of Dr. Andrew Sass-Kortsak, a great scientist and a good friend.

INTRODUCTION

J. Thomas Hindmarsh

DEFINITION

More than 60 elements can be identified in human tissues. They can be classified as "macro elements" (carbon, hydrogen, oxygen, nitrogen, sulfur, calcium, phosphorus, potassium, sodium, chlorine, magnesium) and "trace elements" (human body content less than 2 g — this definition would exclude iron). The approximate human body content of some trace elements is shown in Table 1.[1]

A precise definition for a trace element is not possible. A broad description might be that it is "present in trace amounts and is necessary for optimal functioning of the organism; deficiency consistently results in impairment of a function." A more comprehensive definition might include "its concentration in particular tissues is relatively constant, and its withdrawal produces similar structural and physiological abnormalities in different species; these abnormalities are prevented or reversed by addition of the element to the diet."[2]

CLASSIFICATION

In human nutrition, trace elements can be classified as "essential", "probably essential", "possibly essential", "non-essential", and "toxic".

Essential Trace Elements

A list of these is shown as part of Table 2. These are elements for which a distinct deficiency syndrome has been described in man.

Probably Essential Trace Elements

These are elements that are components of essential processes in man and/or comparable animals, but for which a distinct deficiency syndrome has not been described in man in more than a few cases.

Possibly Essential Trace Elements

These are those whose deficiency has been described to produce some deleterious effect in animals and/or are components of essential processes in man and/or animals and/or higher plants, but for which a distinct deficiency syndrome has not been described in man. The distinction between "possibly" and "probably" essential trace elements is ill defined.

Nonessential Trace Elements

These are trace elements that appear at present to have no role in human metabolism. Arsenic is included in this group rather than as a possible essential trace element as some would classify it (see Arsenic section).

Toxic Trace Elements

This definition is somewhat arbitrary as all elements are toxic if consumed in large excess. However, some elements (particularly those with high molecular weights) are much more toxic than others. Table 2 is not intended to be all inclusive.

Underwood has observed that only 3 of some 27 elements thought to be essential for life have an atomic number greater than 34 and most of the essential trace elements have atomic numbers between 23 and 34.[2] Grandjean has demonstrated that trace elements can be class-

Table 1
HUMAN BODY CONTENT OF SOME TRACE ELEMENTS

Iron	4—5 g	Chromium	<6 mg
Copper	80—120 mg	Selenium	14 mg
Zinc	1—2 g	Molybdenum	<10 mg
Manganese	12—20 mg	Iodine	10—20 mg

Table 2
CLASSIFICATION OF THE ROLES OF TRACE ELEMENTS IN HUMANS

Essential trace elements	
Cobalt[a]	Iron
Copper	Zinc
Iodine	
Probably essential trace elements	
Chromium	Molybdenum
Manganese	Selenium
Possibly essential trace elements	
Boron	Nickel
Bromine	Silicon
Fluorine	Tin
Lithium	Vanadium
Non-essential trace elements	
Aluminum	Mercury
Antimony	Osmium
Arsenic	Platinum
Barium	Rubidium
Beryllium	Silver
Bismuth	Thallium
Cadmium	Titanium
Germanium	Uranium
Gold	Zirconium
Lead	
Toxic trace elements	
Aluminum	Gold
Arsenic	Lead
Barium	Mercury
Beryllium	Silver
Cadmium	Uranium
Cobalt	

[a] Cobalt is a component of cobalamin, but elemental cobalt may also be essential.

ified into two major groups from the perspective of environmental exposure: those to which exposure is relatively constant and relatively high, and those to which exposure is irregular and usually at a low level.[3] During the evolutionary process many elements in the first group seem to have attained useful roles in human metabolism, whereas many of those in the latter are toxic because animals have not needed to evolve tolerance to them. This phenomenon is displayed in Figure 1. Although the heavier metals (density >4 g/cc) tend to be toxic (except for beryllium — density 1.85), toxicity is probably more related to affinity of a metal for reactive groups such as the sulfhydryl groups of enzymes and its ability to replace other elements in essential processes and thereby inactivate them.

RECOMMENDED DIETARY ALLOWANCES

Recommended dietary allowances for humans for essential and some probably essential trace elements are shown in Table 3.

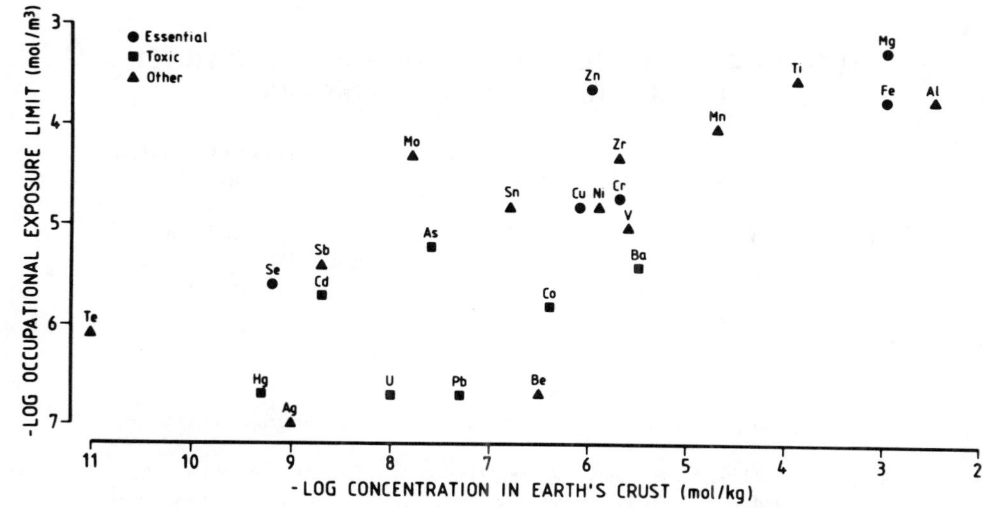

FIGURE 1. Relationship between the molar metal concentraton in the earth's crust and the molar concentration limit for acceptable occupational exposures to nonirritant, inorganic compounds. (Reproduced from Grandjean, P., *Ann. Clin. Lab. Sci.*, 16, 67, 1986. With permission.)

Table 3
RECOMMENDED DIETARY ALLOWANCES FOR SOME TRACE ELEMENTS (mg/d) (ADULT HUMANS)

Chromium	0.05—0.2	Manganese	2.5—5.0
Copper	2.0—3.0	Molybdenum	0.15—0.5
Iodine[a]	0.15	Selenium	0.05—0.2
Iron[b]	10.0	Zinc[c]	15.0

[a] Extra iodine supplement during pregnancy 25 mg; and during lactation 50 mg.
[b] 8 mg iron supplement in women prior to menopause and a supplement of 30—60 mg during pregnancy and lactation.
[c] Extra zinc supplement during pregnancy 5 mg; and during lactation 10 mg.

TRACE ELEMENTS IN PARENTERAL NUTRITION

Patients receiving total parenteral nutrition (TPN) demonstrate some depletion of body reserves of trace elements 2 to 7 weeks after institution of this therapy.[4] Patients requiring TPN frequently already have compromised trace element reserves, because these often have been depleted by protracted illness and, in the case of zinc, the increased urinary losses associated with the metabolic response to trauma and/or starvation. Recommended trace element supplementation is shown in Table 4 and has been reviewed recently.[5]

DEFINING TRACE ELEMENT NEEDS

This is a complex process, as patients who present with features suggestive of trace element deficiencies commonly have multiple deficiencies. Also, the effects of minor imbalances may take months to become clinically apparent and are often subtle in nature. Also, there are complex interactions between trace elements at a variety of sites within the body and, therefore, the dietary intake and body status of one trace element may influence the requirements of another. These interactions may occur during intestinal absorption, during transport in the blood, and when the element is bound at its site of action.[1,6]

Table 4
SUGGESTED DAILY I.V. INTAKE OF ESSENTIAL TRACE ELEMENTS FOR PATIENTS ON TPN

	Stable adult	Acute catabolic state — adult[a]	Stable adult with intestinal losses[a]
Zinc	2.5—4.0 mg	Add 2.0 mg	Add 12.2 mg/l small bowel fluid loss; 17.1 mg/kg of diarrhea or ileostomy fluid loss
Copper	0.5—1.5 mg		
Chromium	10—15 μg		20μg
Manganese	0.15—0.8 mg		

[a] Frequent monitoring of blood levels in these patients is essential for proper dosage.

Adapted from Shils, M. E., Burke, A. W., Greene, H. L., Jeejeebhoy, K. N., Prasad, A. S., and Sandstead, H. H., Guidelines for essential trace element preparations for parental use, JAMA, 241, 2051, 1979.

Table 5
METHODS FOR ASSESSING TRACE ELEMENT STATUS

Measurement of trace element concentration in body fluids or tissues
Assay of metalloenzyme activity in body fluids or tissues
Trace element balance studies
Biochemical and clinical responses to trace element administration
In vitro and *in vivo* assessment of body trace element content by isotopic techniques

ROLE OF ESSENTIAL TRACE ELEMENTS

The roles of trace elements in human metabolism are diverse and will be discussed in the individual sections, but, in general, metals have structural characteristics that particularly fit them for their physiological roles. Thus, they are commonly associated with the active sites of enzymes and other reactive molecules. Their specific functions include stabilization of the tertiary structure of their associated proteins, and participation in redox reactions, electron transfers, substrate binding, and carriage of respiratory gases.

ASSESSMENT OF BODY TRACE ELEMENT STATUS

Methods for assessing human trace element status have been well reviewed by Delves;[7] they are listed in Table 5.

The measurement of trace element concentrations in body fluids such as serum may be helpful in the diagnosis and management of trace element excess or deficiency. Thus, blood lead is probably still the most useful parameter for assessing the severity of lead poisoning and serum zinc is still a useful index in cases of zinc deficiency. When using the serum level of a trace element to assess body status, problems may occur if the serum is rapidly cleared of the element. The level then may reflect only the trace element intake over the recent time; this is the case with arsenic when serum levels reflect the status of only the recent few days. In these circumstances, the investigator has to analyze a tissue whose content is more reflective of the long-term status of the element, for example, hair in the case of arsenic. In contrast, measurement of the tissue content of a trace element may be a less sensitive indicator of deficiency, as the tissue reserves may take longer to become depleted; some have claimed this to be the case when leukocyte zinc is compared with plasma zinc in the assessment of zinc deficiency.[7]

Despite earlier optimism, hair analysis has not proved to be a reliable indicator of body trace element status, except, perhaps, in the detection of chronic arsenic poisoning and then only when external contamination by arsenic can be excluded. Hair is subject to weathering, external contamination (which is often impossible to remove), and adulteration. Reference intervals for normal subjects are extremely wide, as is the inter-hair variation of trace element concentration on an individual's head.[8] Pubic hair analysis does not offer any benefit.[9]

Metalloenzyme activity in the serum and tissues may be reduced in trace element deficiency. Thus, plasma alkaline phosphatase and lactate dehydrogenase activities in serum are reduced in some patients whose dietary zinc intake is restricted. Neutrophil alkaline phosphatase activity may show similar changes.[10] However, enzyme changes, in general, are too non-specific to be of significant use in these circumstances.

Trace element balance studies are very labor intensive and require meticulous supervision; duplicate diets must be prepared and analyzed for their trace element content. These studies require the services of dedicated dietary and nursing staff. They may provide useful information, but cannot be performed outside research establishments.

The measurement of peak plasma levels following dosing with a trace element may provide information about body status. Thus, peak concentrations of zinc are exaggerated in vegetarian humans, implying that this might indicate some clinical zinc deficiency in this group.[7] Rapid clinical improvement after the specific replacement of a suspected deficient trace element in the diet is strong circumstantial evidence that a deficiency existed.

Assessment of tissue trace element status by assessing the uptake of stable or radioactive isotopes *in vitro* or *in vivo* may prove useful in some circumstances. Bone lead and calcium contents can be measured *in vivo* using isotope-induced X-ray emission,[7] but these techniques are feasible only in research establishments.

REFERENCE INTERVALS IN BODY FLUIDS AND TISSUES

These have been extensively documented.[8,11] Normal levels for some trace elements in body fluids (cobalt, chromium, manganese, molybdenum, vanadium) are close to or below the limits of sensitivity for many analytical methods and cannot be measured easily. Also, blood and/or tissue levels may not be representative of the body status of the element. This has led to the development of standards for maximum exposure to toxic metals from the environment and in the workplace.[3] Thus, maximum levels of these toxic metals in drinking water and air have been established, usually by determining a toxic threshold for the metal in laboratory animals and extrapolating these data to man while building a safety factor into the calculation.

SPECIMEN COLLECTION AND STORAGE

Many early trace element determinations on body fluids and tissues were subject to contamination during collection and processing, and meticulous efforts must be made to avoid this problem. This is demonstrated by the progressive reduction in the so-called normal ranges for some trace elements over recent years. Contamination may occur from stainless steel scalpels and venipuncture needles, from skin antiseptics, and from contaminated specimen containers and stoppers, to mention a few. Acid washing of specimen containers, and low-metal stoppers are essential. Often, storage in polycarbonate tubes is the only safe method for long-term storage. Versieck recommends blood collection using a Teflon® catheter through which 20 to 40 ml of blood is allowed to pass before sample aliquots are taken.[11] Less extreme procedures are recommended by Erickson et al., but only time will tell if these are adequate.[12] Analysts must consider all possible sources of contamination for their analyte and make every effort to avoid such contamination; ideally, guidelines should be published for each trace element.

ANALYSIS

Earlier and imprecise colorimetric methods have now largely been replaced by more accurate and precise techniques such as electrothermal atomic atomization, electronically induced plasma emission, voltammetry, and neutron activation. All of these have added immensely to our knowledge of trace element metabolism and have allowed this to become one of the major growth areas in clinical chemistry over the last decade. Because of the increased public interest in environmental pollution and nutrition, this growth is certain to continue.

REFERENCES

1. **Aggett, P. J.,** Physiology and metabolism of essential trace elements: an outline, *Clin. Endocrinol. Metab.*, 14, 513, 1985.
2. **Underwood, E. J.,** *Trace Elements in Human and Animal Nutrition,* 4th ed., Underwood, E. J., Ed., Academic Press, New York, 1977, 2.
3. **Grandjean, P.,** Reference intervals for toxic metals: problems and prospects, *Ann. Clin. Lab. Sci.*, 16, 67, 1986.
4. **Jeejeebhoy, K. N.,** Trace element requirements during total parenteral nutrition, in *Clinical, Biochemical, and Nutritional Aspects of Trace Elements,* Prasad, A. S., Ed., Alan R. Liss, New York, 1982, 469.
5. Working Conference in Parenteral Trace Elements, *Bull. N.Y. Acad. Med.*, 60, 115, 1984.
6. **Kirchgessner, M., Schwarz, F. J., and Schnegg, A.,** Interactions of essential metals in human physiology, in *Clinical, Biochemical, and Nutritional Aspects of Trace Elements,* Prasad, A. S., Ed., Alan R. Liss, New York, 1982, 477.
7. **Delves, H. T.,** Assessment of trace element status, *Clin. Endocrinol. Metab.*, 14, 725, 1985.
8. **Sky-Peck, H. H. and Joseph, B. J.,** The use and misuse of human hair in trace metal analysis, in *Chemical Toxicology and Clinical Chemistry of Metals,* Brown, S. S. and Savory, J., Eds., Academic Press, London, 1983, 159.
9. **Taylor, A.,** Usefulness of measurements of trace elements in hair, *Ann. Clin. Biochem.*, 23, 364, 1986.
10. **Prasad, A. S.,** Zinc, in *Clinical, Biochemical, and Nutritional Aspects of Trace Elements,* Prasad, A. S., Ed., Alan R. Liss, New York, 1982, 3.
11. **Versieck, J.,** Trace elements in human body fluids and tissues, *CRC Crit. Rev. Clin. Lab. Sci.*, 22, 97, 1985.
12. **Ericson, S. P., McHalsky, M. L., Rabinow, B. E., Kronholm, K. G., Arceo, C. S., Weltzer, J. A., and Ayd, S. W.,** Sampling and analysis techniques for monitoring serum for trace elements, *Clin. Chem.*, 32, 1350, 1986.

ALUMINUM

John Savory and Michael R. Wills

Exposure to aluminum is unavoidable since it is the third most common element in the Earth's crust. In certain situations, most notably individuals on hemodialysis treatment, aluminum toxicity presents a serious hazard to the patient.

It is only 100 years since a practical procedure for the large-scale production of aluminum was developed. Since that time, aluminum has found applications in construction, electrical equipment, furnishings, transportation, containers, pharmaceuticals, and many other items encountered in daily life.

One of the first reports of poisoning attributed to aluminum was in 1921.[1] The patient, a metal worker, had memory loss, tremor, jerking movements, and incoordination. Five years later attention was drawn to the potential health hazards of the aluminum present in city drinking water and various medicines.[2]

Industrial aluminum exposure has not been extensively studied. There was one report over 20 years ago[3] of a ball-mill room worker who developed encephalopathy and pulmonary fibrosis associated with inhalation of aluminum dust. Increased aluminum concentrations in urine of industrially exposed individuals have been documented.[4] A higher incidence of respiratory complaints has been reported among aluminum smelter workers.[5] No significant effects have been found on musculoskeletal or hematological parameters in these workers.[6] Studies of the incidence of cancers among aluminum industry workers have been inconclusive.[7,8] Most reviews of industrial aluminum exposure have concluded that the aluminum industry entails no greater occupational health hazard than other comparable industries.[9]

CLINICAL CONSEQUENCES OF ALUMINUM TOXICITY

There is now substantial evidence that in patients with end-stage chronic renal failure managed by long-term intermittent hemodialysis, there is hyperaluminemia with accumulation of aluminum in various tissues. In the absence of renal function, aluminum cannot be eliminated in the urine. The excess in serum and tissue results from intestinal absorption of aluminum salts taken by mouth and from passage of aluminum across the dialysis membrane. The aluminum content of the dialysate depends on the content of the water with which it is made. Aluminum is normally present in raw water; the concentrations are usually low in ground waters and are almost always high in surface waters.[10]

DIALYSIS ENCEPHALOPATHY

A progressive fatal neurological syndrome in patients on long-term intermittent hemodialysis treatment for chronic renal failure was reported in 1972.[11] The first manifestation in these patients was a speech disorder, followed by dementia, convulsions, and myoclonus. The syndrome was considered to be a metabolic disorder due to accumulation of trace metals. Alfrey and colleagues[12] subsequently proposed that this syndrome ("dialysis encephalopathy" or "dialysis dementia"), which arises after 3 to 7 years of treatment, may be due to aluminum intoxication. Their proposal was based on the findings of increased aluminum content in brain, muscle, and bone tissues in affected patients. Dialysis encephalopathy was the major cause of death in one dialysis center.[13]

It was originally proposed that the increased tissue aluminum content in patients with the dialysis encephalopathy syndrome was derived from the oral aluminum hydroxide which had been administered therapeutically to control their serum phosphate concentration. In support of this proposal was the evidence of a positive aluminum balance in patients with chronic renal failure on hemodialysis treatment who were taking oral aluminum-containing phosphate-binding gels.[14] In one hemodialysis center the dialysis fluid was specifically shown to be the source of the aluminum.[15] Most of the existing evidence is consistent with the notion that the excess brain tissue aluminum of dialysis encephalopathy stems from dialysate with some contribution from the gastrointestinal tract. That aluminum is the major toxic factor in the dialysis encephalopathy syndrome is now widely accepted.[16-18] It is also accepted that this syndrome has a common etiology with one of the osteomalacic components of dialysis osteodystrophy.[19,20] Hyperaluminemia and dialysis encephalopathy may occur not only in patients on hemodialysis treatment, but also in those on peritoneal dialysis[21] and in some patients who have not been dialyzed.[22,23] The nondialyzed patients are usually children with renal failure who were on oral treatment with aluminum hydroxide.

DIALYSIS OSTEODYSTROPHY

Bone pain, as a consequence of metabolic bone disease, is a common symptom in chronic renal failure patients on long-term intermittent hemodialysis. The progressive metabolic bone disease in these patients should be called "dialysis osteodystrophy" — a term that distinguishes it, and some aspects of its pathogenesis, from renal osteodystrophy in the undialyzed patient. The osteomalacic component of dialysis osteodystrophy is particularly troublesome in being associated with a high incidence of fractures. The reported incidence and rate of progression of the various components of dialysis osteodystrophy has varied not only between countries, but also between dialysis centers within a country, despite apparently similar dialysis techniques.[24,25] The characteristic of dialysis osteodystrophy is that it progresses or develops despite the maintenance of plasma calcium and magnesium and their fractions, at concentrations which in a healthy person would not interfere with bone mineralization.[26] In the bone mineralization process, the availability of phosphate for the formation of hydroxyapatite crystals, is of crucial importance. Phosphate deficiency, with hypophosphatemia, is a recognized cause of osteomalacia; administration of aluminum-containing medications can induce a state of phosphate deficiency.

Attention was drawn to the possible role of aluminum in renal osteodystrophy in 1971 when Parsons and co-workers reported an increased bone content of the element in patients with end-stage chronic renal failure.[27] The aluminum content tended to be highest in patients who had been longest on hemodialysis. The bulk of the evidence[20,28-31] now supports the notion that aluminum is at least a very important etiological factor in one type of osteomalacic dialysis osteodystrophy. The particular type of osteomalacia induced by aluminum is progressive, associated with a myopathy, and relatively resistant to treatment with either vitamin D or its biologically active metabolites; in some series it has been associated with a high incidence of hypercalcemia.[32] The usual source of the aluminum seems to be the tap water from which the dialysate is made. In addition, there is undoubtedly some intestinal absorption from the phosphate-binding gels, and in some patients this seems to have been the dominant factor.[23,33] The mechanism for the disordered bone formation induced by an excess of aluminum remains to be clarified.

Nearly all patients with chronic renal failure have a normochromic normocytic anemia. The etiology of the anemia has several components; it is not simply the end-result of a decrease in renal production of erythropoietin.[34] In 1978, Elliot and MacDougall reported that anemia was associated with osetomalacic dialysis osteodystrophy and with dialysis encephalopathy. In the encephalopathy patients the hemoglobin fell before onset of neurological symptoms and in the osteodystrophy patients before bone complications; therefore, anemia may be a useful early indication of aluminum toxicity. With regard to heme synthesis, aluminum toxicity has been linked with porphyria in laboratory animals, particularly in those with renal failure, and with hemodialysis-related porphyria cutanea tarda.[35]

Soft-tissue or metastatic extraskeletal calcification is common in chronic renal failure. Calcification may occur in the eyes, in or around joints, in viscera, in the skin, and in arteries.[34] The mechanism of metastatic calcification and, in particular, arterial calcification needs clarification. In view of the observations that aluminum can initiate the precipitation of calcium apatite,[36] aluminum may have a role in the pathogenesis of the metastatic calcification of chronic renal failure.

In patients with normal renal function aluminum has recently been implicated as a factor in the etiology of the amyotrophic lateral sclerosis and Parkinsonism-dementia found in the indigenous (Chamorro) population of Guam.[37] In brain tissue from two patients with the disease, an accumulation of aluminum was demonstrated within the nuclear region and perikaryal cytoplasm of neurofibrillary tangle-bearing hippocampal neurons.[37] Soil and drinking water from areas in which there was a high incidence of these disorders had a high aluminum content with low concentrations of calcium and magnesium.

Aluminum has also been implicated as a neurotoxic agent in the pathogenesis of Alzheimer's disease. In brain tissue from patients with this disorder accumulations of aluminum have been identified within the nuclear region of a high percentage of neurons containing neurofibrillary tangles.[38,39]

ALUMINUM MONITORING

Recommendations on aluminum monitoring in hemodialysis programs have been made in a report of an International Workshop held to study the hazards of aluminum toxicity related to renal insufficiency.[40]

Reverse osmosis is the recommended method for water treatment, since it provides water with a low aluminum content (<10 μg/l), a low content of other cations, and eliminates organic contaminants which may contribute to the problems associated with hemodialysis.

Deionization of the water is the next preferred mode of treatment, but it is recognized that it requires more careful monitoring, since aluminum loading of the resin can occur with the danger of subsequent erratic elution.

Concentrations of water aluminum not exceeding 10 to 15 μg/l are necessary to minimize significant body uptake of aluminum during dialysis. These concentrations can be achieved using reverse osmosis and are recommended.

Between-batch differences in the aluminum concentration of dialysate concentrate exist. The recommendation is that the final concentration of the dialysate after dilution with treated water should be less than 15 μg/l and preferably less than 10 μg/l.

There is evidence that CAPD treatment can remove some of the aluminum present in the plasma. The effectiveness of transfer of aluminum species during this type of treatment is of importance and needs further investigation.

Because of these factors, the concentration of aluminum in the CAPD fluid should be less than 10 μg/1.

In view of the problems with maintaining extremely low serum aluminum concentrations, especially with patients taking aluminum binders, the following criteria are recommended for the management of dialysis patients:

- Serum aluminum concentrations should never exceed 200 μg/l.
- Concentrations over 100 μg/l should be viewed with concern and require careful monitoring.
- Concentrations form 60 to 100 μg/l appear to cause no problems to the patient in the short term.

FREQUENCY OF MONITORING

Hospital Hemodialysis

It is considered appropriate for the main water supply, the treated water, and the dialysis fluid to be monitored weekly. If a reverse osmosis system is used and has been shown to be reliable, then monthly monitoring should be adequate. The dialysis fluid to be monitored should be sampled immediately before it comes into contact with the membrane, since contamination by metal parts of the system is a possibility. Serum aluminum should be monitored in all dialysis patients four to six times each year.

Home Hemodialysis

It is considered appropriate for the main water supply, the treated water, the dialysis fluid, and serum aluminum to be monitored when the patient comes for checkup (usually two or three times per year). If there are important fluctuations in the aluminum content of the main water supply and if reverse osmosis is not employed, then more frequent monitoring is desirable.

Peritoneal Dialysis and CAPD

Serum aluminum should be monitored every 2 or 3 months.

ANALYTICAL METHODS FOR THE DETERMINATION OF ALUMINUM

The key to the success in aluminum monitoring lies in the availability of accurate and precise analytical methods together with guidelines for contamination-free specimen collection. Many problems have existed in the past with aluminum methodology and these have been detailed in the study of Versieck and Cornelis.[41] on the variability in reported normal plasma concentrations. Even with reliable methods mean normal values varied between 2 and 42 μg/l. Other reports have listed normal mean concentrations ranging from 72 to 1460 μg/l, although obvious interferences were present in these methods.[41]

Several methods are available for the determination of total aluminum in biological materials. Chemical and physiochemical methods are insensitive and inaccurate; X-ray fluorescence is specific but lacks sensitivity; neutron activation analysis is complex and subject to interferences, although it is a very sensitive technique. Electrothermal atomic absorption spectrometry (EAAS) is the most widely used technique and can produce reliable results provided the matrix effects on standardization are recognized and corrected.

Recently, Bettinelli et al.[42] have made a thorough evaluation of the direct measurement of aluminum in serum using EAAS with the stabilized temperature furnace with the L'vov platform or stabilized temperature platform furnace. Their recommendation was to use pyrolytically coated graphite tubes and to keep the method as simple as possible with minimal sample pretreatment. Most EAAS procedures have been developed on Perkin-Elmer instruments with the use of auto-sampling to improve precision. Pyrolytically coated graphite tubes are recommended together with a pyrolytic graphite platform. These tubes minimize reactions between the analyte and the graphite tube. Argon is preferred over nitrogen as the purge gas, since argon produces a larger and less variable signal. Furnace signals occur rapidly and in order to follow the absorbance profile, it is necessary to incorporate fast digital electronics into the instrument. The absorbance signal must be integrated to provide accurate results; peak absorbance signals mostly will produce inaccurate quantitative measurements. Measurements also require elimination of matrix effects between samples and standards usually accomplished by some type of background correction. The Zeeman correction system probably provides the most sensitive and reliable results.

Matrix modifiers are an important part of the modern atomic absorption technology. Most analytes require a matrix modifier in order to stabilize the analyte until thermal conditions have become constant. For aluminum, magnesium nitrate permits the use of high char temperatures without losses. The procedure recommended by the present authors is based on that recommended by Bettinelli et al.[42] and uses a stabilized temperature platform with pyrolytically coated graphite tubes. In this procedure serum analysis standards and serum samples are diluted with an equal volume of an aqueous solution containing $Mg(NO_3)_2$ (2 g/l). The autosampler is programmed to deliver a 10-μl aliquot of the sample or standard onto the platform for final analysis. The EAAS program used in the authors' laboratory on the Perkin-Elmer Model 5000 instrument is given in Table 2. Aqueous standards are prepared fresh daily. Linearity ranges are from 0 to 125 μg/l. Standard and serum samples are diluted with an equal volume of an aqueous solution containing 1.4 g/l $Mg(NO_3)_2$. Concentration values are calculated using linear least-square regression.

Leung and Henderson[43] prepare their standards using a serum pool containing a minimal amount of endogenous aluminum, whereas Bettinelli et al.[42] recommend aqueous standards. The use of simple aqueous standards is to be preferred provided that there are no matrix effects from serum.

The present authors have developed a procedure which minimizes any potential matrix effects by protein precipitation.[44] This protein precipitation technique was originally developed for serum nickel[45] and markedly reduces matrix effects in the final atomic spectroscopic analysis.

The sensitivity of all of these serum methods only extends down to approximately 1 μg/l. Probably the upper level of the physiological range for serum aluminum is 2 μg/l, which makes present techniques of limited value in investigating physiological levels.

For urine analysis, sample aliquots are diluted 1:1 with distilled water before application to the atomic absorption spectrometric (AAS) stabilized temperature platform.[43] Fecal analysis requires considerably more complicated preparation steps than serum or urine. The procedure developed in the authors' laboratory[46] is summarized as follows: frozen specimens are thawed and weighed in the orginal plastic container used for collection. Distilled water is added (1 ml/2 g feces) and the sample is homogenized in a sealed can on a paint shaker. A 10-ml aliquot is ashed at 550° in a muffle furnace, dissolved in dilute HNO_3, and analyzed by EAAS.

Solid tissues must be homogenized, dried, ashed, and/or dissolved to produce a liquid sample prior to EAAS analysis. Bone samples for analysis are washed free of marrow by a strong stream of distilled water. Any existing fat or muscle is scraped with an aluminum-free obsidian scapel or blade found to be aluminum free. The bone sample can then be processed by several methods to obtain a solution for injection into the graphite furnace.

Soft tissue samples, such as brain, liver, muscle, etc., must be homogenized before processing, and this can be accomplished easily by pummeling the tissue in a "Stomacher" blender (Fisher Scientific). Distilled water (5 ml) is added to the bag with the tissue, the bag is sealed, and placed in the blender and blended 5 to 15 min, which completely homogenizes the sample.[47] The homogenate can then be processed as described for the fecal samples. Brain samples are dried and ashed, then extracted with EDTA. Muscle samples are dried, grounded, extracted in solvents, redried, and extracted in EDTA.[48] Tissue samples can be processed (after homogenization) with hot nitric acid in a Kjeldahl flask. Soft tissue also can be processed for aluminum analysis by using tetramethyl ammonium hydroxide (TMAH) as a dissolving solution.

Aluminum can be localized in tissue by histochemical staining, electron probe X-ray microanalysis, and by laser microprobe mass spectrometery. These last two are sophisticated techniques which will be powerful tools in future studies of aluminum toxicity.

SPECIMEN COLLECTION AND CONTROL OF CONTAMINATION

Blood Collection

Sample collection is, potentially, a major source of possible contamination. Needles, glass vacuum tubes, plastic syringes, anticoagulants, plastic tubes, and other equipment used in standard venipuncture methods are all potential sources of aluminum contamination and must be checked before introduction into routine use for specimen collection.

The blood collection procedure that has been in use in the authors' laboratory involves drawing blood with a stainless steel needle into a plain glass vacuum tube (No. 6430, Becton-Dickinson and Co., Rutherford, NJ 07070). The blood is allowed to clot for approximately 20 min before it is centrifuged; the serum is then transferred to a 17 × 100-mm polypropylene tube (Falcon, Oxnard, California).

Transport and Storage

Blood drawn into glass vacuum tubes should be separated within 1 h after collection. Variable results may occur if blood is allowed to remain in contact with glass. The cap on the plastic tube should be tight fitting. If the tube is to be shipped, the tube's cap must be wrapped in such a way as to prevent any leakage during handling. There are no data available in the literature that storage of the specimen at room temperature results in any loss of aluminum when compared with storage at 4°C. Specimens for aluminum analysis have been received in the authors' laboratory both at room temperature and on ice packs. Several of these samples were from normal volunteers and the resulting aluminum values were all within our reference range. This finding suggests that there is no problem with shipping the specimen at room temperature.

Tissue samples should be placed into aluminum-free plastic containers. Brain tissue needs to be frozen until analysis. Bone can be kept at room temperature.

Each step of the specimen collection procedure for aluminum analysis may potentially cause contamination and must be scrutinized to determine if it is adding any aluminum to the sample. Each laboratory should check all materials used in the collection process before sampling patients and reporting any results. Once the collection procedure has been established, regular scheduled checks should be employed to verify that little or negligible contamination is being contributed by the technique. A quality control check for the blood collection technique could involve drawing blood from healthy individuals once a week. The individual chosen should be resident in the same area as the patients who are being drawn for aluminum levels; this would enable monitoring of both the sample collection procedure and also the transport system for any spurious aluminum contamination. An aluminum value within the reference range would be acceptable for this specimen. Tissue samples from healthy, nondiseased persons at autopsy can serve as controls for bone, brain, and muscle aluminum determinations.

SOURCES OF CONTAMINATION IN ANALYSIS

As in specimen collection, every item used during analysis should be regarded as a potential source of aluminum contamination. Glassware, pipet tips, plastic tubes, sample cups, the working environment, and the water utilized must all be checked to ensure that they add negligible amounts, if any, of aluminum to the procedure. The room chosen for the analyses should have a limited access to ensure that dust in the working environment is being circulated as little as possible. Sample preparation should be carried out in an environmental laminar flow hood; this precaution helps minimize contamination by dust particles.

Water utilized for standard curve preparation, rinsing of glassware, and sample dilutions must be of high purity. The water should produce a resistivity of at least 18 MΩ; this characteristic should be continuously monitored.

All glassware should be made aluminum free and this can be accomplished by using either acid solutions or saturated disodium EDTA. The acids used should be of ultrapure quality. The method used in our laboratory is based on the acid cleaning technique described by Moody and Lindstrom.[49]

Contamination is a major problem encountered in the performance of aluminum analysis. All items utilized during collection, transport, and assay should be checked for unwanted aluminum contribution to the procedure. Only by taking these stringent precautions will one be able to produce results with confidence.

STANDARD REFERENCE MATERIALS AND QUALITY ASSURANCE

The availability of standard reference materials is of considerable importance in the application of reliable assays for aluminum. Aqueous standards are available for commercial sources, one being Fisher Scientific Company (Fair Lawn, New Jersey). This material is in acid solution and is a stock solution of 1 g/l. Dilution to working standards must incorporate an intermediate standard if accuracy is to be maintained. The National Bureau of Standards (Washington, D.C.) also provides an aqueous standard solution which contains 10 mg/l of aluminum in a 10% acid medium.

Quality assurance in aluminum measurements follows established principles used in all areas of clinical chemistry. Internal quality control materials usually pooled plasma at two or three different concentrations which we used on a daily basis. The concentrations of aluminum in these materials should coincide with various decision levels. Interlaboratory control materials also are an important part of any quality assurance scheme. Presently, such a progress is conducted by Dr. Andrew Taylor through the Robens Institute (University of Surrey, Guildford). The institute mails three water and three serum samples monthly to laboratories for aluminum analysis. The data and statistical information of the 38 participating laboratories are then made available for review.

One useful material to use for monitoring the performance of aluminum assays is plasma from normal individuals. Such a pool would have an aluminum concentration of approximately 5 μg/l and is useful for examining the recovery of aluminum added to normal serum.

THERAPY OF ALUMINUM OVERLOAD

The only method available for removing substantial quantities of aluminum is by chelation with desferrioxamine (DFO).[50-54] Aluminum is bound strongly to DFO which has a molecular weight of 600 Da and can be dialyzed readily. There is an acute rise in serum aluminum concentrations following DFO administration indicating mobilization of aluminum from the tissues. It has been demonstrated that aluminum can be removed from both bone trabeculae and bone marrow by repeated DFO administration.[51] Treatment of fracturing renal osteodystrophy by DFO provides symptomatic relief of dialysis encephalopathy and of osteomalacia.

The infusion of DFO recently has been proposed as a diagnostic test for aluminum-related osteodystrophy.[55,56] These workers[55] evaluated the effect of a standard i.v. dose of DFO on plasma aluminum concentrations in 54 patients on hemodialysis. Stainable bone aluminum, bone histologic findings, and bone aluminum content were studied. Baseline plasma aluminum concentrations of greater than 200 μg/l were associated with aluminum-related osteodystrophy (specificity, 93%), but concentrations of less than 200 μg/l did not exclude the diagnosis (sensitivity, 43%). After administration of DFO the increase in plasma aluminum concentration was 534 $\pm$ 260 (SD) and 214 $\pm$ 92 μg/l in patients with and without aluminum-related bone disease, respectively ($p < 0.001$), and correlated with the bone aluminum content ($r = 0.64$). An increment in plasma aluminum concentration of greater than 200 μg/l identified 35 of the 37 patients with aluminum-related osteodystrophy; sensitivity was 94% and specificity, 50%. These workers concluded that the DFO infusion test was noninvasive, well tolerated, and of value particularly in excluding the diagnosis of aluminum-related osteodystrophy.

SUMMARY

Aluminum is a ubiquitous element in the environment and has been demonstrated to be toxic, especially in individuals with impaired renal function. Not much is known about the biochemistry of aluminum and the mechanisms of its toxic effects. Most of the interest in aluminum has been in the clinical setting of the hemodialysis unit. In the latter aluminum toxicity occurs due to contamination of dialysis solutions, and treatment of the patients with aluminum-containing phosphate binding gels. Aluminum has been shown to be the major contributor to the dialysis encephalopathy syndrome and an osteomalacic component of dialysis osteodystrophy. Other clinical disturbances associated with aluminum toxicity are a microcytic anemia and metastatic extraskeletal calcification.

Major problems exist with the analytical methods for measuring aluminum which result from inaccurate techniques and contamination difficulties. The most widely used analytical technique is EAAS which can provide reliable measurements in the hands of a careful analyst. Considerable care must be taken in the collection, processing, and storage of specimens. Aluminum overload can be treated effectively by chelation therapy with DFO and hemodialysis.

REFERENCES

1. **Spofforth, J.,** Case of aluminum poisoning, *Lancet,* 1, 1301, 1921.
2. **Betts, C. T.,** *Aluminum Poisoning,* Research Publishing, Toledo, 1926.
3. **McLaughlin, A. I. G., Kazantzis, G., King, E., Teare, D., Porter, R. J., and Owen, R.,** Pulmonary fibrosis and encephalopathy associated with the inhalation of aluminum dust, *Br. J. Ind. Med.,* 19, 253, 1962.
4. **Sjorgren, B., Lundberg, I., and Lidums, V.,** Aluminum in the blood and urine of industrially exposed workers, *Br. J. Ind. Med.,* 40, 301, 1983.
5. **Chan-Yeung, M., Wong, R., Maclean, L., Tan, F., Schulzer, M., Enarson, E., Martin, A., Dennis, R., and Grzybowski, S.,** Epidemiologic health study of workers in an aluminum smelter in British Columbia: effects on the respiratory system, *Annu. Rev. Respir. Dis.,* 127, 465, 1983.
6. **Chan-Yeung, M., Wong, R., Tan, F., Enarson, D., Schulzer, M., Subbarao, K., Knickerbocker, J., and Grzybowski, S.,** Epidemilogic health study of workers in an aluminum smelter in Kitimat, B.C. II. Effects on musculoskeletal and other systems, *Arch. Environ. Health,* 38, 34, 1983.
7. **Anderson, A., Dahlberg, B. E., Magnus, K., and Wannag, A.,** Risk of cancer in the Norwegian aluminum industry, *Int. J. Cancer,* 29, 295, 1982.
8. **Rockette, H. E. and Arena, A. C.,** Mortality studies of aluminum reduction plant workers: potroom and carbon department, *J. Occup. Med.,* 25, 549, 1983.
9. **DeHamel, F.,** Aluminum smelters and health, *N.Z. Med. J.,* 1, 123, 1983.
10. **Miller, R. G., Kopfler, F. C., Kelty, K. C., Stober, J. A., and Ulner, N. S.,** The occurrence of aluminum in drinking water, *J. Am. Water Works Assoc.,* 76, 84, 1984.
11. **Alfrey, A. C., Mishell, J. M., Burks, J., Contiguglia, S. R., Rudolph, H., Lewin, E., and Holmes, J. H.,** Syndrome of dyspraxia and multifocal seizures associated with chronic haemodialysis, *Trans. Am. Soc. Artif. Intern. Org.,* 18, 257, 1972.
12. **Alfrey, A. C., Legendre, G. R., and Kaehny, W. D.,** The dialysis encephalopathy syndrome. Possible aluminum intoxication, *N. Engl. J. Med.,* 294, 184, 1976.
13. **Burks, J. S., Alfrey, A. C., Huddlestone, J., Norenberg, M. D., and Lewin, E.,** A fatal encephalopathy in chronic haemodialysis patients, *Lancet,* 1, 764, 1976.
14. **Clarkson, E. M., Luck, V. A., Hynson, W. V., Bailey, R. R., Eastwood, J. B., Woodhead, J. S., Clements, V. R., O'Riordan, J. L. H., and DeWardener, H. E.,** The effect of aluminum hydroxide on calcium, phosphorus and aluminum balances, the serum parathyroid hormone concentration and the aluminum content of bone in patients with chronic renal failure, *Clin. Sci.,* 43, 519, 1972.
15. **Flendrig, J. A., Kruis, H., and Das, H. A.,** Aluminum and dialysis dementia, *Lancet,* 1, 1235, 1976.
16. **Siderman, S. and Manor, D.,** The dialysis dementia syndrome and aluminum intoxication, *Nephron,* 31, 1, 1982.

17. **McKinney, T. D., Basinger, M., Dawson, E., and Jones, M. M.,** Serum aluminum levels in dialysis dementia, *Nephron,* 32, 53, 1982.
18. **Davison, A. M., Walker, G. S., Ott, H., and Lewis, A. M.,** Water supply aluminum concentration, dialysis dementia, and effect of reverse-osmosis water treatment, *Lancet,* 2, 785, 1982.
19. **Alfrey, A. C.,** Dialysis encephalopathy syndrome, *Annu. Rev. Med.,* 29, 93, 1978.
20. **Parkinson, I. S., Ward, M. K., and Kerr, D. N. S.,** Dialysis encephalopathy, bone disease and anemia: the aluminum intoxication syndrome during regular haemodialysis, *J. Clin. Pathol.,* 34, 1285, 1981.
21. **Smith, D. B., Lewis, J. A., Burks, J. S., and Alfrey, A. C.,** Dialysis encephalopathy in peritoneal dialysis, *JAMA,* 244, 365, 1980.
22. **Nathan, E. and Pedersen, S.,** Dialysis encephalopathy in a non-dialysed uraemic boy treated with aluminum hydroxide orally, *Acta Paediatr. Scand.,* 69, 793, 1980.
23. **Griswold, W. R., Reznik, V., Mendoza, A., Trauner, D., and Alfrey, A. C.,** Accumulation of aluminum in a non-dialyzed uremic child receiving aluminum hydroxide, *Pediatrics,* 71, 56, 1983.
24. **Simpson, W., Kerr, D. N. S., Hill, A. V. L., and Siddiqui, J. Y.,** Skeletal changes in patients on regular hemodialysis, *Radiology,* 107, 313, 1973.
25. **Tatler, G. L. V., Baillod, R. A., Varghese, Z., Yong, W. B., Farrow, S., Wills, M. R., and Moorhead, J. F.,** Evolution of bone disease over 10 years in 135 patients with terminal renal failure, *Br. Med. J.,* 3, 315, 1973.
26. **Varghese, Z., Moorhead, J. F., and Wills, M. R.,** Plasma calcium and magnesium fractions in chronic renal failure patients on maintenance haemodialysis, *Lancet,* 2, 985, 1973.
27. **Parsons, V., Davies, C., Goode, C., Ogg, C., and Siddiqui, J.,** Aluminium in bone from patients with renal failure, *Br. Med. J.,* 4, 273, 1971.
28. **Drueke, T.,** Dialysis osteomalacia and aluminum intoxication, *Nephron,* 26, 207, 1980.
29. **Kanis, J. A.,** Osteomalacia and chronic renal failure, *J. Clin. Pathol.,* 34, 1295, 1981.
30. **Hodsman, A. B., Sherrard, D. H., Alfrey, A. C., Ott, S., Brickman, A. S., Miller, N. L., and Coburn, J. W.,** Bone aluminum and histomorphometric features of renal osteodystrophy, *J. Clin. Endocrinol. Metab.,* 54, 539, 1982.
31. **Robertson, J. A., Felsenfeld, A. J., Haygood, C. C., Wilson, P., Clarke, C., and Llach, F.,** Animal model of aluminum-induced osteomalacia: role of chronic renal failure, *Kidney Int.,* 23, 327, 1983.
32. **Boyce, B. F., Fell, G. S., Elder, H. Y., Junor, B. J., Elliot, H. L., Beastall, G., Fogelman, I., and Boyle, I. T.,** Hypercalcaemic osteomalacia due to aluminum toxicity, *Lancet,* 2, 1009, 1982.
33. **Kingswood, C., Banks, R. A., Bunker, T., Harrison, P., and MacKenzie, C.,** Fracture osteomalacia, CAPD and aluminum, *Lancet,* 1, 70, 1983.
34. **Wills, M. R.,** *Metabolic Consequence of Chronic Renal Failure,* 2nd ed., Harvey Miller and Medcalf, Aylesbury, 1978.
35. **Disler, P., Day, R., Burman, N., Bickkenhorst, G., and Eales, L.,** Treatment of hemodialysis-related porphyria cutanea tarda with plasma exchange, *Am. J. Med.,* 72, 989, 1982.
36. **Bachra, B. N. and VanHarskamp, G. A.,** The effect of polyvalent metal ions on the stability of a buffer system for calcification *in vitro, Calcif. Tissue Res.,* 4, 359, 1970.
37. **Perl, D. P., Gajdusek, D. C., Garruto, R. M., Yanagihara, R. T., and Gibbs, C. J., Jr.,** Intraneuronal aluminum accumulation in amyotrophic lateral sclerosis and Parkinsonism-dementia of Guam, *Science,* 217, 1053, 1982.
38. **Perl, D. P. and Brody, A. R.,** Alzheimer's disease: X-ray spectrometric evidence of aluminum accumulation in neurofibrillary tangle-bearing neurons, *Science,* 208, 297, 1980.
39. **Perl, D. P. and Pendlebury, W. W.,** Aluminum (Al) accumulation in neurofibrillary tangle (NFT) bearing neurons of senile dementia Alzheimer's type (SDAT): detection by intraneuronal X-ray spectrometry studies of unstained tissue sections, *J. Neuropathol. Exp. Pathol.,* 43, 349, 1984.
40. **Savory, J., Berlin, A., Courtoux, C., Yeoman, B., and Wills, M. R.,** Summary report of an international workshop on "the role of biological monitoring in the prevention of aluminium toxicity in man: aluminium analysis in biological fluids", *Ann. Clin. Lab. Sci.,* 13, 444, 1983.
41. **Versieck, J. and Cornelis, R.,** Normal levels of trace elements in human blood plasma or serum, *Anal. Chim. Acta,* 116, 217, 1980.
42. **Bettinelli, M., Baroni, U., Fontana, F., and Poisetti, P.,** Evaluation of the L'vou platform and matrix modification for the determination of aluminum in serum, *Analyst,* 110, 19, 1985.
43. **Leung, F. Y. and Henderson, A. R.,** Improved determination of aluminum in serum and urine with use of a stabilized temperature platform furnace, *Clin. Chem.,* 28, 2139, 1982.
44. **Brown, S., Bertholf, R. L., Wills, M. R., and Savory, J.,** Electrothermal atomic absorption spectrometric determination of aluminum in serum with a new technique for protein precipitation, *Clin. Chem.,* 30, 1216, 1984.
45. **Sunderman, F. W., Crisostomo, C., Reid, M. C., Hopfer, S. M., and Nomoto, S.,** Rapid analysis of nickel in serum and whole blood by electrothermal atomic absorption spectrophotometry, *Ann. Clin. Lab. Sci.,* 14, 232, 1984.

46. **Brown, S., Mendoza, N., Bertholf, R. L., Ross, R., Wills, M. R., and Savory, J.,** A study of the metabolic balance of aluminum in healthy adult males administered aceglutamid aluminum, *Res. Commun. Chem. Pathol. Pharmacol.,* in press.
47. **Sunderman, F. W., Jr., Marzouk, A., Crisostomo, M. C., and Weatherby, D. R.,** Electrothermal atomic absorption spectrophotometry of nickel in tissue homogenates, *Ann. Clin. Lab. Sci.,* 15, 299, 1985.
48. **LeGendre, G. R. and Alfrey, A.C.,** Measuring picogram amounts of aluminium biological tissue by flameless atomic absorption analysis of a chelate, *Clin. Chem.,* 22, 53, 1976.
49. **Moody, J. R. and Lindstrom, R. M.,** Selection and cleaning of plastic containers for storage of trace element samples, *Anal. Chem.,* 49, 2264, 1977.
50. **Ackrill, P., Ralston, A. J., Day, J. P., and Hodge, K. C.,** Successful removal of aluminum from a patient with dialysis encephalopathy, *Lancet,* 1, 692, 1980.
51. **Ackrill, P., Day, J. P., Garstang, F. M., Hodge, K. C., Metcalfe, P. J., Benzo, Z., Hill, K., Ralston, A. J., Ball, J., and Denton, J.,** Treatment of fracturing renal osteodystrophy by desferrioxamine, *Proc. Eur. Dial. Transplant Assoc.,* 19, 203, 1982.
52. **Brown, D. J., Dawborn, J. K., Ham, K. N., and Xipell, J. M.,** Treatment of dialysis osteomalacia with desferrioxamine, *Lancet,* 2, 343, 1982.
53. **Day, J. P., Ackrill, P., Garstand, F. M., Hodge, K. C., and Metcalfe, P. J.,** Reduction of the body burden of aluminum in renal patients by desferrioxamine chelation therapy, in *Chemical Toxicology and Clinical Chemistry of Metals,* Brown, S. S. and Savory, J., Eds., Academic Press, London, 1983, 353.
54. **Bertholf, R. L., Roman, J. M., Brown, S., Savory, J., and Wills, M. R.,** Aluminum hydroxide-induced osteomalacia, encephalopathy, and hyperaluminemia in CAPD. Treatment with desferrioxamine, *Periton Dial. Bull.,* 4, 30, 1984.
55. **Milliner, D. S., Nebeker, H. G., Ott, S. M., Andress, D. L., Sherrard, D. J., Alfrey, A. C., Spatopolsky, E. A., and Coburn, J. W.,** Use of the deferoxomine infusion test in the diagnosis of aluminum-related osteodystrophy, *Am. Int. Med.,* 101, 775, 1984.
56. **Berland, Y., Chorhon, S. A., Olmer, M., and Meunier, P. S.,** Predictive value of desferrioxamine infusion test for bone aluminum deposits in hemodialyzed patients, *Nephron,* 40, 433, 1985.

ARSENIC

J. T. Hindmarsh and R. F. McCurdy

INTRODUCTION

Arsenic (As) is a metalloid belonging to the group V(A) elements (N, P, As, Sb, and Bi). It may exist in three possible oxidation states: the metalloid (O), the trivalent (−3 or +3), and the pentavalent (+5) states. It binds covalently with most nonmetals and metals and forms stable organic compounds in both the trivalent and pentavalent states. Arsenic resembles phosphorus chemically and competes with phosphorous analogues for chemical binding sites. Its toxicity follows the following order: arsines > arsenites (inorganic trivalent compounds) > arsenoxides (organic trivalent compounds) > arsenates (inorganic pentavalent compounds) > arsonium compounds > metallic arsenic. The structures in common arsenicals are shown in Table 1.

ARSENIC EXPOSURE

Historical Review

The poisonous and medicinal properties of arsenic have been known since ancient times. Its use as an agricultural insecticide, herbicide, and fungicide became widespread in the 19th Century and reached a peak in the first half of the present century; chronic poisoning was a hazard of working in orchards and vineyards. Since 1950 the hydrocarbon insecticides have progressively replaced it in this role. Probably, as a consequence of its ubiquity, it was a popular homicidal and suicidal poison in the 19th Century. The development of the Marsh test, which allowed its easy detection in body fluids in the latter part of the 19th Century, is probably responsible for its decline in popularity as a homicidal agent and now it appears only occasionally in this role.

Fowler's solution (1% potassium arsenite) was used for many years as a treatment for psoriasis and asthma, and as a tonic. Its prolonged use produces chronic arsenic poisoning and it is no longer available. However, it occasionally reappears from time to time and arsenic is still widely used in patient medicines in the Far East. It was also used for many years as a growth stimulant for farm animals prior to antibiotics being used for this purpose.[1]

Arsenic is commonly found in trace amounts in drinking water and can occasionally reach toxic levels. Arsenopyrites (FeAsS) occurs naturally in gold-bearing rocks, but is relatively insoluble in water. Weathering of the rock may oxidize the insoluble sulfides to more soluble forms which can then enter the ground water, and most cases of poisoning from this source have occurred in populations whose water supply is contaminated by the drainage from gold mine waste rock tailings. This phenomenon is uncommon, however, in North America.[2-4] Weathering of the rock is not essential for arsenic to enter the ground water, however.[4] The ores of many metals (copper, iron, silver, cobalt, nickel, lead, gold, zinc, manganese, tin) often contain some arsenic and workers in the smelting industry may develop poisoning.[5] Shellfish and some fish may have arsenic content, but this appears to be in a relatively nontoxic organic form and is rapidly excreted in the urine.

Current Uses

Small amounts of arsenic are still used in agriculture.[6] It is also used in the glass industry as a clarifier, and in the electronics industry as gallium arsenide and its alloys in semiconductors. Large amounts are still used as wood preservatives.

Table 1
NAMES AND FORMULAS OF SOME ARSENIC COMPOUNDS

Name	Formula
Arsenic trioxide	As_2O_3
Arsenous acid	H_3AsO_3
Arsenite	$H_2AsO_3^{1-}$, $HAsO_3^{2-}$, AsO_3^{3-}
Arsenic pentoxide	As_2O_5
Arsenic acid	H_3AsO_4
Arsenate	$H_2AsO_4^{1-}$, $HAsO_4^{2-}$, AsO_4^{3-}
Arsanilic acid	$C_6H_4NH_2AsO(OH)_2$
Arsenobetaine	$(CH_3)_3As^+CH_2COOH$
Dimethylarsinic acid	$(CH_3)_2AsO(OH)$
Methylarsonic acid	$CH_3AsO(OH)_2$

PHYSIOLOGY

Absorption from the Gastrointestinal Tract

The efficiency of absorption of inorganic compounds from the gastrointestinal tract appears to be related to their water solubilities. Organic arsenicals such as arsanilic acid are poorly absorbed, whereas the organic form of arsenic found in fish and crustaceans is readily absorbed.

Absorption from the Lungs

It is always difficult to determine whether inhaled arsenic is absorbed directly from the alveolar surfaces or is regurgitated by ciliary action in the bronchi and subsequently swallowed. However, appreciable amounts of arsenic appear in the liver and spleen of rats 1 h after intratracheal administration, implying that direct absorption from the lung surfaces can occur.[7] Two human subjects who inhaled a nebulized solution of ^{74}As arsenite retained 32 and 62% of the dose, respectively, in the lung tissue. Only 20% remained after 4 d.[8]

Skin Absorption

Dutkiewicz has shown by immersing rat tails in solutions of ^{74}As-labeled sodium arsenite that some arsenic can be absorbed through the intact skin.[7] However, whether a sufficient amount can be absorbed through the intact skin in humans to produce toxicity is not known. Systemic poisoning has been reported in a patient who applied arsenical paste to his face; however, the skin may have been ulcerated prior to the application of the medication.[9] Death has been reported after the absorption of arsenic from the vagina.[10]

Pharmacokinetics

Arsenic leaves the blood rapidly and 1 d following s.c. injection of potassium arsenite in man, most has left the blood.[11] In man given arsenite intravenously, plasma-level decay fitted a three compartment model. The half-time of the first phase was very short (a few hours) and most (more than 90%) was removed during this phase. Twenty-four hours after dosing, the plasma level was less than 1% of its initial value. The second phase had a half-time of about 30 h, while the third phase showed a very slow disappearance rate. Erythrocyte levels exhibited similar kinetics but at a higher level. The distribution of arsenic between erythrocytes and plasma varies according to species and the valence of the administered arsenic. In humans, the erythrocyte-to-plasma ratio increased to 3:1 from 1:1 as the blood level increased from 2.5 μg/l (normal level for unexposed person) to 62 μg/l.[12] The rat is unusual because the clearance of arsenic from its blood is slower than for other species because of the avid binding of arsenic to an intracorpuscular protein.

Distribution in Tissues

In the normal human exposed to trace amounts of arsenic in his environment, the highest concentrations are found in tissues rich in sulfhydryl groups such as the skin, hair, and nails. When larger amounts of arsenic are given, the pattern of tissue distribution changes and is partly dependent on the time interval between exposure and study. In mice given arsenite or arsenate, highest concentrations were initially found in the bile, kidneys, and liver (2 h after oral dose). Clearance from these tissues was rapid, however, and several days later the highest concentrations were found in the skin, epithelium of the upper gastrointestinal tract, epididymis, and stomach wall.[13] In one human subject given ^{74}As by s.c. injection, the greatest concentrations of arsenic were found in the liver and kidneys 4 d later.[14] The tissue distribution of ^{74}As-labeled trivalent arsenic was similar to that of labeled pentavalent arsenic in mice, except for somewhat higher levels found in the former in the liver and bile 2 h after dosing and in the skin 24 h after dosing.[15]

Detoxification and Excretion

Inorganic arsenicals are rapidly methylated by man and excreted largely in the urine. Methylation probably renders a compound less reactive and facilitates renal excretion.[13] The site of methylation is not known.

The organic arsenicals found in shellfish and fish appear to be rapidly and almost completely absorbed from the gastrointestinal tract and equally rapidly excreted by the kidneys. They appear to be relatively nontoxic. The exact chemical nature of these compounds is not known, but includes methylated species such as arsenobetaine.[16]

Essential

Some authors have claimed that arsenic is an essential trace element for some species; thus, rats fed an arsenic-depleted diet grew slower than those fed the same diet with arsenic supplementation.[17] These growth experiments are difficult to interpret; many farm animals grow better when small amounts of antibiotics are added to their diet, yet we would not claim that these are essential dietary components. Indeed, arsenic was used for many years as a growth stimulant in agriculture before the advent of antibiotics.

BIOCHEMISTRY

Inorganic Arsenicals

Trivalent inorganic arsenicals are generally more toxic than pentavalent compounds. Trivalent arsenic (arsenite) readily combines with thiol groups in tissues; it inhibits the oxidative decarboxylation of pyruvate and of α-ketoglutarate by combining with the di-thiols of the intermediary metabolite lipoic acid.[18] Thus, it impairs the functioning of the tricarboxylic acid cycle. It also inhibits the action of many other sulfhydryl-group-containing enzymes.

Pentavalent inorganic arsenicals uncouple oxidative phosphorylation from the mitochondrial respiratory chain by replacing phosphate in high energy compounds; the arsenic analogue is readily hydrolyzed so that energy is lost rather than conserved and respiration proceeds without concomitant phosphorylation.[18] They exert a similar uncoupling action in the glycolysis pathway at the step where 1:3 diphosphoglycerate is formed; arsenate replaces inorganic phosphate to form arseno-3-phosphoglycerate which hydrolyzes spontaneously to form 3-phosphoglycerate without forming ATP.[18]

Pentavalent inorganic arsenic can readily be reduced to trivalent compounds *in vivo* by humans[19] and animals.[20] It is not known whether there is significant oxidation of arsenite to arsenate in man; it has been demonstrated in mice.[20]

Inorganic arsenic is readily methylated in man (largely to dimethyl arsenic acid)[21] and animals.[22] The methylation appears to render the arsenic relatively nontoxic and facilitates its rapid excretion in the urine. There are few data available regarding the metabolism of organic arsenical compounds.

TOXICOLOGY

Acute Arsenic Poisoning

The fatal dose of arsenic trioxide for man is probably between 100 and 200 mg, although people have survived larger doses; victims suffer severe vomiting and diarrhea (which may be blood stained) and rapidly proceed to peripheral circulatory collapse and death. Neuropathy may follow nonfatal acute poisoning.

Arsine Poisoning

Arsine is formed when nascent hydrogen is formed in the presence of arsenic; thus, it can be formed by the action of acids or even water on metal arsenides and accidental poisoning occurs occasionally in the smelting industry by the action of water on dross (furnace waste). It produces severe hemolysis. The war gases commonly were halogenated arsines.

Chronic Arsenic Poisoning

The general effects of chronic arsenic poisoning are weakness, debility, lassitude, loss of appetite and energy, loss of hair, hoarseness of the voice, loss of weight, and occasionally dementia. Anemia, probably due to bone marrow suppression (normochromic and normocytic), is common and leukopenia may occur. Basophilic stippling of the erythrocytes may be present. Megaloblastic changes have also been reported as have disorders of heme synthesis. Chronic renal and liver damage may occur and presinusoidal portal hypertension without cirrhosis, apparently due to portal tract fibrosis, appears to be an associated rare complication of arsenic poisoning.[23]

More specific features of chronic arsenic poisoning include dermatologic, cardiovascular, and neurologic changes. The dermatologic features of chronic arsenic poisoning are pigmentation and hyperkeratoses; hyperkeratoses occur on the palms of the hands and the soles of the feet and are usually surmounted by small corn-like excrescences, particularly on the thenar and hypothenar aspects of the hands and on the sides of the fingers. Brown pigmentation may be found around the eyelids, nipples, axillae and groin, and on the temples and neck. In severe cases, pigmentation may extend over the chest, back, and abdomen and a mottled appearance (like raindrops on a dusty road) may be seen on the chest and back. The pigmentation is due to melanin deposits. Mees lines (transverse white bands) may be seen on the nails. Peripheral vascular lesions sometimes producing gangrene of the extremities and premature coronary artery disease have been described in chronic arsenic poisoning.[24]

Peripheral neuropathy is a common complication of chronic arsenic poisoning. The symptomatology is particularly pronounced in the arms and legs and is more severe distally. Sensory impairment is often more severe than motor effects and the legs are often more severely affected than the arms.[25] Typical features are numbness, paresthesia, diminished sensations of touch, pain, heat, and cold, and reduced muscle power; ataxia and mental confusion may occur. The neuropathy is thought to be a distal axonopathy possibly related to the effect of arsenic on nerve pyruvate-oxidase activity.

Management of arsenic poisoning is related to the toxic compound encountered. In arsine poisoning, therapy may include exchange transfusion for the hemolysis and hemodialysis for renal failure; chelation therapy does not appear to be helpful. In acute arsenic poisoning, supportive therapy and chelation treatment are indicated; dimercaprol or penicillamine may be used. Chronic poisoning can also be treated with chelating agents. However, provided renal function is good, arsenic appears to be rapidly cleared from body tissues (except for skin, hair, and nails) and damaged tissues regenerate at their own rates.

Table 2
ARSENIC LEVELS IN BODY COMPONENTS OF NORMAL UNEXPOSED HUMANS[24]

Whole blood	2.5 μg/l
Plasma	2.4 μg/l (SD 1.9)
Erythrocytes	2.7 μg/l (SD 1.3)
Urine	<50 μg/l
Hair	<1 ppm

ASSESSMENT OF SEVERITY OF HUMAN POISONING

Thorough clinical examination is probably the best index of the severity of chronic arsenic poisoning. If there is evidence of impaired renal or liver function appropriate tests should be performed. Clinical examination and electromyography may help in the assessment of the neuropathy.

Because of the rapid clearance of arsenic from the blood (little remains after 10 h),[14] blood levels are of little help in assessing human arsenic poisoning, except for perhaps identifying arsenic as a cause of acute poisoning in patients whose renal function is seriously impaired. Normal levels are seen in Table 2. Because of the rapid and effecient excretion of arsenic in the urine, analysis of this is a reliable index of recent (1 to 3 d previous) exposure. Arsenic is found in high concentration in nails and hair because of binding to the sulfhydryl groups in these tissues. Analysis of these is useful in identifying chronic arsenic exposure even months after this has ceased. However, arsenic may bind avidly to hair and nails when it is applied externally, and therefore the measurement is of value in assessing chronic arsenic poisoning only if external contamination can be excluded.[24] In the authors' study in Nova Scotia, the level of arsenic in the hair was only approximately related to the severity of the clinical findings.[4]

CARCINOGENESIS

The association between skin cancer and the ingestion of soluble inorganic arsenicals is well proven.[26] In a recent review of cancer statistics in 478 patients who ingested Fowler's solution for periods of time varying from 2 weeks to 12 years, the authors reported an excess prevalence of skin cancer (11% of a subgroup of 142 patients).[27] They could find no convincing evidence of cancers at other body sites. Skin cancers are more common in patients with hyperkeratoses; they are commonly multiple and occur mostly on the trunk and limbs. The pattern of lesions varies among different reports, but a large series from Taiwan reported intraepidermal carcinomas (Bowen's disease) to be the most common (58%), followed by squamous cell carcinomas (19%) and basal cell carcinomas — superficial multicentric, not the common papulonodular variety (15%).[28] The remaining lesions (8%) were described as combined forms.

Exposure to inorganic arsenicals, particularly arsenic trioxide and sulfides, may occur in smelter workers particularly when copper ores are being processed, and numerous studies have confirmed the increased prevalance of lung cancer in this industry, presumably from inhalation of arsenic-containing dust.[24] Adenocarcinomas of the lung constituted 38% of the tumors in arsenic-exposed patients compared with 12% in controls.

Many reports claim that arsenic may be a carcinogen at other sites. The evidence is unconvincing, however, except in the case of hepatic angiosarcoma.[29] Researchers have been confounded by their inability to find a susceptible animal model for arsenic carcinogenesis. However, Pershagen appears recently to have produced lung cancer in hamsters by the intratracheal injection of arsenic trioxide mixed with charcoal carbon as a carrier.[30]

MUTAGENICITY AND TERATOGENICITY

An increased frequency of chromosome aberrations has been reported among people exposed to inorganic arsenic in the working environment; it appears to impair DNA repair mechanisms. Also an increased rate of spontaneous abortions has been reported in women working at a Swedish copper smelter and several investigators have reported inorganic arsenic to be teratogenic in animals.[24]

ANALYSIS

The type of analytical technique employed is dictated by the sensitivity and the precision required. The most commonly used methods rely on the conversion of arsenic to its corresponding hydride, followed by colorimetric, hydrogen flame atomic absorption, or electrothermal atomic absorption quantitation. For greater sensitivity, neutron activation can be employed. Regardless of the technique used, care must be taken to ensure that the arsenic does not change state either prior to or during the analysis. High-pressure liquid chromatography can be useful for quantitation of arsenic species in a mixture of these.

REFERENCES

1. **Anderson, C. E.,** Arsenicals as feed additives for poultry and swine, in *Arsenic, Industrial, Biomedical and Environmental Perspectives,* Lederer, W. H. and Fensterheim, R. J., Eds., Van Nostrand, New York, 1983, 89.
2. **Morton, W., Starr, G., Pohl, J., and Stoner, S.,** Skin cancer and water arsenic in Lane County, Oregon, *Cancer,* 37, 2523, 1976.
3. **Wyllie, J.,** An investigation of the source of arsenic in a well water, *Can. Public Health J.,* 28, 128, 1937.
4. **Hindmarsh, J. T., McLetchie, O. R., Heffernan, L. P. M., Hayne, O. A., Ellenberger, H. A., McCurdy, R. F., and Thiebaux, H. J.,** Electromyographic abnormalities in chronic environmental arsenicalism, *J. Anal. Toxicol.,* 1, 270, 1977.
5. **Pershagen, G.,** The epidemiology of human arsenic exposure, *Top. Environ. Health,* 6, 199, 1983.
6. **Alden, J. C.,** The continuing need for inorganic arsenical pesticides, in *Arsenic, Industrial, Biomedical and Environmental Perspectives,* Lederer, W. H. and Fensterheim, R. J., Eds., Van Nostrand, New York, 1983, 63.
7. **Dutkiewicz, T.,** Experimental studies on arsenic absorption routes in rats, *Environ. Health Perspect.,* 19, 173, 1977.
8. **Holland, R. H., McCall, M. S., and Lanz, H. C.,** A study of inhaled ^{74}As in man, *Cancer Res.,* 19, 1154, 1959.
9. **Robinson, T. J.,** Arsenic polyneuropathy due to a caustic arsenic paste, *Br. Med. J.,* 3, 139, 1975.
10. **Bowen, D. A. L., Lewis, T. L. T., and Edwards, W. R.,** Acute arsenical poisoning due to acetarsol pessaries, *Br. Med. J.,* 1, 1282, 1961.
11. **Hunter, F. T., Kip, A. F., and Irvine, J. W.,** Radioactive tracer studies on arsenic injected as potassium arsenite, *J. Pharmacol. Exp. Ther.,* 76, 207, 1942.
12. **Heydorn, K.,** Environmental variation of arsenic levels in human blood determined by neutron activation analysis, *Clin. Chim. Acta,* 28, 349, 1970.
13. **Vahter, M.,** Metabolism of arsenic, *Top. Environ. Health,* 6, 171, 1983.
14. **Mealey, J., Brownell, G. L., and Sweet, W. H.,** Radioarsenic in plasma, urine, normal tissues, and intracranial neoplasms, *Arch. Neurol. Psychiatry,* 81, 310, 1959.
15. **Vahter, M. and Norin, H.,** Metabolism of ^{74}As-labelled trivalent and pentavalent inorganic arsenic in mice, *Environ. Res.,* 21, 446, 1980.
16. **Zingaro, R. A. and Bottino, N. R.,** Biochemistry of arsenic, recent developments, in *Arsenic, Industrial, Biomedical and Environmental Perspectives,* Lederer, W. H. and Fensterheim, R. J., Eds., Van Nostrand, New York, 1983, 327.
17. **Schwarz, K.,** Essentiality versus toxicity in metals, in *Clinical Chemistry and Chemical Toxicology of Metals,* Brown, S. S., Ed., Elsevier/North-Holland, Amsterdam, 1977.

18. **Mayes, P. A.,** *Harper's Review of Biochemistry,* 19th ed., Martin, D. W., Mayes, P. A., and Rodwell, V. W., Eds., Lange, Los Altos, CA, 1983, 165.
19. **Yamauchi, H. and Yamamura, Y.,** Urinary inorganic arsenic and methyl arsenic excretion following arsenate-rich seaweed ingestion, *Jpn. J. Ind. Health,* 21, 47, 1979.
20. **Vahter, M. and Envall, J.,** In vivo reduction of arsenate in mice and rabbits, *Environ. Res.,* 32, 14, 1983.
21. **Crecelius, E. A.,** Changes in the chemical speciation of arsenic following ingestion by man, *Environ. Health Perspect.,* 19, 147, 1977.
22. **Vahter, M.,** Biotransformation of trivalent and pentavalent inorganic arsenic in mice and rats, *Environ. Res.,* 25, 286, 1981.
23. **Szuler, I. M., Williams, C. N., Hindmarsh, J. T., and Park-Dincsoy, H.,** Massive variceal hemorrhage secondary to presinusoidal portal hypertension due to arsenic poisoning, *Can. Med. Assoc. J.,* 120, 168, 1979.
24. **Hindmarsh, J. T. and McCurdy, R. F.,** Clinical and environmental aspects of arsenic toxicology, *CRC Crit. Rev. Clin. Lab. Sci.,* 23, 315, 1986.
25. **Goldstein, N. P., McCall, J. T., and Dyck, P. J.,** Metal neuropathy, in *Peripheral Neuropathy,* Vol. 2, Dyck, P. J., Thomas, P. K., and Lambert, E. H., Eds., W. B. Saunders, Philadelphia, 1975, 1227.
26. **Neubauer, O.,** Arsenical cancer, a review, *Br. J. Cancer,* 1, 192, 1947.
27. **Cuzick, J., Evans, S., Gillman, M., and Price-Evans, D. A.,** Medicinal arsenic and internal malignancies, *Br. J. Cancer,* 45, 904, 1982.
28. **Yeh, S.,** Skin cancer in chronic arsenicism, *Hum. Pathol.,* 4, 469, 1973.
29. **Sunderman, F. W., Jr.,** Recent advances in metal carcinogenesis, *Ann. Clin. Lab. Sci.,* 14, 93, 1984.
30. **Pershagen, G.,** Arsenic and Lung Cancer with Special Reference to Interacting Factors, Epidemiological and Experimental Evidence, Ph.D. thesis, Department of Environmental Hygiene, Karolinska Institute, Stockholm, Sweden, 1982.

BERYLLIUM*

H. P. Witschi and L. E. Sendelbach

HISTORICAL CONTEXT

To charge such an admirable metal with having poisonous properties is about as distasteful as accusing a trusted butler of stealing the family plate.[1]

The first reported discovery of beryllium (Be), by Vaquelin in 1797, claimed that the substance was biologically inert and of little medical significance. Later work attributed to the element the general physiological properties of aluminum and iron. In 1930, it was shown that a s.c. injection of beryllium nitrate was lethal in mice. European literature reports of the late 1930s continued to provide evidence as to the toxic nature of Be. Patients from a German Be extraction plant showed clinical signs of bronchitis and pneumonitis with cyanosis. Russian workers from extraction plants were found to have similar lesions. These reports of Be toxicity in Europe gained very little attention within the scientific and medical community of the U.S. until the outbreak of World War II increased demand for Be used in the Manhattan Project. Be was used for slowing down the speed of neutrons and as a neutron reflector. In the early 1940s, zinc beryllium silicate began to be used as an ingredient in the making of phosphors, which give fluorescent tubes their glow. The first reported cases of symptoms experienced within the fluorescent tube industry were essentially chemical pneumonias, due to the inhalation of high doses of soluble Be compounds with acidic pH. In a different plant, workers developed chronic rather than acute forms of Be-induced lung disease.

In 1947 the staff of the Saranac Laboratory in New York called for a symposium to discuss all the available information on Be. It was recognized that Be was the cause of a disease that occurred in the lungs in both acute and chronic forms. The Beryllium Case Registry was established in 1952 to review all relevant data and material concerning this disease.[2]

CHEMICAL AND PHYSICAL PROPERTIES

Be is one of the lightest elements of the periodic table, being the smallest of the Group II metals. The s-oribal of its outer shell is occupied by two valence electrons with the 3 p-orbitals being vacant. The small radius of the Be ion permits unusually close approach of its ion centers. The compounds of Be with the smallest anions, F^{-2}(r = 1.36 Å) and O^{-2}(r = 1.40 Å), have great stability, since the bond strength increases directly as the product of the charges but inversely as the square of the distance between charges.[3]

Be has such a strong tendency to coordinate with oxygen that Be compounds are generally hydrolyzed in solution, with liberation of hydrogen ions. Thus, aqueous solutions of Be salts are strongly acidic. Furthermore, coordination of Be with the hydroxyl ion converts the highly insoluble beryllium hydroxide to a soluble anion. In this regard, Be resembles the Group III element, aluminum, with the same high ratio of charge to radius. Be resembles trivalent aluminum so closely that separation of the two elements is often difficult (Table 1). Be occurs naturally as the ^{9}Be isotope with commercially prepared isotopes of ^{6}Be, ^{7}Be, ^{8}Be, and ^{10}Be available.

Be salts dissociate readily in aqueous solution. At pH values less than 5, Be acts as a

* Research sponsored by the Office of Health and Environmental Research, U.S. Department of Energy under contract DE-AC05-840R21400 with the Martin Marietta Energy Systems, Inc., Georgia.

Table 1
PHYSICAL PROPERTIES OF BERYLLIUM[a]

Property	Value
Atomic number	4
Atomic weight	9.013
Atomic radius	1.123
Atomic volume (cm^3/mol) 25°C	4.877
Electron configuration	$1s^2 2s^2$
First ionization potential (ev)	9.320
Second ionization potential (ev)	18.206
Ionic radius (Be^{+2}) in Angstroms	0.31
Electronegative (Pauling's)	1.5
Density (g/cm^3) 25°C	1.8477 ± 0.0007
Melting point	1283°C
Boiling point	2970°C

[a] Adapted from Krejci and Scheel (1966).

cation. With pH values ranging from 5 to 8, Be tends to form insoluble hydroxides, while at values above a pH of 8, Be complexes form. In general, at pH 7, the strong acid salts of Be hydrolyze to the same extent at any particular concentration.

In the dissociation of water, the equilibrium obtained is expressed in the following reaction:

$$HOH \rightarrow H^+ + OH^-$$

With Be salts, the accumulation of H^+ in solution in large quantities can occur only if the OH^- ion concentration is reduced below its normal equilibrium in water.[3]

For a biunivalent salt (e.g., $Be[NO_3]_2$), hydrolysis may proceed by:

$$Be(NO_3)_2 + HOH \rightarrow Be(OH)NO_3 + H^+ + NO_3^-$$

In the case of a bi-bivalent salt (e.g., $BeSO_4$), complete hydrolysis of a large portion of the salt may occur:

$$BeSO_4 + 2HOH \rightarrow Be(OH)_2 + 2H^+ + SO_4^{-2}$$

Because of the amphoteric character of the Be atom, in aqueous solution its common salts are not neutral. Hydrolysis of Be^{2+} solutions are strongly acidic. Vorwald et al.[4] state that these circumstances have a number of important implications with respect to biological systems and are briefly summarized as follows:

1. Solutions of Be sulfate, chloride, or nitrate have a strongly acidic character capable of causing tissue injury by virtue of the pH effect alone.
2. Continuing hydrolysis of dissolved Be salts causes these solutions to have a high buffer reserve. Hydrogen ions consumed in the course of neutralization are constantly replenished by a shifting ion equilibrium, making the acidity of Be^{2+} solutions more persistent and damaging to tissues than a solution of hydrochloric acid of equal pH.
3. Due to its acidity, Be^{2+} administered to an animal will either overwhelm the buffer systems of the organism or be overwhelmed by them. Overwhelming the buffering capacity will often kill the biological organism. Overwhelming the Be administered by the buffering systems of the organism will cause *in situ* precipitation of Be allowing development of the chronic effects.

ANIMAL TOXICITY

The acute form of berylliosis is readily reproduced in animal experiments. Sprague, in 1947, exposed dogs, rabbits, rats, guinea pigs, mice, and hamsters to $BeSO_4$ dusts at a concentration of 83.3 mg $BeSO_4/m^3$.[5] The results showed a variation in toxicity among the different species with mice being the most sensitive. Acute damage was observed in the lungs, liver, and kidneys.

Stokinger[6] described the acute inhalation effects of beryllium sulfate mist in rats, mice, monkeys, dogs, cats, goats, and chickens. The response among species varied with rats being the most susceptible, showing an LC_{50} at 10 mg $BeSO_4/m^3$. The authors concluded that an acute pulmonary pneumonitis was produced which closely resembled that seen in man. Yet, even after 3 months of exposure, there was no evidence of the chronic granulomatous pneumonitis seen in humans.

The chronic form of the disease, i.e., the granulomatous reaction, has been reproduced with only partial success in experimental animals. Vorwald[7] reported the occurrence of granulomas in lungs of rats continually exposed to an aerosol of $BeSO_4$. Policard[8] reported the formation of granulomas in the lungs of guinea pigs following inhalation and intratracheal exposures to BeO. Elias et al.[9] produced pulmonary lesions, similar to the human respiratory disease, in outbred guinea pigs following intratracheal instillation of 0.05 *M* $BeSO_4$, buffered to a pH of 5.0. More recently, Barna et al.[10] observed the granulomatous disease in guinea pigs following the intratracheal injection of 5 mg BeO.

Exposure to Be and its salts by routes other than inhalation has been shown to produce multiple signs of systemic toxicity. Ingestion of Be salts leads to a depletion of the skeletal system in calcium (''beryllium rickets''). I.V. injection of Be produces acute liver necrosis.[4] Be has been found to interfere with various elements of the immune system. It is thought that the chronic form of the human disease may be mediated by impaired immune function.[11,12]

Be is a proven animal carcinogen.[13] It was discovered as early as 1946 that the i.v. injection of zinc beryllium silicate would produce osteogenic sarcomas in rabbits.[14] The observation was confirmed and expanded by several investigators: sarcoma could be produced by i.v. or intramedullary injection. As little as 20 μg of zinc beryllium silicate implanted under the periosteum would elicit the growth of malignant tumors.[15] In rats chronically exposed to an aerosol of beryllium sulfate tetrahydrate metaplastic changes in the alveolar epithelium developed and, later, alveolar adenocarcinomas were found.[16] Lung tumors were also produced in rats by intratracheal (simple or repeated dose) instillation of Be compounds. Inhalation of beryl or bertrandite ore and dusts produced bronchial alveolar hyperplasia, but no neoplasms, in hamsters. Malignant pulmonary neoplasms also developed in rhesus monkeys exposed to a beryllium sulfate aerosol or given a single intrabronchial injection of beryllium oxide.[4] In 1979 the IARC concluded that there was sufficient evidence that Be metal and several Be compounds are carcinogenic to three animal species (rats, rabbits, and monkeys).[17]

TOXICOKINETICS

The distribution pattern of Be within the body is dependent on the solubility of the compound and the route of administration. Dogs exposed to an aerosol of $BeSO_4$ for 100 d at 40 μg Be/m^3 accumulated 200 times more Be in the lungs than in the kidney, and 20 times that found within bone.[18] Administration of BeF_2 to dogs showed similar distribution after 87 d at 100 μg Be/m^3.[19] The lungs contained 40 times the amount of Be found in the kidney and 4 times that in bone. In rats exposed to $BeSO_4$, pulmonary Be concentrations reached a plateau after 36 weeks of continual exposure. Reeves and Vorwald[20] measured the clearance of Be from the lungs of rats exposed to a $BeSO_4$ aerosol of 34 μg Be/m^3. Be was cleared in a multiphasic pattern and after discontinuation of exposure, half the dose remained in the lungs after 2 weeks with a smaller percentage remaining for many months.

The i.v. injection of Be compounds causes liver toxicity. Be metal, when injected i.v. into rabbits at 40 mg Be per animal, resulted in 19 deaths out of 24 animals within 1 month due to acute liver necrosis.[21] Vacher et al.[22,23] showed that beryllium phosphate blocks the functioning of the reticuloendothelial system in the liver of mice following i.v. injection of 12 μg Be/kg. Beryllium sulfate exerted the same effect at 75 μg Be/kg. Skilleter and Price[24] injected rats with beryllium phosphate, at 12.5 to 520 μmol Be/kg, or beryllium sulfate at 12.5 to 85 μmol Be/kg. Beryllium phosphate was removed from the blood predominantly by the nonparenchymal sinusoidal cells and later distributed to the parenchymal cells.

Oral administration of Be will induce hemorrhagic necrosis of the gastric mucosa, but the metal is poorly absorbed. Repeated administration of Be in the diet will ultimately result in a rachitic condition in rats. Absorption of Be by the gastrointestinal tract is minimal. Following administration of $BeSO_4$ at 0.16 and 1.66 mg Be/ml in the drinking water, rats absorbed only 20% with 60 to 80% passing unabsorbed through the feces.[25] Following absorption, transport of Be is thought to be mainly in the form of a colloidal phosphate, absorbed on plasma α-globulin.[26] Because of the colloidal binding of Be within the plasma, Be is only partially excreted in the urine. Van Cleave and Taylor[27] showed that 20 to 60% of intratracheally injected Be was excreted in the urine. Since Be is bound to colloids within the plasma, Be does not cross the glomerular membrane and is excreted via the tubules.

MECHANISM OF ACTION

In 1949 it was discovered that Be is a potent inhibitor of alkaline phosphatase.[28,29] Non-competitive inhibition of alkaline phosphatase was measured at 10^{-7} *M*, being maximal at 10^{-5} *M*. Subsequent work showed that Be was a highly specific inhibitor at micrometer concentrations for two more enzymes: phosphoglucomutase and NaK-ATPase.[30-32] Effects on other enzymes have been described; however, enzyme inhibition occurred usually at millimolar concentrations only, rather than at micrometer concentrations and was probably an unspecific effect due to protein precipitation.

More recently, Joshi et al. observed that purified phosphoglucomutase from the liver or muscle of rat and rabbit was rapidly and completely inhibited by Be(II), but that activity in a crude homogenate was inhibited only partially.[33] The protecting factor which bound Be(II) was identified to be ferritin.[34] Indeed, this protein, which can amass *in vivo* up to 4500 g of atoms of iron as ferric hydroxyphosphate, can bind up to 1000 g of atoms of Be(II) *in vitro*. Holoferritin, therefore, is the only protein known to date which can bind such a large quantity of Be(II). Consistent with this observation, *in vitro*, ferritin protected against and reversed the inhibition of Be(II) of enzymes such as phosphoglucomutase, alkaline phosphatase, and NaK-ATPase.[35] In *in vivo* experiments, rats pretreated with iron injections, a treatment that increased their ferritin levels, survived otherwise toxic levels of intravenously injected or of inhaled $BeSO_4$.[36]

Be toxicity cannot be explained, however, in terms of inhibition of specific enzymes. In 1968 it was found that Be accumulated in the nuclear fraction of liver cells.[37] While the precise nuclear structure to which Be is attached is not yet identified, it has been shown that Be inhibits DNA synthesis[38] and the synthesis of nucleolar proteins.[39] RNA and protein synthesis are not adversely affected.[38,40] Be prevents appearance and presumably synthesis of thymidine kinase and DNA polymerase in regenerating rat liver.[41] It was concluded that Be would influence events leading to DNA synthesis by interfering with transcription. Evidence for support of this hypothesis was provided by a study which examined the effects of Be on gene expression in cell culture system.[42] Be, under conditions where it did not affect cell growth, inhibited the induction of tyrosine aminotransferase by glucocorticoids.

It was concluded that low Be concentration would interfere with regulatory mechanisms which control the step of transcription in gene expression. More recently, it has also been found that Be, in dividing cells, is capable of producing a selective block of the cell cycle at the G_1-S transition, possibly by inhibiting nuclear protein phosphorylation.[43]

HUMAN TOXICITY

Human Be disease occurs following inhalation or dermal exposure to Be compounds. Almost all cases were observed in industries processing or using Be in manufacturing processes. Berylliosis is recognized to be an occupational disease.[44]

Acute forms of Be disease have been described as chemical-type inflammatory changes in the upper respiratory tract and in the lung. If exposure conditions are high enough, the disease can be quite fulminating, leading to death within a few days. Contact of the skin with Be compounds may produce irritation (dermatitis). If Be salts become implanted under the skin, a chronic ulcerating lesion may develop.[2]

Chronic Be disease is more common than beryllium pneumonitis. It now seems established that it is an idiosyncratic disorder; only about 5% of all exposed workers may be affected. This suggests that some ''host factor'' in addition to Be exposure is required. Chronic Be disease is a granulomatous disease of the lung. Newer investigations have given evidence of disturbed T-lymphocyte function.[45] Histopathological analysis of lung tissue reveals an interstitial pneumonitis with histiocytes as the predominant infiltrating cell and lesser numbers of lymphocyte and plasma cells. Histiocytes may aggregate in large circumscribed granulomas; and giant cells are commonly seen within granulomas. Pulmonary function tests reveal restrictive ventilatory impairment and decreased diffusion capacity. Before the disease was recognized as a clinical entity as such, patients with Be disease were diagnosed as suffering from Boeck's sarcoidosis. Once established, Be disease is irreversible, although it may not always be progressive. Patients suffering from berylliosis may not necessarily die because of the disease, but they will die with the disease. The diagnosis is not easily made, and elements of diagnosis include a working history, chest X-rays, reduced pulmonary function, and, if feasible, proof of Be present in lung tissue.[11]

A Beryllium Case Registry was established in 1952 at the Massachusetts Institute of Technology (MIT) and later moved to the Massachusetts General Hospital. In 1978, the Registry was transferred to the National Institute for Occupational Health and Safety (NIOSH). Since its inception, the Registry collected fewer than 1000 cases of Be disease in the U.S. In a recent analysis it was pointed out that no cases of acute chemical pneumonitis due to Be were reported since 1968 and only 15 cases were known to occur in the U.S. since 1950. It was thought that this is caused thanks to adherence to strict standards for airborne Be: 2 $\mu g/m^3$ for in-plant atmospheric concentration and 0.01 $\mu g/m^3$ for community air. The standards are among the most stringent for any airborne metal compounds.[46,47] A different analysis of chronic Be disease came to the opposite conclusion: that the problem had not been eradicated, that new severe cases continued to be observed, and that a substantial worker population is still at risk.[45]

Experimental data clearly show that Be is an animal carcinogen. Several epidemiological studies examined cancer incidence in cohorts of workers exposed to Be. The results of the studies remain inconclusive and controversial. In 1985 the IARC concluded that there was only limited information which would link occupational exposure to Be, to an increased risk of developing lung cancer.[48]

ANALYSIS

In biological materials, Be may be determined by flame and flameless atomic absorption spectroscopy or with gas chromatography. When gas chromatography is coupled with mass spectroscopy, sensitivities in the 0.04- to 10-pg Be per sample may be obtained. Polarography, enzyme inhibition, and various types of activation techniques may be used under appropriate circumstances.[49]

REFERENCES

1. Editorial, The toxicity of beryllium, *Lancet,* 260, 1357, 1951.
2. **Tepper, L. B., Hardy, H. L., and Chamberlin, R. I.,** *Toxicity of Beryllium Compounds,* Elsevier, Amsterdam, 1961, chap. 1.
3. **Krejci, L. E. and Scheel, L. D.,** The chemistry of beryllium, in *Beryllium: Its Industrial Hygiene Aspects,* Stokinger, H. E., Ed., Academic Press, New York, 1966, 45.
4. **Vorwald, A. J., Reeves, A. L., and Urban, E. C. J.,** Experimental beryllium toxicity, in *Beryllium: Its Industrial Hygiene Aspects,* Stokinger, H. E., Ed., Academic Press, New York, 1966, 201.
5. **Sprague, G. F., LaBelle, C. W., Pettingill, A. G., and Stokinger, H. E.,** Initial studies of the toxicity of inhaled beryllium sulfate dusts and beryllium fume, in *6th Sarranac Symp. Pneumoconiosis: Beryllium, Bauxite Fumes, Compensation,* Vorwald, A. J., Ed., Paul B. Hoeber, New York, 1947.
6. **Stokinger, H. E.,** The Metals (excluding lead), in *Industrial Hygiene and Toxicology,* Vol. 2, Patty, F., Ed., Interscience, New York, 1958.
7. **Vorwald, A. J.,** Adenocarcinoma in the lung of albino rats exposed to compounds of beryllium: cancer of the lung: an evaluation of the problem, in Proc. Sci. Session, Am. Cancer Soc. Annu. Meeting, Detroit, November 3 and 4, 1953.
8. **Policard, A.,** Histological studies of the effects of beryllium oxide (glucine) on animal tissues, *Br. J. Ind. Med.,* 7, 117, 1950.
9. **Elias, J. A., Rossman, M. D., Atkinson, B. J., Phillips, S. M., and Danielle, R. P.,** Experimental berylliosis and beryllium-induced interstitial fibrosis, *Am. Rev. Resp. Dis.,* 123(4), 141, 1981.
10. **Barna, B. P., Deodhar, S. D., Chiang, T., Gautam, S., and Edinger, M.,** Experimental beryllium-induced lung disease. I. Differences in immunological responses to beryllium compounds in strain 2 and 13 guinea pigs, *Int. Arch. Allery Appl. Immunol.,* 73, 42, 1984.
11. **Tepper, L. B.,** Beryllium, *CRC Crit. Rev. Toxicol.,* 1, 235, 1971.
12. **Reeves, A. L. and Preuss, O.,** The immunotoxicity of beryllium, in *Immunotoxicology and Immunopharmacology,* Dean, J. et al., Eds., Raven Press, New York, 1985, 441.
13. **Groth, D. H.,** Carcinogenicity of beryllium: review of the literature, *Environ. Res.,* 21, 56, 1980.
14. **Gardner, L. U. and Heslington, H. F.,** Osteosarcoma from intravenous beryllium compounds in rabbits, *Fed. Proc.,* 5, 721, 1946.
15. **Tapp, E.,** Beryllium induced sarcomas of the rabbit tibia, *Br. J. Cancer,* 20, 778, 1969.
16. **Reeves, A. L., Deitch, D., and Vorwald, A. J.,** Beryllium carcinogenesis. I. Inhalation exposure of rats to beryllium sulfate aerosol, *Cancer Res.,* 27, 439, 1967.
17. IARC Monogr. on the evaluation of the carcinogenic risk of chemicals to humans, Vol. 23, Some metals and metallic compounds, Lyon, 1980.
18. **Stokinger, H. E., Ashenburg, N. J., DeVoldre, J., Scott, J. K., and Smith, F. A.,** Acute inhalation toxicity of beryllium. II. The enhancing effect of the inhalation of hydrogen fluoride vapor in beryllium sulfate poisoning in animals, *Arch. Ind. Hyg. Occup. Med.,* 1, 398, 1950.
19. **Stokinger, H. E., Spiegel, C. F., Root, R. E., Hall, R. H., Steadman, L. T., Stroud, C. A., Scott, J. K., Smith, F. A., and Gardner, D. F.,** Acute inhalation toxicity of beryllium. IV. Beryllium fluoride at exposure concentrations of one and ten milligrams per cubic meter, *Arch. Ind. Hyg. Occup. Med.,* 8, 493, 1953.
20. **Reeves, A. L. and Vorwald, A. J.,** Beryllium carcinogenesis. II. Pulmonary deposition and clearance of inhaled beryllium sulfate in the rat, *Cancer Res.,* 27, 446, 1967.
21. **Aldridge, W. N., Barnes, J. M., and Denz, F. A.,** Experimental beryllium poisoning, *Br. J. Exp. Pathol.,* 30, 375, 1949.
22. **Vacher, J. and Stoner, H. R.,** The removal of injected beryllium from the blood of the rat: the role of the reticuloendothelial system, *Br. J. Exp. Pathol.,* 49, 315, 1968.
23. **Vacher, J., Deraedt, R., and Berzoni, J.,** Compared effects of two beryllium salts (soluble and insoluble): toxicity and blockage of the reticuloendothelial system, *Toxicol. Appl. Pharmacol.,* 24, 497, 1973.
24. **Skilleter, D. N. and Price, R. J.,** The uptake and subsequent loss of beryllium by rat liver parenchymal and non-parenchymal cells after the intravenous administration of particulate and soluble forms, *Chem. Biol. Interact.,* 20, 383, 1978.

25. **Reeves, A. L.,** The absorption of beryllium from the gastrointestinal tract, *Arch. Environ. Health,* 11, 209, 1965.
26. **Vacher, J. and Stoner, H. B.,** The transport of beryllium in rat blood, *Biochem. Pharmacol.,* 17, 93, 1968.
27. **Van Cleave, C. D. and Taylor, C. T.,** Distribution, retention and elimination of Be in the rat after intratracheal injection, *Arch. Ind. Health,* 11, 375, 1955.
28. **DuBois, K. P., Cochran, K. W., and Mazur, M.,** Inhibition of phosphatase by beryllium and antagonism of the inhibition by manganese, *Science,* 110, 420, 1949.
29. **Grier, R. S., Hood, D. B., and Hoagland, M. B.,** Observations on the effects of beryllium on alkaline phosphatase, *J. Biol. Chem.,* 180, 289, 1949.
30. **Thomas, M. and Aldridge, W. N.,** The inhibition of enzymes by beryllium, *Biochem. J.,* 98, 94, 1966.
31. **Aldridge, W. N. and Thomas, M.,** The inhibition of phosphoglucomutase by beryllium, *Biochem. J.,* 98, 100, 1966.
32. **Hashimoto, T., Joshi, J. G., Del Rio, C., and Handler, P.,** Phosophoglucomutase. IV. Inactivation by beryllium, *J. Biol. Chem.,* 242, 1671, 1967.
33. **Joshi, J. G., Price, D. J., and Fleming, J.,** Ferritin and metal toxicity, *Protides Biol. Fluids Proc. Colloq.,* 31, 183, 1984.
34. **Price, D. J. and Joshi, J. G.,** Ferritin: binding of beryllium and other divalent metal ions, *J. Biol. Chem.,* 258, 10873, 1983.
35. **Price, D. J. and Joshi, J. G.,** Ferritin: protection of enzyme activity against the inhibition of divalent metal ions *in vitro, Toxicology,* 31, 151, 1984.
36. **Lindenschmidt, R. C., Sendelbach, L. E., Witschi, H. P., Price, D. J., Fleming, J., and Joshi, J. G.,** Ferritin and *in vivo* beryllium toxicity, *Toxicol. Appl. Pharmacol.,* 82, 344, 1986.
37. **Witschi, H. P. and Aldridge, W. N.,** Uptake distribution and binding of beryllium to organelles of rat liver cells, *Biochem. J.,* 106, 811, 1968.
38. **Witschi, H. P.,** Inhibition of deoxyribonucleic acid synthesis in regenerating rat liver by beryllium, *Lab. Invest.,* 19, 67, 1968.
39. **Bassleer, R.,** Effects of beryllium on synthesis of nucleoproteins and DNA in fibroblasts cultivated in vitro, *C. R. Soc. Biol.,* 159, 1620, 1965.
40. **Witschi, H. P. and Aldridge, W. N.,** Biochemical changes in rat liver after acute beryllium poisoning, *Biochem. Pharmacol.,* 16, 263, 1967.
41. **Witschi, H. P.,** Effects of beryllium on deoxyribonucleic acid synthesizing enzymes in regenerating rat liver, *Biochem. J.,* 120, 623, 1970.
42. **Perry, S. J., Kulkarni, S. B., Lee, K. L., and Kenney, F. T.,** Selective effect of the metallocarcinogen beryllium on hormonal regeneration of gene expression in cultured cells, *Cancer Res.,* 42, 473, 1982.
43. **Skilleter, D., Price, R. J., and Legg, R. F.,** Specific G_1-S phase cell cycle block by beryllium as demonstrated by cytofluorometric analysis, *Biochem. J.,* 216, 773, 1983.
44. **Hardy, H. L.,** Beryllium disease: a clinical perspective, *Environ. Res.,* 21, 1, 1980.
45. **Cullen, M. R., Cherniack, M. G., and Kominsky, J. R.,** Chronic beryllium disease in the United States, *Semin. Respir. Med.,* 7, 203, 1986.
46. **Eisenbud, M. and Lisson, J.,** Epidemiological aspects of beryllium-induced nonmalignant lung disease: a 30 year update, *J. Occup. Med.,* JOMO 25, 198, 1983.
47. **Eisenbud, M.,** Origins of the standard for control of beryllium disease (1947—1949), *Environ. Res.,* 27, 79, 1982.
48. **Weld, N. J. and Doll, R.,** Interpretation of negative epidemiological evidence for carcinogenicity, IARC Scientific Publ. No. 65, Lyons, 1985, 199.
49. **Fishbein, L.,** Overview of analysis of carcinogenic and/or mutagenic metal in biological and environmental samples. I. Arsenic, beryllium, cadmium, chromium and selenium, *Int. J. Environ. Anal. Chem.,* 17, 113, 1984.

CADMIUM

Zahir A. Shaikh and Karen J. McVety

CHEMISTRY, USES, AND SOURCES OF POLLUTION

Cadmium is a divalent metal with many physical properties similar to zinc and is found in nature associated with zinc. It is used as a pigment in paints and plastics, as plating to prevent corrosion, as a stabilizer in plastics such as PVC, and in nickel-cadmium batteries. The main source of environmental pollution of cadmium is from its production from the refinement of zinc, copper, and lead and in the iron and steel industries. Annual production of cadmium exceeds 17,000 tons/year. The cadmium which is released either in the form of atmospheric emissions or as sewage sludge becomes deposited in the soil and water and taken up by plants. Pollution of the estuaries also leads to contamination of shellfish. Workers in cadmium-producing factories are exposed mainly to cadmium dust and fumes.

THRESHOLD VALUES

Occupational safety standards for cadmium exposure have been set in many countries. The American Conference of Governmental Industrial Hygienists has set the threshold limit value at 50 $\mu g/m^3$, which is the value used by most countries.[1] Lower limits have been set by Finland (20 $\mu g/m^3$ for dust and 10 $\mu g/m^3$ for fumes) and Sweden (20 $\mu g/m^3$ for existing industries and 10 $\mu g/m^3$ for new industries). A proposed new limit of 10 $\mu g/m^3$ has been suggested by the World Health Organization (WHO).

The limit for cadmium in foodstuffs has been set at 1 mg/kg in the U.S.; Japan also uses this value for cadmium limit in rice. A recommended limit on human intake of cadmium has been set at 400 to 500 μg/week by the WHO.[3] The limit suggested by this organization for drinking water cadmium is 5 μg/l[3].

BIOCHEMISTRY

Due to its high affinity for sulfhydryl groups, the most important biochemical interaction of cadmium is with metallothionein (MT). MT is a low-molecular-weight sulfhydryl-rich protein which binds metals and appears to act in the detoxification of cadmium and mercury and as a storage site for zinc and copper.[4-8] Its synthesis can be induced by copper, zinc, cadmium, mercury, bismuth, and other metals.[7-11] Cadmium binds to MT in the plasma and this appears to be one of the forms in which cadmium is taken up by the kidneys.[12-14] Protection against cadmium toxicity in the liver can be produced by a prior induction of MT.[15,16] This is probably the mechanism by which the liver is protected against toxicity of long-term low-level cadmium exposure. Up to 80% of liver and kidney cadmium from long-term exposure is MT bound.[5,17] The kidney is exposed to cadmium in the form of Cd-MT complexes which are freely filterable and which are reabsorbed by the proximal tubules.[12-14] However, Cd-MT has a half-life of a few days while the retention half-time of cadmium is in the order of years.[11] Thus, new MT must be synthesized in order to bind the released cadmium during MT turnover to prevent ensuing toxicity. When the cadmium content of the kidney exceeds about 200 μg/g, apparently a saturation of these binding sites occurs and the non-MT-bound cadmium is then able to induce toxic reactions.[18]

Cadmium also interacts with other sulfhydryl-containing enzymes. It has been shown to inhibit intestinal ALPase and ATPase[19] and to depress both RNA and protein synthesis.[20] It is also believed that cadmium may substitute for zinc in zinc-containing enzymes, thus, decreasing their activity.[21]

UPTAKE AND METABOLISM

Gastrointestinal absorption of cadmium is less than 10% in humans.[22,23] Exposure via inhalation can result in absorption of up to 20% of the dose of cadmium depending on the size of the suspended particles.[24] Cadmium which enters the blood by either route of exposure becomes bound to proteins and is taken up slowly by the red blood cells. Plasma concentrations of cadmium following acute exposure are higher than whole blood concentrations on day 1 of exposure, but decrease to below whole blood concentrations by the second day of exposure. Chronic cadmium exposure in animals results in 94 to 99% red blood cell binding of cadmium within 3 weeks and in all of the blood cadmium in the red blood cells within 10 weeks.[26] After 10 weeks of exposure to 5 μmol $CdCl_2$/kg/day, 5 d/week, plasma cadmium levels begin to rise concurrent with a drop in cadmium levels in the liver and kidney of rats due to possible tissue damage.[27,28]

Initial protein binding of cadmium in animals and humans in the plasma is to albumin and other high-molecular-weight proteins.[29] Within 4 d of a single injection, binding of cadmium to MT in the plasma can be detected.[30] Cadmium binding to this protein is also seen following occupational exposure in humans.[31]

From the blood, cadmium is taken up primarily by the liver and kidneys.[29] The highest concentrations following a single injection in rats are found in the liver. A gradual redistribution occurs during the months following exposure in which hepatic cadmium levels fall and renal cadmium levels increase, until, by the fifth month, the concentration in the renal cortex is twice that in the liver.[32] If the exposure is chronic the cadmium levels in the liver will remain above those in the kidney.[26] A release of cadmium-bound MT from the liver is believed to be responsible for the redistribution of cadmium to the kidneys seen after long-term exposure.[12-14] The renal cortex contains more cadmium than the medulla at a ratio of about 2:1 between the cortex and medulla content of cadmium.[33,34] The kidneys contain up to 50% of the total body burden, the liver about 18%, and all other organs combined make up the remainder.[35]

Other organs which accumulate cadmium are the pancreas, spleen, lungs, heart, and intestine.[29] In humans exposed to normal concentrations of cadmium, the muscles accumulate the cadmium throughout the lifetime of the subject while the kidneys and liver reach maximums by about age 40 and 30, respectively.[35]

Following a single dose of cadmium, two thirds of the cytosolic cadmium is bound to MT.[36] This is also the case after long-term oral exposure in rats.[37] Binding to MT accounts for the very long retention time for cadmium in the tissues.

Elimination of cadmium occurs mainly through urine and feces. Following a single s.c. dose of cadmium, urinary excretion only accounts for a small percentage of the dose; less than 1% of an injected dose in rats is excreted in the urine after 4 to 7 d.[29,38] About 15% of the dose is excreted via the feces within the first 4 d after s.c. injection in rats.[29] Thereafter the fecal excretion accounts for 1% of the body burden per day. In humans, daily urinary excretion varies from 0.2 to 2.0 μg/l.[35] The retention time of cadmium is very long. In one study using radiolabeled cadmium a minimum retention half-time of 26.3 years was reported.[22] Chronic exposure to cadmium in rats results in a linear increase in urinary excretion of cadmium as body burden increases.[30] Tubular reabsorption in rabbits is complete when the amount of filtered cadmium is below 1.2 μg/min.[12] Once toxicity to the kidney occurs a sharp rise in urinary cadmium can be seen due to a release of cadmium from injured cells and to a decrease in tubular reabsorption.[26-28] Cadmium-exposed individuals who exhibit signs of renal toxicity also show higher urinary cadmium levels than those with no signs of renal damage.[40-42] Other routes of excretion which make minor contributions to the elimination of cadmium are the hair,[43] milk,[44] saliva,[45] and sweat.[46]

TOXIC EFFECTS

Acute Effects

Upon acute exposure to cadmium, the target organ depends on the route of exposure. Exposure to cadmium dust or fumes leads to nasopharyngeal irritation and may be followed by gastrointestinal disturbances, chest pain, and dyspnea if the exposure is high.[47] Exposure which exceeds the lethal dose (2600 to 2900 min/mg/m^3)[47] results in death due to pulmonary edema leading to cardiopulmonary insufficiency.

Oral intake of cadmium results in damage primarily to the intestinal mucosa and typically manifests as nausea, vomiting, and diarrhea. Damage to the kidneys occurs when the acute exposure is oral or parenteral and presents mainly as proteinuria or anuria in humans.[48] Studies conducted on rabbits showed renal tubular lesions after acute cadmium exposure, but these were considered secondary to the vascular effects.[49]

Severe endothelial cell damage in small vessels occurs upon acute exposure to a high dose of cadmium in animals. An impairment of microcirculation and cellular respiration leads to secondary damage to the kidneys,[49] liver,[50] and testes.[51] This type of cadmium-induced damage has not been documented in humans.

Probably the most pronounced type of damage is the liver injury. Rats given high doses of cadmium (3.9 to 4.0 mg/kg)[52,53] showed severe liver damage within hours; parenchymal cell necrosis and mitochondrial degeneration were the main morphological changes noted. Data on human hepatic injury upon acute cadmium exposure is limited, although one case study showed pronounced liver damage and metabolic acidosis following ingestion of 5 g CdI_2.[48]

Another effect which has been documented only in animals is the damage to the testes. Following injection of 0.02 mmol cadmium chloride the testes decrease in size and weight, interstitial edema and hemorrhage occur, and they progress to total necrosis within 24 to 48 h.[54] The testicular damage may be related to the vascular changes seen following cadmium treatment, since a decrease in capillary blood flow within the testes can be measured within hours following cadmium injection.[55]

S.c. or i.m. injection of cadmium to rats produces sarcoma at the site of injection and in some cases has caused distant tumor development in the testes.[55,56] Some attempt has been made to correlate cadmium pollution with an increased prevalence of prostatic cancer in an area of Alberta, Canada,[58] and exposure to cadmium in factories with high incidences of deaths due to lung and prostatic cancer.[59] A correlation does appear to exist, but these results may be insufficient to suggest a direct causative effect of cadmium upon carcinogenesis.

Teratogenesis from cadmium has not been seen in man, although many malformations can be produced in rodents if the exposure is prior to the closure of the vitelline duct. Transfer of cadmium to the fetus early in development produces mainly skeletal and facial abnormalities.[60] In rats, internal organ defects such as hydronephrosis, hydroureter, bladder defects, and undescended testes can be induced by cadmium.[61]

Chronic Effects

Chronic exposure to cadmium can produce quite a different spectrum of effects than acute exposure. Liver injury in humans chronically exposed to cadmium appears to be minimal. Some increases in serum gammaglobulin levels and an elevation in aspartate aminotransferase have been shown in patients suffering from long-term cadmium exposure.[62] Morphological changes similar to those seen upon acute exposure are seen in animals and include degeneration of rough endoplasmic reticulum and swollen mitochondria.[63] Increased activity of liver adenyl cyclase, aspartate aminotransferase, alkaline phosphatase, and gluconeogenic enzymes as well as a drop in glycogen levels also occurs.[63,64]

The most dramatic result of chronic cadmium exposure is the damage seen in the kidneys.

Renal effects do not manifest until the renal cadmium concentration reaches about 200 μg/g in the renal cortex. Urinary cadmium levels also do not increase until this renal level is reached.[27,65] Morphological changes include proximal tubular necrosis, mitochondrial injury, interstitial nephrosclerosis, swelling, and vacuolization of epithelial cells.[66] Some alteration of the glomeruli may also be involved, but may be secondary to the changes in microcirculation seen in the kidney.[49] Characteristic functional changes in the kidney are proteinuria, aminoaciduria, and enzymuria.[67,69] A decreased concentrating capacity also leads to polyuria, hypercalciuria, and increased urinary uric acid.[70] These changes point to the proximal tubule as the source of the renal dysfunction. Cadmium has been found to inhibit the carriers of amino acid reuptake in the proximal tubule.[71] High-molecular-weight proteinuria seen upon chronic cadmium exposure has suggested glomerular damage as well.[72] It is believed that the proximal tubular damage may be indicative of early cadmium toxicity, since even when no change in glomerular filtration rate can be detected, an increased level of urinary β_2-microglobulin (freely filtered at the glomerulus) can be shown.[73] Increased urinary levels of glutamic oxaloacetic transaminase, glutamic-pyruvic transaminase, and lactate dehydrogenase as well as other enzymes present in the epithelium of renal tubules can be detected early in chronic cadmium poisoning and may be the most sensitive indicators of early proximal tubular damage.[74]

Vascular changes within the kidney have also been noted in animals. A non-dose-dependent thickening of arteriole walls and a dilation of arteries are the main changes observed.[75] Rats receiving 5 mg/l cadmium in the drinking water exhibit hypertension and show signs of renal arterial lesions similar to those seen in hypertension due to other causes.[76-78] In humans, some suggested evidence that cadmium exposure is linked to a higher incidence of death from cardiovascular disease has been presented,[79-81] but other reports indicate the contrary.[82,83] Since smoking habits of the subjects are a confounding factor in most of these reports, more evaluation of this subject needs to be done before conclusive correlations can be drawn.

The only incidence of a large-scale human poisoning has occurred in Japan, where soon after World War II Itai-Itai disease was discovered.[84] This disease, seen almost exclusively in the Jinzu River Basin, an area in the Toyama Prefecture of Japan which has been contaminated by cadmium discharged from a mine, is caused by consumption of rice grown in the polluted soil. This disease presents as lumbago, rheumatic pain, decrease in stature, bone decalcification, and thickening of bone with a thinning of the cortex. All of the patients also show increased urinary protein which indicates kidney involvement, although advanced nephrosis is usually not seen.[86] Clinically, the disease appears to be limited to women over the age of 40 with a history of multiple pregnancies. One possible etiology is that Itai-Itai disease is linked directly to the renal impairment.[87] A calcium loss due to tubular dysfunction leads to redistribution of body calcium, a loss of bone calcium, and osteomalacia. Cadmium also inhibits the activation of vitamin D_3 by the kidney.[47] Since males show cadmium-induced renal impairment but do not manifest the bone involvement, some factor other than just cadmium exposure must be involved. Administration of vitamin D appears to be helpful in easing the bone deformities while also causing an increase in the urinary excretion of cadmium.[85] It is believed that cadmium exposure may also accelerate the normal course of senile osteoporosis, since in some suspected cases of Itai-Itai disease the bone deformities are difficult to discern from senile osteoporosis.[47] The metabolism of bone collagen may also be impaired by cadmium exposure. Increased amounts of proline and hydroxyproline (indicators of collagen metabolism) are found in the urine of Itai-Itai patients.

RETENTION AND BODY BURDEN

Since cadmium has a long half-life in the body ($t^1/_2$ = 30 years in the kidney) it tends to accumulate over time.[35,86-88] The body burden depends upon the diet, cigarette smoking, and occupation. In the U.S. the average body burden at 45 years of age is 8.7 mg, while in Japan it is 21 mg.[35] Cadmium concentration in liver and kidney of the Japanese population is higher than the American population at all ages.[35] Average values of 50 μg/g wet renal cortex weight in the U.S. and 100 μg/g in Japan are found.[87] The accumulation of cadmium in the kidney, however, does not appear to be linear over time, but begins to drop past the age of 50.[87]

Cigarette smoking also affects the body burden of cadmium.[35,86,89] It causes faster accumulation of cadmium with age in both sexes.[89] Up to 110% more cadmium can be seen in smokers as compared to the nonsmokers.[35] Although the major route of exposure to cadmium in the general population is through the foodstuffs, smoking is a major contributor to cadmium exposure.

METHODS OF ASSESSING TOXICITY

Urinary abnormalities are indicative of the tubular dysfunction seen upon prolonged cadmium exposure. Low-molecular-weight proteinuria relates to the tubular damage. In particular, the increased excretion of 10,000 molecular weight β_2-microglobulin is commonly used as a test to assess early tubular damage.[73,90] When the urine pH is low and alkalinization of the urine cannot be performed, degradation of the β_2-microglobulin occurs. α_1-Microglobulin, which also correlates with tubular damage, may be a better indicator for use in field testing.[91,92]

The concentration of cadmium in urine correlates with renal impairment as measured by proteinuria,[93,94] but may not be useful as an indicator on an individual basis due to high subject variability and the risk of external sample contamination during collection and analysis.[94] Nevertheless, urinary cadmium can be used as an indicator of recent exposure when the exposure has been high.[88] Blood cadmium may also be used as an indicator of recent exposure.[88]

MT in the urine also relates to cadmium exposure and renal dysfunction as measured by urinary protein, glucose, and β_2-microglobulin.[42,95] From the available literature it appears that a combination of urinary cadmium, α_1- and β_2-microglobulins, and MT would be the best method to assess cadmium exposure on an individual basis.

ACKNOWLEDGMENT

Some of the work cited here from the author's laboratory was supported by U.S. Public Health Service Research Grants ES 01448 and ES 03187.

REFERENCES

1. International Labor Office, Occupational exposure limits for airborne toxic substances, *Occup. Saf. Health Ser.*, p. 37, 1977.
2. World Health Organization, Recommended health-based limits in occupational exposure to heavy metals. Report of a study group, *Tech. Rep. Ser.*, p. 647, 1980.
3. World Health Organization, Evaluation of certain food additives and the contaminants mercury, lead and cadmium, *Tech. Rep. Ser.*, p. 505, 1972.
4. **Kimura, M., Otaki, N., Yoskiki, S., Suzuki, M., and Suda, T.,** The isolation of metallothionein and its protective role in cadmium poisoning, *Arch. Biochem. Biophys.*, 165, 340, 1974.
5. **Shaikh, Z. A.,** Metallothionein as a storage protein for cadmium: its toxicological implications, in *Biological Roles of Metallothionein*, Foulkes, E. C., Ed., Elsevier, Amsterdam, 1982, 69.
6. **Webb, M. and Cain, K.,** Functions of metallothonein, *Biochem. Pharmacol.*, 31, 137, 1982.

7. **Kagi, J. H. R. and Nordberg, M., Eds.,** *Metallothionein,* Birkhauser, Basel, 1979, 41.
8. **Cherian, M. G. and Goyer, R. A.,** Metallothioneins and their role in the metabolism and toxicity of metals, *Life Sci.,* 23, 1, 1978.
9. **Bremner, I. and Davies, N. T.,** The induction of metallothionein in rat liver by zinc injection and restriction of food intake, *Biochem. J.,* 149, 733, 1975.
10. **Onosaka, S. and Cherian, M. G.,** The induced synthesis of metallothionein in various tissues of rats in response to metals. I. Effect of repeated injection of cadmium salts, *Toxicology,* 22, 91, 1981.
11. **Shaikh, Z. A. and Smith, J. C.,** The biosynthesis of metallothionein in rat liver and kidney after administration of cadmium, *Chem. Biol. Interact.,* 15, 327, 1976.
12. **Nimiyama, K. and Foulkes, E. C.,** Reabsorption of filtered cadmium-metallothionein in the rabbit kidney, *Proc. Soc. Exp. Biol. Med.,* 156, 97, 1977.
13. **Cherian, M. G. and Shaikh, Z. A.,** Metabolism of intravenously injected cadmium-binding protein, *Biochem. Biophys. Res. Commun.,* 65, 863, 1975.
14. **Johnson, D. R. and Foulkes, E. C.,** On the proposed role of metallothionein in the transport of cadmium, *Environ. Res.,* 21, 360, 1980.
15. **Morita, S.,** Defense mechanisms against cadmium toxicity. II. Effects of pretreatment with a small oral dose of cadmium on absorption, distribution and excretion of cadmium after a large oral dose in mice, *Jpn. J. Pharmacol.,* 35, 143, 1984.
16. **Webb, M.,** Protection by zinc against cadmium toxicity, *Biochem. Pharmacol.,* 21, 2767, 1972.
17. **Webb, M.,** The metallothioneins, in *The Chemistry, Biochemistry and Biology of Cadmium,* Webb, M., Ed., Elsevier, Amsterdam, 1979, 195.
18. **Nomiyama, K. and Nomiyama, H.,** Critical concentration of 'unbound' cadmium in the rabbit renal cortex, *Experientia,* 42, 149, 1986.
19. **Sugawara, N. and Sugawara, C.,** Effect of cadmium, *in vivo* and *in vitro* on intestinal brush border ALPase and ATPase, *Bull. Environ. Contam. Toxicol.,* 14, 653, 1975.
20. **Hidalgo, H. A., Koppa, V., and Bryan, S. E.,** Effect of cadmium on RNA-polymerase and protein synthesis in rat liver, *FEBS Lett.,* 64, 159, 1976.
21. **Petering, D. H., Loftsgaarden, J., Schneider, J., and Fowler, B. A.,** Metabolism of cadmium, zinc and copper in the rat kidney: the role of metallothionein and other binding sites, *Environ. Health Perspect.,* 54, 73, 1984.
22. **Shaikh, Z. A. and Smith, J. C.,** Metabolism of orally ingested cadmium in humans, in *Mechanisms of Toxicity and Hazard Evaluation,* Holmstedt, B., Lauwerys, R., Mercier, M., and Roberfroid, M., Eds., Elsevier, Amsterdam, 1980, 569.
23. **McLellan, J. S., Flanagan, P. R., Chamberlain, M. J., and Valberg, L. S.,** Measurements of dietary cadmium absorption in humans, *J. Toxicol. Environ. Health,* 4, 131, 1978.
24. **Hadley, J. G., Conklin, A. S., and Sanders, C. L.,** Rapid solubilization and translocation of ^{109}Cd following pulmonary deposition, *Toxicol. Appl. Pharmacol.,* 54, 156, 1980.
25. **Carlson, L. A. and Friberg, L.,** The distribution of cadmium in blood after repeated exposure, *Scand. J. Clin. Lab. Invest.,* 9, 1, 1957.
26. **Friberg, L.,** Further investigations on chronic cadmium poisoning; a study on rabbits with radioactive cadmium, *Arch. Ind. Hyg. Occup. Med.,* 5, 30, 1952.
27. **Shaikh, Z. A. and Hirayama, K.,** Metallothionein in the extracellular fluids as an index of cadmium toxicity, *Environ. Health Perspect.,* 28, 267, 1979.
28. **Suzuki, Y.,** Cadmium metabolism and toxicity in rats after long term subcutaneous administration, *J. Toxicol. Environ. Health,* 6, 469, 1980.
29. **Shaikh, Z. A. and Lucis, O. J.,** Biological differences in cadmium and zinc turnover, *Arch. Environ. Health,* 24, 410, 1972.
30. **Nordberg, G. F.,** Cadmium metabolism and toxicity. Experimental studies on mice with special reference to the use of biological materials as indices of retention and the possible role of metallothionein in transport and detoxification of cadmium, *Environ. Physiol. Biochem.,* 2, 7, 1972.
31. **Nordberg, G. F., Garvey, J. S., and Chang, C. C.,** Metallothionein in plasma and urine of cadmium workers, *Environ. Res.,* 28, 179, 1982.
32. **Gunn, S. A. and Gould, T. C.,** Selective accumulation of ^{115}Cd by cortex of rat kidney, *Proc. Soc. Exp. Biol. Med.,* 96, 820, 1957.
33. **Ishizaki, A., Fukushima, M., and Sakamoto, M.,** On the accumulation of cadmium in the bodies of Itai-Itai patients, *Jpn. J. Hyg.,* 25, 86, 1970.
34. **Livingston, H. D.,** Measurement and distribution of zinc, cadmium and mercury in human kidney tissue, *Clin. Chem.,* 18, 67, 1972.
35. **Kjellstrom, T.,** Exposure and accumulation of cadmium in populations from Japan, the United States, and Sweden, *Environ. Health Perspect.,* 28, 169, 1979.
36. **Shaikh, Z. A. and Lucis, O. J.,** Cadmium and zinc binding in mammalian liver and kidneys, *Arch. Environ. Health,* 24, 419, 1972.

37. **Waku, K.,** The chemical form of cadmium in subcellular fractions following cadmium exposure, *Environ. Health Perspect.,* 54, 37, 1984.
38. **Lucis, O. J., Lynk, M. E., and Lucis, R.,** Turnover of cadmium-109 in rats, *Arch. Environ. Health,* 18, 307, 1969.
39. **Suzuki, Y. and Yoshikawa, H.,** Cadmium excretion in urine and feces of rats at different levels of cadmium toxicity, *Ind. Health,* 21, 43, 1983.
40. **Lauwerys, R. R., Buchet, J.-P., and Roels, H. A.,** The relationship between cadmium exposure or body burden and the concentration of cadmium in blood and urine in man, *Int. Arch. Occup. Environ. Health,* 36, 275, 1976.
41. **Roels, H. A., Lauwerys, R. R., Bucket, J.-P., Bernard, A., Chettle, D. R., Harvey, T. C., and Al Haddad, I. K.,** *In vivo* measurement of liver and kidney cadmium in workers exposed to this metal: its significance with respect to cadmium in blood and urine, *Environ. Res.,* 26, 217, 1981.
42. **Tohyama, C., Shaikh, Z. A., Nogawa, K., Kobayashi, E., and Honda, R.,** Urinary metallothionein as a new index of renal dysfunction in Itai-Itai disease patients and other Japanese women environmentally exposed to cadmium, *Arch. Toxicol.,* 50, 159, 1982.
43. **Nordberg, G. F. and Nishiyama, K.,** Whole-body and hair retention of cadmium in mice, including an autoradiographic study on organ distribution, *Arch. Environ. Health,* 24, 209, 1972.
44. **Lucis, O. J., Lucis, R., and Shaikh, Z. A.,** Cadmium and zinc in pregnancy and lactation, *Arch. Environ. Health,* 25, 14, 1972.
45. **Smith, L. W. and Borzelleca, J. F.,** Excretion of cadmium and mercury in rat saliva, *Toxicol. Appl. Pharmacol.,* 54, 134, 1980.
46. **Cohn, J. R. and Emett, E. A.,** The excretion of trace metals in human sweat, *Ann. Clin. Lab. Sci.,* 8, 270, 1978.
47. Japan Public Health Association, *Effects of Cadmium on Human Health,* Tokyo, December 1976.
48. **Wisniewska-Knypl, J. M., Jablonska, J., and Myslak, Z.,** Binding of cadmium on metallothionein in man: an analysis of a fatal poisoning by cadmium iodide, *Arch. Toxicol.,* 28, 46, 1971.
49. **Foster, C. L. and Cameron, E.,** Observations on the histological effects of sub-lethal doses of cadmium chloride in the rabbit. II. The effect on the kidney cortex, *J. Anat.,* 97, 281, 1963.
50. **Gabbiani, G., Badonnel, M.-C., Mathewson, S. M., and Ryan, G. B.,** Acute cadmium intoxication. Early selective lesions of endothelial clefts, *Lab. Invest.,* 30, 686, 1974.
51. **Parizek, J.,** Effect of cadmium salts on testicular tissue, *Nature,* 177, 1036, 1956.
52. **Dudley, R. E., Svoboda, D. J., and Klaassen, C. D.,** Acute exposure to cadmium causes severe liver injury in rats, *Toxicol. Appl. Pharmacol.,* 65, 302, 1982.
53. **Hoffman, E. O., Cook, J. A., DiLuzio, N. R., and Coover, J. A.,** The effects of acute cadmium administration in the liver and kidney of the rat, *Lab. Invest.,* 32, 655, 1975.
54. **Gunn, S. A., Gould, T. C., and Anderson, W. A. D.,** The selective injurious response of testicular and epididymal blood vessels to cadmium and its prevention by zinc, *Am. J. Pathol.,* 42, 685, 1963.
55. **Aoki, A. and Hoffer, A. P.,** Reexamination of the lesions in rat testis caused by cadmium, *Biol. Reprod.,* 18, 579, 1978.
56. **Gunn, S., Gould, T., and Anderson, W.,** Cadmium-induced interstitial cell tumors in rats and mice and their prevention by zinc, *J. Natl. Cancer Inst.,* 31, 745, 1963.
57. **Kazantsis, G. and Hanbury, W.,** Induction of sarcoma in the rat by cadmium sulfide and by cadmium oxide, *Br. J. Cancer,* 20, 190, 1966.
58. **Bako, G., Smith, E. S. O., Hanson, J., Dewar, R., and Hill, G. B.,** Cadmium and prostatic cancer in Alberta, *Environ. Toxicol.,* 124, 121, 1981.
59. **Kipling, M. D. and Waterhouse, J. A. H.,** Cadmium and prostatic carcinoma, *Lancet,* 1, 730, 1967.
60. **Mulvihill, J., Gramm, S., and Ferm, V.,** Facial formations in normal and cadmium treated golden hamsters, *J. Embryol. Exp. Morphol.,* 24, 393, 1970.
61. **Nolen, G., Buehler, E., Geil, R., and Goldenthal, E.,** Effects of tri sodium nitrolo tri acetate, tri sodium citrate and tri sodium nitrilo tri acetate ferric chloride mixture on cadmium and methyl in rats, *Toxicol. Appl. Pharmacol.,* 23, 238, 1972.
62. **Nogawa, K.,** Itai-Itai disease and follow-up studies, in *Cadmium in the Environment,* Nriagu, J. D., Ed., John Wiley & Sons, New York, 1981, 1.
63. **Faeder, E. J., Chaney, S. Q., King, L. C., Hinners, T. A., Bruce, R., and Fowler, B. A.,** Biochemical and ultrastructural changes in livers of cadmium-treated rats, *Toxicol. Appl. Pharmacol.,* 39, 473, 1977.
64. **Merali, Z., Kacew, S., and Singhal, R. L.,** Response of hepatic carbohydrate and cyclic AMP metabolism to cadmium treatment in rats, *Can. J. Physiol. Pharmacol.,* 53, 174m, 1974.
65. **Goyer, R. A., Cherian, M. G., and Richardson, L. D.,** Renal effects of cadmium, in *Cadmium 77,* Metal Bulletin Ltd., London, 1978, 183.
66. **Chan, W. Y. and Kennert, O. M.,** Cadmium nephropathy, *Ann. Clin. Lab. Sci.,* 11, 229, 1981.
67. **Saito, H., Shioji, R., Hurukawa, Y., Nagai, K., Arikawa, T., Saito, T., Sasaki, Y., Furuyama, T., and Yoshinaga, K.,** Cadmium-induced proximal tubular dysfunction in a cadmium-polluted area, *Contr. Nephrol.,* 6, 1, 1977.

68. **Nogawa, K., Kobayashi, E., Honda, R., Ishizaki, A., Kawano, S., and Matsuda, H.,** Renal dysfunctions of inhabitants in a cadmium-polluted area, *Environ. Res.*, 23, 13, 1980.
69. **Nomiyama, K. and Sugata, Y.,** Urinary low-molecular-weight proteins in Itai-Itai disease, *Environ. Res.*, 6, 373, 1973.
70. **Kazantzis, G., Flynn, F. V., Spowage, J. S., and Trott, D. G.,** Renal tubular malfunction and pulmonary emphysema in cadmium pigment workers, *Q. J. Med.*, 32, 165, 1963.
71. **Samarawickrama, G. P.,** Biological effects of cadmium in mammals, in *The Chemistry, Biochemistry and Biology of Cadmium,* Webb, M., Ed., Elsevier, Amsterdam, 1979, 341.
72. **Lauwerys, R., Bucket, J. P., Roels, H. A., Brouwers, J., and Stanescu, D.,** Epidemiological survey of workers exposed to cadmium, *Arch. Environ. Health,* 28, 145, 1974.
73. **Nogawa, K.,** Biologic indicators of cadmium nephrotoxicity in persons with low-level cadmium exposure, *Environ. Health Perspect.*, 54, 163, 1984.
74. **Nomiyama, K., Sato, C., and Yamamoto, A.,** Early signs of cadmium intoxication in rabbits, *Toxicol. Appl. Pharmacol.*, 4, 625, 1973.
75. **Fowler, B. A., Jones, H. S., Brown, H. W., and Haseman, J. K.,** The morphologic effects of chronic cadmium administration on the renal vasculature of rats given low and normal calcium diets, *Toxicol. Appl. Pharmacol.*, 34, 233, 1975.
76. **Kanisawa, M. and Schroeder, H. A.,** Renal arteriolar changes in hypertensive rats given cadmium in drinking water, *Exp. Mol. Pathol.*, 10, 81, 1969.
77. **Perry, H. M., Jr., Erlanger, M., and Perry, E. F.,** Elevated systolic pressure following chronic low level cadmium feeding, *Am. J. Physiol.*, 232, 114, 1977.
78. **Thind, G. S., Biery, D. N., and Bovee, K. C.,** Production of arterial hypertension by cadmium in the dog, *J. Lab. Clin. Med.*, 81, 549, 1973.
79. **Perry, H. M., Jr. and Kopp, S. J.,** Does cadmium contribute to human hypertension?, *Sci. Total Environ.*, 26, 223, 1983.
80. **Tulley, R. T. and Lehmann, H. P.,** Method for the simultaneous determination of cadmium and zinc in whole blood by atomic absorption spectrophotometry and measurement in normotensive and hypertensive humans, *Clin. Chim. Acta,* 122, 189, 1982.
81. **Inskip, H., Beral, V., and McDonall, M.,** Mortality of Shipham residents: 40-year follow-up, *Lancet,* 1, 896, 1982.
82. **Sandifer, S. H., Keil, J. E., Preister, L. E., and Stara, J. F.,** Cadmium exposure and human health effects, in *Trace Substances in Environmental Health — V,* Hemphill, D. D., Ed., University of Missouri, Columbia, 1972, 269.
83. **Nogawa, K. and Kawano, S. A.,** Survey of the blood pressure of women suspected of Itai-Itai disease, *J. Juzen Med. Soc.*, 77, 357, 1969.
84. **Tsuchiya, K.,** Causation of Ouch-Ouch disease, I. Nature of the disease, *Keio J. Med.*, 18, 181, 1969.
85. **Tsuchiya, K.,** Causation of Ouch-Ouch disease. II. Epidemiology and evaluation, *Keio J. Med.*, 18, 195, 1969.
86. **deGroot, G. and Sangster, B.,** Determination of urinary cadmium excretion as diagnostic procedure to evaluate cadmium exposure of inhabitants of a polluted quarter of Stadskanaal, *Fresenius Z. Anal. Chem.*, 317, 433, 1984.
87. **Elinder, C., Kjellstrom, T., and Friberg, L.,** Cadmium in kidney cortex, liver and pancreas from Swedish autopsies, *Arch. Environ. Health,* 31, 292, 1976.
88. **Shaikh, Z. A. and Smith, L. M.,** Biologic indicators of cadmium exposure and toxicity, *Experientia,* 40, 36, 1984.
89. **Lewis, G. P., Tusko, W. J., and Coughlin, L. L.,** Cadmium accumulation in man: influence of smoking, occupation, alcoholic habit and disease, *J. Chronic Dis.*, 25, 717, 1972.
90. **Nogawa, K., Yamada, Y., Honda, R., Tsuritani, I., Kobayashi, E., and Ishizaki, M.,** Copper and zinc levels in serum and urine of cadmium-exposed people with special reference to renal tubular damage, *Environ. Res.*, 33, 29, 1984.
91. **Nogawa, K., Kido, T., Yamada, Y., Tsuritani, I., Honda, R., Ishizaki, M., and Terahata, K.,** α_1-Microglobulin in urine as an indicator of renal tubular damage caused by environmental cadmium exposure, *Toxicol. Lett.*, 22, 63, 1984.
92. **Kido, T., Honda, R., Yamada, Y., Tsuritani, I., Ishizaki, M., and Nogawa, K.,** α_1-Microglobulin determination in urine for the early detection of renal tubular dysfunctions caused by exposure to cadmium, *Toxicol. Lett.*, 24, 195, 1985.
93. **Nogawa, K., Kobayashi, E., Inaoka, H., and Ishizaki, A.,** The relationship between the renal effects of cadmium and cadmium concentration in urine among the inhabitants of cadmium-polluted areas, *Environ. Res.*, 14, 391, 1977.

94. **Nogawa K., Ishizaki, A., and Kobayashi, E.,** A comparison between health effects of cadmium and cadmium concentration in urine among inhabitants of the Itai-Ita disease endemic district, *Environ. Res.*, 18, 397, 1979.
95. **Shaikh, Z. A., Tohyama, C., and Nolan, C. V.,** Occupational exposure to cadmium: effect on metallothionein and other biological indices of exposure and renal function, *Arch. Toxicol.*, in press.

CHROMIUM

Paul O'Brien

INTRODUCTION

The work which led to the suggestion that chromium was an essential trace element started in 1955.[1] Rats fed on a "tolura" yeast diet, then believed to be chromium deficient, developed an impaired tolerance for i.v. glucose. This could be reversed by a nutritional factor present in brewer's yeast, meat, and various other foods. The unknown species present in such foodstuffs was termed the glucose tolerance factor (GTF). GTF-deficient rats were subsequently treated with 47 elements[2] of which only chromium(III) restored glucose tolerance. Unfortunately, nearly 30 years of work has failed to lead to the isolation or synthesis of a chemically pure compound corresponding to GTF.

However, *vide infra,* there is now unequivocal evidence that chromium(III) is implicated in the glucose metabolism of higher animals. The long-standing association of chromium with glucose intolerance has led many workers to seek a link with diabetes, and later with diseases associated with diabetes, such as arteriosclerosis and cardiovascular disease.

Chromate has a number of specific biological effects, including carcinogenicity,[3] mutagenicity,[4] and inducing a variety of immunological responses.[5] The compounds involved in mediating chromium toxicity are probably chromium(III), chromium(V), or chromium(VI) species formed *in vivo.* Chromate readily penetrates cells via anion channels; ^{51}Cr-labeled chromate is, hence, extensively used both to mark cells and to assess their viability.[6,7]

This article will concentrate on more recent studies of chromium in humans. A number of important reviews of chromium in biological systems are available and include: a comprehensive survey of the early literature,[8] a number of recent articles,[9-12] and two particularly useful and critical reviews of chromium in nutrition[13] and disease states.[14] Specific articles dealing with relevant chromium chemistry,[15,16] prenatal and neonatal nutrition,[17] cardiac disease,[18] and lipid metabolism[19] have also been published.

CHROMIUM ANALYSIS

The analysis of chromium in biological fluids presents a challenge to the analyst. Erroneous values for tissue and bio-fluid levels of the element abound in the older literature. The best method for the routine analysis of biological samples seems to be a high-sensitivity atomic absorption method, such as electrothermal-AA or a related technique.[12-14,20,21] It is essential that the method of additions be used in such assays. Interlaboratory comparisons confirm this,[22,23] but as yet there are no suitable standard reference materials[24] for chromium in biological samples. Neutron activation analysis is also capable of providing very accurate analytical results,[13,25] however, the expense and time involved mean that this method is unlikely to be chosen for routine clinical analysis.

CHROMIUM(III) AS AN ESSENTIAL TRACE NUTRIENT

Chromium Levels in Humans

Over the last 20 years the reported "normal" levels of chromium in readily available human samples such as serum, plasma, urine, or hair have varied by orders of magnitude. This now appears to be mainly because of the contamination of samples, particularly in the earlier studies, with exogenous chromium from a variety of sources, but including such common items as stainless steel syringe needles.[12,13,26]

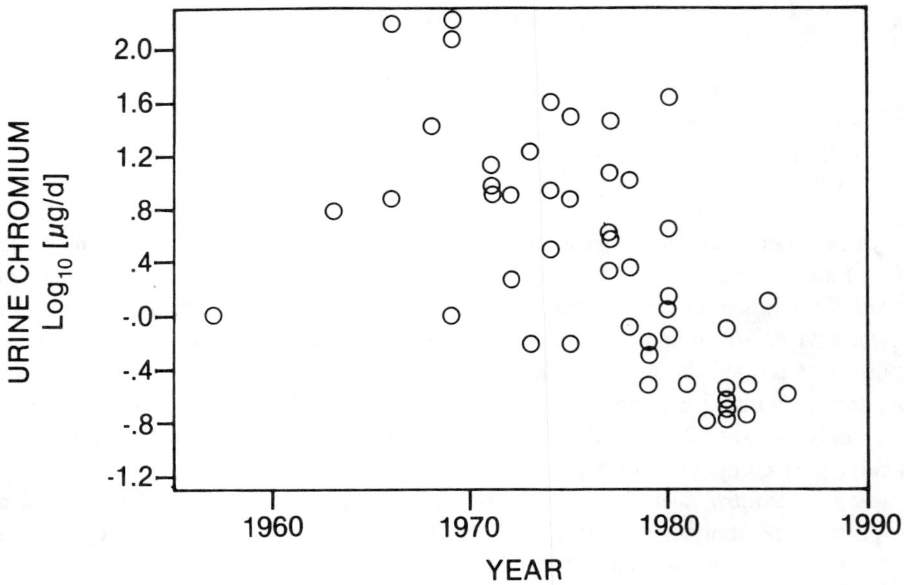

FIGURE 1. Twenty-four-hour urine chromium levels for normal subjects. Values are taken from the compilations in reviews by Guthrie[13] and Anderson,[12] and also from References 23, 29, 33, and 40. If no urine volume was reported, a 24-h volume was approximated as 1.5 l.

This article will emphasize the more recent results, as there is now a growing consensus as to the range of values acceptable. Chromium in serum is now held to normally fall in the range 0.1 to 0.2 ng/ml,[11-13,27] and urine levels are typically[11-13,28] <1 ng/ml (corresponding to 0.3 to 3.0 μg/d). Figure 1 illustrates (on a logarithmic scale!) typical 24-h urine levels of chromium reported in the last 26 years. It is apparent that since about 1980 there has been an increasing international agreement on the normal level.

Samples of hair are somewhat easier to analyze for chromium than those of serum or urine, chromium levels are considerably higher, and samples are collected by a noninvasive method. Levels in "normal" subjects generally fall between 200 and 800 ng/g.[12,13] There is little certainty as to the biochemical significance of chromium in hair, but its concentration may correlate with diabetes (383 ppb normal, 317 ppb diabetics) and does increase on chromium supplementation (average increase 117 ppb).[30] A correlation with arteriosclerosis and diabetes was suggested in an earlier study; the average in both these groups was 449 ppb compared to a normal level of 819 ppb.[31]

The uncertainty about the levels of chromium normal in serum and urine makes an assessment of chromium in diseases hard to make. Many studies report such high "normal" levels of chromium that, at present, the change in a disease state is almost impossible to assess.[12-14]

Chromium levels in urine increase with supplementation and are a reliable indicator of dietary intake, but not status.[23,32,33] The absorption of chromium from dietary sources is estimated at ~0.5%. Urinary excretion of chromium after a glucose load is unpredictable.[32] Serum chromium levels in 76 adult subjects were recently shown to be elevated by chromium supplementation (0.13 to 0.38 ng/ml),[34] but serum chromium following a glucose load showed no significant changes. The situation with regard to chromium levels in patients with arteriosclerosis and/or coronary heart disease is even worse; totally contradictory results are reported by different workers.[14,35]

Reliable results indicate that chromium levels are elevated in patients with terminal renal insufficiency treated by continuous ambulatory peritoneal dialysis (CAPD). Serum levels as high as 4.20 ± 1.73 ng/ml with a normal value of 0.160 ± 0.083 ng/ml have been reported.[13,25,36] Similarly, reliable results show that urine chromium (0.16 vs. 0.3 μg/d)[37] and serum chromium (0.12 vs. 0.17 ng/ml)[38] are elevated by strenuous exercise. Urinary chromium excretion is also elevated in trauma patients (to ~2.5 ng/ml, average for seven patients).[39] Anderson has recently found that diets high in sugars stimulate the urinary excretion of chromium.[40] Patients who have received chromium-containing metallic implants also have elevated serum levels of chromium.[14,41]

The Biological Effects of Chromium Supplementation in Humans

Studies[42-44] of three patients who developed glucose intolerance during total parenteral nutrition (TPN) provide unequivocal evidence of a marked response to chromium supplementation in humans. The results of these studies are summarized in Table 1; a number of points emerge from these papers. In all cases, after some time on a regime of TPN, the patients developed marked glucose intolerance; considerable quantities of insulin were ineffective in controlling the condition. However, after the addition of chromium chloride (presumably, *trans*-$[Cr(H_2O)_4Cl_2]^+Cl{\cdot}2H_2O$), the patients made a rapid return to more normal glucose tolerance and eventually insulin was not required. These results strongly support the link between glucose metabolism and chromium in man. There has been a recent tendency[13,14] to doubt this link. Mertz, in his early studies,[8] found that inorganic chromium salts could have a marked effect on glucose tolerance in a variety of animal models. This work was an important influence on the workers who supplemented TPN formulations with chromium for glucose-intolerant patients. The American Medical Association now recommends that TPN formulations should contain 10 to 15 μg of chromium per day.[45]

Other results in humans and animals are less convincing; many of the earlier studies used yeasts or yeast extracts high in chromium. There will always remain the possibility that in these studies the effect of some other substance contained in the yeast was being studied.

In a 12-week study of 23 healthy adult males supplemented with 200 μg of chromium chloride solution per day (water pH = 4), Riales and Albrink[46] found that high-density lipoprotein increased, weight decreased, and that insulin levels decreased. It is interesting that the results in this study seemed to indicate a falloff in response with time (12 vs. 6 weeks); this could be connected with the hydrolysis of the chromium solutions at pH 4 over the 12-week period. In marked contrast, no effect was observed[47] on glucose tolerance or lipid variables in a group of ten Finnish maturity-onset diabetic males receiving 200 μg of chromium per day (in 0.02 *M* HCl) over a 6-week period.

Anderson has recently reported studies of chromium supplementation on glucose-intolerant older people,[48] a number of hyperglycemic, hypoglycemic, and normal subjects,[49] and insulin and lipid variables in 76 free-living adults.[50] In general, modest but positive effects have been observed in the double-blind 3-month study.[50] Subjects with a serum level of glucose greater than 100 mg/dl, 90 min after glucose challenge, had this level lowered by 200 μg/d of chomium chloride; similar results were reported for both hypo- and hyperglycemic subjects.[49] The interpretation of these data has been challenged[14] on the basis that if all subjects were considered, no statistically significant improvement in glucose tolerance is observed. There is probably a real problem in interpreting such studies, as it may well be that only a subset of subjects in any study will respond to chromium. In the study of five older patients glucose tolerance was only very marginally improved.[48]

Additional evidence that chromium can affect glucose metabolism has been provided by Vinson.[51] He has shown that a high chromium yeast produces a marked decrease in the fasting serum glucose level. Fairly substantial effects were observed for high chromium yeasts and a chromium EDTA complex, but not with a control low chromium yeast. The group size was, however, very small (seven). In a related[52] study a high chromium yeast was shown to improve glucose control and lower serum lipids in a group of hyperglycemics.

Table 1
CHROMIUM DEFICIENCY IN TOTAL PARENTERAL NUTRITION (TPN)

Time on TPN	Chromium levels	Symptoms	Insulin	Final state	Ref.
3 1/2 years	0.55 ng/ml (blood) 154 ng/ml (hair)	15% weight loss Peripheral neuropathy, glucose intolerance	45 U/d ineffective	No additional insulin required; chromium supplementation initially 250 μg/d, then 20 μg/d. Glucose requirement lowered	42
6 months	0.5 μg/l (serum)	Hyperosmolar, nonketonic coma, metabolic encephalopathy, blood glucose up to 1400 mg/dl, glucosurea	20—30 U/d ineffective	No insulin requirement 150 μg/d chromium No glucosurea	43
6 1/2 months	0.1 μg/dl (plasma) (reference 1.8—3.8)	2 kg weight loss, serum glucose 200 mg/dl, glucosurea	Up to 30 U/d ineffective	No insulin requirement 26 μg/d chromium; no glucosurea; chromium still ~0.1 μg/dl (plasma)	44

There is some evidence of chromium deficiency in protein-calorie malnutrition.[12,13] Epidemiological work has also been reported to suggest a link between environmental levels of chromium and those diseases — diabetes etc. — that may relate to chromium deficiency.[12]

Chromium(III) and the Glucose Tolerance Factor

Early work on the isolation of chromium(III) complexes from yeasts indicated that GTF might be a chromium(III) complex of nicotinic acid together with glycine, cysteine, and glutamic acid[8,53] (the constituent amino acids of glutathione). To date, nobody has succeeded in isolating a chemically pure chromium complex from yeast that potentiates insulin activity. Indeed, Blackwell et al. have gone so far as to suggest that chromium(III) may not be an essential component of GTF.[54,55] A very thorough chromatographic purification of GTF from brewer's yeast led to the separation of the Mertz "GTF" into several fractions, including biologically active fractions which contained no chromium(III). Similar results have been reported by Holdsworth.[56] However, in these studies[54,55] biological activity was only assayed by monitoring effects on yeast metabolism; hence, the results do not preclude the biological activity of a chromium(III) complex in mammalian systems. Second, all work on the extraction of GTF from yeasts is hampered by the many substances present and the low levels of chromium found; e.g., Elwood[57] has recently shown, by high performance liquid chromatography (HPLC), that potent antagonists of GTF activity are found together with GTF-active fractions in yeast extracts.

It is not clear at present that yeast is a good source of biologically available or active chromium. Some caution is, hence, needed in interpreting studies in which a yeast is used as a chromium supplement. However, some suggestions as to the possible composition of active complexes have been provided by work[53] on yeast extracts, and work on well-defined chromium complexes is needed to clarify the situation.

Mode of Action of GTF

Various *in vitro* assays of chromium activity have been developed. Many involve isolated adipose tissue or adipocytes.[8,9,15] The most recent assay system reported[58] allows up to 75 assays to be performed using the fat cells of a single rat. Adipocytes were prepared from the epididymal fat pads of chromium(III)-deficient rats by the method of Rodbell.[59] Small quantities of these cells are then incubated with the material to be tested and radiolabeled glucose. Carbon dioxide production after 30 min of incubation was then measured by trapping the gas in hyamine and scintillation counting. Such systems have been used to screen both synthetic complexes and compounds isolated from foodstuffs.[58]

Other systems involve microbiological assays[60] and studies of effects of chromium on yeast metabolism.[55,61] Although these assays are facile, their relevance to mammalian systems is unclear.

One unequivocal action of the hormone insulin is its ability to facilitate the transport of glucose across the plasma membranes of muscle and fat cells. There are numerous reviews of the effect and mechanism of action of this hormone.[62,63] As chromium(III) appears to affect glucose metabolism, many workers have sought a link between chromium(III) and insulin activity.

Insulin is believed to exert its primary action (i.e., interact with the receptor) at the membrane of the target cell. In the isolated adipocyte model,[64,65] isotopically labeled insulin covalently bound to Sepharose® beads (which are larger than the isolated fat cells), is capable of exerting its full hormonal activity. This kind of work may be relevant to the mode of action of GTF. Detailed studies of GTF or chromium/cell/insulin interactions, using well-defined complexes, will be needed for a full understanding of the mode of action of GTF in this model and of the relevance of the assay to the *in vivo* situation. The various ways in which a chromium complex might interact with the surface of an adipocyte and influence

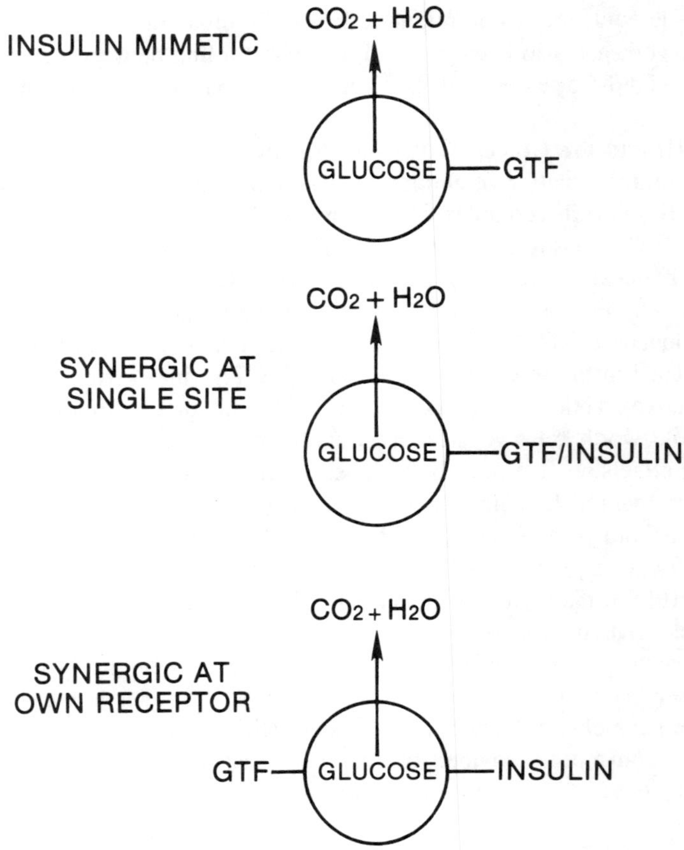

FIGURE 2. Chromium complexes and isolated adipocytes shown as circles. Schematic representation of how a chromium complex might interact with the surface of an isolated adipocyte.

metabolism are summarized in Figure 2. Such work is hampered by the fact that no chemically pure chromium(III) complexes have yet shown high activity in such test systems.

Nicotinic acid was believed to be present in the more active chromium-containing GTF preparations isolated from yeasts.[8,53,58] The coordination chemistry of this ligand is particularly relevant to glucose tolerance and the presence of this substance is apparently essential for the maximal activity of *synthetic* complexes in *in vitro* tests.[8,15] Anderson et al. have reported a complex of glutathione and nicotinic acid with chromium(III) to be active in their *in vitro* assay.[58] The complex was purified by HPLC, however, the details of the preparation are not available.[66,67] This is unfortunate as this is a species of quite remarkable *in vitro* biological activity. This work does demonstrate high specific activity unequivocally associated with a chromium complex. Related observation can be found in earlier work from the Mertz group.[8] We have recently prepared and characterized a number of chromium(III) complexes of glutathione.[68] Ultimately, we hope to synthesize well-defined ternary GSH/nicotinate complexes of the kind with maximal *in vivo* activity.

The lack of structural information on GTF means that any suggested structure must be highly speculative. Mertz originally suggested[8,69] that two *trans* nicotinic acids are coordinated to chromium as illustrated below (I) the first coordination sphere being completed by the various amino acids involved. One product of the reaction of nicotinic acid and chromium(III) (mole ratio 3:1, pH 3.2) has been suggested to be the tetra-aquo *trans*-bis-nicotinate complex[69] (II), which showed biological activity in solution. However, this procedure would be more likely to lead to a solution containing complexes containing coordinated chloride ions and *O*-bonded nicotinate.

Blackwell has recently suggested that biological activity may relate to the ability of the relatively hard chromium(III) center to form stable complexes with the carboxylate group of nicotinic acid.[70,71] The similar effect on yeast metabolism of 1,4-diguanidinobutane to the chromium(III) complexes with *trans* nicotinates may be explained on this basis.

The coordination chemistry of nicotinic acid and chromium has been developed over the last few years and well-defined *O*- and *N*-bonded (III) complexes of chromium(III) are now known.[72-74] Perhaps the most interesting and recent of these compounds is an air-stable chromium(II) complex of nicotinic acid (IV).[75] This most unusual complex, with a biologically relevant ligand, opens up the possibility of a redox role for chromium in biological systems. Crystal structures of chromium(III) peptide complexes have been reported[76] for bis-prolyl-histidine and glycyl-glycine complexes, both involving two facially coordinated peptides, and dimeric complex of β-alanyl-histidine. All the complexes contain chromium-to-peptide (imine) bonds.

CHROMIUM(VI) AND TOXICITY

Mutagenicity and Carcinogenicity

Chromium was first suggested to be the cause of a cancer over 50 years ago.[3,77] The historical aspects of chromates as carcinogens have been reviewed.[78] Chromate is particularly implicated in lung and nasal cancers. Chromates are mutagenic in the Ames' and related tests and work in this area has also been extensively reviewed.[4,79]

The carcinogenicity and mutagenicity of chromium(VI) are usually considered in terms of the uptake/reduction model.[80] Chromium(VI) readily passes into the cell, via anion channels, and once within the cell it is eventually reduced to chromium(III) species. Wetterhahn has recently comprehensively reviewed the effect of chromate on cells.[80] Complexes trapped within the cell are the agents responsible for the toxic effects of chromate.

Glutathione is important in the maintenance of redox status;[81,82] it is found at millimolar concentrations in typical mammalian cells and is implicated, in general, in the defense of cells challenged with oxidizing agents. However, in the case of chromate toxicity, cells with elevated levels of GSH have been shown to have an increased susceptibility to damage,[83,84] which may mean that a toxic metabolite is generated by a reaction with GSH. We and others have recently shown that quite stable chromium(V) species can be generated in the reaction of GSH with chromate at neutral pH.[85,86]

Microsomes[87] are capable of reducing chromium(VI), as are various nucleotides[88] and free sulhydryls.[80,85,86] In these cases, chromium(V) species of considerable stability have been observed using electron paramagnetic resonance spectroscopy. Another step in the reduction of chromate by GSH is the formation of thiolate esters,[89] which may also be implicated in the toxicity.

Chromium(VI) and Immunological Response

The ability of chromium compounds to cause contact dermatitis was first demonstrated in 1925,[90] even though chromates are only present in very low concentrations (typically 0.01 to 0.1 mg%). Cements are one of the two major causes of occupational contact dermatitis.[90] Many subjects who have a positive response to chromate are also affected by chromium(III) compounds. Clearly, similar factors — bioavailability, chemical reactivity, species formed *in vivo* to those detailed above for carcinogenicity — will control immunological response to the different oxidation states. An uptake reduction/pathway is again generally accepted as the mechanism of action of chromate, but even less is known of the chemical species involved than in mutagenicity/carcinogenicity. Oxidation states other than

(VI) or (III) have generally been discounted in explaining immunological response; this may be unwise in view of the increasing number of relatively stable chromium(V) species becoming known.

REFERENCES

1. **Mertz, W. and Schwarz, K.,** *Arch. Biochem. Biophys.,* 58, 504, 1955.
2. **Mertz, W. and Schwarz, K.,** *Arch. Biochem. Biophys.,* 85, 292, 1959.
3. **Hayes, R. B.,** *Biological and Environmental Aspects of Chromium,* Langard, S., Ed., Elsevier, Amsterdam, 1982, 199.
4. **Leonard, A. and Lauwerys, R. R.,** *Mutat. Res.,* 1, 293, 1980.
5. **Burrows, D.,** *Chromium Metabolism and Toxicity,* CRC Press, Boca Raton, FL, 1983.
6. **Sanderson, C. J.,** *Biological and Environmental Aspects of Chromium,* Langard, S., Ed., Elsevier, Amsterdam, 1982, 102.
7. **Anon.,** *Br. J. Haematol.,* 21, 241, 1971.
8. **Mertz, W.,** *Physiol. Rev.,* 49, 163, 1969.
9. **Anderson, R. A.,** *Sci. Total Environ.,* 17, 13, 1981.
10. **Wallach, S.,** *J. Am. Coll. Nutr.,* 4, 107, 1985.
11. **Doisy, R. J., Streeten, J. H. P., Freiberg, J. M., and Schneider, A. J.,** *Trace Elements in Human Health and Disease,* Praesad, A. S., Ed., Academic Press, New York, 1976, 79.
12. **Borel, J. S. and Anderson, R. A.,** *The Biochemistry of the Essential Ultratrace Elements,* Frieden, E., Ed., Plenum Press, New York, 1984, 2175.
13. **Guthrie, B. E.,** *Biological and Environmental Aspects of Chromium,* Langard, S., Ed., Elsevier, Amsterdam, 1982, 117.
14. **Cornelis, R. and Wallaeys, B.,** *Trace Element — Analytical Chemistry in Biology and Medicine,* Vol. 3, Bratter, P. and Schramel, P., Eds., Walter de Gruyter, Berlin, 1984, 219.
15. **Barrett, J., Pedrosa de Jesus, J., and O'Brien, P.,** *Polyhedron,* 4, 1, 1985.
16. **Larkworthy, L. F., Nolan, K. B., and O'Brien, P.,** *Comprehensive Coordination Chemistry,* Gillard, R. D. and Mcleverty, A. J., Eds., in press.
17. **Seelig, M. S.,** *Clin. Disord. Pediatr. Nutr.,* 2, 167, 1982.
18. **Simonoff, M.,** *Cardiovasc. Res.,* 18, 591, 1984.
19. **Anon.,** *Nutr. Rev.,* 41, 307, 1983.
20. **Jons, D., Jensen, A., and Persson, P.,** *Metal Ions in Biological Systems,* Vol. 16, Sigel, H., Ed., Marcel Dekker, New York, 1983, 201.
21. **Torgrimsen, T.,** *Biological and Environmental Aspects of Chromium,* Langard, S., Ed., Elsevier, Amsterdam, 1982, 65.
22. **Veillon, C., Patterson, K. Y., and Bryden, N. A.,** *Anal. Chim. Acta,* in press.
23. **Kumpulainen, J., Lehto, J., Koivistoinen, P., Uusitupa, M., and Vuori, E.,** *Sci. Total Environ.,* 31, 71, 1983.
24. **Kumpulainen, J., Anderson, R. A., Polansky, M. M., and Wolf, W. R.,** *Chromium in Nutrition and Medicine,* Shapcott, D. and Hubert, J., Eds., 1979, 79.
25. **Cornelis, R., Van Renterghem, D., Wallaeys, B., Mees, L., Lamiere, N., and Ringoir, S.,** *Trace Elements in Man and Animals,* TEMA-5, Mills, C. F., Bremner, I., and Chesters, J. K., Eds., Commonwealth Agricultural Bureaux, 1985, 819.
26. **Cornelis, R., Speecke, A., and Hoste, J.,** *Anal. Chim. Acta,* 78, 317, 1975.
27. **Anderson, R. A., Bryden, N. A., and Polansky, M. M.,** *Am. J. Clin. Nutr.,* 41, 571, 1985.
28. **Veillon, C., Patterson, K. Y., and Bryden, N. A.,** *Anal. Chim. Acta,* 136, 233, 1982.
29. **Harnly, J. M., Patterson, K. Y., Veillon, C., Wolf, W R., Marshall, J., Littlejohn, D., Ottaway, J. M., Miller-Ihli, N. J., and O'Haver, T. C.,** *Anal. Chem.,* 55, 1417, 1983.
30. **Hunt, A. E., Allen, K. G. D., and Smith, A. B.,** *Nutr. Res. (N.Y.),* 5, 131, 1985.
31. **Cote, M., Munan, L., Gagne-Billon, M., Kelly, A., Pietri, O.Di., and Shapcott, D.,** *Chromium in Nutrition Metabolism,* Shapcott, D. and Hubert, J., Eds., Elsevier/North-Holland, Amsterdam, 1979, 223.
32. **Anderson, R. A., Polansky, M. M., Bryden, N. A., Roginski, E. E., Patterson, K. Y., Veillon, C., and Glinsman, W. H.,** *Am. J. Clin. Nutr.,* 36, 1184, 1982.
33. **Anderson, R. A., Polansky, M. M., Bryden, N. A., Patterson, K. Y., Veillon, C., and Glinsman, W. H.,** *J. Nutr.,* p. 276, 1983.
34. **Anderson, R. A., Polansky, M. M., and Bryden, N. A.,** *Am. J. Clin. Nutr.,* 41, 571, 1985.

35. **Von Versiek, J.,** *Z. Gesamte Inn. Med. Jahrg.*, 41, 111, 1986.
36. **Wallaeys, B., Cornelis, R., and Lameire, N.,** *Frontiers in Peritoneal Dialysis,* Field Rich and Associates, New York, 1986, 478.
37. **Anderson, R. A., Polansky, M. M., Bryden, N. A., Patterson, K. Y., and Reamer, D. C.,** *Diabetes,* 31, 212, 1982.
38. **Anderson, R. A., Polansky, M. M., and Bryden, N. A.,** *Biol. Trace Element Res.*, 6, 327, 1984.
39. **Borel, J. S., Majerus, T. C., Polansky, M. M., Moser, P. B., and Anderson, R. A.,** *Biol. Trace Element Res.*, 6, 317, 1984.
40. **Kozlovsky, A. S., Moser, P. B., and Anderson, R. A.,** *Metabolism,* 35, 515, 1986.
41. **Black, J., Maitin, E. C., Gelman, H., and Morris, D. M.,** *Biomaterials,* 4, 1604, 1983.
42. **Jeejeebhoy, K. N. R. C., Chu, E. B., Marliss, G. R., and Bruce-Robinson, A.,** *Am. J. Clin. Nutr.*, 30, 531, 1977.
43. **Freund, H., Atamian, S., and Fischer, J. E.,** *J.A.M.A.*, 241, 496, 1979.
44. **Brown, R. O., Forloines-Lynn, S., Cross, R. E., and Heitzer, W. D.,** *Dig. Dis. Sci.*, 31, 661, 1986.
45. **Anon.,** *J.A.M.A.*, 241, 2051, 1979.
46. **Riales, R. and Albrink, M. J.,** *Am. J. Clin. Nutr.*, 34, 2670, 1981.
47. **Uusitupa, M. I. J., Kumpulainen, J. T., Voutilainen, E., Hersio, K., Sarlund, H., Pyorala, K., Koivistoienen, P., and Lehto, J. T.,** *Am. J. Clin. Nutr.*, 38, 404, 1983.
48. **Potter, J. F., Levin, P., Anderson, R. A., Freiberg, M. J., Andres, R., and Elahi, D.,** *Metabolism,* 34, 199, 1985.
49. **Anderson, R. A., Polansky, M. M., Bryden, N. A., and Kozlovsky, A. S.,** *Trace Elements in Man and Animals,* TEMA-5, Mills, C. F., Bremner, I., and Chesters, J. K., Eds., Commonwealth Agricultural Bureaux, 1985, 811.
50. **Anderson, R. A., Polansky, M. M., Bryden, N. A., Roginski, E. E., Mertz, W., and Ginsmann, W.,** *Metabolism,* 32, 894, 1983.
51. **Vinson, J. A. and Hsiao, K.-H.,** *Nutr. Rep. Int.*, 32, 1, 1985.
52. **Vinson, J. A. and Bose, P.,** *Nutr. Rep. Int.*, 30, 911, 1984.
53. **Toepfer, E. W., Mertz, W., Polansky, M. M., Roginsky, E. E., and Wolf, W. R.,** *J. Agric. Food Chem.*, 25, 162, 1977.
54. **Haylock, S. J., Buckley, P. D., and Blackwell, L. F.,** *J. Inorg. Biochem.*, 18, 195, 1983.
55. **Haylock, S. J., Buckley, P. D., and Blackwell, L. F.,** *J. Inorg. Biochem.*, 19, 105, 1983.
56. **Davies, D. M., Holdsworth, E. S., and Sherriff, J. L.,** *Biochem. Med.*, 33, 297, 1985.
57. **Elwood, J. C.,** personal communication, 1986.
58. **Anderson, R. A., Branter, J. H., and Polansky, M. M.,** *J. Agric. Food Chem.*, 26, 1219, 1978.
59. **Rodbell, M.,** *J. Biol. Chem.*, 239, 375, 1964.
60. **Gutierrez, J., Mertz, W., Toepfer, E. W., Polansky, M. M., and Roginsky, E. E.,** *J. Agric. Food Chem.*, 22, 100, 1974.
61. **Haylock, S. J., Thornton, R. J., Buckley, P. D., and Blackwell, L. F.,** *Exp. Mycol.*, 6, 335, 1982.
62. **Perry, M. C.,** *Companion to Biochemistry,* Bull, A. T., Lagnado, J. R., Thomas, J. O., and Tipton, K. F., Eds., Longman, Chicago 1976, 587.
63. **Blundell, T., Dodson, G., Hodkin, D., and Mercola, D.,** *Adv. Protein Chem.*, 26, 279, 1972.
64. **Cuatrecasas, P. and Hollenberg, M. D.,** *Adv. Protein Chem.*, 30, 252, 1976.
65. **Cuatrecasas, P.,** *Proc. Natl. Acad. Sci. U.S.A.*, 68, 1264, 1971.
66. **Anderson, R. A., Polansky, M. M., Branter, J. H., and Roginsky, E. E.,** *Trace Element Metabolism in Man and Animals,* Vol. 3, Kirchgessner, M., Ed., Institut für Ernahrungsphysiologie der Universität München, Freising-Weihenstephen, 1978, 272.
67. **Anderson, R. A. and Branter, J. H.,** *Fed. Proc.*, 36, 1123, 1977.
68. **Abdullah, M., Barrett, J., and O'Brien, P.,** *J. Chem. Soc. Dalton Trans.*, p. 2085, 1985.
69. **Mertz, W., Toepfer, E. W., Roginsky, E. E., and Polansky, M. M.,** *Fed. Proc.*, 33, 2275, 1974.
70. **Cooper, J. A., Anderson, B. F., Buckley, P. D., and Blackwell, L. F.,** *Inorg. Chim. Acta,* 91, 1, 1984.
71. **Cooper, J. A., Blackwell, L. F., and Buckley, P. D.,** *Inorg. Chim. Acta,* 92, 23, 1984.
72. **Chang, J. C., Gerdom, L. E., Baeziger, N. C., and Goff, H. M.,** *Inorg. Chem.*, 22, 1739, 1983.
73. **Green, C. A., Bianchini, R. J., and Legg, J. I.,** *Inorg. Chem.*, 23, 2713, 1984.
74. **Broderick, W. E. and Legg, J. I.,** *Inorg. Chem.*, 24, 3725, 1985.
75. **Broderick, W. E., Pressprich, M. R., Geiser, U., Willet, R. D., and Legg, J. I.,** *Inorg. Chem.*, 25, 3372, 1986.
76. **Murdoch, C. M., Cooper, M. K., Hambley, T. W., Hunter, W. N., and Freeman, H. C.,** *J. Chem. Soc. Chem. Commun.*, p. 1329, 1986.
77. **Langard, S.,** *Chromium Metabolism and Toxicity,* Burrows, D., Ed., CRC Press, Boca Raton, FL, 1984, 13.
78. **Maclure, K. M. and McMahon, B.,** *Epidemiol. Rev.*, 2, 19, 1980.

79. **Levis, A. G. and Bianchi, V.,** *Biological and Environmental Aspects of Chromium,* Langard, S., Ed., Elsevier, Amsterdam, 1982, 171.
80. **Wetterhahn, K. E. and Connett, P. H.,** *Struct. Bonding (Berlin),* 54, 93, 1983.
81. **Rabenstein, D. L., Guevremont, G., and Evans, C.,** *Metal Ions in Biological Systems,* Vol. 9, Sigel, H., Ed., 1979, 103.
82. **Meister, A.,** *Science,* 220, 472, 1983.
83. **Wetterhahn, K. E. and Cupo, D. Y.,** Abstr. of the 23rd Int. Conf. on Coordination Chemistry, Boulder, CO, 1984, 474.
84. **Wetterhahn, K. E. and Cupo, D. Y.,** *Carcinogenesis,* 5, 1705, 1984.
85. **O'Brien, P., Barrett, J., and Swanson, F.,** *Inorg. Chim. Acta,* 108, L19, 1985.
86. **Goodgame, D. M. L. and Joy, A. M.,** *J. Bioinorg. Chem.,* 26, 219, 1986.
87. **Jennette, K. W.,** *J. Am. Chem. Soc.,* 104, 874, 1982.
88. **Goodgame, D. M. L., Hayman, P. B., and Hathaway, D. E.,** *Polyhedron,* 1, 497, 1982.
89. **Connett, P. H. and Wetterhahn, K. E.,** *J. Am. Chem. Soc.,* 107, 4282, 1985.
90. **Polak, I.,** *Chromium Metabolism and Toxicity,* Burrows, D., Ed., CRC Press, Boca Raton, FL, 1983, 51.

COBALT

Lindsay G. Morgan

INTRODUCTION

Cobalt metal is used extensively in the metallurgical industry, especially in the manufacture of high-duty alloys. It is a constituent of hard metal which is used in the manufacture of cutting tools. Cobalt salts and oxides are used in the enameling, paint, and pottery industries as drying agents and pigments, and in agriculture as animal-feed additives.

Chemistry

Cobalt (Co) is a silver-white metal. It is a transition element belonging to Group VIII of the periodic table between iron and nickel, with both of which it shares biochemical and toxicological properties. The cobaltic ion is a strong oxidizing agent and reacts with organic and inorganic acids to form pink or blue soluble and insoluble salts.

Analytical Chemistry

A variety of techniques for analyzing cobalt in biological tissues have been used (see Table 1) and their sensitivities have improved dramatically in recent years.[1]

Biochemistry

Cobalt is an essential element for man. Apart from being a constituent of vitamin B_{12}, cobalt contained in low-molecular-weight complexes is essential to mammals. It is a co-factor for at least one mammalian enzyme, glycylglycinedipeptidase, and is essential to certain bacteria involved in the synthesis of vitamin B_{12} in ruminants. Divalent cobalt can replace other divalent cations in metal-activated enzymes, for example, zinc in carbonic anhydrase and manganese in phosphoglucomutase.

Cobalt has been found to have antidotal effects against certain poisons and its action in detoxifying cyanide has been used therapeutically. Certain cobalt coordination complexes are important because they can bind oxygen reversibly. Cobalt is known to interfere with the Krebs cycle because it binds the S H group in dihydrolipoic acid, preventing the oxidative decarboxylation of pyruvate to acetyl-coenzyme A (CoA) and ketoglutarate to succinyl CoA. This depresses energy uptake and lowers tissue oxygen concentration.[2]

OCCURRENCE

Cobalt is widely distributed in the earth's crust, but there are only a few areas where the concentrations are sufficient to be commercially recoverable. Principal deposits are in the nickel ores of the Sudbury Basin, Ontario and the cobalt/copper deposits of Zaire and Zambia. Some coals contain cobalt which can be a constituent of fly ash and it is also a minor constituent of crude oils. Soils normally contain cobalt in concentrations of 1 to 40 mg/kg; cobalt-deficient soil in New Zealand has resulted in Pining disease in sheep. Cobalt content of seawater varies from 0.01 to 4.6 μg/l, while the content of fresh water has even greater variability from 0.1 to 10 μg/l, depending on local geology and human activity.[3]

The principal dietary sources are seafood and caffeinated beverages (100 to 300 μg/100 g), and vegetables and grains (10 to 100 μg/100 g).[4]

Table 1
INSTRUMENTAL ANALYSIS OF COBALT: BIOLOGICAL TISSUES

Method	Detection limit	Ref.
DC arc emission		
Spectrometry	2—5 μg/g	54
Spectrography	20 μg/g	55
Stallwood jet	5 μg/g	55
DC plasma emission	0.001 μg/ml	56
Spark emission	2 μg/ml	54
Radiofrequency plasma emission	0.003 μg/ml	54
Flame atomic emission	0.03 μg/ml	54
Flame atomic absorption	0.002 μg/ml	56
Electrothermal atomic absorption		
Carbon rod atomizer	5×10^{-12}	5
	34 *n* mol/l	52
Graphite furnace	80×10^{-12}	58
Molecular fluorescence	0.0001 μg/ml	60
Flame atomic fluorescence	1.5 μg/ml	54
Neutron activation analysis	0.0002 μg	55
	0.04 μg	55
X-ray fluorescence	100—1000 μg/l	59
Polarography	50—400 μg/g	55

Based on Carson, B. L. and Smith, I. C., *Cobalt An Appraisal of Environmental Exposure,* Vol. 11, Table 111-15, Midwest Research Institute for National Institute for Environmental Health Science, 1979.

INTAKE

Gastrointestinal

Cobalt generally enters the body via the gastrointestinal tract. Absorption is variable but is higher for the organic form than the free ionic state. Normal intake is in the order of 140 to 580 μg/d.[4].

Dermal

Significant dermal absorption of cobalt may occur. Wahlberg[5] found that painting cobalt chloride solution to the skin of guinea pigs caused a high mortality, but only 1.7% of a dose of radioactive cobalt metal applied to the skin appeared in the urine.

Respiratory

Long-term inhalation of 2 to 160 μg/l cobaltous oxide by hamsters resulted in less than 1% retention.[6] Occupational studies suggest that cobalt metal and cobaltous oxide are ingested via the lungs, and they showed that a hard metal worker with pulmonary fibrosis had retained excess cobalt in the lung 12 years after cessation of exposure.

METABOLISM

No tissue selectively concentrates cobalt. The highest levels are found in the blood, liver, kidneys, and gastric mucosa.[9] (See Table 2).

EXCRETION

Cobalt is excreted in the urine and feces in humans. Although 80% of an administered dose is excreted in the feces, much has been absorbed and reexcreted by the liver. Onkelinx[10]

Table 2
NORMAL VALUES OF COBALT IN MAN (VARIOUS AUTHORS)

Matrix	Normal value	Ref.
Whole blood	1.5 mg/70-kg man	61
Whole blood	0.35 μg/l	62
Serum	0.29 μg/l	63
Blood total	0.7 μg	4
Plasma total	1.4 μg	64
RBC total	0.34 μg	4
Lungs	0.03 mg/kg	61
Bone marrow	18.5 μg/g	4
Urine	260 μg/d (= 87% of intake)	4
	160 μg/l	4
	0.2—0.8 μg/l	64
Sweat	17 μg/d	4
Hair	0.17—0.28 μg/g	4
	0.04—0.15 ppm	64
Feces	0.19—1.21 μg/d	65
Saliva	0.125 μg/l	66
Daily intake	140—580 μg/l	4
Milk	0.6 μg/l	67

administered Co_{57} Cl intravenously to rats and found that 66% of the administered dose was excreted in the urine in 3 d. Retention is increased in the presence of antibiotics and high copper intake,[11] while excretion can be facilitated by chelation with EDTA, or glutathione which is less toxic.[12]

TOXICITY

Acute

Acute cobalt poisoning in man does not appear to have been recorded. Cobalt metal has a low oral toxicity of LD_{50} 7000 mg/kg,[13] but was found to be lethal intratracheally.[14] Speijers et al.[15] studied the acute toxicity of eight cobalt compounds. The LD_{50} varied from 150 mg/kg for cobalt chloride to 708 mg/kg for cobalt acetate. The addition of cobalt chloride, 500 ppm, to the drinking water of rats for a 3-month period caused weight loss and polycythemia,[16] and a similar dosage schedule was noted to cause testicular degeneration[17] (see Table 3).

Chronic

Therapeutic use and occupational exposure to cobalt metal and compounds have led to considerable clinical and pathological experience (see Table 4).

Polycythemia

In 1955, Davies and Fields[18] showed that cobalt compounds could induce polycythemia in man when hemopoiesis is depressed such as in toxic marrow disease, but not in iron deficiency or pernicious anemia.[19]

Probably, cobalt stimulates the production of renal hemopoietin and this is independent of any vitamin B_{12} effect. The effect has also been noted in two studies of cobalt-exposed workers.[20] Cobalt was used in the treatment of refractory anemia for a while in the late 1950s, but because of the associated toxic effects on the thyroid, nervous system, and gastrointestinal tract it has now been discontinued.

Table 3
ACUTE ORAL TOXICITY OF COBALT COMPOUNDS IN EXPERIMENTAL ANIMALS

Route form	Species	LD_{50} (mg/kg)	Ref.
Oral cobalt metal	Rat	7000	13
Cobalt oxide Co_3O_4	Rat	5000	13
Cobalt(II) oxide	Rat	1700	68
Cobalt chloride	Rat	180	69
CoF	Rat	150	15
CoO	Rat	202	15
$Co_3(PO_4)_2$ $8H_2O$	Rat	539	15
$CoBr_2$	Rat	406	15
$CoCl_2 \cdot 6H_2O$	Rat	766	15
$CoSO_4 \cdot 7H_2O$	Rat	768	15
$Co(NO_3)$ $2 \cdot 6H_2O$	Rat	691	15
$Co(CH_3CO_2)$ $2 \cdot 4H_2O$	Rat	708	15
$CoCl_2$	Mouse	40.8	70
I.p. cobalt chloride	Mouse	1000	71
Cobalt chloride	Mouse	36.7	70
Metal powder	Mouse	10—20	
S.c. cobalt salts	Mouse	12—67	72

Table 4
CHRONIC TOXICITY OF COBALT AND COBALT COMPOUND IN MAN

Organ or tissue	Effect	Ref.
Blood	Polycythemia	73
	Clotting time increased	74
	Plasma lipids raised	75
	Plasma proteins raised	76
	Glucose increased	75
Thyroid	Goiter	77
Skin	Contact dermatitis	78
	Vasodilatation	79
Heart	Cardiomyopathy	30
		33
Gastric	Indigestion	80
Respiratory		
Upper	Anosmia	81
Lower	Hard metal disease	34
		8
	Asthma	42
	Cobalt carbonyl poisoning	82
Implants	Necrosis	83

Hypersensitivity

Dermatological — Cobalt allergy is seen in the pottery, printing, and metallurgical industries,[21] especially hard metal manufacturing.[22] Cobalt chloride is a standard patch-test material for dermatologists, and in a multinational study Malten et al.[23] reported that 4.9% of patients attending European dermatology clinics for nonoccupational dermatitis had positive patch tests to cobalt sulfate, but 14% of those with occupational dermatitis were positive. Some cement workers develop sensitivity to cobalt as well as to chrome. Cobalt sensitivity unassociated with sensitivity to other substances is a rare event.[24] Urticaria and papuloerythematous rashes have been described at the wrists, antecubital fossae, neck, and ankles, and may be aggravated by sunlight.[25] Cobalt salts may also cause primary irritant dermatitis.

Asthma — While there is ample evidence that cobalt is capable of forming ligands with proteins and becoming allergenic to both skin and respiratory tissue, there are few reports of asthma being attributed specifically to cobalt. Bruckner[26] and Morgan[7] each reported two possible cases in men manufacturing cobalt salts. Asthma frequently occurs in hard metal disease.

Metal implants — Many metal implants contain cobalt alloyed with other metals. Some individuals have developed sensitivity reactions to the cobalt in these implants and specific patch testing is recommended before their use.[27] Evans[28] used a fibroblast tissue culture technique to study the toxicity of cobalt-chrome-molybdenum alloys compared with cobalt and nickel separately and found the alloy to be less toxic.

Heart disease — Kocetkova[29] reported the case of a machinist exposed to high concentrations of cobalt dust who had a diseased myocardium. However, the earliest recognized cases of cobalt cardiomyopathy were those that occurred in the Quebec episode of 1965, when fatal cardiomyopathy in heavy beer drinkers was eventually traced to cobalt added as an antifrothing agent.[30] Once cobalt ceased to be used for this purpose no further cases occurred.[31] Kennedy et al.[32] consider the pathology to be similar to the heart disease of beriberi because cobalt depressed the metabolism of thiamin. They drew this conclusion as the result of an interesting study of a cobalt worker who went into cardiogenic shock under anesthesia. Unverth[33] used cobalt cardiomyopathy in dogs to study heart failure.

Hard Metal Disease

Hard metal is manufactured by heating a mixture of finely powdered cobalt metal and tungsten carbide in an electric arc furnace. The process can be very dusty if not properly controlled and significant exposure may also occur in the use and sharpening of hard metal tools.

Pulmonary reaction to cobalt was reported by Jobs and Ballhausen in 1940.[34] The early history of hard metal disease was reviewed by Bech[35] and the disease is extensively documented. It presents as a form of progressive interstitial fibrosis and emphysema with or without an asthmatic element. Patients characteristically have a chronic cough with shortness of breath and in the later stages they may develop right heart failure. The chest X-ray shows soft shadowing with loss of definition of the cardiac shadow. Employees in the hard metal industry have a high prevalence of chronic bronchitis.

The continuing controversy as to whether tungsten carbide, cobalt, or the mixture causes the problem was recently reviewed by Cullen.[36] Schepers[37,38] investigated the problem in a series of intratracheal studies in rats and considered that cobalt metal was the culprit. Kerfoot,[40] using inhalation studies, showed ultrastructural changes in miniature swine at very low exposure levels. The National Institute of Safety and Health in the U.S.[41] reviewed the possibility that all cobalt compounds are capable of causing fibrosis and deemed it prudent to consider that they do. Davison et al.[42] reported the presence of multinucleated giant cells comprising type 2 alveolar epithelial cells and alveolar macrophages and their bronchial provocation tests indicated cobalt as the responsible agent.

Recently, suspicion has fallen upon cobalt as the causative agent in a condition of interstitial pneumonitis and asthma, clinically similar to hard metal disease, occurring in diamond polishers who use high-speed grinding tools and have no exposure to tungsten carbide.[43] Rom et al.[44] have reported a similar condition in dental technicians, but they have a mixed exposure.

Cancer and Mutagenicity

Cobalt and its salts are not generally recognized as carcinogens, except that the M.A.K. Commission in West Germany has classified cobalt (as respirable dust/aerosols from cobalt metal and cobalt salts of low solubility) as a class A2 carcinogen, i.e., proven in animals. Heath[45] induced metastasizing rhabdomyosarcomas in rats after i.m. injections of metallic cobalt powder, and Shabaan[46] observed similar results with cobalt chloride injected into the abdominal wall. Safiotti[47] noted hyperplasia of bronchial epithelium after instillation of cobalt chloride into the trachea, and other workers have produced injection site tumors with cobalt powders or other cobalt compounds.[41]

Cobalt can be shown to affect the accuracy of DNA synthesis.[48] No inhalation studies have been specifically directed at the carcinogenesis question, however, Wehner et al.[49] exposed hamsters to cobalt oxide, 10 mg/m^3 each day of the working week for a life span. They observed fibrosis and other changes, but no increase in cancer specifically. Similarly in man, although none of the studies reported have been directed at investigating the incidence of cancer, none have reported an increased incidence as an incidental finding.

There is no clear evidence which suggests that metal implants containing alloys, when correctly used, are carcinogenic to man.

Industrial "Overexposure"

In respect to occupational exposure there are a number of reports available.[7,20,31,50] Clinical problems that were recorded included nasal irritation anosmia, pulmonary and gastrointestinal symptoms, polycythemia, anorexia, and dermatitis. Other papers reported similar findings, but the atmospheric conditions are not well defined in their papers. Morgan[7] found elevated urinary cobalt, but no relevant clinical problems despite exposures in of excess of 0.1 mg/m^3.

Monitoring of Occupational Cobalt Exposure

Atmospheric Sampling

Workplace sampling of cobalt may be by static instruments or the use of personal samplers worn on the lapel. These generally sample at the rate of 2 l/min and are worn for the complete 8-h shift. Dust may be collected on a glass fiber or cellulose filter paper which is subsequently analyzed for cobalt content. Analysis is generally by atomic absorption spectrophotometry (AAS), but if X-ray fluorescence is available then this can provide a quick and simple analytical technique after appropriate calibration. The American Conference of Governmental and Industrial Hygienists (ACGIH) is currently recommending a reduction of the maximum 8-h time-weighted average exposure from 0.10 to 0.05 mg/m^3.[51]

Biological Monitoring

The level of cobalt in blood and urine is dependent upon a variety of factors, including the chemical nature and composition of the process material and its particle size as well as the route of exposure and individual rates of absorption and excretion. Cobalt levels in both matrices have both diurnal and temporal variations and urinary cobalt concentrations also related to the specific gravity of urine.

The analysis of urine and serum cobalt are generally by AAS after appropriate sample preparation and great care has to be taken to avoid contamination. Schroeder's values[4] for nonoccupational exposed personnel are reported in Table 2. Morgan[7] reported urinary cobalt values ranging from 0.1 to 0.9 mg/ml in a group of cobalt-exposed workers.

Insoluble cobalt is more likely to remain in the lung and exert a toxic effect than soluble cobalt, from which it is rapidly excreted with a half-life of approximately 30 h.

Pellet et al.[52] studied the value of monitoring urinary cobalt in hard metal workers on a routine basis using flame atomic absorption, and concluded that it has greater value in assessing the working environment than predicting the occurrence of occupational disease.

REFERENCES

1. **Smith, I. C. and Carson, B. L.,** *Trace Metals in the Environment,* Vol. 6, Ann Arbor Science, Ann Arbor, MI, 1979, 15.
2. **Alexander, C. S.,** Cobalt — beer cardiomyopathy. A study of 28 cases (Minnesota), *Am. J. Med.,* 53, 395, 1972.
3. **Young, R. S.,** *Cobalt in Biology and Biochemistry,* Academic Press, London, 1979, 31.
4. **Schroeder, H. A., Nason, A. P., and Tipton, I. H.,** Essential trace element metals in man. Cobalt, *J. Chronic Dis.,* 4, 20, 869, 1967.
5. **Wahlberg, J. E.,** A method of studying percutaneous toxicity of metal compounds in the guineapig, *Acta Derm. Venereol.,* 45, 171, 1965.
6. **Wehner, A. P. and Craig, D. K.,** Toxicology of inhaled NiO, and CoO, in Syrian golden hamsters, *Am. Ind. Hyg. Assoc. J.,* 33, 145, 1972.
7. **Morgan, L. G.,** A study into the health and mortality of men exposed to cobalt and oxides, *J. Soc. Occup. Med.,* 33, 181, 1983.
8. **Hillerden, G. and Hartung, M.,** On cobalt in tissues from hard metal workers, *Int. Arch. Occup. Environ. Health,* 53, 89, 1983.
9. **Flodh, H.,** Autoradiographic studies on distribution of radio-cobalt chloride in pregnant mice, *Acta Radiol. Ther. Phys. Biol.,* 7, 121, 1968.
10. **Onkelinx, C.,** Compartment analysis of cobalt (ll) metabolites in rats of various ages, *Toxicol. Appl. Pharmacol.,* 38, 425, 1976.
11. **Cocozza, E. P. and Ferguson, A.,** *Appl. Spectrosc.,* 21, 286, 1967.
12. **Llobet, J. M. and Domingo, J. L.,** Comparison of antidotal efficacy of chelating agents upon acute toxicity of Co(II) in mice, *Res. Commun. Chem. Pathol. Pharmacol.,* 50 (2), November 1985.
13. Food and Drugs Research Laboratories, Acute Oral Toxicity of Seventeen Nickel-Cobalt Compounds in Sprague-Dawley Rats, commissioned report for Nipera, Food and Drugs Research Laboratories, Inc., unpublished, 1983.
14. **Harding, H. E.,** Notes on the toxicology of cobalt metal, *Br. J. Ind. Med.,* 7, 78, 1950.
15. **Speijers, G. J. A., Krank, E. I., and Berkvens, J. M.,** Acute oral toxicity of inorganic cobalt compounds in rats, *Food Chem. Toxicol.,* 20, 311, 1982.
16. **Domingo, J. L., Llobet J. M., and Bernat, R.,** A study of the effects of cobalt administered orally to rats, *Arch. Farmacol. Toxicol.,* 10, 13, 1984.
17. **Corrier, D. E., Mollenhauer, H. H., Clark, D. E., Hare, M. F., and Elissalde, M. H.,** Testicular degeneration and necrosis induced by dietary cobalt, *Vet. Pathol.,* 22, 610, 1985.
18. **Davies, J. E. and Field, J. P.,** Cobalt polycythaemia in humans, *Fed. Proc.,* 14, 13, 1955.
19. **Goldwasser, E., Jacobson, L. O., Fried, W., and Plzak, L. F.,** Studies on erythropoiesis. V. The effect of cobalt on the production of erythropoietin, *Blood,* 13, 55, 1958.
20. **Cau, G., Hollard, D., Gimmert, E., and Zarb, R.,** Troubles respiratoires et generaux conseintifs à l'inhalation de ponssières de cobalt. "Respiratory and general disorders following the inhalation of cobalt dust", *Rev. Lyon. Med.,* 12, 491, 1963.
21. **Fischer, T. and Rystedt, I.,** Cobalt allergy in hard metal workers, *Contact Derm.,* 9, 115, 1983.
22. **Rystedt, I. and Fischer, T.,** Relationship between nickel and cobalt sensitization in hard metal workers, *Contact Derm.,* 9, 195, 1983.
23. **Malten, K. E., Fregert, S., Bandmann, H. J., Calnan, C. D., Cronin, E., Hjorth, N., Magnusson, B., Maibach, H. I., Meneyhini, C. L., Pirila, V., and Wilkinson, D. S.,** *Berufs Dermatosen,* 19, 51, 1971.
24. **Everall, J., Truter, M., and Truter, E.,** Epidermal sensitivity to Cr, Co, and Ni, *Acta Derm. Venereol.,* 34, 477, 1954.
25. **Romaguera, C., Lecha, M., Grimalt, F., Muniesa, A. M., and Mascaro, J. M.,** Photocontact dermatitis to cobalt salts, *Contact Derm.,* 8, 383, 1982.
26. **Bruckner, H. C.,** Extrinsic asthma in a tungsten carbide worker, *J. Occup. Med.,* 9, 518, 1967.
27. **B. M. J.,** Editorial 281, 1303, 1980.
28. **Evans, E. J. and Thomas I. T.,** The "in vitro" toxicity of cobalt-chrome-molybdenum alloy and its constituent metals, *Biomaterials,* 7, 1, 1986.
29. **Kocetkova, T. A.,** The pathological effects of cobalt dust (in Russian), *Gig. Tr. Prof. Zabol.,* 4, 11, 34, 1960.
30. **Morin, Y. and Daniel, J.,** Quebec beer drinkers cardiomyopathy. Etiological consideration, *Can. Med. Assoc. J.,* 97, 926, 1967.
31. **Barborik, M. and Dusek, J.,** Cardiomyopathy accompanying industrial cobalt exposure, *Br. Heart J.,* 34, 113, 1972.

32. **Kennedy, A., Dornin, J. D., and King, R.,** Fatal myocardial disease associated with industrial exposure to cobalt, *Lancet,* 1, 412, 1981.
33. **Unverth, D. V. et al.,** Canine cobalt cardiomyopathy: a model for the study of heart failure, *Am. J. Vet. Res.,* 44 (6), 989, 1983.
34. **Jobs, H. and Ballhausen, C.,** The medical and technical view points of metal ceramics as a source of dust (in German), *Vertravensarzt Krankenkasse,* 8, 142, 1940.
35. **Bech, A. O., Kipling, M. D., and Heather, J. C.,** Hard metal disease, *Br. J. Ind. Med.,* 19, 239, 1962.
36. **Cullen, M. R.,** Exposure to hard metal (letter to Editor), *Chest,* 87, 4, 554, 1984.
37. **Schepers, G. H. W.,** The biological action of particulate cobalt metal, *AMA Arch. Ind. Health,* 12, 127, 1955.
38. **Schepers, G. H. W.,** The biological action of tungsten carbide and cobalt, *AMA Arch. Ind. Health,* 12, 140, 1955.
39. **Schepers, G. H. W.,** Biological action of cobaltic oxide, *AMA Arch. Ind. Health,* 12, 124, 1955.
40. **Kerfoot, E. J., Fredrick, W G., and Domier, E.,** Cobalt metal inhalation studies on miniature swine, *Am. Ind. Hyg. Assoc.,* 36, 17, 1975.
41. **NIOSH,** Criteria for Controlling Occupational Exposure to Cobalt, D.M.S.S. National Institute of Safety and Health Publ. No. 82/107, 1981.
42. **Davison, A. G., Haslam, P. L., Corrin, B., Coutts, I. I., Dewar, A., Riding, W. D., Studdy, P. R., and Newman-Taylor, A. J.,** Interstitial lung disease and asthma in hard-metal workers: bronchoalveolar lavage, ultrastructural, and analytical findings and results of bronchial provocation tests, *Thorax,* 38, 119, 1983.
43. **Gheyseurs, B., Auwerx, J., Van den Eeckhout, A., and Demedts, M.,** Cobalt-induced bronchial asthma in diamond polishers, *Chest,* 88, 5, 1985.
44. **Rom, W. N., Lockey, J. E., Lee, J. S., Kimball, A. C., Bang, K. M., Leaman, H., Johns, R. E., Jr., Perrota, D., and Gibbons, H. L.,** Pneumoconiosis and exposures of dental laboratory technicians, 74, 11, 1984.
45. **Heath, J. C.,** The production of malignant tumours by cobalt in the rat, *Br. J. Cancer,* 10, 668, 1956.
46. **Shabaan, A. A., Marks, V., Lancaster, M. C., and Dufeau, G. R.,** Fibrosarcomas induced by cobalt chloride ($CoCl_2$) in rats, *Lab. Anim.,* 11, 43, 1977.
47. **Saffiotti, V. F., Ceffis, L. H., Kolb, L. H., and Grote, M. J.,** Intertracheal injection of particulate carginogens into hamster lungs, *Proc. Am. Assoc. Cancer Res.,* 4, 59, 1963.
48. **Sirover, M. A. and Loeb, L. A.,** Infidelity of DNA synthesis, *Science,* 194, 1434, 1976.
49. **Wehner, A. P., Busch, R. H., Olsen, R. J., and Craig, D. K.,** Chronic inhalation of cobalt oxide and cigarette smoke by hamsters, *Am. Ind. Hyg. Assoc. J.,* 38, 338, 1977.
50. **Verhamme, E. N.,** Contribution to the evaluation of the toxicity of cobalt, *Cobalt,* 2, 29, 1973.
51. American Conference of Governmental and Industrial Hygienists, Threshold Limit Value for Chemical Substances in the Work Environment, adopted by the ACGIH for 1984—1985 Publ. ACGIH.
52. **Pellet, F., Perdrix, A., et al.,** Biological Monitoring of Urinary Cobalt. Significance, Methodology, Preliminary Results, Toxicovigilance Industrielle, No. 125,
53. **Carson, B. L. and Smith, I. C.,** Cobalt. An Appraisal of Environmental Exposure, Vol. 11, report for Niehs by Mid West Research Institute, 1979.
54. **Pinta, M.,** *Modern Methods for Trace Element Analysis,* translated from the French by STS, Inc., Ann Arbor Science Publishers, Ann Arbor, MI, 1978.
55. **Reeves, R. D. and Brooks, R. R.,** *Trace Element Analysis of Geological Materials,* John Wiley & Sons, New York, 1978.
56. **Veillon, C.,** Optical atomic spectroscopic methods, in *Trace Analysis Spectroscopic Methods for Elements,* Winefordner, J. D., Ed., John Wiley & Sons, New York, 1976, 123.
57. **Culver, B. R.,** Analytical Methods for Carbon Rods, Atomizers, Varain Techtron Pty., Springvale, Victoria, Australia, 1975.
58. **Anon.,** Analytical Methods for Atomic Absorption Spectrophotometry, Perkin Elmer Corp., Norwalk, CT, April 1974.
59. **Anon.,** Simultaneous multi-element analysis of aqueous solutions, *Notes Water Res. (U.K.),* No. 8, 1, 1977.
60. **St. John, P. A.,** Fluorometric methods for traces of elements, in *Trace Analysis Spectroscopic Methods for Elements,* Winefordner, J. D., Ed., John Wiley & Sons, New York, 1976.
61. **Widdowson, E. M. and Dickerson, J. W.,** *Mineral Metabolism,* Vol. 2 (Part A), Academic Press, New York, 1964.
62. **Haerdi, A.,** Cobalt in blood, *Helv. Chem. Acta,* 43, 869, 1960.
63. **Parr, R. M. and Taylor, D. M.** Serum cobalt, *Biochem. J.,* 91, 424, 1964.
64. **Coleman, R. F., Herrington, J., and Scales, J.,** Varying cobalt levels in patients with vitallium implants, *Br. Med. J.,* 1, 528, 1973.

65. **Harp, M. J. and Scoular, F. I.,** Cobalt in faeces, *J. Nutr.,* 47, 67, 1952.
66. **Dreizen, B.,** Cobalt in saliva, *J. Dent. Res.,* 31, 137, 1952.
67. **Archibald, J. G.,** Cobalt in milk, 0.6 μg/l, *J. Dairy Sci.,* 30, 293, 1947.
68. **Smythe, H. F., Carpenter, C. P., Weil, C. S., Pozzani, U. C., Striegel, J. N., and Nycum,** Range finding toxicity data list VIII, *Am. Ind. Hyg. J.,* 30, 470, 1969.
69. **Christiansen, H. E. and Luginbyn, T. T.,** Toxic Substances List 218, Public Health Service, NIOSH, Rockville, MA, 1974.
70. **Hasagawa, T.,** Tissue cobalt content and excretion in mice and rats following injection of cobalt chloride, *Jpn. J. Ind. Health,* 13 (1) (Engl. Abstr.), 289, 1971.
71. **Berlin, N. I. and Seri, W. E.,** Rate of absorption of cobalt from the peritoneal cavity and rate of elimination from the blood, *Am. J. Physiol.,* 164, 221, 1951.
72. **Kaerber, R. and Behrens, O.,** Second phase acute toxicology quoted by Caijolle, F., and Meynier, D., *Rev. Pathol. Gen.,* 59, 245, 1957.
73. **Edwards, M. S. and Curtis, J. R.,** Use of cobaltous chloride in anaemia of maintenance haemodialysis patients, *Lancet,* p. 582, 583, 1971.
74. **Krantz, S., Loeber, M., and Fielder, H.,** Investigation on the cleavage of fibrinopeptides from fibrinogen of cobalt treated rabbits, *Biochem. Biophys. Acta,* 230, 630, 1971.
75. **Telib, M. and Schmidt, F. U.,** Effect of cobaltous chloride in laboratory animals. II. Blood sugar, plasma insulin and plasma lipids, *Endokrinol. Band* 61, 3, 395, 1973.
76. **Stokinger, H. E. and Wagner, W. D.,** Early metabolic changes following cobalt exposure, elevations in globulin and serum neurominic acid, *AMA Arch. Ind. Health,* 17, 273, 1958.
77. **Kriss, J. P., Carnes, W. H. and Gross, R. T.,** Hypothyroidism and thyroid hyperplasis in patients treated with cobalt, *JAMA,* 157 (2), 117, 1955.
78. **Spruit, D. and Malten, K. E.,** Occupational cobalt and chromium dermatitis in an offset printing factory, *Dermatologica,* 151, 34, 1975.
79. **LeGoff, J. M.,** Elimination du Cobalt par le Rein Chez L'Homme (quoted by Derville 1961), source unavailable, 1930.
80. **Dervillee, M. M., Heraut, P., and Dervillee, E. T.,** Le Risque d'Intoxication Professionelle par le Cobalt et sa Prevention, in Proc. 14th Congr. Int. de Medicine du Travail, Madrid, September 1973.
81. **Kaplun, J. S.,** Toxicity of industrial dust of cobalt and it's compounds, *Tevetyne Metally,* 30 (9), 42, 1957.
82. **Kincaid, J. F., Strong, J. F., and Sunderman, F. W.,** Toxicology studies of cobalt carbonyls, *Arch. Ind. Hyg. Occup Med.,* 10, 210, 1954.
83. **Evans, M., Freeman, M. A. R., Miller, A. J., and Roberts, V.,** Metal sensitivity as a cause of bone necrosis and loosening of the prosthesis in total joint replacement, *J. Bone Jt. Surg.,* 56B, 626, 1974.

COPPER

B. Sarkar

CHEMISTRY

Copper is a tough, soft, and ductile reddish metal with a high thermal and electrical conductivity. It is widely distributed in nature in the elemental state, in sulfides, arsenites, chlorides, and carbonates. In the sedimentary cycle, copper is concentrated in the clay mineral fractions with a slight enrichment in clays rich in organic carbon. There are three oxidation states of copper. The copper(I) ion has the electronic structure $3d^{10}$. Copper(I) compounds are diagmagnetic and colorless and mostly possess tetrahedral coordination geometry. The copper(II) is the most important form of copper. It has a d^9 electronic configuration and, hence, it is normally paramagnetic with a single unpaired electron. Virtually all complexes and compounds of copper(II) are blue or green. The coordination geometry of copper(II) is either distorted octahedral or square-planar. Copper(III) has a d^8 electronic configuration. Many copper(III) complexes are short lived, although copper(III)-peptide complexes have relatively long lifetimes in aqueous solution.

METABOLISM

Copper content of the normal human adult is 50 to 120 mg. The liver, brain, heart, and kidney contain the highest concentration of copper. Hepatic copper accounts for about 10% of the total. The copper levels of biological fluids are shown in Table 1.

A single day's diet may contain 10 mg or more of copper. However, half of the dietary copper is not absorbed; instead, it is excreted in the feces. The mechanisms by which copper is absorbed and transported by the intestine are unclear. However, the intestinal metallothionein may have some function in maintaining copper homeostasis. It is known that cadmium, mercury, silver, and zinc interfere with copper metabolism. Radioactive copper given by mouth to human subjects appears very rapidly in the blood. Immediately after its absorption, copper is transported in plasma bound to albumin and low molecular compounds including amino acids.[1-3] Free ionic copper cannot exist in appreciable amounts in the living organism. In the body, copper is present in complexed form with proteins, peptides, amino acids, and other organic substances.

In spite of continued accumulation of copper, the hepatic concentration of copper does not increase with age in man. The appearance in blood of some copper after its incorporation into ceruloplasmin during hepatic synthesis and the excretion of copper from lysosomes into the bile, maintain this constant concentration. About 150 to 300 mg of ceruloplasmin containing 0.5 to 1.0 mg of copper is catabolized daily in the adult liver and about 30 mg ceruloplasmin, containing 0.1 mg copper, is excreted into the intestine. Excretion of fecal copper comes from salivary, gastric, pancreatic, and jejunal secretions along with any unabsorbed dietary copper.

BIOCHEMISTRY

Although most of the copper in plasma is bound to ceruloplasmin, results show that albumin is the immediate transport form of copper in blood.[4,5] The copper transport site of human albumin has been characterized. Copper binds to albumin at the NH_2 terminal in 1:1 stoichiometry involving the αNH_2 of aspartyl residue, two intervening peptide nitrogens, imidazole nitrogen of histidine in position 3, and the carboxyl side chain of NH_2-terminal

Table 1
COPPER CONTENT OF BIOLOGICAL FLUIDS IN MAN

Fluid	Copper (μg/100 ml ± SD)
Blood	
Whole	98.0 ± 13.0
Plasma	109.0 ± 17.0
Erythrocytes	115.0 ± 22.0
Saliva	31.7 ± 15.0
Bile	5.6—205.0
Cerebrospinal fluid	1.6 ± 0.4
Urine	18.0 ± 7.2 μg/24 h

Reproduced from Sarkar, B., *An Introduction to Bio-Inorganic Chemistry,* Williams, D. R., Ed., Charles C Thomas, Springfield, IL, 1976, 318. With permission.

Table 2
COPPER PROTEINS IN MAN

Common name	Functional role	Known or expected consequence of deficiency
Ceruloplasmin	Ferroxidase? other roles	Uncertain
Cytochrome oxidase	Electron transport chain	Uncertain
Superoxide dismutase	Free radical detoxification	Uncertain
Metallothionein	Copper and zinc homeostasis in fetus	Uncertain
Dopamine-β-hydroxylase	Catecholamine production	Neurological effects
Lysyl oxidase	Cross-linking of collagen and elastin	Vascular rupture
Tyrosinase	Melanin production	Failure of pigmentation
Enzyme not known	Cross-linking of keratin (disulfide bonds)	Pili torti

aspartyl residue.[6] Besides albumin, there appears to be another protein, transcuprein, which also binds exchangeable copper in plasma.[7] However, there are a few proteins in the body in which copper is an integral part and they are metalloproteins. Copper in these proteins is part of the molecular structure having a characteristic ratio between moles of protein and copper. Many of these copper proteins are enzymes and the copper is part of their active site. Copper enzymes found in humans are listed in Table 2.

Ceruloplasmin — Is a blue plasma glycoprotein containing 0.3% copper and 7% carbohydrate. Depending on the analytical method employed, its molecular weight is between 132,000 and 160,000 Da. There are six to eight atoms of copper in ceruloplasmin which exist in both cupric and cuprous forms. The 8-10 oligosaccharide chains of ceruloplasmin are composed primarily of glucosamine, mannose, and galactose. Almost all chains are terminated by a sialic acid residue which apppears to be essential for the survival of the protein in the circulation. Ceruloplasmin has oxidase activity towards several substrates at pH 4 to 5.9, the best substrate being the *p*-phenylenediamine or its dimethyl derivative. Ceruloplasmin has been shown to oxidize ferrous ion to ferric ion. Various pathological states affect the level of ceruloplasmin. In Wilson's disease, the level of ceruloplasmin is low, although normal levels have also been noted. The levels increase in pregnancy and also in response to administration of estrogens. Several functional roles have been proposed for ceruloplasmin. They include copper transport function, maintenance of the zero copper balance in the body, transfer of copper specifically to cytochrome oxidase, and oxidation of ferrous ion to ferric state.

Cytochrome oxidase — The enzyme is located in the mitochondrial membrane and it provides energy for the cell by coupling electron transport through the cytochrome chain with the process of oxidative phosphorylation. Copper is essential for the catalytic activity. Each monomer of cytochrome oxidase contains one molecule of heme, one atom of iron, and one atom of copper. Severe copper deficiency, whether acquired or hereditary, is generally associated with reduced cytochrome oxidase activity. Cytochrome oxidase appears during the fetal development. Initially, in the first 6 months of gestation, there is a low activity of the enzyme. This is followed by a rise in brain, liver, and heart reaching adult levels at birth.

Superoxidase dismutase — This enzyme has been found in erythrocytes which could catalyze the dismutation of superoxide. Each molecule of the enzyme has two copper atoms. The enzyme exhibits a pale greenish-blue color and contains 0.38% copper. It has an isoelectric point of 4.74 to 4.76 and an approximate molecular weight of 34,000 Da. A high quantity of superoxide radical is generated through various enzymatic processes. The superoxide radical reduces cytochrome oxidase. If the radical is not removed, the enzyme cytochrome oxidase could be fixed in the reduced stage. The role of superoxide radical may lie in protecting the cell from the damaging effects of the superoxide radical.

Metallothionein — Is a colorless protein having a molecular weight of 6000 Da. It has a high sulfhydryl content and the capacity for binding copper, zinc, and cadmium. A large amount of copper in human fetal liver is present in this form. This protein apparently maintains copper and zinc homeostasis in the fetus. However, very little is known about the mechanisms of intracellular copper transport and storage in animal tissues. In Wilson's disease, the liver usually accumulates large quantities of copper, but fails to remove copper through biliary excretion. The presence of a strong copper ligand in the hepatocytes could cause the intracellular accumulation of copper.

Other copper proteins — Dopamine-β-hydroxylase is a copper-containing enzyme. It is a glycoprotein of molecular weight 290,000 Da (tetramer) isolated from adrenal medulla. This enzyme plays a role in the biosynthesis of noradrenaline, a neurotransmitter, and adrenalin, a hormone from tyrosine. Lysyl oxidase is another copper enzyme which is involved in the cross-linking of collagen and elastin. Tyrosinase is a copper-containing enzyme which is essential in the pigmentation process. A genetic abnormality of the absence of tyrosinase results in albinism. This defect is characterized by extreme sensitivity to light.

DEFICIENCY

Copper deficiency is a rare condition. Normal infants are born with a generous store of copper in the liver. However, there are instances of copper deficiency associated with anemia, neutropenia, and severe demineralization of the bone. Appearance of clinically significant copper deficiency seems to require that either severe malnutrition or intestinal malabsorption be present. Additions to the diet or infusion of 1 mg copper per day, except for the premature infant who should receive only 100 μg/kg, is more than adequate supplementation. This is capable of reversing the anemia, neutropenia, and bone lesions that are the most important clinical symptoms of copper deficiency in infants.

Menkes' disease — This is an X-linked fatal genetic disorder first described by Menkes et al.[8] Affected male infants exhibit spirally twisted hair, physical and mental retardation with widespread degeneration of the brain, hypothermia, and death within the first several years of life. Copper deficiency was demonstrated in infants with markedly low serum concentration of copper and ceruloplasmin.[9] The deficiency may be caused by a defect in the intestinal absorption of copper. Although copper deficiency has been found to be a universal characteristic of the disease, evidence points to a more widespread disorder of copper transport. Copper deficiency is linked to hypomyelination which, in turn, may cause lack of development of the nervous system.

Table 3
METABOLIC ABNORMALITIES IN SYMPTOMATIC AND EARLY ASYMPTOMATIC HOMOZYGOUS PATIENTS AND HETEROZYGOTES OF WILSON'S DISEASE

	Homozygotes		
Abnormality	**Symptomatic**	**Asymptomatic**	**Heterozygotes**
Serum copper	Usually low	Usually low	Rarely low
Serum ceruloplasmin	Low	Low	Rarely low
Nonceruloplasmin-bound copper	High	Normal	Normal
Urinary copper	High	Normal	Normal
Copper in tissues	Elevated	Elevated	?
Handling of radiocopper			
Rate of clearance from blood	Slow	Normal	Normal
Incorporation into ceruloplasmin	None	None	Slow
Uptake in liver			
Initial rate	Slow	Normal	Normal
Extent	Low	Normal	Normal
Discharge from liver	Impaired	Impaired	Impaired
Excretion			
Urinary	High	Normal	Normal
Fecal	Low	Low	?
Other abnormalities			
Aminoaciduria	Present	Absent	Absent
Phosphaturia + low serum P	Present	Absent	Absent
Uricosuria + low serum urate	Present	Absent	Absent

Reproduced from Sarkar, B., *Elements in Health and Disease,* Arora, R. B., Vohora, S. B., and Khan, M. S. Y., Eds., Institute of History of Medicine and Medical Research, New Delhi, India, 1984. With permission.

TOXICITY

Symptoms of acute toxicity of copper are gastrointestinal mucosal ulceration, hemolysis, hepatic necrosis, and renal damage from deposition of hemoglobin. Hemolysis has been reported after applying solutions of copper salts to large areas of burned skin and after introducing copper into the circulation during hemodialysis. Chronic toxicosis has been reported in the lungs of some vineyard workers after years of exposure to Bordeaux mixture, 1 to 2% solution of copper sulfate neutralized with hydrated lime which is used widely to prevent mildew of grapevines, particularly in France, Portugal, and southern Italy. They show pulmonary copper deposition and fibrosis occurring in the lungs.

Wilson's disease — This is an inborn error of copper metabolism characterized by severe progressive liver disease, a peculiar neurological syndrome, and the presence of Kayser-Fleisher rings of the cornea.[10] The disease is inherited in an autosomal-recessive manner. Patients fall into two broad categories: one with the symptoms involving mainly the nervous system, the other with the liver. The neurological symptoms stem from disorganization of the lenticular region of the brain. Spastic rigidity is usually observed with the younger patients, while progressively worsening tremor is more common in older patients. Liver damage may occur in these conditions. It has been noticed that the kidney is also frequently involved in Wilson's disease, resulting in generalized amino aciduria and other signs of abnormal renal tubular function. The age of onset of the disease is generally during the first two decades of life. Table 3 shows the metabolic abnormalities in Wilson's disease. There is a high concentration of copper in the body. The blood level of ceruloplasmin is low.[11] Fecal and serum copper levels are found to be below normal. The direct-reacting exchangeable form of copper fraction that is not bound to ceruloplasmin is increased. Several findings suggest that biliary excretion of copper is impaired in Wilson's disease. Excessive amounts of copper are accumulated in many organs in Wilson's disease patients. When the accumulated copper is removed by effective chelation therapy such as by D-penicillamine, the symptoms of the disease regress.[12] The primary defect causing the copper intoxication is not yet known.

ANALYTICAL METHODS

Contamination and loss are prime concerns in estimating copper concentration. Consequently, the collection and treatment of the sample prior to analysis require particular attention. The method of atomic absorption spectroscopy is used frequently to quantitatively determine copper. For microanalysis of biological materials a spectrophotometric method using dicyclohexanoneoxalyldihydrazone is also used. The measurement of ceruloplasmin levels in serum provides an important diagnostic feature of detecting both Wilson's and Menkes' diseases. This is determined by following the rate at which *p*-phenylenediamine is oxidized which is proportional to the concentration of ceruloplasmin in serum.

REFERENCES

1. **Bearn, A. G. and Kunkel, H. G.,** Localization of ^{64}Cu in serum fractions following oral administration: an alteration in Wilson's disease, *Proc. Soc. Exp. Med.,* 85, 44, 1954.
2. **Neuman, P. Z. and Sass-Kortsak, A.,** The state of copper in human serum: evidence for an amino acid-bound fraction, *J. Clin. Invest.,* 46, 646, 1967.
3. **Sarkar, B. and Kruck, T. P. A.,** Copper-amino acid complexes in human serum, in *Biochemistry of Copper,* Peisach, J., Aisen, P., and Blumberg, W., Eds., Academic Press, New York, 1966, 183.
4. **Gubler, C. J., Lahey, M. E. Cartwright, G. E., and Wintrobe, M. M.,** Studies on copper metabolism. IX. The transportation of copper in blood, *J. Clin. Invest.,* 32, 405, 1953.
5. **Sternlieb, I., Morcell, A. G., Tucker, W. D., Greene, M. W., and Scheinberg, I. H.,** The incorporation of copper into ceruloplasmin: studies with ^{64}copper and ^{67}copper, *J. Clin. Invest.,* 40, 1834, 1961.
6. **Laussac, J.-P. and Sarkar, B.,** Characterization of the copper(II) and nickel(II)-transport site of human serum albumin. Studies of copper(II) and nickel(II) binding to peptide 1-24 of human serum albumin by ^{13}C and ^{1}H NMR spectroscopy, *Biochemistry,* 23, 2832, 1984.
7. **Weiss, K. C. and Linder, M C.,** Copper transport in rats involving a new plasma protein, *Am. J. Physiol.,* 249, E77, 1985.
8. **Menkes, J. H., Alter, M., Steigleder, G. K., Weakly, D. R., and Sung, G. H.,** A sex-linked recessive disorder with retardation of growth, peculiar hair, and focal cerebral and cerebellar degeneration, *Pediatrics,* 29, 764, 1962.
9. **Danks, D. M., Campbell, P. E., Walker-Smith, J., Stevens, B. J., Gillespie, J. M., Blomfield, J., and Turner, B.,** Menkes' kinky-hair syndrome, *Lancet,* 1, 1100, 1972.
10. **Wilson, S. A. K.,** Progressive lenticular degeneration: a familial nervous disease associated with cirrhosis of the liver, *Brain,* 34, 295, 1912.
11. **Scheinberg, I. H. and Gitlin, D.,** Deficiency of ceruloplasmin in patients with hepatolenticular degeneration (Wilson's disease), *Science,* 116, 484, 1952.
12. **Walshe, J. M.,** Penicillamine, a new oral therapy for Wilson's disease, *Am J. Med.,* 21, 487, 1956.
13. **Sarkar, B.,** Miscoordination in a metal-dependent disease: Wilson's disease, in *An Introduction to Bio-Inorganic Chemistry,* Williams, D. R., Ed., Charles C. Thomas, Springifield, IL, 1976, 318.
14. **Sarkar, B.,** Specificity of metal-binding in normal physiology and biochemical alterations in disease conditions: Wilson's and Menkes' diseases, in *Elements in Health and Disease,* Arora, R. B., Vohora, S. B., and Khan, M. S. Y., Eds., Institute of History of Medicine and Medical Research, New Delhi, India, 1984.

LEAD

Robert A. Goyer

INTRODUCTION

Lead is one of the oldest metals known to man; history of its use dates back to antiquity. Awareness of its potential toxicity to living things also dates back to ancient times. Because of its physical properties, malleability, and resistance to corrosion, as well as abundance, it has been one of the most useful metals known to man. The major modern uses have been for electric storage batteries, paint pigments, gasoline additives, various metal products, and as cable sheathing. Total usage in Europe, the U.S., and Japan in 1983 was about 1800 metric tons.

These uses have been influenced in recent years by substitution with alternative materials and new technologies, as well as new knowledge about the environmental and health effects of lead. Consumption of lead-containing products decreased by 10% in Europe and 15% in the U.S. between 1979 and 1983. In 1983 batteries accounted for 55% of lead usage in the western world; 70% in the U.S.,; various pigments and other compounds, 13%; gasoline additives, 5%.

Lead is ubiquitous in the environment and is detectable in all biologic systems. Since it is toxic to most living things at high exposures and there is no demonstrated biologic need, the major issue is at what dose does it become toxic.

CHEMISTRY

Lead (atomic number 82; atomic weight 207.19; specific gravity 11.34) is a soft, malleable, gray-white metal with a low melting point (327.5°C). It is the most abundant of the natural heavy elements with an atomic number greater than 60. It occurs in nature as four stable isotopes in varying relative amounts: ^{204}Pb, 1.2 to 1.6%; ^{206}Pb, 20 to 28%; ^{207}Pb, 20 to 23%; and ^{208}Pb, 50 to 54%; and four short-lived radioactive isotopes: ^{210}Pb, ^{211}Pb, ^{212}Pb, and ^{214}Pb, which are decay products of uranium and thorium.

Metallic lead is readily smelted from its major natural source, the mineral galena, in which it occurs as the sulfide PbS. It is also found in cerussite (the carbonate), anglesite (the sulfate), as well as in a number of other metal-bearing minerals. The inorganic salts of lead, lead sulfide, and oxides of lead are generally poorly soluble. Lead acetate is the most soluble common form of lead; the nitrate, chlorate, and, to a lesser degree, the chloride are less soluble in descending order. Toxicity of inorganic lead salts is attributable to lead per se and the chemical form of the salt is thought to be of significance only in that it influences solubility and bioavailability.

The toxicity of organolead compounds, tetraethyl- and tetramethyl lead, that are used as petrol fuel additives differs from that of inorganic compounds and is only of importance from potential occupational exposure and abusive use of leaded gasoline such as "sniffing" for hallucinogenic effects.

Although the natural biomethylation of lead has been postulated, it has not been possible to methylate lead by bacterial species known to alkylate mercury and other heavy metals. Lead forms complexes with humic and fulvic acids in soils and surface waters, but they are not known to have any specific biologic significance.

SOURCES OF HUMAN EXPOSURE

Lead may enter the environment at any point during mining, smelting and refining, manufacturing processes, and use of lead-containing products, and from disposal. The largest emission source by far is lead from gasoline combustion, but because of the introduction of lower compression engines, lower octane fuel requirements, and the availability of low-lead or no-lead gasoline, sales of lead fuel derivatives declined nearly 40% in the U.S. from 1970 to 1975 and was only 61,000 tons in 1979. Still, this accounted for 91.4% of total atmospheric lead emissions for the U.S. in 1979. Global emissions were 273,000 tons in 1979.[5]

For persons in the general population, exposure to lead occurs from inhaled air, dusts of various types, and food and water. Additional exposures may be related to particular environments (urban), occupational sources, and certain habits and activities such as pica in toddlers. Concentration of lead in air varies greatly in different parts of the world and is generally between 0.3 and 1.1 $\mu g/m^3$ for urban areas and about 0.15 and 0.3 $\mu g/m^3$ in rural areas, although extremes above and below these levels do occur.

Lead in surface soil is influenced by atmospheric lead, proximity to roadsides and other emission sources, and composition of the soil. Retention is greatest in soils with at least 5% organic matter and a pH of 5 or above. Air lead deposited on soil surfaces mostly complexes with humic substances and dissolves in soil moisture. Concentration of natural soil lead typically ranges from 10 to 30 μg/g, but the upper layer of soil along roadsides may contain 30 to 2000 μg/g of lead within 25 m of the road due to automobile emissions.

Lead occurs in water in either dissolved or particulate form. The most important factors determining lead in water are the pH and hardness of the water and the stream flow. At low pH, lead is dissolved from sediments and particles. For lead in ground water the geochemical composition of the water-bearing bedrock is the most important factor, so that well water is little influenced by acid rain, surface runoff, or deposition of airborne lead. Chemical treatment to soften water increases the solubility of lead, increasing lead uptake from distribution systems.

Lead in food is derived from root absorption from soil and atmospheric deposition on above-ground parts. Lead in raw vegetables is generally under 0.05 μg/g and is reported to be 0.2 μg/g in pork (muscle) and 0.06 μg/g in beef (muscle). Lead is added to food in all steps of transport and processing including packaging and preparation. Several studies report average dietary lead intake from 100 to 500 μg/d for adults.[3] About 90% of lead in food is derived directly or indirectly from atmospheric deposition so that the lowest possible consumption of dietary lead is probably in the order of 10 to 15 μg/d. The U.S. Food and Drug Administration has an intensive program to reduce lead intake in children to less than 100 μg/d by 1988. Considerable reduction has already been achieved by the removal of soldered cans used for infant formula.

METABOLISM

Atmospheric lead and lead in food are the two major routes of exposure for people in the general population. For the contemporary urbanite, the amount of respired air is about one half the intake from the diet.[4]

Atmospheric lead is largely in the form of dust or particles of lead oxide. Absorption by the lungs is dependent on size distribution of lead-containing particles and volume of air respired per day. Particles are characterized by their mean mass external diameter (MMED) and only a minor fraction of particles over 0.5 μm in MMED is retained. The percentage of particles less than 0.5 μm retained in the lung increases with reduction in particle size.[5] Nearly 90% of lead particles in ambient air that are deposited in the lung are small enough to be retained.

Gastrointestinal absorption of lead is influenced by a large number of factors, particularly age and nutrition. Adults absorb about 5 to 15% of ingested lead and retain less than 5%.[6]

Alexander[7] found that children between 3 months and 8 years absorbed 50% or more of ingested lead, and Ziegler and co-workers[8] found that infants 14 to 746 d of age with daily lead intake of 5 μg/kg/d absorbed 41.5% and net retention averaged 31.8% of intake. The highly efficient absorption and retention of ingested lead by young children may contribute to the apparent increase in susceptibility of young children to lead toxicity. Dietary lipids and certain milk components, particularly lactose, increase absorption of lead. Low dietary calcium and phosphate and deficiencies of other essential minerals, including zinc and copper, may also enhance gastrointestinal absorption of lead.[9]

Blood Lead

More than 95% of lead in blood is associated with red blood cells. There are at least two compartments for lead in red cells, one associated with the membrane and the other with red cell proteins, particularly hemoglobin. Whole blood lead has a biological half-life of about 25 to 28 d and is in equilibrium with plasma lead and, in turn, other tissue compartments as well as excretory pathways. Plasma lead is mostly bound to albumin; other plasma ligands are not well defined, but may increase with increasing blood levels.

Blood lead levels reflect recent exposure to lead and are influenced by levels of intake by inhalation and by ingestion. About 1 μg lead/m^3 air lead will increase the blood lead concentration of an adult by approximately 1 μg lead/dl blood when air lead concentrations are in the range of 1 to 5 μg/m^3. For higher levels of air lead the relationship is nonlinear and the slope tends to decrease as exposures increase. Nevertheless, estimates of impact on blood lead are dependent on the choice of model.[10] Studies on the relationship of dietary lead intake to blood lead indicate a range of 5.4 to 18.3 for men and 4.4 to 13 μg/dl per 100 dietary lead intake per day for women.[11] The relationship between lead in water and blood lead is even less well defined, but may be linear at lower water lead values typical for U.S. populations.[12]

Tissue Lead

Total lead in body tissues for a child less than 10 years of age is less than 2 mg, but increases with age and may be over 200 mg for persons in the later decades of life.[13] From estimates made a few years ago, Thompson[14] thought that the body accumulation of lead could be accounted for from a retention of less than 0.4% of the average daily intake or about 10 μg/d. Using results of more recent studies, Rosen estimates percent retention for adults to be more in the order of 1% and as high as 34% for infants.[15] Persons with excessive occupational exposure to lead may have a total body lead retention exceeding 500 mg. The largest concentration of lead is in bone where the major fraction is bound in a nondiffusible form. This fraction of bone lead may have a biological half-life of greater than 20 years. However, this pool equilibrates with bound blood lead in less than a day so that bone lead may contribute to the persistence of elevated blood lead in persons with excessive exposure for years after exposure ceases.

Whole blood lead has a biological half-life of 25 to 28 d, whereas the diffusible or labile plasma pool biological half-life may be as short as 1 to 3 d. Only the diffusible lead fraction from plasma passes in and out of capillaries, permeates cell membranes, and enters parenchymal cells of the central nervous system, liver, kidney, and other organs. The relatively large content of lead in liver and kidney may be related to the excretory function of these two organs, whereas only trace amounts of lead are present in muscle and brain. Lead in teeth increases with age and is proportional to blood lead.

Renal Excretion of Lead

Renal excretion of lead is with glomerular filtrate and enhanced or decreased by secretion or resorption by renal tubular lining cells. Renal excretion is influenced by a number of factors including level of lead exposure, age, renal function, and competing cations. Rate of glomerular filtration and tubular resorption must be the major determinants of renal excretion in persons in the general population with exposure to lead limited to minor levels in the ambient environment. With higher blood levels, tubular cell secretion by way of transcellular transport from peritubular capillaries must assume a greater role. However, precise measurement of rate of renal clearance has been hampered by difficulties in the accurate measurement of the filtrable component of plasma lead. Using the assumption that 5% of blood lead is filtrable, Campbell and co-workers[16] found a mean clearance of lead from blood of 2.08 ml/min with wide variation (1.44—7.92 ml/min). Renal tubular secretion of lead into the tubular lumen may occur by transtubular transport from peritubular capillaries,[17] and as glomerular filtration decreases, renal tubular secretion increases. Cellular transport of lead is a passive process and is probably dependent on the capillary-lumenal gradient.

Transplacental Transfer of Lead

Although it is well known that lead passes through the placenta from mother to fetus, the mechanisms and factors affecting kinetics are not known. Placental transfer of lead has been demonstrated as early as the 12th week of gestation and lead content of the fetus increases throughout pregnancy.[18] It probably begins very early in fetal life since lead can be found in embryonic and fetal tissue obtained from early legal abortions.[19] The pattern of tissue distribution is similar to adults: highest levels are in bone and liver. Umbilical cord blood lead levels tend to be 20 or 30% lower than maternal blood lead levels.[20]

Interaction with Essential Metals

Lead has been shown to interact with essential metals, particularly calcium, iron, and zinc.[21,22] The alterations in lead metabolism induced by dietary calcium are complex and illustrate the metabolic relationships between lead and calcium metabolism. Animals maintained on a low calcium diet (0.03%) exhibit decreased rather than increased lead absorption.[23] Mahaffey and co-workers[24] suggest that the increased susceptibility to lead toxicity found in animals on a low calcium regime is not a direct effect of calcium intake, but of increased lead retention brought about by decreases in renal lead excretion.

Low dietary calcium (0.1%) releases lead from bone and increases the proportion of body lead found in soft tissues.[25,26] Calcium deficiency also appears to cause wealing rats to increase their intake of lead-containing solutions,[27] but no related studies exist for human populations.

The interaction of lead and calcium metabolism is of particular interest during pregnancy and lactation, since changes in calcium metabolism might influence both the amount of lead transported to the fetus in utero and the lead content of milk. In studies with rats, transplacental transport of lead was found to be eight times lower than that of calcium,[28] consistent with the hypothesis that the mechanism for placental transfer of lead is passive, whereas there is an active placental transport mechanism for calcium.[29]

Iron deficiency enhances the intestinal absoption of lead in rats[30] and in children there is a negative correlation between blood lead and serum iron levels.[31] Iron-deficient adults may absorb about 24% of dietary lead instead of the normal level of 10%.[32]

Lead levels in bone are increased in rats fed a diet low in zinc. Soft tissue lead levels are not changed and bone lead is reduced as diet zinc is increased.[33] Zinc also provides protection to the clinical manifestations of lead toxicity in horses grazing on pastures contaminated with lead and zinc from refinery effluents.[34] The mechanism of these effects at the cellular level is not known. However, lead does inhibit zinc-containing enzymes in heme synthesis. Blood lead levels in rats are associated with a dose-related increase in urinary zinc excretion beginning at 40 μg/dl, the lowest level of blood lead measured in lead-exposed animals.[35]

CELLULAR TOXICITY

Toxicological effects of lead have their basis in perturbations in cell function of various organ systems. Effects of lead at the cell level are multiple and complex as is the diversity of effects at the tissue and organ system level, particularly heme synthesis, interactions with calcium metabolism, and binding to mitochondrial membranes and enzymes involved in energy metabolism.

Lead has particular affinity for the carboxyl groups of glutamic and aspartic acids, the sulfhydryl groups of cysteine, and the phenoxy groups of tyrosine whether present in proteins of unknown function such as acidic proteins in inclusion bodies or on cell membranes or enzyme systems.

Lead interacts with calcium at the molecular level. Calcium provides cells with important homeostatic mechanisms and mediates intracellular communication or messenger activities so that lead-calcium interactions may impact on diverse cellular processes. Pounds,[36] in a thorough review of calcium-lead interactions, has subdivided lead effects into direct and indirect effects. Direct effects of lead on calcium metabolism result from the displacement of Ca_2^{2+} by lead ions at functionally significant binding sites. This may result in inhibition of calcium transport or function. Examples of these binding sites include mitochondrial Ca_2^{2+} transport proteins where lead competitively inhibits the uptake of calcium.[37] Lead may also substitute for Ca^{2+} in the activation of phosphodiesterase by calmodulin[38] and lead may compete with calcium for binding to the triphosphate chain of ATP.[39] Extracellular presence of lead inhibits movement of calcium ions through voltage-dependent calcium channels, perhaps by interfering with an active recognition site on the outside of the membrane. These interactions with calcium may provide the molecular basis for effects on neural development and function.[40] In an even more general way, effects of lead on (Na^+-K^+)-ATPase may alter cellular Na^+ transport affecting both the influx of Ca^{2+} and efflux of calcium into and out of cells and altering calcium homeostasis and affecting a number of other metabolic processes.

Lead impairment of mitochondrial respiration and oxidative phosphorylation was demonstrated in kidneys of lead-exposed rats many years ago and lead effects on mitochondrial function have been demonstrated in intestinal lining cells,[41] hepatocytes,[42] and brain tissue.[43]

CLINICAL MANIFESTATIONS

The major clinical manifestations of lead toxicity relate to specific organ systems. Effects of major importance involve the nervous system, hematologic effects, and renal effects. Effects on male reproductive organs and possibly the immune system are also important but less well understood. Nonspecific symptoms and signs include loss of appetite, metallic taste in the mouth, constipation and obstipation, pallor (from anemia), malaise, weakness, insomnia, headache, irritability, muscle and joint pains, fine tremors, and colic.

Lead lines (Burton's lines) or purple-blue discoloration of gingiva is a classical feature of severe lead toxicity in children with lead encephalopathy. However, this feature of lead toxicity, as well as the presence of lead lines at the epiphyseal margins of long bones seen on X-rays of children with severe lead exposure, is uncommon today.

Neurological Effects

The central nervous system is probably the most clinically significant in terms of human health and performance.[44] Children with lead intoxication who are brought to hospitals or emergency rooms for medical aid invariably manifest some central nervous system abnormality. Symptoms vary from ataxia to stupor, coma, and convulsions. This form of lead intoxication has decreased appreciably in North America over the past 20 years with improved control of major sources of lead poisoning for children, namely, ingestion of lead-containing paint chips. Also, there is now increased awareness of other potential sources of lead poisoning although episodes of acute lead toxicity from burning of battery casings still occur.

Morphological effects of lead on the brain are nonspecific. Lead encephalopathy is accompanied by severe cerebral edema, increase in cerebral spinal fluid pressure, proliferation, and swelling of endothelial cells in capillaries and arterioles, proliferation of glial cells, neuronal degeneration, and areas of focal cortical necrosis in fetal cases. The pathogenesis of neuronal damage in lead encephalopathy is not well understood. In severe cases, there are obvious changes in hemodynamics (cerebral edema) and cellular hypoxia, but it is now apparent that there is a direct effect of lead on neuronal and possibly synaptic transmission at levels of lead exposure that do not produce apparent symptoms of intoxication. These effects are termed "low-level lead toxicity" because they are believed to be associated with blood lead levels of approximately 30 to 50 μg/dl and they can only be detected by assessment of neuropsychological behavior, such as hyperactivity, poor classroom behavior (decreased attention span), and even small decrements (point average in group studies) of two to five IQ points.[45] In some studies of young children (toddlers), there seem to be one to two IQ point differences with blood leads in the 15- to 30-μg/dl range.[46] Also, EEG alterations may be detected with blood lead levels as low as 15 μg/dl.[47] For further critique of existent studies and new studies currently under way the reader is referred to the review by Yule and Rutter.[48] For the present it is uncertain what is the lowest level of blood lead below which no effect occurs.

Neurochemical studies in experimental models have shown that lead in the absence of morphologic changes does produce deficits in neurotransmission through inhibition of cholinergic function, possibly by reduction of extracellular calcium. Other noted changes in neurotransmitter function include impairment of dopamine uptake by synaptosomes and impairment of the function of inhibitory neurotransmitter, gamma-aminobutyric acid.[40]

Lead enters the brain in rough proportion to the circulating blood lead concentrations. The higher the blood lead the lower the blood/brain lead ratio, suggesting that lead tends to be tightly bound to certain brain structures.[49,50] Also, lead in the brain is not homogeneous, but tends to localize in particular parts of the brain. Greatest concentrations are in cortical and cerebellar gray matter and brain stem nuclei[51] and seem to be particularly concentrated in the hippocampal gyrus.[52]

Peripheral neuropathy is a classic manifestation of lead toxicity, particularly the foot-drop and wrist-drop that characterized the house painter and other workers with excessive occupational exposure to lead more than a half century ago.[53] Motor nerve function, assessed clinically by electrophysiologic measurement of nerve conduction velocities, has been shown to occur with blood lead levels in the 30- to 50-μg/dl range.[54] Segmental demyelination and possibly axonal degeneration follow lead-induced Schwann cell degeneration.[55] However, histological studies of nerves of humans with relatively low level of lead exposure show no overt evidence of segmental demyelination, but there may be changes in the axon membrane itself.[56] Wallerian degeneration of posterior roots of sciatic and tibial nerves is possible, but sensory nerves are less sensitive to lead than motor nerve structure and function.[57]

Hematologic Effects

Lead has multiple hematological effects. In lead-induced anemia, the red blood cells are microcytic and hypochronic as in iron deficiency and usually there are increased numbers of reticulocytes with basophilic stippling. Basophilic stippling results from the inhibition of the enzyme pyrimidine-5-nucleotidase, an enzyme that cleaves residual nucleotide chains remaining in erythrocytes after extrusion of the nucleus.[58] The activity of this enzyme is decreased in persons with elevated blood lead levels even when stippling is not morphologically evident.

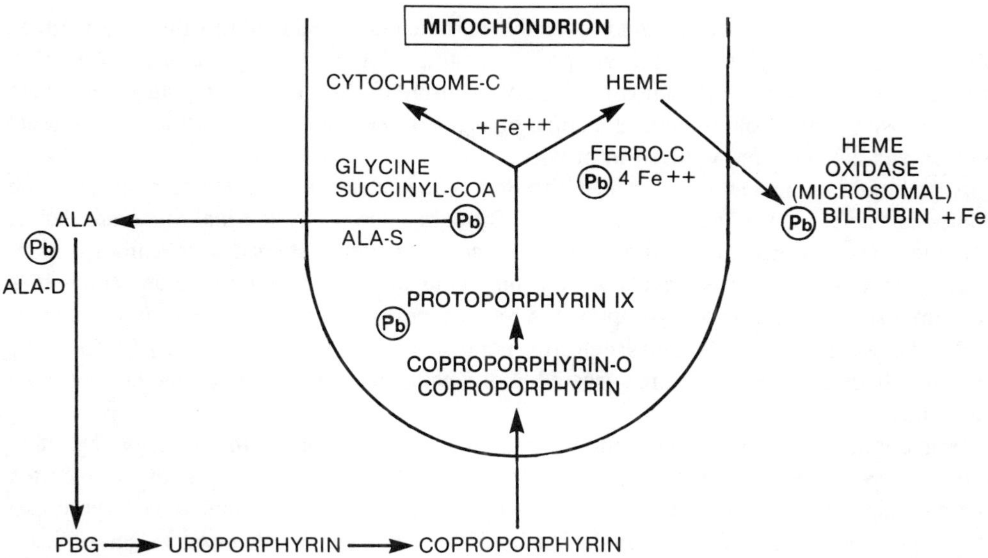

FIGURE 1. Scheme of heme synthesis showing sites where lead has an effect. COA, coenzyme A; ALA-S, aminolevulinic acid synthetase; ALA, *d*-aminolevulinic acid; ALA-D, aminolevulinic acid dehydratase; PBG, porphobilinogen; UROPOR, uroporphyrinogen; COPRO, coproporphyrinogen; COPRO-O, coproporphyrinogen oxidase; FERRO-C, ferrochelatase; CYTOCH-C cytochrome *c*; Pb, site for lead effect.

The anemia that occurs in lead poisoning results from two basic defects: shortened erythrocyte life span and impairment of heme synthesis. Shortened life span of the red blood cell is thought to be due to increased mechanical fragility of the cell membrane. The biochemical basis for this effect is not known, but is accompanied by inhibition of sodium- and potassium-dependent ATPase.[59]

A schematic presentation of effects of lead on heme synthesis is shown in Figure 1. Probably the most sensitive effect is inhibition of δ-aminolevulinic acid dehydratase (ALA-D) resulting in a negative exponential relationship between ALA-D and blood lead. There is also depression of coproporphyrinogen oxidase, resulting in increased coproporphyrin activity. Lead also decreases ferrochelatase activity. This enzyme catalyzes the incorporation of the ferrous ion into the porphyrin ring structure. Failure to insert iron into protoporphyrin results in depressed heme formation. The excess protoporphyrin takes the place of heme in the hemoglobin molecule and, as the red blood cells containing protoporphyrin circulate, zinc is chelated at the center of the molecule at the site usually occupied by iron. Red blood cells containing zinc-protoporphyrin are intensely fluorescent and may be used to diagnose lead toxicity. Depressed heme synthesis is thought to be the stimulus for increasing the rate of activity of the first step in the heme synthetic pathway, δ-aminolevulinic acid synthetase by virtue of negative feedback control as proposed by Granick and Levere,[60] As a consequence, the increased production of *d*-aminolevulinic acid and decreased activity of ALA-D results in a marked increase in circulating blood levels and urinary excretion of *d*-ALA. Perinatal feeding of lead to experimental animals also raises heme oxygenase activity resulting in some increase in bilirubin formation. The changes in rates of activity of these enzymes by lead produce a dose-related alteration in activity of affected enzymes, but anemia only occurs in very marked lead toxicity. The changes in enzyme activities, particularly ALA-D in peripheral blood and excretion of ALA in urine, correlate very closely to actual blood lead levels and serve as early biochemical indices of lead effect.

Renal Effects of Lead

Toxicologic effects of lead on the kidney consist of reversible renal tubular dysfunction that occurs mostly in children with acute exposure to lead, usually associated with overt central nervous system effects, and irreversible chronic interstitial nephropathy characterized by vascular sclerosis, tubular cell atrophy, interstitial fibrosis, and glomerular sclerosis.[61] It is most often seen in workmen with years of exposure to lead. In the early stages of excess lead exposure, morphologic and functional changes in the kidney are confined to the renal tubules and are most pronounced in proximal tubular cells.

A pathognomonic feature of lead poisoning is the presence of characteristic inclusion bodies in nuclei of proximal tubular cells.[62] By light microscopy the inclusions are dense, homogeneous, eosinophilic bodies. They are acid fast when stained with carbolfuchsin. Ultrastructurally, the bodies have a dense central core and outer fibrillary region. The bodies are composed of a lead-protein complex.[63] A recent report suggests the protein has an apparent molecular weight of 32,000 and isoelectric point of 6.3.[64] The sequestering of lead in the inclusion body may protect more susceptible organelles like mitochondria and endoplasmic reticulum.[62]

With continued exposure to lead there is a gradual change in kidney morphology beginning with the appearance of peritubular and periglomerular fibrosis, particularly in the deep cortex of the juxtamedullary zone.[61] This is accompanied by atrophy of some tubules and hyperplasia of others. There are also fewer inclusion bodies present in the advanced stages of lead-induced nephrosclerosis and it may be impossible to find any inclusion bodies in lead-induced nephrosclerosis. Recognition of interstitial fibrosis induced by lead, therefore, from any other forms of interstitial fibrosis is not possible morphologically, but must be made from history and knowledge of progression of the disease if this is available.

Gout

Gout occurs in about 50% of persons with chronic lead nephropathy,[65] and Campbell and co-workers[66] report that among gout patients in Scotland without known lead exposure, blood lead levels are higher than among non-gout controls. Lead does reduce uric acid excretion,[67] so that lead at the least alters susceptibility to gout. Beyond the renal effects, lead can disrupt purine metabolism per se. Farkas et al.,[68] have shown that lead inhibits quanine aminohydrolase which catalyzes the hydrolytic deamination of quanine to xanthine.

Hypertension

The relationship between lead and hypertension has long been suspected, but has been tenuous at best, even among workmen with excessive exposure to lead.[69] Recently, however, the potential relationship between lead and elevated blood pressure received new attention, with the finding of a statistically significant relationship ($p < 0.01$) between systolic and diastolic blood pressure and blood lead levels[70] from data obtained from the second National Health and Nutrition Examination Survey. Experimental support for this finding is largely from animals exposed to small increases in lead for long periods of time. Perry and Erlanger[71] found that chronically feeding rats lead at doses 0.1, 1.0, or 5.0 ppm produces statistically significant increases in systolic blood pressure. The cellular basis for this effect of lead is not known, but most studies have focused on the renin angiotensin system,[72] and effects of lead on calcium flux in endothelial cells of arterioles[74] or juxtaglomerular cells[74] may also be important.

Reproduction

Severe lead toxicity has long been known to cause sterility, abortion, and neonatal mortality and morbidity. In the 19th and early 20th centuries, women in lead occupations recognized lead as an abortifacient and that male workers employed in lead-related industries had a high incidence of sterile marriages. Lancranjan[75] found that fertility of lead workers with blood lead levels above 40 μg/dl was decreased and this was correlated with an increased frequency of oligospermia and astherospermia or teratospermia.

Table 1
INDICATORS OF LEAD TOXICITY IN TERMS OF BLOOD LEAD CONCENTRATION (μg/dl)

Effect	Adults	Children	Pregnant women and fetuses
Gastrointestinal colic	>80	>80—100	
Acute encephalopathy	>80	>80	
Anemia	>80	>70	
Chronic renal disease	>60		
Cognitive and behavioral	>40	>30	>15—20?
Peripheral neuropathy	>30—50		
Reproductive (testicular)	>40		
Increased urinary aminolevulinic acid	>40		
Increased erythrocyte protoporphyrin	>25 males >20 females	>15	
Decreased blood aminolevulinic acid			
Dehydratase	>20	>10	
Pyridine-5-nucleosidase		>10	
Calcium-mediated cellular functions	Unknown	Unknown	Unknown

Immunological Effects

Lead induces suppression of the three major components of the immune system, humoral-, cellular-, and macrophage-related immunity. These effects have been shown to occur in animal models at dose levels in which there is no apparent evidence of toxicity.[76] The significance of the experimental findings is not clear, but experimental animals have increased susceptibility to endotoxins and infectious agents. Whether related or not, it has also been noted that children with an asymptomatic increase in blood lead levels appear to have more frequent febrile illness.[77]

Carcinogenesis

The possible carcinogenic effects of lead have been receiving increasing attention.[78] It is clear that lead can induce cancer in kidneys of rodents when fed high doses of lead.[79] On the other hand, the evidence that lead is carcinogenic to man is very limited. A study of causes of mortality in 7000 lead workers in the U.S. showed a slight excess of deaths from cancer.[80] The most common tumors found were of the respiratory and digestive systems, not the kidney, although case reports of renal adenocarcinoma in workmen with prolonged occupational exposure to lead have been reported.[81,82]

ASSESSMENT OF TOXICITY

Adults

Health effects of lead are related to blood lead in Table 1. The predominant effect depends on intensity and duration of exposure and other factors that determine susceptibility to lead effect. Gastrointestinal colic is usually not seen unless the exposure to lead is extreme; that is, it produces blood lead levels over 80 μg/dl. Because of the severity of pain and discomfort, lead colic has been regarded as an acute effect and usually stops with cessation of exposure and chelation therapy. Severity of anemia also depends on intensity and duration of exposure, and reflects balance between impairment of lead on heme-synthesizing enzymes and compensatory enhancement of rate of heme synthesis. Clinical encephalopathy is rare in adults and probably will not occur with blood lead levels under 80 dl, but behavioral effects and peripheral neuropathy manifested by impairment in motor function and nerve conduction velocity may occur at blood lead levels of between 30 and 50 μg/dl. In adults, lead-induced renal disease is most commonly discovered after many years of exposure and often is diagnosed as chronic interstitial nephropathy. How much lead over what period of time produces chronic renal disease is still uncertain, but chronic renal disease is unlikely to occur if blood lead levels do not exceed 60 μg/dl. Oligospermia and possibly morphologic aberrations in sperm have been noted at blood levels near 40 μg/dl.

Subclinical effects on heme enzymes, particularly inhibition of ALA-D, occur at blood lead levels as low as 20 μg/dl, although an increase in urinary ALA is not apparent until blood lead levels are 40 μg/dl or above.

Children

Children are less tolerant to lead than adults, particularly with effects on the central nervous system and on heme synthesis. As previously discussed, decreased nerve conduction velocity may occur with blood lead levels as low as 40 μg/dl and impairment of cognitive development and behavior at blood lead levels as low as 30 μg/dl. Inhibition of ALA-D and pyridine-5-nucleosidase occurs in children with blood lead levels as low as 10 μg/dl, although these effects are generally regarded as biochemical effects without meaningful adverse health effects. However, this interpretation may merely reflect present lack of understanding of these effects. Also, the minimum level of tolerance to adverse calcium-lead interactions at the level of cellular membranes is not known.

Pregnant Women and Fetuses

It is not the pregnant women per se, but the fetuses that have a higher risk of toxicity to lead than other people, particularly effects on the developing nervous system. Precise levels of risk in terms of maternal or fetal blood lead levels have not been determined with precision. Slight effects (one to two IQ points) are suggested with blood lead levels in the 15- to 25-μg/dl range. Furthermore, the uptake of lead by the fetal brain is not readily reversed by lowering blood lead levels.

TREATMENT

Chelation Therapy

Treatment of excess lead exposure is removal of lead source and removal of lead from the body using chelating agents. For adults, the most widely accepted procedure is the continuous infusion of the calcium salt of disodium ethylenediamine tetraacetate ($CaNa_2$-EDTA) 1 to 2 g/d for 4 or 5 consecutive days. The lead chelate formed by exchange of calcium for lead is promptly excreted in the urine. Indications for treatment with $CaNa_2$-EDTA is a blood lead level of 70 μg/dl.

Indications for treatment of children with excess lead exposure have been recently detailed in a statement by the Centers for Disease Control.[83] The dose of EDTA for children is 1000 $mg/m^2/24$ according to body size. For children with encephalopathy chelation therapy is begun with BAL alone, 75 mg/m^2, for one dose, 450 $mg/m^2/24$ h, followed by the i.v. infusion of $CaNa_2$-EDTA, 1500 $mg/m^2/24$ h, after the priming dose of BAL.

Asymptomatic children with blood lead greater than 70 μg/dl should also be treated with BAL and $CaNa_2$-EDTA. Children with blood lead between 55 and 70 μg/dl should be treated with a single course of $CaNa_2$-EDTA, 1000 $mg/m^2/d$. The choice of a second course of treatment is based on blood lead level after 4 or 5 d following initial treatment.

A decision to administer chelating agents to children with blood lead levels between the action level of 25 and 55 μg/dl must be based on results of a Ca_2Na-EDTA provocative test as an estimate of body burden. The Centers for Disease Control (CDC) statement[83] provides details for therapeutic procedures involving EDTA. Removal of the person from the lead-containing environment may be the only measure required for persons with low-level exposures and blood lead levels near the action level of 25 μg/dl who do not have excess body burden.

ANALYTICAL METHODS

Procedures for Sampling of Body Fluids and Tissue

For collection of biological samples particular attention must be given to avoidance of contamination and obtaining a sample that is representative of the material to be analyzed. Blood lead can be analyzed from venous or capillary samples collected with anticoagulant (heparin). Skin surface at the sampling site should be thoroughly washed with soap and brush, and diluted aqueous solutions of HNO_3 and EDTA. The collection devices, e.g., evacuated tubes, rubber stoppers, anticoagulants, and urine conservants, should always be checked for their lead blank contribution. A sample should never be analyzed in which the lead content is less than the blank contribution.

Tissue samples should be excised with a titanium-bladed knife, placed on a teflon surface, and washed with high-purity water. The tissue should only be handled with talc-free teflon gloves. Hair samples should be washed with a nonionic detergent and/or organic solvent. The first 1-cm segment of hair close to the hair root is preferred for cleanliness and reflection of recent exposure to lead.

Storage of Samples

Blood samples stored by room temperature should be analyzed within 1 or 2 d, but may be kept up to 2 weeks at 4°C. Samples stored at 20°C are stable for several weeks. The best approach for even longer term storage is to lyophilize specimens (blood, urine, tissue). Blood samples can also be dried on filter paper and stored until analyzed.

Determination Methods

Literature describing various methods to determine metal content of lead in biologic samples is extensive.[84] Atomic absorption spectroscopy is the most popular method for routine determination. There are a variety of specific methods dependent on the instrument to be used and method of sample preparation.[85] In general, samples must be ashed using a wet digestion or dry ash technique. Wet digestion techniques are generally acceptable to trace lead analyses and a number of applicable approaches are available. Dry ashing techniques may be more suitable for large and fatty samples and for samples with considerable inorganic material such as bone, teeth, or milk. Lead atoms are vaporized in a controlled flame or in a furnace and a light beam is directed through the flame into a monochromator and onto a detector that measures the amount of light absorbed. Furnace techniques provide high sensitivity as well as the ability to measure small samples. The design and materials used for furnaces are still evolving. Improvements in the graphite furnace include more rapid furnace heating and matrix modification techniques.

Anodal stripping voltammetry is an analytical method utilizing electrochemical phenomena and is based on the relationship between current, charge, potential, and time for electrolyte reactions. The electric potential from the oxidation of lead that is adhered to a mercury electrode is scanned and measured.

X-ray fluorescence is attractive for certain uses because it is nondisruptive and allows simultaneous identification of several elements, including rods using a high-energy irradiation source.

A number of other technically feasible methods are available including optical emission spectroscopy, neutron activation, and isotope dilution mass spectrometry. It should be pointed out that all of the methods mentioned only measure total lead and do not identify specific compounds of lead. There are methods, however, such as gas chromatography using the electron capture detector that may be used to measure organolead compounds.

REFERENCES

1. **Reisinger, K., Stoeppler, M., and Nurnberg, H. W.,** Evidence for the absence of biological methylation of lead in the environment, *Nature,* 291, 228, 1981.
2. **Nriagu, J. O.,** Global inventory of natural and anthropogenic emissions of trace metals to the atmosphere, *Nature,* 279, 409, 1979.
3. **Elias, R. W.,** Lead exposures in the human environment, in *Dietary and Environmental Lead: Human Health Effects,* Mahaffey, K. R., Ed., Elsevier, New York, 1985, 79.
4. **Rabinowitz, M. D., Wetherill, G. W., and Kopple, J. D.,** Magnitude of lead intake from respiration by normal man, *J. Lab. Clin Med.,* 90, 238, 1977.
5. **Nozaki, K.,** Methods for studies on inhaled particles in human respiratory system and retention of lead fume, *Ind. Health,* 4, 118, 1966.
6. **Rabinowitz, M. D., Wetherill, G. W., and Kopple, J. D.,** Studies of human lead metabolism by use of stable isotope traces, *Environ. Health Perspect.,* 7, 145, 1974.
7. **Alexander, F. W.,** The uptake of lead by children in differing environments, *Environ. Health Perspect.,* 7, 156, 1974.
8. **Ziegler, E. E., Edwards, B. B., Jensen, R. L., Mahaffey, K. R., and Fomon, S. J.,** Absorption and retention of lead by infants, *Pediatr. Res.,* 12, 29, 1978.
9. **Mahaffey, K. R. and Michaelson, I. A.,** The interaction between lead and nutrition, in *Low Level Lead Exposure,* Needleman, H. L., Ed., Raven Press, New York, 1980, 159.
10. **Azar, A., Snee, R. D., and Habibi, K.,** An epidemiologic approach to community air lead exposure using personal air samples, *Environ. Qual. Saf. Suppl.,* 2, 254, 1975.
11. WHO, Environmental Health Criteria, 3, Lead, World Health Organization, Geneva, 1977, 78.
12. **Pocock, S. J., Shaper, A. G., Walker, M., Wale, C. J., Clayton, B., Delves, H. T., Lacey, R. F., Packham, R. F., and Powell, P.,** Effects of tap water lead, water hardness, alcohol, and cigarettes on blood lead concentrations, *J. Epidemiol. Commun. Health,* 37, 1, 1983.
13. **Barry, P. S. I. and Mossman, D. B.,** Lead concentrations in human tissues, *Br. J. Ind. Med.,* 27, 339, 1970.
14. **Thompson, J. A.,** Balance between intake and output of lead in normal individuals, *Br. J. Ind. Med.,* 28, 189, 1971.
15. **Rosen, J.,** Metabolic and cellular effects of lead: a guide to low level lead toxicity in children, in *Dietary and Environmental Lead: Human Health Effects,* Mahaffey, K. R., Ed., Elsevier, New York, 1985, 157.
16. **Campbell, B. C., Elliott, H. L., and Meredith, P. A.,** Lead exposure and renal failure: does renal insufficiency influence lead kinetics?, *Toxicol. Lett.,* 9, 121, 1981.
17. **Vostal, J. and Heller, J.,** Renal excretory mechanisms of heavy metals. I. Transtubular transport of heavy metal ions in the kidney, *Environ. Res.,* 21, 1, 1968.
18. **Barltrop, D.,** Transfer of lead to the human foetus, in *Mineral Metabolism in Pediatrics,* Barltrop, D. and Burland, W. L., Eds., Blackwell Scientific, Oxford, 1969, 139.
19. **Chambe, S., Swinyard, C. A., and Nishimura, H.,** A quantitative study of human embryonic and fetal lead with considerations of maternal-fetal lead gradients and the effect of lead on human reproduction, *Teratology,* 5, 253, 1972.
20. **Bell, J. U. and Thomas, J. A.,** Effects of lead on mammalian reproduction, in *Lead Toxicity,* Singhal, R. L. and Thomas, J. A., Eds., Urban and Schwarzanberg, Baltimore, 1980, 169.
21. **Six, K. M. and Goyer, R. A.,** The influence of iron deficiency on tissue content and toxicity of ingested lead in the rat, *J. Lab. Clin. Med.,* 79, 128, 1972.
22. **Mahaffey, K. R., Goyer, R. A., and Haseman, J. K.,** Dose-response to lead ingestion in rats fed low dietary calcium, *J. Lab. Clin. Med.,* 82, 92, 1973.
23. **Gerber, G. B. and Deroo, J.,** Absorption of radioactive lead (^{210}Pb) by different parts of the intestine in young and adult rats, *Environ. Physiol. Biochem.,* 5, 314, 1975.
24. **Mahaffey, K. R., Banks, T. A., Stone, C. L., Capar, S., Compton, J., and Glubik, M.,** Effect of varying levels of dietary calcium on susceptibility to lead toxicity, *Proc. Int. Conf. Heavy Metals Environ,* 3, 155, 1975.
25. **Six, K. M. and Goyer, R. A.,** Experimental enhancement of lead toxicity by low dietary calcium, *J. Lab. Clin. Med.,* 76, 933, 1970.

26. **Rosen, J. F. and Wexler, E. E.,** Studies on lead transport in bone organ culture, *Fed. Proc.*, 34, 267, 1975.
27. **Snowdon, C. T. and Sanderson, B. A.,** Lead pica produced in rats, *Science,* 183, 92, 1974.
28. **Kostial, K. and Momcilovic, B.,** Transport of lead-203 and calcium-47 from mother to offspring, *Arch. Environ. Health,* 29, 28, 1974.
29. **Harris, P. and Holley, M. R.,** Lead levels in cord blood, *Pediatrics,* 49, 606, 1972.
30. **Six, K. M. and Goyer, R. A.,** The influence of iron deficiency on tissue content and toxicity of ingested lead in the rat, *J. Lab. Clin. Med.*, 79, 128, 1972.
31. **Delves, H., Bicknell, J., and Clayton, B.,** The excessive ingestion of lead and other metals by children, *Comm. Eur. Communities (Rep.) Eur EUR-5004,* p. 345, 1973.
32. **Watson, W. S., Hume, R., and Moore, M. R.,** Oral absorption of lead and iron, *Lancet,* 2, 236, 1980.
33. **Clerklewski, F. L. and Forbes, R. M.,** Influence of dietary zinc on lead toxicity in the rat, *J. Nutr.*, 106, 689, 1976.
34. **Willoughby, R. A., MacDonald, E., McSherry, B. J., and Brown, G.,** Lead and zinc poisoning and the interactions between the lead and zinc poisoning in the foal, *Can. J. Comp. Med.*, 36, 348, 1972.
35. **Victery, W., Miller, C., and Goyer, R. A.,** Essential trace metal excretion from rats with lead exposure and during chelation therapy, *J. Lab. Clin. Med.*, 107, 129, 1986.
36. **Pounds, J. G.,** Effect of lead intoxication on calcium homeostasis and calcium-mediated cell functions: a review, *Neurotoxicology,* 5, 295, 1984.
37. **Kapoor, S. C. and van Rossum, G. O.,** Effects of Pb^{2+} added *in vitro* on Ca^{2+} movements in isolated mitochondria and slices of rat kidney cortex, *Biochem. Pharmacol.*, 33, 1771, 1983.
38. **Goldstein, G. W. and Ar, D.,** Lead activates calmodulin sensitive processes, *Life Sci.*, 33, 1001, 1983.
39. **Bock, J. L.,** The binding of metal ions to ATP: a proton and phosphorus NMR investigation of diamagnetic metal-ATP complexes, *J. Inorg. Biochem.*, 12, 119, 1980.
40. **Audesirk, G.,** Effects of lead exposure on physiology of neurons, *Prog. Neurobiol.*, 24, 199, 1985.
41. **Wapnir, R. A., Exeni, R. A., McVicar, M., and Lifshitz, F.,** Experimental lead poisoning and intestinal transport of glucose, amino acid and sodium, *Pediatr. Res.*, 11, 153, 1977.
42. **Pounds, J. G., Morrison, D., Wright, R., Casciano, D. A., and Shaddock, J. G.,** Effect of lead on calcium-mediated cell function in the isolated rat hepatocyte, *Toxicol. Appl. Pharmacol.*, 63, 402, 1982.
43. **Silbergeld, E. K. and Hruska, R. E.,** Neurochemical investigations of low level lead exposures, in *Low Level Lead Exposure,* Needlemen, H. L., Ed., Raven Press, New York, 1980, 135.
44. **Rutter, M.,** Lead versus health sources and effects of low level lead exposure, in *Lead Versus Health,* Rutter, M. and Jones, R. R., Eds., John Wiley & Sons, New York, 1983, 333.
45. **Needleman, H. L., Gunnoe, C., Leviton, A. Reed, R., Peresie, H., Maher, C., and Barrett, P.,** Deficits in psychologic and classroom performance of children with elevated blood lead levels, *N. Engl. J. Med.*, 300, 689, 1979.
46. **Burchfield, J. L., Duffy, F. H., Bartels, P. H., and Needleman, H. L.,** The combined discriminating power of quantitative electroencephalography and neuropsychologic measures in evaluating central nervous system effects of lead at low levels, in *Low Level Lead Exposure,* Needleman, H. L., Ed., Raven Press, New York, 1980, 75.
47. **Benignus, V. A., Otto, D. A., Muller, K. E., and Seipk, J. J.,** Effects of age and body lead burden on CNS function in young children. II, *Electroencephalogr. Clin. Neurophysiol.*, 52, 240, 1981.
48. **Yule, W. and Rutter, M.,** Effect of lead on children's EEG spectra behavior and cognitive performance: a critical review, in *Dietary and Environmental Lead: Human Health Effects,* Mahaffey, K. R., Ed., Elsevier, New York, 1985, 211.
49. **Grant, L. D., Kimmel, C. A., West, G. L., Martinez-Vargus, C. M., and Howard, J. L.,** Chronic low-level lead toxicity in the rat. II. Effects on postnatal physical and behavioral development, *Toxicol. Appl. Pharmacol.*, 56, 42, 1980.
50. **Govani, S., Memo, M., Lucchi, L., Spano, P. R., and Trabucchi, M.,** Brain neurotransmitter systems and chronic lead intoxication, *Pharmacol. Res. Commun.*, 12, 447, 1980.
51. **Stowe, H. D., Goyer, R. A., and Krigman, M. R.,** Experimental oral lead toxicity in young dogs, *Arch. Pathol.*, 95, 106, 1973.
52. **Fjerdingstad, E. J., Danscher, G., and Fjerdingstad, E.,** Hippocampus: selective concentration of lead in the normal rat brain, *Brain Res.*, 80, 350, 1974.
53. **Thomas, H. M.,** A case of generalized lead paralysis, with a review of the cases of lead palsy, *Bull. Johns Hopkins Hosp.*, 15, 209, 1904.
54. **Seppalaine, A. M. and Hernberg, S.,** A follow-up study of nerve conduction velocities in lead exposed workers, *Neurobehav. Toxicol. Teratol.*, 4, 721, 1982.
55. **Lampert, P. W. and Schochet, S. S.,** Demyelination and remyelination in lead neuropathy, *J. Neuropathol. Exp. Neurol.*, 27, 527, 1968.
56. **Buchthal, F. and Behse, F.,** Electrophysiology and nerve biopsy in men exposed to lead, *Br. J. Ind. Med.*, 36, 135, 1979.

57. **Schlaepfer, W. W.,** Experimental lead neuropathy: a disease of the supporting cells in the peripheral system, *J. Neuropathol. Exp. Neurol.*, 28, 401, 1969.
58. **Paglia, D. E., Valentine, W. N., and Dahlgner, J. G.,** Effects of low level lead exposure on pyrimidine-5′-nucleotidase and other erythrocyte enzymes. Possible role of pyrimidine-5′-nucleotidase in the pathogenesis of lead-induced anemia, *J. Clin. Invest.*, 56, 1164, 1975.
59. **Hernberg, S., Nurminen,M., and Hasan, H.,** Nonrandom shortening of red cell survival times in men exposed to lead, *Environ. Res.*, 1, 247, 1967.
60. **Granick, S. and Levere, R. D.,** Heme synthesis in erythroid cells, *Prog. Hamatol.*, 4, 1, 1964.
61. **Goyer, R. A.,** Lead toxicity: a problem in environmental pathology, *Am. J. Pathol.*, 64, 167, 1971.
62. **Goyer, R. A.,** Lead and the kidney, *Curr. Top. Pathol.*, 55, 147, 1971.
63. **Moore, J. F., Goyer, R. A., and Wilson, M.,** Lead-induced inclusion bodies solubility, amino acid content and relationship to residual acidic nuclear proteins, *Lab. Invest.*, 29, 488, 1973.
64. **Shelton, K. R. and Egle, P. M.,** The proteins of lead-induced intranuclear inclusion bodies, *J. Biol. Chem.*, 257, 11802, 1982.
65. **Emmerson, B. T.,** Chronic lead nephropathy, *Kidney Int.*, 4, 1, 1973.
66. **Campbell, B. C., Moore, M. R., Goldberg, A., Hernandex, L. A., and Dick, W. C.,** Subclinical lead exposure: a possible cause of gout, *Br. Med. J.*, 2 (6149), 1403, 1978.
67. **Emmerson, B. T., Mirosch, W., and Douglas, J. B.,** The relative contributions of tubular reabsorption and secretion to urate excretion in lead nephropathy, *Aust. N.Z. Med.*, 4, 353, 1971.
68. **Farkas, W. R., Stanawitz, T., and Schneider, M.,** Saturninc gout: lead-induced formation of quanine crystals, *Science*, 199, 786, 1978.
69. **Cramer, K. and Dahlberg, L.,** Incidence of hypertension among lead workers: a follow-up study based on regular control over 20 years, *Br. J. Med.*, 23, 101, 1966.
70. **Pirkle, J. L., Schwartz, J., Landis, J. R., and Harlan, W. R.,** The relationship between blood lead levels and blood pressure and its cardiovascular risk implications, *Am. J. Epidemiol.*, 121, 1985.
71. **Perry, H. M. and Erlanger, M. W.,** Pressor effects of chronically feeding lead and cadmium together. Trace substances in environmental health, in *Proc. 12th Conf. on Trace Substances in the Environment*, Hemphill, D. D., Ed., University of Missouri, Columbia, 1978, 268.
72. **Victery, W., Schoeps, P., Shulak, J. M., and Vander, A. J.,** Perinatal exposure to lead alters the renin-angiotensin system in rats, *Proc. Soc. Exp. Biol. Med.*, 172, 1, 1983.
73. **Revis, N. W., Zinmeister, A. R., and Bull, R.,** Atherosclerosis and hypertension induction by lead and cadmium ions: an effect prevented by calcium ion, *Proc. Natl. Acad. Sci. U.S.A.*, 78, 6494, 1981.
74. **Meredith, P. A., Campbell, C. B., Blower, A., Derkx, F. H. M.,** and Reid, J. L., The effects of lead on the renin-angiotensin system, *Xenobiotica*, 15, 521, 1985.
75. **Lancranjan, I., Popescu, H. I., Gavanescu, O., Klepsch, I., and Servanerscu, M.,** Reproductive ability of workmen occupationally exposed to lead, *Arch. Environ. Health*, 30, 396, 1975.
76. **Koller, L. D.,** Immunological effects of lead, in *Dietary and Environmental Lead*, Mahaffey, K. R., Ed., Elsevier, New York, 1985, 339.
77. **Perlstein, M. A. and Attala, R.,** Neurologic sequelae of plumbism in children, *Clin. Pediatr.*, 5, 292, 1966.
78. International Agency for Research on Cancer, Monogr. on the Evaluation of the Carcinogenic Risk of Chemicals to Humans: Some Metals and Metallic Compounds, Vol. 23, IARC, World Health Organization, Lyon, France, 1980, 325.
79. **Moore, M. R. and Meredith, P. A.,** The carcinogenicity of lead (review), *Arch. Environ. Health*, 34, 158, 1979.
80. **Cooper, W. C. and Gaffey, W. R.,** Mortality of lead workers, *J. Occup. Med.*, 17, 100, 1975.
81. **Baker, E. L., Goyer, R. A., Fowler, B. A., Khettry, U., Bernard, D. B., Adler, S., White, R. D., Babayan, R., and Feldman, R. G.,** Occupational lead exposure, nephropathy and renal cancer, *Am. J. Ind. Med.*, 1, 139, 1980.
82. **Lilis, R.,** Long-term occupational lead exposure: chronic nephropathy and renal cancer: a case report, *Am. J. Ind. Med.*, 2, 293, 1981.
83. Centers for Disease Control (CDC), Preventing Lead Poisoning in Children, U.S. Department of Health and Human Services, Atlanta, 1985.
84. **Facchetti, S., Ed.,** *Analytical Techniques for Heavy Metals in Biological Fluids, Commission of the European Communities*, Elsevier, Amsterdam, 1983.
85. **Tsalev, D. L.,** *Atomic Absorption Spectroscopy in Occupational and Environmental Health Practice*, Vol. 1 and 2, CRC Press, Boca Raton, FL, 1984.

MANGANESE

B. Sarkar

CHEMISTRY

Manganese (Mn) can have ten oxidation states. Despite its variable valence states Mn is not directly involved in biological electron exchange reactions. Although it can be easily oxidized in an alkaline solution, divalent Mn is the most stable form. Mn(II) compounds are often pale pink. Complexes of Mn(II) have a coordination number of 6 and thus form an octahedral coordination complex. Mn(II) chemistry resembles that of Mg(II) in that is prefers weaker ligand domains such as carboxylate or phosphate groups. Mn(III) is a powerful-enough oxidizing agent to react with water to give Mn(II) and oxygen. However, it is the Mn(III) which is present in Mn superoxide dismutase and it is this form which binds transferrin. It appears to be easier for Mn(III) to be chelated in biological systems. Injected Mn(II) is thought to be converted to Mn(III) in the alkaline medium of the duodenum. Free Mn(II) shows a sextet hyperfine electron spin resonance spectrum, while bound Mn(II) does not. This property is utilized for studying enzyme-metal ion-substrate complexes.[1]

METABOLISM

Absorption of Mn occurs throughout the length of the small intestine.[2] The absorption of dietary Mn is relatively poor and is not under any homeostatic control. The percentage of Mn absorbed, which is approximately 3 to 4% of dietary Mn, is independent of the body burden of the element. Mn absorbed from the gastrointestinal tract enters into the portal blood. Free and protein-bound Mn in the portal blood is rapidly taken up by the liver where it is incorporated into the Mn-containing metalloenzymes. Absorbed Mn which is not retained by the body is almost totally excreted with the feces. The excretion of Mn through bile is very rapid. In contrast to the absorptive process, the excretion of Mn is not influenced by other dietary metal ions. Mn is involved in carbohydrate and lipid metabolism, and brain function.

Turnover of Mn in mammalian tissues is rapid, depending upon exogenous supply. The absorbed Mn is retained initially by mitochondria in the liver, pancreas, and kidney. Approximately 40% of the total body content of Mn is retained in the bone marrow and the rest is in a freely exchangeable, labile pool. Mn is retained in erythrocytes in a porphyrin complex. Absorption of Mn is increased by Fe deficiency, but under normal conditions other metal ions do not affect Mn metabolism, suggesting little or no interaction between Mn and other metal ions in biologic tissues.[3] However, manganese-vanadium antagonism is known in mammals. While V inhibits, Cr and Mn stimulate the hepatic synthesis of cholesterol and fatty acids in rats.

BIOCHEMISTRY

Mn can function both as an enzyme activator and as a constituent of metalloenzymes. The role of Mn as cofactor is not specific for most enzymes which can be activated by this metal. Examples are kinases, decarboxylases, hydrolases, and transferases.[4] In many cases Mg(II) can take the place of Mn. Iron(II) has also been reported to compete with Mn(II) as an activator in phosphoenolpyruvate carboxylase. There are three known Mn metalloenzymes: pyruvate carboxylase, superoxide dismutase, and arginase.

Pyruvate carboxylase is an enzyme which is responsible for the first step of gluconeogenesis, catalyzing the direct carboxylation of pyruvate to yield oxaloacetate. The enzyme has a wide tissue and species distribution. Its activity has been detected in most mammalian tissues examined, although maximal catalytic activities are confined to liver, kidney cortex, and adipose tissues. The enzyme is a tetramer with molecular weight of 500,000. Each molecule contains four Mn(II) which are on the outer surface of the enzyme and four biotins which are necessary for the activity of the enzyme. Mn contained in the enzyme appears to play a role in activating the methyl group of pyruvate by virtue of its electron-withdrawing properties. The methyl group of pyruvate then serves as the acceptor for carbon dioxide in the formation of oxaloacetate. Although the enzyme was first identified as a Mn metalloenzyme, several other metals have also been shown to exist as a constituent of the enzyme resulting in Mg, Zn, and Co containing pyruvate carboxylases.

Superoxide dismutase is another Mn metalloenzyme. The reaction mechanism of superoxide dismutase involves the alternate reduction and oxidation of the metal contained at its active site by superoxide ions. The biological role of this enzyme is to combat potentially toxic effects of superoxide ions by catalyzing the dismutation of these ions. This enzyme was first identified in 1939 as a copper-containing protein in bovine erythrocytes and liver and named erythrocuprein and hepatocuprein, respectively. At that time these enzymes were thought to be Cu storage proteins because they did not seem to show any enzyme activity. The first protein showing superoxide dismutase activity was isolated from bovine blood in 1969, and it was not until then that Mn- or Fe-containing superoxide dismutase was discovered and isolated from prokaryotes and mitochondria.[5] Generally, superoxide dismutase is classified into three groups based on the type of metal ion present in the enzyme. These are Cu/Zn superoxide dismutase, Mn-superoxide dismutase, and Fe-superoxide dismutase. Properties of Mn-containing superoxide dismutases are similar to those of Fe-containing superoxide dismutases. They contain either two Mn(III) ions or two Fe(III) ions having a reddish-purple or yellow color, respectively. The active form of superoxide dismutase found in *Escherichia coli* is a dimer with a molecular weight of 40,000, while in chicken liver mitochondria it is a tetramer with a molecular weight of 80,000. Although Mn-superoxide dismutase and Fe-superoxide dismutase are quite similar in many respects, their amino acid sequences show only 44% homology. The apoprotein which originally contains Mn(III) has been successfully reconstituted with Mn, Co, Fe, Ni, and Cu. Only the Mn-reconstituted enzyme is active and this activity is indistinguishable from that of the native enzyme.[6]

Arginase is a cytosolic enzyme involved in urea formation with a molecular weight of 120,000. The molecule is a tetramer containing four Mn(II) ions, of which two cannot be removed without irreversible loss of activity, while the other two can be removed by extensive dialysis, resulting in a 50% loss of activity. Addition of Mn(II) to the system restores activity to 100%. Although the activity of arginase is affected by dietary factors, the functional significance of lower activity of arginase in Mn-deficient animals is not known.

DEFICIENCY

Mn deficiency causes skeletal abnormalities, impaired growth, depressed reproductive function, ataxia of the newborn, and defects in lipid and carbohydrate metabolism. Mn deficiency in humans is rather rare. However, the clinical features include hypocholesterolemia, weight loss, transient dermatitis, occasional nausea, and vomiting. Whole blood Mn is significantly lower in epileptics and Mn-deficient conditions are more susceptible to convulsions.

TOXICITY

Mn is toxic to varying degrees depending upon the type of Mn ion and its oxidation states. The toxicity of some Mn salts is summarized in Table 1. Growth retardation, nonspecific

Table 1
MANGANESE TOXICITY

Compound	Animal	Route	Toxicity	Dosage/kg body weight: Compound (mg)	Metal: mg	Metal: m*M*
Manganese dioxide	Mouse	s.c.	LD_{100}	500	316	5.75
(MnO_2)	Rabbit	i.v.	LD_{100}	45	28	0.51
Manganese fluoride	Guinea pig	p.o.	MLD	200	118	2.15
(MnF_2)	Guinea pig	s.c.	MLD	700	414	7.54
Manganese chloride	Mouse	s.c.	LD_{100}	210	59	1.07
($MnCl_2$)	Guinea pig	s.c.	LD_{100}	180	50	0.91
	Rabbit	s.c.	LD_{100}	180	50	0.91
	Rabbit	i.v.	MLD	65	18	0.33
	Dog	i.v.	LD_{50}	201	56	1.02
Manganese perchlorate [$Mn(ClO_4)_2 \cdot 6H_2O$]	Rat	i.p.	LD_{50}	410	62	1.13
Manganese sulfate	Mouse	i.p.	LD_{50}	120	44	0.80
($MnSO_4$)	Mouse	i.p.	LD_{50}	120	44	0.80
Potassium permanganate	Mouse	s.c.	MLD	500	174	3.17
($KMnO_4$)	Rat	p.o.	LD_{50}	1090	380	6.92
	Rabbit	i.v.	LD_{100}	70	24	0.44
Manganese acetate	Mouse	i.v.	LD_{50}	16	6.19	0.11
[$Mn(C_2H_3O_2)_2 \cdot 4H_2O$]	Rat	p.o.	LD_{50}	3730	3730	15.2

Reproduced from Venugopal, B. and Luckey, T. D., Eds., *Metal Toxicity in Mammals,* Vol. 2, Plenum Press, New York, 1978. With permission.

anemia, ''metal fume fever'', and psychic and neurological disorders associated with ''manganism'' are some of the symptoms of Mn intoxication. Chronic Mn toxicity in humans following chronic exposure to Mn through inhalation, ingestion, or parenteral administration results in a disease of the central nervous system involving psychic and neurological disorders. A peculiar slopping gait, cramps or tremors of the body and extremities, slurred speech, hallucinations, insomnia, and mental confusion are some of the symptoms. Mn toxicity decreases the level of catecholamines and serotonin in the corpus striatum and injures the dopaminergic neurons, for which the biogenic amines act as neurotransmitters. The derangement of amine metabolism is accompanied by a disturbance in functional reaction among these amines, cyclic AMP and Mn(II). In spite of the severity of the symptoms, chronic Mn intoxication (or manganism) is not a fatal disease, although it causes permanent disability. Intradermal or i.m. injection of large doses of soluble Mn salts can result in calcified matrix, owing to the reaction of Mn with tissue proteins. Mn is also a potential carcinogen. Chronic parenteral injections of $MnCl_2$ induce an increased frequency of lymphosarcomas of mice. Leukemia and mammary adenosarcomas have also been noted.

ANALYTICAL METHOD

Mn can be determined by atomic absorption spectrometry. It has also been measured by neutron activation analysis and emission spectroscopy.

REFERENCES

1. **Taylor, A., Sawas, S., and James, T. L.,** Structural aspects of the inhibitor complex formed by *N*-(leucyl)-*O*-aminobenzenesulfonate and manganese with Zn^{2+}-Mn^{2+} leucine aminopeptidase (EC3.4.11.1), *J. Biol. Chem.*, 257, 11571, 1982.
2. **Underwood, E. J.,** *Trace Elements in Human and Animal Nutrition,* Academic Press, New York, 1971, 416.
3. **Pollack, S., Georg, J. N., Reba, R. C., Kaufman, R. N., and Crosby, W. H.,** The absorption of non-ferrous metals in iron deficiency, *J. Clin. Invest.*, 44, 1470, 1965.
4. **Vallee, B. L. and Coleman, J. E.,** *Comprehensive Biochemistry,* Vol. 12, Florkin, M. and Stotz, E., Eds., Elsevier, New York 1964, 165.
5. **McCord, J. M. and Fridovich, I.,** Superoxide dismutase: an enzymic function for erythrocuprein (hemocuprein), *J. Biol. Chem.*, 244, 6049, 1969.
6. **Brock, C. J., Harris, J. I., and Sato, S.,** Superoxide dismutase from Bacillus stearothermophilus. Preparation of stable apoprotein and reconstitution of fully active Mn enzyme, *J. Mol. Biol.*, 107, 1975, 1976.
7. **Venugopal, B. and Luckey, T. D., Eds.,** *Metal Toxicity in Mammals,* Vol. 2, Plenum Press, New York, 1978.

MERCURY

Sherry L. Perkins

CHEMISTRY

The unique physicochemical properties of mercury distinguish this metal from other proximal transition metals of the periodic table. The elemental state, Hg(O), is liquid at room temperature, boils at 357°C, has a vapor pressure of 1.2×10^{-3} mmHg, and demonstrates minimal aqueous solubility (6.4×10^{-5} g/l @ 25°C). Two other stable oxidation states are known: mercury(I) and mercury(II). The mercury(I) (mercurous) form is novel in that it exists as the binuclear cation Hg_2^{2+}, solutions of which contain about 1% Hg^{2+} (mercuric) ions, corresponding to the equilibrium:

$$Hg_2^{2+} \rightleftharpoons Hg^{2+} + Hg^0$$

The magnitude of the equilibrium constant for this disproportionation reaction has been the subject of considerable controversy owing to difficulties in quantitating all three oxidation states without disrupting the equilibrium during analysis. Values of 0.0120 to 10^{-8} have been reported depending on the initial concentration of mercurous ion and the analytical technique employed.[1-3]

Cinnabar or mercury(II) sulfide is the most common natural ore of mercury. Deposits are found throughout the world, with the earth's crust containing an average of 0.5 ppm mercury. Small droplets of elemental mercury can often be found with cinnabar deposits.

Normal atmospheric concentrations of mercury are 5 $\mu g/m^3$, increasing to 200 $\mu g/m^3$ near areas containing high soil levels.[4] Burning of fossil fuels releases mercury and accounts for higher atmospheric levels over urban centers. Groundwater concentrations in the U.S. and Canada are usually below 1 ppb, with background levels of around 0.1 ppb.[5]

Mercury compounds can be divided into two major classes: organic and inorganic. The inorganic forms include mercury(O) and salts of mercury(I) and mercury(II). Elemental mercury is much less reactive than cadmium, its electronically similar neighbor in the periodic table. All of the organic forms of mercury contain the element in the 2^+ oxidation state which is covalently bound to aryl, alkyl, or alkoxy alkyl groups and may be classified as such. These organic forms are generally unreactive to air and water and are often utilized as reagents for the preparation of other organometallic compounds.

TOXICOLOGY AND CLINICAL MANIFESTATIONS

Sources of Human Exposure

Poisonings have been documented for centuries from the widespread and indiscriminate use of mercury and mercury compounds. The earliest use of a mercury compound may have been the employment of cinnabar as a pigment by prehistoric man.[5] From the 4th century B.C., when mercury compounds became popular among Arab and Greek physicians, elemental mercury and mercury salts were used in medicine for a variety of purposes including dermal antiseptics, laxatives, the treatment of syphilis, as diuretics, in dental preparations, as cytotoxic agents, and for peritoneal lavage.[4-7]

Contemporary commercial uses include: electrical apparatus, thermometers, barometers, switches, mercury vapor lamps, neon lamps, in the paint industry for the production of anticorrosive paints, in the cellulose industry as a preservative for wet pulp from bacterial and fungal degradation, in metal refinement processes, in electrolyte cells for the production of chlorine and sodium hydroxide, as fungicides, and as catalysts in the manufacture of plastics.[6,8]

Inorganic Mercury Toxicity

Elemental Mercury

Elemental mercury presents a unique toxicological hazard among trace metals due to its appreciable vapor pressure at room temperature; inhalation represents the major route of uptake, especially in chronically exposed subjects. Approximately 80% of vapor inhaled is retained with subsequent distribution throughout the body. The main target organs include the kidneys and the central nervous system, the latter being particularly susceptible due to the high lipid solubility of the elemental form and the resultant ease with which it crosses the blood-brain barrier.[9] The toxic manifestations include: numbness in the fingers and lips, gingivitis, hearing impairment, tremor in the hands and tongue, headache, fatigue, weakness, ataxia, loss of hair, anorexia, excessive salivation, and erethism (excessive shyness, irritability, depression, impatience, intolerance, loss of memory, insomnia). Proteinuria often occurs with progression to renal failure in severe cases. In contrast to poisoning with the inorganic salts, elemental mercury produces no adverse effects when ingested provided it is not retained in the gastrointestinal tract; absorption is estimated at less than 2%.

At the biochemical level, there appears to be little difference between the effects of exposure to Hg(O) vapor and Hg(I) or Hg(II) compounds. This can be explained by the hypothesis that elemental mercury, once distributed in the bloodstream and, hence, throughout the body, is probably ultimately oxidized to Hg^{2+}, which produces toxic effects by forming covalent bonds with sulfhydryl groups.[10] The mechanism by which this *in vivo* oxidation of Hg(O) occurs has been extensively investigated over the last two decades, with most data suggesting an enzymatic oxidation involving the catalase-hydrogen peroxide system.[9,11-15] A unique characteristic of this oxidation pathway has been the observation that ethanol ingestion provides a protective effect, presumably through competitive inhibition of the catalase-hydrogen peroxide complex.[16-19]

Mercury(II) Salts

The first documented cases of the toxic effects of mercury(II) compounds may have been the reports of mental deterioration and ataxia among Roman quicksilver mine workers.[20] During the 17th Century, the introduction of mercuric nitrate to the felting process of hat making initiated an insidious era of chronic industrial mercury intoxication. Mercury poisoning became so common among hatters in Victorian Britain that the expression "as mad as a hatter" has become part of our language and has been portrayed in literature (the Mad Hatter in Lewis Carroll's *Alice's Adventures in Wonderland*).[21]

Acute poisoning due to inorganic mercuric salts usually arises from accidental or intentional ingestion. Although mercuric chloride (corrosive sublimate) taken orally is an unpleasant means of death, it once ranked sixth in frequency for use as a suicidal agent.[20]

The oral LD_{50} for mercuric chloride is less than 3 to 6 mg/kg with target organs affected in the following order: kidney > liver > blood > brain. Death is associated with blood mercury concentrations of 75 μmol/l. The clinical picture of acute mercuric chloride poisoning includes: sudden profound circulatory collapse with tachycardia, hypotension, peripheral vasoconstriction, and vomiting; hemorrhagic colitis may occur, with bloody diarrhea.[22] Renal failure usually develops within 24 h, progressing from oliguria to complete anuria. Uptake of mercury by other body tissues results in nonspecific changes including liver necrosis, cerebral petechial hemorrhages, respiratory distress, and/or bronchopneumonia.[7]

The nephrotoxic effects of mercuric compounds are thought to be the consequence of concentration of mercuric ions in the terminal part of the proximal renal tubule, resulting in tubular necrosis.[23] Tubular epithelial cells show degeneration, fragmentation, and necrosis.

The basement membrane may lose functional integrity while casts and granular debris often block the tubular lumen.[7] Glomerular disease resulting from exposure to mercurials is uncommon and occurs independently of acute proximal tubule damage and in the absence of neurological manifestations.[24]

At the biochemical level, Hg^{2+} acts as a potent, nonspecific enzyme poison. It inactivates metabolic enzymes, denatures structural proteins, disrupts membranes in the cells of target organs, complexes nitrogenous bases of DNA, and inhibits protein synthesis.[25-27] According to recent reports, the final mechanism for cell death in Hg^{2+} poisoning may be an uncontrolled influx of calcium ions into the cell which is toxic to the mitochondria and activates lysosomal enzymes.[20]

Mercury(I) Salts

Toxicity from mercurous salts has largely been attributed to the use of calomel, Hg_2Cl_2, in children's teething powders, and as a vermicide. Poisoning produces acrodynia (pink disease), the symptoms of which include: edema and blistering of skin, photophobia, excessive salivation, insomnia, profuse sweating, pink coloration of the extremities and cheeks, anorexia, irritability, apathy, and hypotonia.[26] Toxicity is thought to arise from the rapid disproportionation of the Hg_2^{2+} ions *in vivo* into Hg^0 and Hg^{2+}. As described before, the latter inactivates biologically important macromolecules through chelation of sulfhydryl groups. This sequestration of Hg^{2+} provides a driving force for the disproportionation reaction. Direct binding, and toxicity due to the Hg_2^{2+} itself, probably play a minor role.

Excretion

The excretion of inorganic mercury is complex and multiphasic. Over 85% of an acute single dose of inorganic mercury becomes concentrated in the kidney. Elimination occurs largely in the urine and is a consequence of tubular secretion and the exfoliation of renal tubular cells. Three phases of excretion have been described: (1) the initial phase where 35% of a single dose is excreted within a few days, (2) the second phase which has a half-life of 30 d, where 50% is lost, and (3) the slow third phase which has a half-life of 100 d.[23]

Organic Mercury Toxicity

The 20th Century's worst toxicological disaster involving mercury began in 1950 when a mysterious disease of the nervous system appeared among inhabitants, especially fishermen and their families, of Minamata, Japan. Three years passed before authorities traced the problem to local contamination of fish by mercurial compounds dumped into the bay by a plastics factory. The problem reached epidemic proportions in 1956 and by 1968, when the final Minamata report was written, 46 out of 121 official reported cases of methyl mercury poisoning had died.[28]

Methyl mercury and other organomercurial compounds produced by industry and by biotransformation of inorganic mercury compounds are particularly hazardous due to their high volatility and lipid solubility. Inhalation, ingestion, or cutaneous absorption of these species may result in toxicity. The three most common classes of organomercurials are (1) the short-chain alkyl compounds, (2) aryl compounds, and (3) alkoxy alkyl compounds.

Short-Chain Alkyl Compounds

Methyl mercury has been responsible for most incidences of poisoning from this group. Sources of exposure have included consumption of contaminated fish, as was the case in Minamata, and consumption of homemade bread prepared from wheat treated with methyl (or ethyl) mercury fungicides. Over 90% of the ingested dose is absorbed and distributed throughout the body. The principal target organ of methyl mercury is the central nervous system.

The clinical effects of exposure to monomethyl, dimethyl, or ethyl mercuric compounds are similar and are termed the Hunter-Russell syndrome. A latent period both during and after the cessation of exposure is often observed. This period, during which the patient feels no ill effects, may vary from a few weeks to a year. Typically, the first feature is paresthesia followed by progression to ataxia, dysarthria, hearing impairment, and constriction of visual fields.[26] In severe cases, widespread damage to the brain may occur resulting in mental derangement, coma, and death. Since the use of short-chain alkyl mercury compounds in industry has been discontinued, few cases of poisoning due to organomercurials have been reported.

Aryl and Alkoxy Alkyl Compounds

These compounds are rapidly converted to inorganic mercury in the body by cleavage of the carbon-mercury bond. Excretion then occurs in the urine. They are rapidly absorbed from the gastrointestinal tract and phenyl mercury can be absorbed by the skin.[10] These compounds are extensively used in industry as seed fungicides and as components of paint, yet reports of poisoning have been few, suggesting a lower relative toxicity compared to the alkyl compounds.

ASSESSMENT OF HUMAN TOXICITY

The half-life of methyl mercury in blood is long (70 d). Since erythrocyte levels are ten times greater than those of plasma, whole blood mercury levels are the preferred measure of exposure to this species and concentrations are reported to reflect body burden and the brain content. Normal levels should be less than 30 nmol/l, except in populations where fish consumption is high and values of 150 nmol/l can be seen.[29]

Elemental mercury is rapidly cleared from the circulation, its half-life being approximately 3 d; therefore, blood levels reflect only very recent exposure.[29] Thereafter, it is excreted in the urine and measurement of urine mercury levels is probably the preferred parameter for identifying toxicity from this and inorganic mercurials.[30,31] How long urine excretion remains elevated once exposure to elemental mercury or its inorganic salts has ceased is uncertain, and the investigator may have to rely upon clinical features to make the diagnosis in these circumstances. Normal urine excretion should be less than 50 nmol/d (less than 10 μg/d).[29] Symptomatic patients have levels in excess of 200 μg/l (1 μmol/l, approximately 1.5 μmol/d) and they are often much higher.[24] However, minor clinical and biochemical evidence of mercury toxicity (changes in hand tremor measured electronically, increased urinary excretion of high-molecular-weight proteins, and increased level of plasma and urine β galactosidase activity) can be detected in workers exposed to mercury vapor when urine mercury levels exceed 50 μg/g of creatinine (approximately 65 μg/d).[32]

Studies from Iraq of patients exposed to methyl mercury show that there is good correlation between the location of high mercury levels in hair and blood levels taken during the period of exposure. Hair mercury levels in pregnant women were also shown to be related to the frequency of abnormalities in their infants.[10] Thus, hair mercury levels can apparently be used to assess the time course of methyl mercury exposure and perhaps the approximate extent of toxicity.[33] There is little evidence to support its more extensive use in evaluating inorganic mercury toxicity, particularly in view of the susceptibility of hair to external contamination and the uncertain normal range for mercury.[34]

ANALYTICAL METHODS

Sample Preparation and Preservation

Preservation of mercury in biological samples presents major difficulties due to the disproportionation of Hg(I) to Hg(II) and Hg(O) and the subsequent loss of Hg(O) to the air.

In addition to considerations of the chemical form of mercury, the type of container and method of cleaning it, the biological matrix, and the type of preservative have all been reported to affect mercury preservation.[35] Aqueous samples should be brought to low pH and treated with oxidizing agents such as potassium dichromate in 10% nitric or sulfuric acid. Polypropylene, Pyrex®, Teflon®, and soft glass are all satisfactory for specimen preservation provided the oxidizing agent has been added.

Untreated urine, even when stored at $-10°C$, loses up to 20% of its mercury during the first few days of storage.[36] The most convenient storage mechanism appears to be the addition of sulfuric acid and Triton X-100 detergent and storage at 4°C. This method is compatible with most analytical techniques and digestion, preconcentration, and extraction are not required prior to analysis.

Determination of mercury in whole blood is fraught with difficulty due to the strong protein binding of mercury. Digestion is necessary. Techniques include: cysteine-trichloracetic acid[37] and sulfuric acid-potassium permanganate.[38] Direct cold vapor atomic absorption analysis without predigestion has also been reported.[39,40] The addition of cysteine to biological matrices preferentially binds mercury, thereby facilitating release during the reduction phase of the cold vapor atomic absorption analysis. EDTA vacutainers should be used.

The atomic absorption technique employing cold vapor analysis is the preferred method in most clinical laboratories.[41-43] In this technique, mercury compounds are chemically reduced using stannous chloride and cadmium chloride to the metallic form, which is then vaporized from aqueous solution into an inert gas stream. The mercury vapor is carried into a 20- to 30-cm gas cell in which the quantity of mercury present is measured by UV absorbance at 254 nm. This technique can be modified to differentiate between total organic and total inorganic mercury.[44]

Direct anodic stripping polarography using a glassy carbon electrode has been developed to quantitate mercury in the range of 0.1 to 0.005 ppm. However, this sensitivity is possible only if interferences can be eliminated.[45] Quantitation of the change in dithizone concentration after extraction of the mercury dithizone complex by AC polarography is an accurate indirect measurement of mercury concentration, but lacks the sensitivity of the anodic stripping polarographic technique.[8]

Neutron activation analysis requires no chemical manipulation of the specimen and provides the only suitable means of nondestructive analysis at the 0.01-ppm level. However, long irradiation times and expensive instrumentation requirements have limited the role of this assay to a reference method. Concentration of the sample can be achieved by lyophilization or physical or chemical extraction of the mercury from the sample.

The majority of analytical techniques for mercury are based on quantitation after it has been converted to one chemical oxidation state. Speciation of Hg(O), Hg(I), Hg(II), and CH_3Hg^+ has presented an analytical challenge. Quantitative differentiation of total organic and total inorganic mercury has been accomplished through selective reduction of the inorganic form with stannous chloride and total reduction of inorganic and organic mercury with a stannous chloride-cadmium chloride mixture.[36,44] A triple speciation of Hg(O), total inorganic and total organic mercury may be accomplished by omitting the reducing agent, in which case only the Hg^0 species is measured. This approach provides a rapid and accurate speciation of Hg^0, Hg^{2+}, and CH_3Hg^+. However, in the presence of equimolar concentrations of Hg_2^{2+}, the analysis has been found to be ambiguous and extreme caution should be exercised in interpreting the nonreducing signal.[12]

REFERENCES

1. **Moser, H. C. and Voigt, A. F.,** Dismutation of the mercurous dimer in dilute solutions, *J. Am. Chem. Soc.,* 79, 1837, 1957.
2. **Toribara, T. Y., Shields, C. P., and Koval, L.,** Behavior of dilute solutions of mercury, *Talanta,* 17, 1025, 1970.
3. **Hildebrand, D. A.,** Ph.D. thesis, University of North Dakota, Grand Forks, 1977.
4. **Casarett, L. J. and Doull, J., Eds.,** *Toxicology: The Basic Science of Poisons,* Macmillan, New York, 1975.
5. **Saha, J. G. and McKinlay, K. S.,** Use of mercury in agriculture and its relationship to environmental pollution, *Toxicol. Environ. Chem. Rev.,* 1, 271, 1973.
6. **McAuliff, Ed.,** *The Chemistry of Mercury,* 1st ed., Macmillan, New York, 1977.
7. **Laundy, T., Adam, A. E., Kershaw, J. B., and Rainford, D. J.,** Deaths after peritoneal lavage with mercuric chloride solutions: case report and review of the literature, *Br. J. Med.,* 289, 96, 1984.
8. **Shrivastava, A. K. and Tandon, S. G.,** The determination of mercury: a mini review, *Toxicol. Environ. Chem.,* 5, 311, 1982.
9. **Clarkson, T. W., Halbach, S., Magos, L., and Sugata, Y.,** On the mechanism of oxidation of inhaled mercury vapour, in *Molecular Basis of Environmental Toxicity,* Bhatnager, R. S., Ed., Ann Arbor Science, Ann Arbor, MI, 1980.
10. **Clarkson, T. W. and Marsh, D. O.,** Mercury toxicity in man, in *Clinical, Biochemical and Nutritional Aspects of Trace Elements,* Alan R. Liss, New York, 1982, 549.
11. **Wigfield, D. C. and Perkins, S. L.,** Oxidation of mercury by catalase and peroxidase in homogeneous solution, *J. Appl. Toxicol.,* 3(4), 185, 1983.
12. **Perkins, S. L.,** Studies of the Mechanism of the Metabolic Oxidation of Elemental Mercury, Ph.D. thesis, Carleton University, Ottawa, Canada, 1984.
13. **Magos, L., Halbach, S., and Clarkson, T. W.,** Role of catalase in the oxidation of mercury vapour, *Biochem. Pharmacol.,* 27, 1373, 1978.
14. **Ogata, M. and Ikeda, M.,** In vitro mercury uptake by human acatalasemic erythrocytes, *Arch. Environ. Health,* 34, 218, 1979.
15. **Eide, I. and Syversen, T. L. M.,** Relationship between catalase activity and uptake of elemental mercury by rat brain, *Acta Pharmacol. Toxicol.,* 52, 217, 1983.
16. **Magos, L., Clarkson, T. W., and Greenwood, M. R.,** The depression of pulmonary retention of mercury vapour by ethanol: identification of the site of action, *Toxicol. Appl. Pharmacol.,* 26, 180, 1973.
17. **Dunn, J. D., Clarkson, T. W., and Magos, L.,** Interaction of ethanol and inorganic mercury: generation of mercury vapour *in vivo, J. Pharmacol. Exp. Ther.,* 216, 19, 1981.
18. **Dunn, J. D., Clarkson, T. W., and Magos, L.,** Ethanol reveals novel mercury detoxification step in tissue, *Science,* 213(4), 1123, 1981.
19. **Dunn, J. D., Clarkson, T. W., and Magos, L.,** Ethanol-increased exhalation of mercury in mice, *Br. J. Ind. Med.,* 35, 241, 1978.
20. **Winek, C. L., Fochtman, F. W., Bricker, J. D., and Wecht, C. H.,** Fatal mercuric chloride ingestion, *Clin. Toxicol.,* 18(3), 261, 1981.
21. **Waldron, H. A.,** Did the Mad Hatter have mercury poisoning, *Br. Med. J.,* 287, 24, 1983.
22. **Datyner, M. E. and Cox, P. A.,** Inorganic mercurial poisoning, *Anaesth. Intensive Care,* 9(3), 266, 1981.
23. **Worth, D. P., Lewins, A. M., Davison, A. M., Ledgerwood, M. J., and Taylor, A.,** Hemodialysis and charcoal haemoperfusion in acute inorganic mercury poisoning, *Postgrad. Med. J.,* 60, 636, 1984.
24. **Wedeen, R. P.,** Lead, mercury and cadmium nephropathy, *Neurotoxicology,* 4(3), 134, 1983.
25. **Hughes, W. L.,** A physicochemical rationale for the biological activity of mercury and its compounds, *Ann. N.Y. Acad. Sci.,* 65(5), 454, 1957.
26. **Clarkson, T. W.,** Recent advances in the toxicology of mercury with emphasis on the alkylmercurials, *CRC Crit. Rev. Toxicol.,* March 1972.
27. **Stacey, N. H. and Kappus, H.,** Cellular toxicity and lipid peroxidation in response to mercury, *Toxicol. Appl. Pharmacol.,* 63, 29, 1982.
28. **Takeuchi, T.,** Pathology of Minamata disease with special reference to its pathogenesis, *Acta Pathol. Jpn.,* 32 (Suppl. 1), 73, 1982.
29. **Taylor, A.,** The biochemistry and toxicology of heavy metals, in *Scientific Foundation of Clinical Biochemistry — Biochemistry in Clinical Practice,* Williams, D. L. and Marks, V., Eds., Heinemann Medical Books, 1983, 351.
30. **Calder, I. M., Kelman, G. R., and Mason, H.,** Diurnal variations in urinary mercury excretion, *Hum. Toxicol.,* 3, 463, 1984.
31. **Yoshida, M.,** Relation of mercury exposure to elemental mercury levels in the urine and blood, *Scand. J. Work Environ. Health,* 11, 33, 1985.

32. **Roels, H., Gennart, J. P., Lauwerys, R., Buchet, J. P., Malchaire, J., and Bernard, A.,** Surveillance of workers exposed to mercury vapour: validation of a previously proposed biological threshold limit value for mercury concentration in urine, *Am. J. Ind. Med.*, 7, 45, 1985.
33. **Phelps, R. W., Clarkson, T. W., Kershaw, T. G., and Wheatley, B.,** Interrelationships of blood and hair mercury concentrations in a North American population exposed to methyl mercury, *Arch. Environ. Health,* 35(3), 161, 1980.
34. **Taylor, A.,** Usefulness of measurements of trace elements in hair, *Ann. Clin. Biochem.*, 23, 364, 1986.
35. **Rosain, R. M. and Wai, C. M.,** The rate of loss of mercury from aqueous solution when stored in various containers, *Anal. Chim. Acta,* 65, 279, 1973.
36. **Littlejohn, D., Fell, G. S., and Ottaway, J. M.,** Modified determination of total and inorganic mercury in urine by cold vapour atomic absorption spectrometry, *Clin. Chem.*, 22(10), 1719, 1976.
37. **Coyle, P. and Hartley, T.,** Automated determination of mercury in urine and blood by the Magos reagent and cold vapor atomic absorption spectrometry, *Anal. Chem.*, 53, 354, 1981.
38. **Kudsk, F. N.,** Uptake of mercury vapour in blood in vivo and in vitro from Hg-containing air, *Acta Pharmacol. Toxicol.*, 27, 149, 1969.
39. **Greenwood, M. R., Dhahir, P., Clarkson, T. W., Farant, J.-P., Chartrand, A., and Khayat, A.,** Epidemiological experience with Magos' reagents in the determination of different forms of mercury in biological samples by flameless atomic absorption, *J. Anal. Toxicol.*, 1, 265, 1977.
40. **Sharma, D. C. and Davis, P. S.,** Direct determination of mercury in blood by use of sodium borohydride reduction and atomic absorption spectrometry, *Clin. Chem.*, 25(5), 769, 1979.
41. **Magos, L.,** Selective atomic-absorption determination of inorganic mercury and methyl mercury in undigested biological samples, *Analyst,* 96, 847, 1971.
42. **Baltisberger, R. J. and Knudson, C. L.,** The differentiation of submicrogram amounts of inorganic and organomercury in water by flameless atomic absorption spectrometry, *Anal. Chim. Acta,* 73, 265, 1974.
43. **Uthe, J. F. and Armstrong, F. A. J.,** The microdetermination of mercury and organomercury compounds in environmental materials, *Toxicol. Environ. Chem. Rev.*, 2, 45, 1974.
44. **Magos, L. and Clarkson, T. W.,** Atomic absorption determination of total, inorganic and organic mercury in blood, *J. Assoc. Off. Anal. Chem.*, 55(5), 966, 1972.
45. **Chilov, S.,** Determination of small amounts of mercury, *Talanta,* 22, 205, 1975.

MOLYBDENUM

Gillian Lockitch

INTRODUCTION

Molybdenum (Mo) is one of nine trace elements currently accepted as essential for man.[1] Reliable data on the metabolism of Mo in healthy humans are scarce. Physiological levels of Mo in human tissues and body fluids are very low (Table 1).[2-11] Sample contamination during collection or processing, lack of analytical sensitivity, and failure to validate assay procedures by use of certified reference material may partially account for the higher Mo concentrations found in earlier studies, levels of which greatly exceeded the concentrations currently proposed as reference norms.[10]

Daily dietary intake probably exceeds normal needs, as spontaneous dietary deficiency has not been reported.[12,13] Recent reports of an inherited disorder, molybdenum co-factor deficiency,[14] and of acquired Mo deficiency in a patient on long-term parenteral nutrition[15] have caused a resurgence of interest in Mo as an essential trace element in human physiology.[16,17]

CHEMISTRY AND INTERACTION WITH OTHER TRACE ELEMENTS

Mo (atomic weight 95.9, atomic number 42) occurs naturally in different oxidation states. The most important are Mo^{3+} to Mo^{6+}. Mo serves as a catalyst in electron transfer reactions by coupling oxygen or proton transfer with electron transfer. Mo forms six-coordinate complexes with oxygen and sulfur and to a lesser extent with nitrogen.[17-19]

Both Mo and sulfur, a Mo antagonist, interfere with absorption and possibly utilization of copper. This was recognized through studies of endemic diseases in grazing animals in areas of high Mo content,[20-22] and confirmed in laboratory studies.[23] Tungsten is another Mo antagonist which potentiates the effect of induced Mo deficiency in animal studies.[20]

BIOCHEMISTRY

Mo is an essential component of several enzymes, of which three are important in human metabolism (Table 2). These are sulfite oxidase (SO) (EC 1.8.3.1), xanthine dehydrogenase (EC 1.2.1.37), and aldehyde oxidase (EC 1.2.3.1.). All molybdoenzymes in nature, except nitrogenase, a nitrogen-fixing enzyme found in bacteria, contain a noncovalently bound Mo cofactor. Molybdopterin, the apoprotein component of this cofactor, is a reduced pterin with a side chain containing four carbon atoms with two sulfur moieties, a hydroxyl group and a terminal phosphate ester. The co-factor of nitrogenase is an iron-molybdenum-sulfur complex.[19]

SO catalyzes the conversion of sulfite, a highly toxic metabolite, to sulfate. Sources of sulfite include methionine and cysteine catabolism, ingestion of bisulfite (a food preservative), and sulfur-dioxide inhalation. A syndrome characterized by severe neuropathy and ectopia lentis is seen in the rare autosomal-recessive disorder of SO deficiency (Table 3). The characteristic changes seen in this disorder may be primarily due to toxicity from excess sulfite or to deficiency of sulfate, an essential component of sulfoproteins and sulfolipids.[19] This issue has not been resolved.

Xanthine dehydrogenase catalyzes the conversion of xanthine, a product of purine catabolism, to uric acid. This enzyme functions both as a dehydrogenase (XD), using NAD^+, and as an oxidase (XO), using oxygen, as electron acceptors *in vivo*. Xanthine oxidase

Table 1
MOLYBDENUM CONCENTRATION IN BIOLOGICAL FLUIDS AND TISSUE

Specimen	Reported means	Reported range	SI units	Ref.
Serum (μg/l)	0.55—0.6	0.19—1.16	2—12 nmol/l	2—4, 10, 11
Urine (μg/24 h)	32—34	11—88	114—917 nmol/24 h	5, 6
Liver (μg/g) wet weight	0.57—0.76	0.20—1.20	2—12.5 nmol/g	7—9

Table 2
MOLYBDOENZYMES IN MAN

Enzymes	Reaction	Substrates
Xanthine oxidase	$RH + H_2O \rightarrow ROH + 2e + 2H^+$	Xanthine hypoxanthine
Sulfite oxidase	$SO_3^{2-} + H_2O \rightarrow SO_4^{2-} + 2e + 2H^+$	Sulfite
Aldehyde oxidase	$RH + H_2O \rightarrow ROH + 2e + 2H^+$	Heterocyclic compounds

Table 3
INHERITED DISORDERS INVOLVING MOLYBDOENZYMES

Disorder	Syndrome	Biochemical abnormalities
Xanthinuria (xanthine oxidase deficiency)	Xanthine nephrolithiasis	Absent XO/XD activity, hypouricemia, hypouricosuria, xanthine, hypoxanthine increased in serum and urine
Sulfite oxidase deficiency	Mental retardation, ocular lens dislocation, multiple neurological problems (ataxia, choreoathetosis)	Increased sulfite, thiosulfite, *S*-sulfocysteine, and decreased sulfate in urine
Molybdenum cofactor deficiency	Feeding difficulties, mental retardation, myoclonus, abnormal facies, dislocation of ocular lenses	Hypouricemia, xanthinuria, increased sulfite, thiosulfite, *S*-sulfocysteine, and decreased sulfate in urine

deficiency, characterized by xanthine nephrolithiasis and hypouricemia, is more common and has less adverse clinical effect than SO deficiency. Patients on allopurinol, an inhibitor of xanthine oxidase, do not usually experience complications of xanthine oxidase inhibition other than xanthine crystalluria. This contrasts with the severe neurological impairment of isolated SO deficiency, suggesting that impaired sulfur metabolism causes the most significant clinical problems in Mo deficiency.

Aldehyde oxidase (AO) hydroxylates heterocyclic nitrogenous compounds such as purines, pyrimidines, aldehydes, and pteridines. No specific disorder has been described for isolated AO deficiency.

Deficiency of Mo cofactor, probably an autosomal-recessive disorder, results in decreased activity of all three enzymes. The symptoms result from reduced sulfite and xanthine oxidase activity. None of the biochemical findings in Mo cofactor deficiency have been specifically attributed to AO deficiency (Table 3).[14]

METABOLISM

Human data are sparse and most information is derived from animal studies.[20,22] A comprehensive study of Mo metabolism and the effect of Mo excess in humans was published in 1979.[13] However, the serum levels reported for controls in this study are ten times greater than the levels now considered reference norms.[10,11]

Gastrointestinal absorption of molybdate and water-soluble inorganic forms is high.[13] Almost 100% of a single ingested dose was absorbed in the stomach and small intestine in rats. Clearance is equally efficient; over 70% of a single ingested dose is excreted within 3 h and within 24 h almost 90% is excreted.[13] A common carrier system has been proposed for molybdate and sulfate in both intestine and kidney.[20] Sulfate competitively inhibits molybdate absorption in rat intestine[24] and renal tubules.[25]

Mo is rapidly taken up by the liver,[26] reaching maximum uptake 1 h after a single dose. Blood is almost completely cleared of Mo within 6 h after a single i.v. injection of ^{99}Mo.[26] Excess Mo taken up by liver is excreted in bile and can be reabsorbed resulting in an enterohepatic circulation. Mo is also distributed to peripheral tissues, particularly to the thyroid and the adrenal which have a high Mo content.[13]

It is possible that the apoprotein component of the Mo cofactor is needed to bind Mo in order to retain Mo in tissues.[27] Without apoprotein, inorganic molybdate may not be retained in the liver. Acute viral hepatitis as well as other forms of liver disease is associated with higher Mo levels in blood.[28] This may reflect decreased uptake by liver cells or release from damaged tissue. In tissue obtained at time of autopsy, the highest Mo concentrations were found in liver, kidney, small intestinal mucosa, and adrenals.[22] In chronic overload, Mo also accumulates in kidney, testes, and ovary. Infertility and reproductive problems have been reported in Mo toxicity in animals.[13]

The main route of excretion is urinary, loss through the gut being normally less than one third of urine excretion. This can substantially alter in intestinal diseases such as Crohn's disease, where substantial fecal losses can occur.[16] Urinary excretion is extremely efficient and appears to provide a good defense against toxicity.[13,22] The efficiency of clearance is such that progressive accumulation is not severe in chronic industrial overexposure.[13] Urine Mo concentration in smelter workers ranged between 200 μg Mo/l and 11 mg/l while in a control group; Mo concentration was 20 to 230 μg Mo/l urine.[13] The mean serum Mo content was similar in both the control group and exposed workers, which appeared to support the concept that increased excretion removed the excess Mo. However, the mean serum values in both groups were ten times above the currently accepted normal values.[10,11]

HUMAN NUTRITION

Tipton et al.[12] described dietary Mo intakes of 200 to 500 μg/d, while Chappell et al.[13] found intakes to be between 120 and 240 μg Mo/d depending on age, sex, and income group. Estimates ranged from 8 to 10 μg/kg body weight at 1 year to 2 to 3 μg/kg at 65.[13] Intake was higher in males than females, and higher in lower-income families. The daily intake probably greatly exceeds physiological requirements, because no spontaneous nutritional deficiency under natural conditions has been reported for any species.[22]

Rich food sources of Mo include meats, especially organ meats, cereal grains, legumes, and dairy products.[29] Milk Mo content varies substantially with Mo intake of cows.[29] Natural water Mo content in areas not contaminated by industrial exposure is less than 10 to 20 μg Mo per liter.[13]

HUMAN DEFICIENCY

Acquired

The first case of probable acquired Mo deficiency was reported in a 24-year-old man with extensive Crohn's disease who developed symptoms after 12 months of total parenteral nutrition. Initial symptoms were tachycardia, tachypnea, severe bifrontal headache, night blindness, central scotomas, nausea, vomiting, lethargy, disorientation, and coma Blood findings included very high plasma methionine levels and marked hypouricemia. Urine uric acid and sulfate excretion were extremely low, while excess urinary thiosulfate, sulfite, xanthine, and hypoxanthine were present. I.v. ammonium molybdate completely reversed the clinical and biochemical abnormalities.[15]

Genetic

Molybdenum co-factor deficiency was first described in 1978,[30] and there have now been eight reported cases.[16,30] Inherited enzyme deficiency disorders have been reported independently for both xanthine oxidase (Xanthinuria) and SO (Table 3). Recognition of Mo co-factor deficiency occurred when a child appeared to have symptoms suggestive of a combined deficiency state.[30] The symptoms were early feeding difficulties, leading to severe dehydration, dysmorphic features, abnormal muscle tone, mental retardation, dislocated lenses, and excretion of xanthine concretions in the urine. Hepatic SO and xanthine oxidase activities were markedly decreased compared to controls. Neither clinical nor biochemical abnormalities responded to up to 200 μg/d of i.v. molybdate. Defective Mo co-factor formation or utilization was suspected. No active Mo cofactor could be detected in liver from three patients.[14] The detection of SO activity in normal cultured fibroblasts and absent SO activity in fibroblasts from SO-deficient patients means that antenatal detection of SO deficiency and Mo co-factor deficiency is possible.[14]

TOXICITY IN MAN

Mo was considered to be relatively nontoxic, with urinary excretion increasing in response to increased intake. However, symptoms have been described in mine and smelter workers in Armenia and India, Mo-rich areas, where intakes of up to 15 mg Mo per day have been reported.[18] This is almost 100 times greater than normal Western intakes.[13] A syndrome of gout-like symptoms and joint deformities, together with hyperuricemia, high plasma Mo, and increased Mo and Cu excretion, was described in these workers.[18]

MOLYBDENUM IN ANIMALS

In contrast to other elements such as zinc, there appear to be marked interspecies differences in manifestations of disordered Mo metabolism.[20]

Toxicity

Molybdenosis in cattle is characterized by severe diarrhea, loss of weight, loss of hair, and anemia. This is probably the result of induced copper deficiency, since symptoms respond to copper supplements. Fertility and reproduction are also affected.[20] Mandibular exostoses, growth retardation, and anemia are found in rats on excessively high Mo intake.

Deficiency

Naturally occurring Mo deficiency has not been reported. Experimental Mo deficiency in rats and chicks on a low Mo diet alone did not result in symptoms unless tungsten, a MO antagonist, was added.[20] Prenatal Mo deficiency in goats[31] and chicks[32] results in abnormalities of conception, morphogenesis, and fetal survival, indicating the essentiality of Mo in these species.

ANALYTICAL METHODOLOGY AND REPORTED REFERENCE VALUES

Recent excellent reviews have emphasized the problems of analysis of trace elements such as Mo which may be present at physiological concentrations close to the lower limits of sensitivity for some analytical methods.[10,11] Electrothermal atomization-atomic absorption spectroscopy (EAAS), which can measure elements such as manganese or chromium on a 20-μl sample, requires pretreatment and concentration to measure Mo accurately in serum or urine.[11] Severe matrix effects, poor sensitivity, carbide formation, and very low physiological levels preclude direct analysis of Mo by EAAS in these fluids. Acceptable reference values for Mo concentration in serum or plasma, urine, and tissues have come from studies using neutron activation analysis,[10] a technology unavailable to most laboratories. Studies in which control-subject serum values greatly exceed currently accepted reference values and where reference standards have not been used to validate the analyses, should be read with caution.

Currently, the main use for measurement of Mo in serum, urine, or tissues is in diagnosis of Mo toxicity states or investigation of Mo metabolic pathways. The normally low level of Mo in blood precludes diagnosis of deficiency by blood levels. Diagnosis of suspected Mo deficiency, as in long-term parenteral nutrition, is best made by looking for evidence of decreased activity of the molybdoenzymes xanthine oxidase and SO. The characteristic changes in uric acid and sulfur metabolism result in the biochemical findings described in Table 3.[15,33] Ultimate proof of acquired Mo deficiency as a cause of symptoms in a specific case will come from response to molybdate therapy.

The rarity of reports of acquired Mo deficiency in man may partially be due to a low physiological requirement for this element, but may also reflect a low index of suspicion. Careful studies of patients with chronic gastrointestinal disease, on elemental diets or on long-term parenteral nutrition, may reveal more cases and provide further information about daily requirements and normal Mo metabolism in man.

REFERENCES

1. **Aggett, P. J. and Davies, N. T.,** Some nutritional aspects of trace metals, *J. Inher. Metab. Dis.*, 6 (Suppl. 1), 22, 1983.
2. **Baert, N., Cornelis, R., and Hoste, J.,** Molybdenum in human blood, *Clin. Chim. Acta,* 68, 355, 1976.
3. **Versieck, J., Hoste, J., Barbier, F., Vanballenberghe, L., De Rudder, J., and Cornelis, R.,** Determination of molybdenum in human serum by neutron activation analysis, *Clin. Chim. Acta,* 87, 35, 1978.
4. **Kasparek, K., Iyengar, G. V., Kiem, J., Borberg, H., and Feinendegen, L. E.,** Elemental composition of platelets. III. Determination of Ag, Au, Cd, Co, Cr, Cs, Mo, Rb, Sb, and Se in normal human platelets by neutron activation analysis, *Clin. Chem.*, 25, 711, 1979.
5. **Robinson, M. F., McKenzie, J. M., Thomson, C. D., and van Rij, A. L.,** Metabolic balance of zinc, copper, cadmium, iron, molybdenum and selenium in young New Zealand women, *Br. J. Nutr.*, 30, 195, 1973.
6. **Cornelis, R., Versieck, J., Desmet, A., Mees, L., and Vanballenberghe, L.,** Neutron activation analysis of the trace element molybdenum in the urine of healthy persons, *Bull. Soc. Chim. Belg.*, 90, 281, 1981.
7. **Brune, D., Nordberg, G., and Wester, P. O.,** Distribution of 23 elements in the kidney, liver and lungs of workers from a smeltery and refinery in northern Sweden exposed to a number of elements and of a control group, *Sci. Total Environ.*, 16, 13, 1980.
8. **Sumino, K., Hayakawa, K., Shibata, T., and Kitamura, S.,** Heavy metals in normal Japanese tissues, *Arch. Environ. Health,* 30, 487, 1975.
9. **Meinel, B., Bode, J. C., Koenig, W., and Richter, F. W.,** Contents of trace elements in the human liver before birth, *Biol Neonate,* 36, 225, 1979.
10. **Versieck, J.,** Trace elements in human body tissues and fluids, *CRC Crit. Rev. Clin. Lab. Sci.*, 22, 97, 1985.
11. **Delves, H. T.,** The analysis of biological and clinical materials, *Prog. Anal. At. Spectrosc.*, 4, 1, 1981.
12. **Tipton, I. H., Stewart, P. L., and Dickson, J.,** Patterns of elemental excretion in long term balance studies, *Health Phys.*, 16, 455, 1969.
13. **Chappell, W. R., Meglen, R. R., Moure-Eraso, R., Tsongas, T. A., Walravens, P. A., and Winston, P. W.,** Human Health Effects of Molybdenum in Drinking Water, DPA-600/1-79-006, U.S. Environmental Protection Agency, 1979.

14. **Wadman, S. K., Duran, M., Beemer, F. A., Cats, B. P., Johnson, J. L., Rajagopalan, K. V., Saudubray, J. M., Ogier, H., Charpentier, C., Berger, R., Smit, G. P. A., Wilson, J., and Kryawawych, S.,** Absence of hepatic molybdenum cofactor: an inborn error of metabolism leading to a combined deficiency of sulphite oxidase and xanthine dehydrogenase, *J. Inher. Metab. Dis.*, 6 (Suppl. 1), 78, 1983.
15. **Abumrad, N. N., Schneider, A. J., Steel, D., and Rogers, L. S.,** Amino acid intolerance during prolonged total parenteral nutrition reversed by molybdate therapy, *Am. J. Clin. Nutr.*, 34, 2551, 1981.
16. **Abumrad, N. N.,** Molybdenum — is it an essential trace metal?, *Bull. N.Y. Acad. Med.*, 60, 161, 1984.
17. **Coughlan, M. P.,** The role of molybdenum in human biology, *J. Inher. Metab. Dis.*, 6 (Suppl. 1), 70, 1983.
18. **Friberg, L.,** Molybdenum, in *Handbook on the Toxicology of Metals*, Friberg, L., Nordberg, G. F., and Vouk, V. B., Eds., Elsevier, Amsterdam, 1979, 531.
19. **Rajagopalan, K. V.,** Molybdenum, in *Biochemistry of the Essential Ultra Trace Elements*, Frieden, E., Ed., Plenum Press, New York, 1984, 149.
20. **Mills, C. F. and Bremner, I.,** Nutritional aspects of molybdenum in animals, in *Molybdenum and Molybdenum-Containing Enzymes*, Coughlan, M., Ed., Pergamon Press, Oxford, 1980, 517.
21. **Mills, C. F.,** Dietary interactions involving the trace elements, *Annu. Rev. Nutr.*, 5, 173, 1985.
22. **Underwood, F. J.,** Molybdenum, in *Trace Element Metabolism in Man and Animals*, Academic Press, New York, 1977, 116.
23. **Gray, L. F. and Daniel, L. J.,** Effect of the copper status on the copper-molybdenum-sulphate interaction, *J. Nutr.*, 84, 31, 1964.
24. **Cardin, C. J. and Mason, J.,** Molybdate and tungstate transfer by rat ileum; competitive inhibition by inorganic sulphate, *Biochim. Biophys. Acta*, 455, 937, 1976.
25. **Bishara, H. N. and Bray, A. C.,** Competition between molybdate and sulphate for renal tubular reabsorption in the sheep, *Proc. Aust. Soc. Anim. Prod.*, 12, 123, 1978.
26. **Sorensen, L. B.,** Specific uptake of molybdenum by the liver, *J. Lab. Clin. Med.*, 60, 1020, 1962.
27. **Johnson, J. L., Waud, W. R., Rajagopalan, K. V., Duran, M., Beemer, F. A., and Wadman, S. K.,** Inborn errors of molybdenum metabolism. Combined deficiencies of sulphite oxidase and xanthine dehydrogenase in a patient lacking the molybdenum cofactor, *Proc. Natl. Acad. Sci. U.S.A.*, 77, 3175,
28. **Versieck, J., Hoste, J., Vanballenberghe, L., Barbier, F., Cornelis, R., and Waelput, I.,** Serum molybdenum in diseases of the liver and biliary system, *J. Lab. Clin. Med.*, 97, 535, 1981.
29. **Hazell, T.,** Minerals in foods: dietary sources, chemical forms, interactions, bioavailability, *World Rev. Nutr. Diet.*, 46, 1, 1985.
30. **Duran, M., Beemer, A., van der Haiden, C., Korteland, J., de Bree, P. K., Brink, M., and Wadman, S. K.,** Combined deficiency of xanthine oxidase and sulphite oxidase: a defect of molybdenum metabolism or transport?, *J. Inher. Metab. Dis.*, 1, 175, 1978.
31. **Anke, M., Grun, M., Partschfeld, M., and Groppel, B.,** Molybdenum deficiency in ruminants, in *Trace Element Metabolism in Man and Animals — 3*, Kirchgessner, M., Ed., Institut für Ernahrungsphysiologie, Friesing-Weihenstephan, 1978, 230.
32. **Payne, G. C.,** Molybdenum responsive syndromes of poultry, in *Trace Element Metabolism in Man and Animals — 3*, Kirchgessner, M., Ed., Institut für Ernahrungsphysiologie, Friesing-Weihenstephan, 1978, 315.
33. **Wadman, S. K., Cats, B. P., and de Bree, P. K.,** Sulfite oxidase deficiency and the detection of urinary sulfite, *Eur. J. Pediatr.*, 141, 62, 1983.

NICKEL

F. William Sunderman, Jr.

CHEMISTRY, OCCURRENCE, AND USES[1-3]

Nickel, the 24th element in order of natural abundance in the earth's crust, is a malleable, ductile, ferromagnetic metal. Nickel is obtained primarily by refining sulfide and oxide ores. The oxidation states of nickel include −1, 0, +1, +2, +3, or +4, but the most prevalent oxidation state is +2. The nickel salts of greatest commercial importance are $NiCl_2$, $NiSO_4 \cdot 6H_2O$, $Ni(NO_3)_2 \cdot 6H_2O$, $NiCO_3$, $Ni(OH)_2$, $Ni(CH_3COO)_2$, and NiO. Nickel monoxide (NiO) exists in two major forms with different properties: black NiO is chemically reactive and readily yields nickel salts upon contact with mineral acids, whereas green NiO is relatively refractory to solubilization. Crystalline nickel sulfides (e.g., αNi_3S_2, nickel subsulfide; βNiS, nickel monosulfide) are intermediates in the refining of sulfide ores; they possess biological importance because of their potency as carcinogens. Nickel carbonyl ($Ni[CO]_4$), a colorless liquid that boils at 43°C and decomposes at 60°C, has biological importance because of its exceptional acute toxicity. $Ni(CO)_4$ is an intermediate in the Mond process for refining nickel and is also used as a catalyst in the petroleum and chemical industries. Metallic nickel is widely used for production of ferronickel alloys (e.g., stainless steel), cupronickel alloys (e.g., Monel metal), and precious metal alloys (e.g., German silver). Various nickel alloys are used as rods for arc welding, in molds for manufacture of glass and ceramics, in catalysts for hydrogenation of fats and oils, and in surgical and dental prostheses. Nickel compounds are used in electroplating, in electric batteries, in paint pigments, textile dyes, and mordants. Nickel enters the atmosphere from combustion of fossil fuels, industrial emissions, waste incineration, weathering of rocks, and volcanic emissions.

BIOCHEMISTRY AND METABOLISM[4-6]

Nickel-containing enzymes (e.g., ureases, hydrogenases) have been isolated from plants and microorganisms, but not, as yet, from mammalian tissues. The average dietary intake of nickel by adult persons is 165 μg/d, but may reach 900 μg/d in diets rich in oatmeal, cocoa, chocolate, nuts, and soya products. Inhalation of nickel averages 0.4 μg/d for urban dwellers and 0.2 μg/d for rural dwellers. A minor fraction (1 to 5%) of ingested nickel and a major fraction (35%) of inhaled nickel are absorbed, enter the plasma, and are excreted primarily via the urine, and to a limited extent via bile and sweat. Fecal excretion of nickel reflects the unabsorbed fraction of ingested or tracheally cleared nickel, as well as nickel eliminated in bile and intestinal secretions. Ni^{2+} in blood is bound to albumin, ultrafiltrable ligands (e.g., histidine, cysteine, aspartic acid), and a specific macroglobulin ("nickeloplasmin"). In human autopsy tissues, nickel contents are generally highest in lungs, bones, thyroid, and adrenal glands. The estimated body burden of nickel in healthy adults averages 0.5 mg/70 kg (7.3 μg/kg body weight). Nickel concentrations in sweat from healthy adults and in milk from lactating mothers average 10 to 20 times greater than the mean concentration of nickel in urine. Hence, under conditions of profuse sweating or during lactation, appreciable quantities of nickel may be eliminated via sweat or milk. Serum nickel concentrations are oftentimes increased in patients after myocardial infarction, myocardial ischemia, stroke, or thermal burns; the pathophysiological mechanisms for such disturbances of nickel metabolism have not been elucidated. The elimination kinetics of soluble Ni^{2+} compounds administered to rodents by parenteral routes can be described by a two-compartment model. The disappearance half-time for plasma nickel concentrations in human volunteers after oral intake of $NiSO_4$ is approximately 11 h. In electroplating workers who are exposed to inhalation of soluble nickel-containing aerosols, the elimination half-time for nickel in urine is 17 to 39 h; in welders and moldmakers exposed to inhalation of insoluble nickel-containing dusts, the elimination half-time for nickel in urine ranges from 30 to 53 h.

NUTRITIONAL ESSENTIALITY[4-6]

Nutritional essentiality of nickel has been established in chickens, rats, swine, and goats. In rats, nickel deficiency is associated with retarded body growth, anemia, hypoglycemia, and diminished activities of dehydrogenase and aminotransferase enzymes in liver and kidney. There is no direct evidence that nickel is essential for human nutrition; nickel deficiency has not been documented in any human diseases.

TOXIC EFFECTS[7-9]

Health hazards from human exposure to certain nickel compounds and alloys include (1) hypersensitivity, (2) sinonasal and lung cancers, (3) nonmalignant disorders of the respiratory tract, and (4) local or systemic toxicity from iatrogenic exposures. Nickel is a common etiology of allergic dermatitis, resulting from contact with stainless steel utensils, tools, jewelry, coinage, and clothing fasteners. Nickel hypersensitivity also causes pulmonary asthma, conjunctivitis, and inflammation around nickel-containing prostheses. Increased mortality rates for sinonasal and lung cancers have been reported in nickel refinery workers who are exposed primarily to inhalation of nickel compounds with low aqueous solubilities, such as nickel subsulfide (αNi_3S_2) and nickel monoxide (NiO). Whether or not the soluble nickel compounds that are involved in electrolytic nickel refining processes should be viewed as carcinogenic is a controversial matter. Respiratory disorders, including rhinitis, bronchitis, and asthma, have been reported among workers exposed to nickel dusts and aerosols in nickel refineries, welding shops, and electroplating shops. Severe pneumonitis develops in persons accidently exposed to inhalation of nickel carbonyl ($Ni[CO]_4$), oftentimes attended by brain, adrenal, hepatic, and renal damage. Nickel toxicity has been reported following inadvertent parenteral administration of nickel compounds to patients via extracorporeal hemodialysis or nickel-contaminated drugs. Certain oral medications (e.g., disulfiram, used for aversion therapy of alcoholism) cause increased nickel concentrations in blood and urine by enhancing the gastrointestinal uptake of nickel. Sodium diethyldithiocarbamate is considered the chelating agent of choice for therapy of patients with acute nickel carbonyl poisoning. Based on animal experiments, D-penicillamine appears to be the most effective chelating agent for treatment of acute Ni^{2+} poisoning, since it promotes renal excretion of Ni^{2+} without enhancing nickel uptake into the brain and other tissues.

BIOLOGICAL MONITORING[10,11]

Urine and serum are the body fluids with widest applicability and greatest practicality for monitoring human exposures to nickel compounds from occupational, environmental, or iatrogenic sources. Urine is generally preferable to serum, since (1) collection of blood for nickel analysis necessitates venepuncture with plastic i.v. cannulae to avoid nickel contamination from steel needles, (2) in nonexposed subjects, nickel concentrations are considerably higher in urine than serum, enhancing analytical accuracy and precision, and (3) in exposed subjects, nickel concentrations are proportionately much higher in urine than serum. In hemodialysis patients, serum is obviously selected for nickel analysis, since little or no urine is available. Fecal nickel analysis provides the best index of oral exposures to nickel. Analyses of nickel in whole blood, saliva, milk, hair, and tissue specimens are useful for forensic or

Table 1
NICKEL CONCENTRATIONS IN BODY FLUIDS AND EXCRETA OF HEALTHY, NONEXPOSED PERSONS[5]

Specimen	Subjects	Nickel concentrations Mean ± SD	Range	Units
Whole blood	30 (15 m, 15 f)	0.34 ± 0.28	<0.05—1.05	μg/l
Serum	30 (15 m, 15 f)	0.28 ± 0.24	<0.05—1.08	μg/l
Urine (random collection)	34 (18 m, 16 f)	2.0 ± 1.5	0.5—6.1	μg/l
		2.0 ± 1.5	0.4—6.0	μg/g creatinine
Urine (24-h collection)	50 (24 m, 26 f)	2.2 ± 1.2	0.7—5.2	μg/l
		2.6 ± 1.4	0.5—6.4	μg/d
Feces (72-h collection)	10 (6 m, 4 f)	14.2 ± 2.7	10.8—18.7	μg/g dry wt
		258 ± 126	80—540	μg/d

research purposes. Under all circumstances, precautions are needed to minimize nickel contamination of specimens during collection, transport, storage, and processing. Urine should be voided directly into acid-washed plastic containers that have been tested for nickel contamination. Blood specimens should be collected through polyethylene cannulae into plastic acid-washed syringes; steel needles and rubber-stoppered tubes should be avoided.

ANALYTICAL METHODOLOGY[10-13]

Electrothermal atomic absorption spectrophotometry (EAAS) and differential pulse adsorption voltammetry (DPAV) are the analytical approaches that offer the sensitivity required for measurements of nickel concentrations in body fluids of nonexposed persons. The detection limits for nickel determinations by EAAS analysis with Zeeman background correction are approximately 0.45 μg/l for urine, 0.1 μg/l for whole blood, 50 ng/l for serum, and 10 ng/g (dry weight) for tissues. DPAV analysis with a dimethylglyoxime-sensitized mercury electrode provides detection limits as low as 1 ng/l for nickel concentrations in whole blood, urine, saliva, and tissue homogenates. However, DPAV analyses are more cumbersome, difficult, and time-consuming than EAAS procedures. Nickel concentrations in body fluids and tissues from nonexposed persons are listed in Table 1, based upon measurements in the author's laboratory.

REFERENCES

1. **Nriagu, J. O., Ed.,** *Nickel in the Environment,* John Wiley & Sons, New York, 1980, 1.
2. **Sunderman, F. W., Jr., Ed.-in-Chief,** *Nickel in the Human Environment,* Oxford Press, Oxford, 1984, 1.
3. **Merian, E., Ed.-in-Chief,** *Metalle in der Umwelt,* Verlag Chemie, Weinheim, 1984, 1.
4. **Sunderman, F. W., Jr., Nickel,** in *Disorders of Mineral Metabolism,* Vol. 1, Bronner, F. and Coburn, J. W., Eds., Academic Press, New York, 1981, 201.
5. **Sunderman, F. W., Jr., Nickel,** in *Handbook on Toxicity of Inorganic Compounds,* Seiler, H. G. and Siegel, H., Eds., Marcel Dekker, New York, in press.
6. **Anon.,** Health Assessment Document for Nickel, U.S. Environmental Protection Agency, Research Triangle Park, NC, 1985, 1.
7. **Brown, S. S. and Sunderman, F. W., Jr., Eds.,** *Nickel Toxicology,* Academic Press, London, 1980, 1.
8. **Brown, S. S. and Sunderman, F. W., Jr., Eds.,** *Progress in Nickel Toxicology,* Blackwell Scientific, Oxford, 1985, 1.
9. **Norseth, T.,** Nickel, in *Handbook on the Toxicology of Metals,* Vol. 2, Friberg, L., Nordberg, G. F., and Vonk, V. B., Eds., Elsevier, Amsterdam, 1986, 462.

10. **Sunderman, F. W., Jr., Aitio, A., Morgan, L. G., and Norseth, T.,** Biological monitoring of nickel, *Toxicol. Ind. Health,* 2, 17, 1986.
11. **Sunderman, F. W., Jr.,** Nickel, in *Biological Monitoring of Toxic Metals,* Clarkson, T., Friberg, L., Nordberg, G., and Sager, P., Eds., Plenum Press, New York, in press.
12. **Sunderman, F. W., Jr.,** Determination of nickel in body fluids, tissues excreta, and water, in *Environmental Carcinogens: Selected Methods of Analysis,* Vol. 8, O'Neill, I. K., Schuller, P., and Fishbein, L, Eds., IARC, Lyon, 1986, 319.
13. **Sunderman, F. W., Jr., Hopfer, S. M., and Crisostomo, M. C.,** Nickel analysis by atomic absorption spectrometry, *Methods Enzymol.,* in press.

SELENIUM

B. Sarkar

CHEMICAL PROPERTIES

Selenium (Se) is a steel-gray or purplish powder. It resembles sulfur in many respects but is a metalloid. Some of the chemical and physical properties of Se are listed in Table 1. The element exhibits variable valence from −2 to +6. Selenium dioxide, selenous acid, and the alkali metal selenites are water-soluble compounds. When an acid solution of a Se compound is reduced by metals such as Sn and Zn, a highly poisonous hydrogen selenide is generated. Selenic acid reacts vigorously with water. Se compounds of biological importance are selenomethionine and selenocysteine. Selenate can function as a bridging group in polynucleate ions. Selenomercaptides and ethers form complexes with Hg^{2+}. Selenopersulfide linkages have been reported in proteins.[1]

METABOLISM

Se has been recognized as essential in animals.[2] The gastrointestinal absorption of Se in the form of selenites, selenates, and seleniferous organic compounds is rapid and efficient, but metal selenides and elemental Se are poorly absorbed. Retention of Se in tissues depends on the dietary intake and existing body level of the metal. Accumulation in mammalian tissues is owing partially to Se replacing sulfur in cystine and methionine. They may exist in structural and functional proteins or as free selenoamino acids. Se is excreted in feces and urine and exhaled. Se is also excreted in sweat and milk. The elimination of organic Se is slow, while selenite is eliminated rapidly. Sulfate is a competitive antagonist and increases the urinary excretion of selenate, but not selenite. Little is known about Se homeostasis but it appears that within physiological limits animals are able to maintain nontoxic levels of Se by controlled excretory mechanisms. Estimated dietary threshold levels at which Se can cause physiologic and pathologic symptoms are listed in Table 2.

BIOCHEMISTRY

Se is thought to exert its biological effect through its antioxidant properties.[3] Most of the Se in erythrocytes is known to be in the form of glutathione peroxidase and this enzyme, together with glutathione, protects against the oxidative damage of hemoglobin.[4] Glutathione peroxidase prevents the accumulation of toxic peroxides within the cell. Antioxidants, such as tocopherols, are synergistic in preventing peroxide accumulation. Apart from lowering vitamin E requirements, Se has other biochemical functions, such as a sparing effect on biologically active sulfur compounds. It has been suggested that Se, particularly selenide, might have a role in the electron transfer function associated with mitochondria and smooth endoplasmic reticulum.[5] Glutathione peroxidase utilizes equivalents of reduced glutathione in the reduction of H_2O_2, lipid hydroperoxides, or sterol hydroperoxides. Glutathione peroxidase consists of four apparently identical subunits and has a molecular weight of about 80,000 Da. Both ovine and bovine red cells contain 4 g-atoms of Se per mole of glutathione peroxidase. The subject of Se incorporation into this enzyme is still a matter of controversy. There are reports that Se in reduced glutathione peroxidase is present as selenocysteine. Selenocysteine is incorporated into the enzyme by the normal translational pathway using a specific selenocystenyl-t RNA. However, there are reports of Se being incorporated by a post-translational process in which Se is attached enzymatically to an amino acid residue

Table 1
PROPERTIES OF SELENIUM COMPOUNDS

	Formula	Appearance and odor	Melting point (°C)	Boiling point (°C)
Selenium	Se	Black, gray, or red odorless solid	150	585
Sodium selenite solid	Na_2SeO_3	Colorless and odorless	710 (decomposes)	Decomposes
Sodium selenate	Na_2SeO_4	Colorless and odorless solid	Decomposes	Decomposes
Selenium dioxide	SeO_2	Colorless and odorless solid	340	315 (sublimes)
Selenium oxychloride	$SeOCl_2$	Colorless and yellow liquid	10.8	176

Table 2
DIETARY THRESHOLD LEVELS OF SELENIUM

Characterization	ppm
Deficiency	<0.03
Physiological	0.03—0.4
Pathological	0.4—3.0
Clinical	3—20
Lethal	>20

Reproduced from Venugopal, B. and Luckey, T. D., Eds., *Metal Toxicity in Mammals, Vol. 2,* Plenum Press, New York, 1978.

already present in the molecule. This enzyme is thought to reduce hydrogen peroxide which may react with superoxide anion O_2 to form a highly reactive oxygen species such as OH, which may also degrade and damage the phospholipid constituents of biological membranes.

DEFICIENCY

Se deficiency causes liver necrosis in rats. Se-vitamin E deficiency also causes oxidative diathesis in chicks, mulberry heart disease in swine, and some forms of muscular dystrophy in lambs. Animals fed Se-deficient diets showed rapid decreases in glutathione peroxidase activity that are correlated with the appearance of Se deficiency symptoms. Se deficiency is considered to be responsible for Keshan disease in the People's Republic of China, which appears to be the first expression of such deficiency in humans.

TOXICITY

Se toxicity is caused by both organic and inorganic forms of the element. In man, early signs of acute Se poisoning are fever, vomiting, and somnolence. The blood pressure falls and labored breathing is accompanied by tetanic and clonic spasms. Death is caused by respiratory failure from prolonged convulsions caused by the toxic action on the nervous system. In animals blind staggers, ataxia, elevation of body temperature, labored respiration, and death from respiratory failure are the acute symptoms. Se has significant toxic effects on reproduction, calcification, and teratogenesis. There are also reports of Se-caused carcinogenesis. Se toxicity is generally attributed to its interference with sulfur metabolism and function. Se affects enzyme systems associated with cellular respiration and replaces thiol groups by SeH groups in some of the dehydrogenase enzyme systems, with subsequent inhibition of the dehydrogenase. Selenites react with sulfhydryl compounds such as cysteine or coenzyme A, forming stable selenotrisulfides and rendering these essential cofactors unavailable. Functional and structural proteins may undergo conformational changes if selenotrisulfide cross-links are formed between SH groups and selenites. Also, selenoamino acids bind to sulfhydryl groups of functional proteins forming complexes which prevent reversible redox reactions.

ANALYTICAL METHOD

The reference method used to analyze Se in both biological and environmental samples, and the most commonly used, is the fluorometric method.[6] Se can also be measured by flameless atomic absorption procedures.

REFERENCES

1. **Ganther, H. E.,** Reduction of the selenotrisulfide derivative of glutathione to a persulfide analog by glutathione reductase, *Biochemistry,* 10, 4089, 1971.
2. **Schwarz, K. and Foltz, C. M.,** Factor 3 activity of selenium compounds, *J. Biol. Chem.,* 233, 245, 1958.
3. **Tappel, A. L. and Caldwell, K. A.,** Selenium, in *Biomedicine,* Mutts, O. H., Oldfield, J. E., and Weswig, P. H., Eds., AVI Publishing, Westport, CT, 1967, 345.
4. **Rotruck, J. T., Pope, J. T., Ganther, H. E., Swanson, A. B., Hafeman, D. G., and Hoekstra, G. W.,** Selenium: biochemical role as a component of glutathione peroxidase, *Science,* 179, 588, 1973.
5. **Caygill, C. P. J., Lucy, J. A., and Diplock, A. T.,** The effect of vitamin E on the intracellular distribution of the different oxidation states of selenium in rat liver, *Biochem. J.,* 125, 407, 1971.
6. **Olson, O. E., Palmer, I. S., and Whitehead, I. E.,** Determination of selenium in biological materials, *Methods Biochem. Anal.,* 21, 39, 1973.
7. **Venugopal, B. and Luckey, T. D., Eds.,** *Metal Toxicity in Mammals,* Vol. 2, Plenum Press, New York, 1978.

URANIUM

Michael A. Moss

SOURCES

Uranium is ubiquitous in nature, yet it has no known metabolic function. Naturally occurring uranium (^{nat}U) is a mixture of three radionuclides: ^{238}U (~99.25%), ^{235}U (~0.7%), and ^{234}U (~0.005%). These decay by α emission with half-lives ($T^1/_2$) of 4.5 × 10^9, 0.7 × 10^9, and 0.2 × 10^6 years, respectively.[1] Natural uranium in this combination has an activity of about 0.67 pCi/μg.[2] It is found in granites and various other mineral deposits such as uraninite, carnotite, and pitchblende,[3,4] Uranium exists in valence states +2 to +6, but the hexavalent form is most common in nature, existing as the uranyl ion (UO_2^{2+}).

Uranium is chemically extracted from the ore and converted to gaseous uranium hexafluoride (UF_6). $^{235}UF_6$ is enriched through membrane diffusion and is converted to uranium dioxide ($^{235}UO_2$), the final reactor fuel product.[1] Besides its use in nuclear energy plants, uranium is also used in nuclear weapons and isotope production. ''Depleted'' uranium (^{235}U content ~0.2%) is used as a ballast or counterweight, in inertial guidance devices, gyro compasses, and as a shielding material.[4] Uranyl compounds are also used as photographic intensifiers, in dry-copying ink, and as staining pigments in the ceramics/glass industry. Orange pottery, in particular, may contain significant amounts of uranium.[5]

Uranium may enter drinking water from naturally occurring deposits or as a result of human activity, such as mill tailings.[2,6] Phosphate fertilizers may contain an average of 150 μg U per gram and can also contribute uranium to groundwater.[7] Average uranium content of municipal tap waters has been reported as 4.8 to 11.4 ng/l (Japan) to 32 ng/l (New York City).[8,9] However, private domestic supplies have been identified in Canada with uranium levels up to 700 μg/l.[10,11] Typical daily uranium intake from water has been estimated as 0.035 μg for Reference Man in the U.S.[12] Coal and fuel combustion are major sources of uranium contamination in air, but this is a minor source of exposure, amounting to only about 2 ng/d.[13] Vegetables and cereals normally contribute most heavily to the daily intake of uranium.[1] Average per capita intake from food has been reported as 1.3 μg/d (U.S.) and 1.5 μg/d (Japan).[8,9,14] Plant (vegetable) produce generally contains a tenfold higher uranium concentration than animal produce, the highest concentrations being found in root and stem vegetables.[15]

PHARMACOKINETICS

Estimates for the uptake coefficient of uranium from the gastrointestinal (GI) tract to blood have ranged from 0.01 to 30%.[14] Absorption from the GI tract varies with the solubility of the uranium compound,[1] and is also markedly species dependent.[12,16] From a recent review of six animal studies and four human studies, a best estimate for man of 1.4% at environmental levels was derived. It remains to be determined whether uptake of uranium is greater from water than from food.[17] Water-soluble uranium compounds have been shown to be absorbed by the skin to about 0.1%.[18] Mobilization and dissolution of inhaled particles show large species differences.[19] Animal models investigating inhaled uranium dioxide showed halftimes ranging from 300 d for the rat and from 420 to 660 d for the dog and monkey.[20] The dissolution halftimes of uranium compounds in simulated lung fluid have been shown to vary from 0.01 d to more than 10,000 d, depending upon the physical and chemical form of the uranium.[21] Estimates of transfer coefficients have been derived for blood-bone (0.2), and blood-kidney (0.12).[22] Reported values for uranium distribution amongst lung, liver, bone, and kidney in the general population vary widely, and may reflect differences in analytical technique.[23] Estimates of average adult total body uranium content have ranged from 20 to about 100 μg.[17,24] Reference Man has been estimated to contain a total of 38 μg of uranium, of which 66% or 24.9 μg is in the skeleton.[12] Uranium predominates in the mammalian organism as the water-soluble hexavalent (uranyl) form.[14,25] This forms soluble complexes with bicarbonate ions and proteins. A nondiffusible uranyl-albumin complex is formed in equilibrium with a diffusible ionic uranyl bicarbonate complex (UO_2HCO^{3+}). The uranyl compounds have a high affinity for phosphate, carboxyl, and hydroxyl groups, and will combine readily with proteins and nucleotides, forming stable complexes.[1,26] The uranyl bicarbonate complex is stable, but will dissociate under conditions of reduced pH, such as in bone and in the kidney tubule.[5] The uranyl ion replaces calcium in the hydroxy-apatite complex of bone crystals.[1] Hexavalent uranium is rapidly excreted by the kidney; the excretory rate is partly dependent upon the pH of the tubular urine.[1] At high pH, most of the uranyl complex is stable and is excreted in the urine. If the pH is low, the complex dissociates to a variable degree, and the uranyl ion may then interact with cellular proteins in the tubular wall. This results in reduced uranyl excretion, and the protein binding can impair tubular function. Biphasic urinary excretion has been observed following an acute single dose. There is an initial rapid phase during which 70% of the absorbed dose is excreted during the first 24 h, followed by a much slower phase with a half-life of many months.[1,27] The overall elimination half-life of uranium under conditions of normal daily intake has been estimated to be between 180 and 360 d.[1]

TOXICITY

Radiobiological effects of ingested or parenteral ^{nat}U have not been observed; nevertheless, skeletal cancer is regarded as the major potential radiological effect. Experimental evidence of its carcinogenicity is restricted to highly insoluble or enriched uranium compounds delivered by inhalation or injection. The naturally occurring mixture of uranium isotopes has such a long half-life and a correspondingly low specific activity, that its biological action is predominantly that of a nonradioactive element. In the case of ^{nat}U, chemical toxicity is a more restrictive health criterion than the risk of radiation-induced bone cancer, unless the cancer risk limit is set at an unrealistically low level. Lung cancer has been induced in dogs by inhalation of highly insoluble $^{nat}UO_2$ dust for 2 to 5 years. Calculated radiation doses to dog lungs were 200 to 500 rads, while the dose to tracheobronchial lymph nodes was 10,000 rads. Monkeys showed pulmonary fibrosis only at lung doses of ~900 rads.[20,28] Sarcomas were induced in rats (33 animals in each group) following injections of metallic uranium. (Intrafemoral deposition of 50 mg U resulted in a 33% induction rate, and intrapleural injection of 300 mg U a 6% rate.[29]) These observations do not seem relevant to environmental exposures to uranium, where chemical toxicity is still assumed to be the limiting health factor.[12]

Uranium is a classic experimental nephrotoxin, and nephritis is the primary chemically induced health effect of uranium in animals and humans.[14] Lesions are seen in the kidney at both glomerular and tubular levels.[30] These effects have been seen down to a single dose level of ~0.1 mg/kg (i.v.), but not less.[14,31] In experimental animals, the acute renal injury caused by uranium is manifest by morphological changes in the proximal convoluted tubules.[1] Cells of the proximal convoluted tubules show apical swelling, increase and enlargement of cytoplasmic vacuoles, and some loss of the brush-border microvilli. At sublethal doses, regeneration of the injured epithelium begins at 2 to 3 d. The new epithelial lining differs morphologically from normal tubular epithelium.[1] These ultrastructural changes have been described for both glomerular and tubular cells following acute exposure.[32,33] The biochemical functional changes are consistent with damage to the distal two thirds of the proximal tubule. There is evidence from dog studies that the uranyl ion competes with ATP for a binding site on the Ca^{2+} and Mg^{2+} ATPase, and that this may be a factor in its renal toxicity.[34] Diabetes in the rat confers partial protection against the nephrotoxicity of uranyl nitrate; this may be due to decreased binding and impaired transport of the uranyl ion into the tubular cells.[35] Some of the early case histories relate the use of uranyl nitrate during the last century to treat diabetes.[36] There is evidence to suggest that tolerance may develop to repeated uranium exposure,[37-39] and this has been reviewed.[25] The proteinuria of acute uranyl toxicity is characterized by both albumin and low-molecular-weight classes, presumed to reflect a partial glomerular lesion combined with failure of tubular reabsorption.[14,31,39] The threshold concentration for renal injury in man remains uncertain, but is likely to be above 0.6 μg/g, and may be as high as 3 μg/g.[12] For metabolic modeling purposes, a reasonably conservative estimate at this time is 1.0 μg U per gram renal tissue.[12]

INDICES OF EXPOSURE

The chronic health effects of natural environment exposure to uranium have received little attention. Clinical studies from Nova Scotia reported 324 persons exposed to variable amounts of uranium in their drinking water (up to 0.7 mg U per liter) supplied from private wells.[10,11] There was no pattern of overt disease or symptomatic complaints which could be attributed to uranium exposure. Various parameters were studied as potential indices of quantitative exposure or subclinical toxicity. Urinary uranium excretion was studied; this index is used by most authorities to assess accidental exposures in the occupational setting.[40,41] It was shown to be unreliable as an index of quantitative uranium exposure, appearing to reflect only recent intake. Hair uranium content has been reported in the context of human occupational exposure and, although a sensitive potential index, was found to be subject to interference from certain proprietary shampoos.[42] Hair uranium was assessed as an index of quantitative exposure in the Nova Scotia studies and showed a strong association with exposure.[10] This was particularly marked in females, and it was later shown that lifestyle factors such as hairwashing greatly influence the hair uranium content. Hair was found to be capable of adsorbing and accumulating uranium by a factor of 10^3 from water in which it is soaked.[11] Unless washing procedures can be developed to remove exogenous (adsorbed) uranium prior to analysis, this index appears unreliable. Plasma levels would be expected to reflect only recent intake of uranium. The most reliable indices of cumulative exposure are probably skeletal or renal tissue concentrations, which are unfortunately not readily accessible for screening purposes. Whole body counting has been applied in occupational exposure. The technique is relatively insensitive, capable of detecting a burden in the range of 10 mg ^{nat}U *in vivo* (or 1 mg of enriched uranium); exposure to environmental levels may not be reliably assessed.[43]

A sensitive index of renal proximal tubular function might serve as an index of subclinical toxicity. Patterns of aminoaciduria following occupational uranium exposure with urinary excretion up to 50 μg U per liter have been reported.[44] Uranium appears far more potent in producion of aminoaciduria than lead or mercury, but shows no consistent pattern. Urinary β_2-microglobulin excretion (β_2m) was reported in the Nova Scotia studies; a trend effect was observed with increasing uranium exposure, with evidence of reversibility for individuals who had stopped using the uranium-contaminated drinking water.[10,11] This marker deserves further study, but has the disadvantage of rapid degradation under conditions of low urinary pH. Retinol-binding protein is another candidate marker of proximal tubular function.[11]

ANALYTICAL CONSIDERATIONS

Several methods are available for the determination of uranium, but most are not sufficiently sensitive for the levels normally encountered in biological systems. Neutron activation analysis is not widely available. Using an adaptation of a standard fluorometric procedure,[45,46] a method for processing biological samples has been employed.[11] It is based upon the fluorescence produced when uranium, fused into beads of sodium fluoride, is illuminated by UV light. The use of a sample fusion rotary burner allows a number of samples to be processed simultaneously in a uniform manner.[47] Tissue samples can be processed following nitric acid digestion.[11] and amounts as low as 1 μg U per liter can be detected.

REFERENCES

1. **Berlin, M. and Rudell, B.,** Uranium, in *Handbook on the Toxicology of Metals,* Friberg, L. et al., Eds., Elsevier/North-Holland, Amsterdam 1979, chap. 40.
2. **Cothern, C. R. and Lappenbusch, W. L.,** Occurrence of uranium in drinking water in the U.S., *Health Phys.,* 45 (1), 89, 1983.
3. **Roessler, C. E., Smith, Z. A., Bolch, W. E., and Prince, R. J.,** Uranium and radium-226 in Florida phosphate materials, *Health Phys.,* 37, 269, 1979.
4. **Weast, R. C., Ed.,** Uranium, in *CRC Handbook of Chemistry and Physics,* 66th ed., CRC Press, Boca Raton, FL, 1985, B40.
5. **Venugopal, B. and Luckey, T. D.,** Uranium, in *Metal Toxicity in Mammals,* Vol. 2, Plenum Press, New York, 1978, 161.
6. **Dressen, D. R., Williams, J. M., Marple, M. L., Gladney, E. S., and Perrin, D. R.,** Mobility and bioavailability of uranium mill tailings constituents, *Environ. Sci. Technol.,* 16 (10), 702, 1982.
7. **Spalding, R. F. and Sackett, W. M.,** Uranium in runoff from the Gulf of Mexico distributive province: anomalous concentrations, *Science,* 175, 629, 1972.
8. **Nozaki, T., Ichikawa, M., Sasuga, T., and Inarida, M. J.,** Neutron activation analysis of uranium in human bone, drinking water and daily diet, *J. Radioanal. Chem.,* 6, 33, 1970.
9. **Welford, G. A. and Baird, R.,** Uranium levels in human diet and biological materials, *Health Phys.,* 13, 1321, 1967.
10. **Moss, M. A., McCurdy, R. F., Dooley, K. C., Givner, M. L., Dymond, L. C., Slayter, J. M., and Courneya, M. M.,** Uranium in drinking water — report on clinical studies in Nova Scotia, in *Chemical Toxicology and Clinical Chemistry of Metals,* Brown, S. S. and Savory, J., Eds., Academic Press, London, 1983, 149.
11. **Moss, M. A.,** Chronic Low Level Uranium Exposure via Drinking Water — Clinical Investigations in Nova Scotia, MSc. thesis, Dalhousie University, Halifax, Canada, 1985.
12. **Wrenn, M. E., Durbin, P. W., Howard, B., Lipsztein, J., Rundo, J., Still, E. T., and Willis, D. L.,** Metabolism of ingested U and Ra, *Health Phys.,* 48, (5), 601, 1985.
13. United Nations, Sources and Effects of Ionizing Radiation, U.N. Rep. #E.77.IX.1, United Nations, New York, 1977.
14. **Hursh, J. B. and Spoor, N. L.,** Data on man, in *Handbook of Experimental Pharmacology,* Vol. 36, Hodge, H. C., Stannard, J. N., and Hursh, J. B., Eds., Springer-Verlag, Berlin, 1973, 197.
15. **Tracy, B. L., Prantl, F. A., and Quinn, J. M.,** Transfer of ^{226}Ra, ^{210}Pb and uranium from soil to garden produce, *Health Phys.,* 44, 469, 1983.
16. **Tracy, B. L., Quinn, J. M., Gilman, A. P., Villeneuve, D. C., Secours, V. E., Valli, V. E., and Yagminas, A. P.,** The metabolism of ingested uranium in mammals, in *Proc. 7th Int. Congr. of Radiation Research,* Broerse, J. J. et al., Eds., Martinus Nijhoff, The Hague, 1983, E5.
17. **Hamilton, E. I.,** The concentration of uranium in man and his diet, *Health Phys.,* 22, 149, 1972.
18. **Orcutt, J. A.,** *Pharmacology and Toxicology of Uranium Compounds,* Voegtlin, C. and Hodge, H. C., Eds., McGraw-Hill, New York, 1949, 377.
19. **Spoor, N. L. and Hursh, J. B.,** Protection criteria, in *Handbook of Experimental Pharmacology,* Vol. 36, Hodge, H. C., Stannard, J. N., and Hursh, J. B., Eds., Springer-Verlag, Berlin, 1973, 241.
20. **Leach, L. J., Yuile, C. L., Hodge, H. C., Sylvester, G. E., and Wilson, H. B.,** A five year inhalation study with natural uranium dioxide (UO_2) dust. II. Postexposure retention and biologic effects in the monkey, dog and rat, *Health Phys.,* 25, 239, 1973.
21. **Kalkwarf, D. R.,** Dissolution rates of uranium compounds in simulated lung fluid, *Sci. Total Environ.,* 28, 405, 1983.
22. International Commission on Radiological Protection, *ICRP Publ. #30,* Pergamon Press, Elmsford, NY, 1979.

23. **Wrenn, M. E. and Singh, N. P.,** Comparative distribution of uranium, thorium and plutonium in human tissues of the general population, in *Natural Radiation Environment,* Vohra, K.C. et al., Eds., Wiley Eastern, New Delhi, 1982, 144.
24. **Welford, G. A., Baird, R., and Fisenne, I. M.,** Concentration of natural uranium in the human body, in Proc. 10th Midyear Topical Symp. of the Health Physics Society, Troy, NY, 1976, 239.
25. **Yuile, C. L.,** Animal experiments, in *Handbook of Experimental Pharmacology,* Vol. 36, Hodge, H. C., Stannard, J. N., and Hursh, J. B., Eds., Springer-Verlag, Berlin, 1973, 165.
26. **Adams, N. and Spoor, N. L.,** Kidney and bone retention functions in the human metabolism of uranium, *Phys. Med. Biol.,* 19 (4), 460, 1974.
27. **Bassett, S. H., Frenkel, A., Cedars, N., Van Alstine, H., Waterhouse, C., and Cusson, K.,** *Rep. UR* -37, USAEC, 1948.
28. **Leach, L. J., Maynard, E. A., Hodge, H. C., Scott, J. K., Yuile, C. L., Sylvester, G. E., and Wilson, H. B.,** A five year inhalation study with natural uranium dioxide (UO_2) dust. I. Retention and biologic effect in the monkey, dog and rat, *Health Phys.,* 18, 599, 1970.
29. **Hueper, W. C., Zuefle, J. H., Link, A. M., and Johnson, M. G.,** Experimental studies in metal cancerigenesis. II. Experimental uranium cancers in rats, *J. Natl. Cancer Inst.,* 13, 291, 1951.
30. **Cothern, C. R., Lappenbusch, W. L., and Cotruvo, J. A.,** Health effects guidance for uranium in drinking water, *Health Phys.,* 44 (Suppl. 1), 377, 1983.
31. National Academy of Science, *Drinking Water and Health,* Vol. 3, National Academy of Sciences, Washington, D.C., 1980.
32. **Stone, R. S., Bencosme, S. A., Latta, H., and Madden, S. C.,** Renal tubular fine structure studied during reaction to acute uranium injury, *Arch. Pathol.,* 71, 160, 1961.
33. **Haley, D. P.,** Morphologic changes in uranyl nitrate-induced acute renal failure in saline – and water – drinking rats, *Lab. Invest.,* 46 (2), 196, 1982.
34. **Thompson, J. D. and Nechay, B. R.,** Inhibition by metals of a canine renal calcium, magnesium-activated adenosine triphosphatase, *J. Toxicol. Environ. Health,* 7, 901, 1981.
35. **Roth, D., Morales, J., Teixeira, R. B., Kelley, J., Alpert, H., Pardo, V., and Vaamonde, C. A.,** *Kidney Int.,* 21, 223, 1982.
36. **Hodge, H. C.,** A history of uranium poisoning (182401942), in *Handbook of Experimental Pharmacology,* Vol. 36, Hodge, H. C., Stannard, J. N., and Hursh, J. B., Eds., Springer-Verlag, Berlin, 1973, 5.
37. **Maynard, E. A. and Hodge, H. C.,** *Pharmacology and Toxicology of Uranium Compounds,* Veogtlin, C. and Hodge, H. C., Eds., McGraw-Hill, New York, 1949, 309.
38. **Durbin, P. W. and Wrenn, M. E.,** Metabolism and effects of uranium in animals, Conference on Occupational Health Experience with Uranium, Arlington, VA, ERDA 93, U.S. Energy Research and Development Administration, Washington, D.C., 1975, 67.
39. **Campbell, D. C.,** The Development of an Animal Model with Which to Study the Nephrotoxic Effects of Uranium-Contaminated Drinking Water, MSc. thesis, Dalhousie University, Halifax, Canada, 1985.
40. Nuclear Regulatory Commission (U.S.), Regulatory Guide 8.22, Div. of Technical Information Document Control, NRC, Washington, D.C., 1978.
41. Atomic Energy Control Board (Canada), Reg. Guide P-5, 1981.
42. **Bentley, K. W. and Wyatt, J. H.,** Quantitative determination of uranium in human hair, *Environ. Res.,* 21, 407, 1980.
43. **Donoghue, J. K., Dyson, E. E., Hislop, J. S., Leach, A. M., and Spoor, N. L.,** Human exposure to natural uranium, *Br. J. Ind. Med.,* 29, 81, 1972.
44. **Clarkson, T. W. and Kench, J. E.,** Urinary excretion of amino acids by men absorbing heavy metals, *Biochem. J.,* 62, 361, 1956.
45. **Ingles, J. C. and Zimmerman, J. B.,** Uranium-detection by fluorometry in fused fluoride beads, in Energy, Mines and Resources Methods Manual, Government of Canada, 1979, 236.
46. American Society for Testing and Materials, Microquantities of uranium in water by fluorimetry, in *Annual Book of ASTM,* Philadelphia, 1973, 630.
47. **Ashbrook, A. W.,** The construction and operation of a rotary burner for fluorometric uranium analysis, Mines Branch Technical Bull. TB-137, Information Canada, Ottawa, 1971.

ZINC

John Patrick

Zinc is the most studied of the trace elements known to be necessary in man. It has a remarkable range of effects which are summarized in Table 1. As yet there is still no universal agreement about the mechanisms underlying the symptomatology of zinc deficiency, but two mechanisms, not necessarily exclusive, have the most experimental backing. The first invokes changes in zinc metalloenzymes as the mechanism.[1] Examples of zinc metalloenzymes can be found in all enzyme classes. Over 200 zinc metalloenzymes are known. The role of zinc within these enzymes also varies, being either catalytic, structural, regulatory, or unknown. The alternative or complementary mechanism proposes that zinc interacts directly with the cell membrane; the arguments for the two views have been reviewed.[2]

CHARACTERISTICS OF ZINC DEFICIENCY

There are multiple reviews of zinc deficiency available.[3-5] Zinc deficiency is usually a slowly evolving process, but can, on occasion, evolve quite rapidly, especially in the presence of diarrheal illness. The cure of zinc deficiency can be extremely rapid and is one of the most gratifying therapeutic activities that medicine affords. Patients will say that they feel better in hours and the skin rash will almost be visibly improving within 24 h of adequate repletion. Until recently, in practical terms, the clinical setting in which to suspect zinc deficiency was when circumoral and circumanal eczematous rashes developed in patients receiving total parenteral nutrition (TPN), but now that the addition of trace element supplements to TPN are routine, this problem is unlikely to occur. The important therapeutic situation in which to think about deficiency is when fungal or candida infections develop which do not respond to standard therapy. Provision of zinc can dramatically improve the patient's capacity to kill the fungi. The way this knowledge was acquired is quite interesting. For many years researchers had attempted to determine why protein-energy malnutrition was associated with increased risks of infection without any very convincing results. It then became apparent that one of the features of protein-energy malnutrition, thymic atrophy, was rapidly corrected by zinc supplementation, and associated with this there was an improvement in cell-mediated immunity. This had been demonstrated in A46 cattle in 1971 and subsequently in children with acrodermatitis, but its significance went largely unnoticed at that time.[6] Even more remarkable has been the demonstration by Beach et al. that mice, whose mothers were fed a zinc-deficient diet during gestation, showed immune deficiency through to the F3 generation despite adequate dietary zinc.[11]

Rapid weight change can be a precipitating factor for zinc deficiency. The mechanism appears to be that when catabolism ceases and anabolism starts zinc needs escalate acutely. Lean tissue contains approximately 200 mg Zn/kg dry weight or 50 mg/kg wet weight. Absorption from the diet is usually around 60%, but may be as low as 20%. Thus, with rapid weight gain, dietary needs may reach 250 mg/d or 25 times the usual recommended intake. Thus, relative zinc deficiency may develop while positive zinc balance is occurring. As a general principle, the induction of trace element deficiency in animals requires the use of growing animals, so it is not surprising that trace element deficiency in clinical medicine often develops when anabolism begins after a serious illness. Zinc excretion is mainly via the gut; diarrhea may, therefore, be another precipitant of zinc deficiency.[12]

Continued zinc deficiency will ultimately cause anabolism to cease; thus, the other important clinical situation in which to think about zinc deficiency is in unexplained growth failure, particularly with chronic diseases like Crohn's disease, even if diarrhea is not a major feature.

Table 1
EFFECTS OF ZINC DEFICIENCY

	Ref.
Growth failure	3, 4
Skin rash	23
Failure of sexual maturation	3
Taste abnormalities	24
Abnormalities of fetal development	4, 19—21
Impaired cell-mediated immunity	6—10, 17
Impaired vitamin A metabolism	27
Impaired appetite control	28
Altered central nervous system function	29, 30

All these reports concern clinically significant zinc deficiency, but the major health concern is whether marginal zinc deficiency, which certainly exists in North America, is of importance. The group most likely to be vulnerable are pregnant mothers from disadvantaged socioeconomic groups. Recent experimental work from Hurley's group gives some guidance about what to expect.[13-15] Female rhesus monkeys were fed a diet which reduced plasma zinc by 20 to 30% but did not produce overt signs of zinc deficiency in non-pregnant animals. The pregnant animals fed the same zinc-deficient diet (4 g/g) did develop skin changes and anorexia in the second half of pregnancy and post partum. Anorexia, more marked in some animals than others, reduced the fall in plasma zinc, presumably via zinc release from catabolized tissue. Some zinc-deficient animals developed a microcytic anemia and their lymphocyte count dropped. There were variable reductions in response to mitogens, but immunoglobulins and complement were not affected. There was some evidence of impaired transport of vitamin A from the liver in the animals with the lowest plasma zinc level. Pregnancy losses and complications were higher than in controls, but the same as pair-fed zinc-sufficient controls. Symmetrical growth retardation was more common in the zinc-deficient group than in either ad libitum or pair-fed controls. As usual males were most vulnerable; seven of eight were affected. The problems of interpreting plasma zinc were well illustrated, because in the controls there was a positive correlation between birth weight and plasma zinc and in the zinc-deficient group there was a negative correlation. Unlike severe Zn deficiency, there were no teratogenic effects.

Alcoholism is another major problem in which zinc may play a role. Zinc is a constituent of alcohol dehydrogenase and in zinc deficiency alcohol levels may remain higher for longer.[16] The possibility that zinc has an important role in the control of memory could lead to another explosion of interest in zinc.[17]

MEASUREMENT OF ZINC DEFICIENCY

This subject has been recently reviewed.[18] In general, simple dietary deficiency will lead to a fall in plasma zinc as negative balance is induced. However, in human clinical settings plasma zinc is very unreliable, because all sorts of "stress" situations are associated with rapid decrements in plasma zinc caused not by deficiency, but by redistribution.[19] There is some evidence that the coincidental measurement of serum metallothionein will distinguish between "stress"-induced hypozincemia and true deficiency.[20] In the former, metallothionein is still detectable, while in true deficiency it is not. Assays for zinc metallothionein are still confined to research studies at present. Other measures of zinc status have been proposed including hair, muscle, red cell, leukocyte zinc, and tolerance tests. Hair can be useful in survey work, but extremely careful sampling and preparation are essential.[21] Hair is of no value in acute clinical settings, because hair growth rates change in serious illnesses;

hence, changes may reflect the reference measure rather than zinc status.[22] Red cell zinc is more often an inverse reflection of iron status than a measure of zinc status. Muscle and leukocyte assays appear to be useful, but are demanding techniques whose use in a clinical laboratory can hardly be justified. This is especially so because zinc is a very nontoxic metal, so that clinical suspicion followed by a therapeutic trial in suspected zinc deficiency rapidly provides definitive evidence. Although 10 mg is an average intake, 50 times that amount can be taken with impunity for a short period. There is evidence that long-term administration of zinc supplements to zinc-replete subjects may impair the function of the immune system[23] just as zinc deficiency does. This is the usual trace element phenomenon of a normal requirement, with deficiency at one extreme and toxicity at the other.

METHODS OF ANALYSIS

Atomic absorption spectrophotometry is the most frequently used technique. Very great care is needed to ensure that no contamination occurs during sample collection and preparation. Some of the tubes specifically sold for collection of samples for trace element analysis still have enough zinc in them to contaminate the small blood volume obtained from neonates. Almost all rubber products contain large amounts of zinc. The only safe approach is to use a deionized water blank which goes through the same process as the sample. The actual measurement of zinc is straightforward; a double-beam system with background subtraction is best. If very low levels in small specimens must be measured, then a flameless system preferably with Zeeman Background correction is required.[24]

Assays of the activity of zinc metalloenzymes like alkaline phosphatase have also been proposed as measures of zinc status, but they are subject to too many other sources of variation, apart from zinc deficiency, to be of value in most clinical settings. Alkaline phosphatase also changes only slowly with zinc deficiency. Nucleoside phosphorylase changes much more rapidly and may turn out to be clinically useful.[25]

CONCLUSION

Growth failure, typical rash, and impaired cell-mediated immunity are the major clinical situations in which zinc deficiency should be considered. At present, a therapeutic trial ol zinc is the best investigation, plasma zinc being almost uninterpretable in the context of acute illness, although for more chronic situations it may be helpful. If a metallothionein assay becomes clinically feasible, then it should be possible to separate nonspecific hypozincemia from that due to zinc deficiency. Atomic absorption spectrophotometry remains the most useful analytical method.

REFERENCES

1. **Galdes, A. and Vallee, B. L.,** *Metal Ions in Biological Systems,* Sigel, H., Ed., Marcel Dekker, New York, 1983, 1.
2. **Bettger, W. J. and O'Dell, B. L.,** A critical physiological role for zinc in the structure and function of biomembranes, *Life Sci.,* 28, 1425, 1980.
3. **Rennert, O. M. and Chen, W. Y.,** *Metabolism of Trace Metals in Man,* CRC Press, Boca Raton, FL, 1984.
4. **Hurley, L. S.,** Teratogenic aspects of mangenese, zinc and copper nutrition, *Physiol. Rev.,* 61, 249, 1981.
5. **Prasad, A. S.,** *Clinical, Biochemical, and Nutritional Aspects of Trace Elements,* Prasad, A. S., Ed., Alan R. Liss, New York, 1982.
6. **Brummerstedt, E., Flagstad, T., Basse, A., and Andresen, E.,** The effect of zinc on calves with hereditary thymus hyperplasia (lethal trait A46), *Acta Pathol. Microbiol. Scand. Sect.,* 79, 686, 1971.

7. **Golden, M. H. N., Jackson, A. A., and Golden, B. E.,** Effect of zinc deficiency on the immune responses of young adult A/J mouse, *J. Nutr.*, 107, 1889, 1977.
8. **Fraker, P. J., Haas, S. M., and Luecke, R. W.,** Effect of zinc deficiency on the thymus of recently malnourished children, *Lancet,* 2, 1057, 1977.
9. **Golden, M. H. N., Golden, B. E., Harland, P. S. E. G., and Jackson, A. A.,** Zinc and immunocompetence in protein energy malnutrition, *Lancet,* 1, 1226, 1978.
10. **Chandra, R. K.,** Nutrition, immunity, and infections: present knowledge and future directions, *Lancet,* 1, 688, 1983.
11. **Beach, R. S., Gershwin, M. E., and Hurley, L. S.,** Gestational zinc deprivation in mice: persistence of immunodeficiency through three generations, *Science,* 218, 469, 1982.
12. **Wolman, S. L., Anderson, G. M., Marliss, E. B., and Jeejeebhoy, K. N.,** Zinc in total parenteral nutrition: requirements and metabolic effects, *Gastroenterology,* 76, 458, 1979.
13. **Golub, M. S., Gershwin, M. E., Hurley, L. S., Baly, D. L., and Hendricks, A. S.,** Studies of marginal zinc deprivation in rhesus monkeys. I. Influence on pregnant dams, *Am. J. Clin. Nutr.*, 39, 265, 1984.
14. **Golub, M. S., Gershwin, M. E., Hurley, L. S., Baly, D. L., and Hendricks, A. S.,** Studies of marginal zinc deprivation in rhesus monkeys. II. Pregnancy outcome, *Am. J. Clin. Nutr.*, 39, 879, 1984.
15. **Baly, D. L., Golub, M. S., Gershwin, M. E., and Hurley, L. S.,** Studies of marginal zinc deprivation in rhesus monkeys. III. Effects on vitamin A metabolism, *Am. J. Clin. Nutr.*, 40, 199, 1984.
16. **Smith, J. C.,** The vitamin A-zinc connection: a review, *Ann. N.Y. Acad. Sci.*, 62, 1980.
17. **Howell, G. A. and Kaswskis, E. J., Eds.,** *The Neurobiology of Zinc, Parts A and B, Alan R. Liss,* New York, 1984.
18. **Patrick, J. and Dervish, C.,** Leukocyte zinc in the assessment of zinc status, *CRC Crit. Rev. Lab Sci.*, 20, 94, 1984.
19. **Falchuk, K. H.,** Effect of acute disease and ACTH on serum zinc protein, *N. Engl. J. Med.*, 296, 1129, 1977.
20. **Mehra, R. K. and Bremner, J.,** Development of a radioimmunoassay for rat liver metallothionein-I and its application to the analysis of rat plasma and kidneys, *Biochem. J.*, 213, 459, 1983.
21. **Gibson, R. S.,** Hair as a biopsy material for the assessment of trace element status in infancy, *J. Hum. Nutr.*, 34, 405, 1980.
22. **Bradield, R. B. and Hambidge, K. M.,** Problems with hair zinc as an indicator of body zinc status, *Lancet,* 1, 363, 1980.
23. **Chandra, R. K.,** Excessive uptake of zinc impairs immune responses, *JAMA,* 252, 1443, 1984.
24. **Slavin, W., Carnrick, G. R., Manning, D. C., and Pruszkowska, E.,** Recent experiences with the stabilized temperature platform furnace and Zeeman Background connection, *At. Spectrosc.*, 4, 1047, 1983.
25. **Prasad, A. S. and Rabbani, P.,** Nucleoside phosphorylase in zinc deficiency, *Trans. Assoc. Am. Physicians,* 94, 314, 1981.
26. **Moynahan, E. J.,** Acrodematitis enteropathics: a lethal inherited human zinc deficiency disorder, *Lancet,* 2, 399, 1974.
27. **Henkin, R. I., Schechter, P. J., Hoye, R., and Mattern, C. F. T.,** Idiopathic hypogenusia with dysgeusia, hyposmia and dysosomia. A new syndrome, *JAMA,* p. 434, 1971.
28. **Essatara, M. B., Levine, A. S., Morley, J. E., and McClain, C. J.,** Zinc deficiency and anorexia in rats: normal feeding patterns and stress induced feeding, *Physiol. Behav.*, 32, 469, 1984.
29. **Sandstead, H. H.,** Zinc essentiality for brain development and function, *Nutr. Rev.*, 43, 129, 1985.
30. **Anon.,** *Nutr. Rev.*, 42, 279, 1984.

Nutrients and Vitamins

J. Thomas Hindmarsh, M.D.

INTRODUCTION

J. Thomas Hindmarsh

The clinical biochemist must be able to help in the management of nutritional problems. Whereas in the Third World undernutrition is widespread, in the West it is largely confined to specific patient groups. Protein calorie malnutrition is common in hospitalized patients, particularly those with debilitating diseases. Vitamin deficiencies are uncommon except in patients with alcoholism, and, except for iron deficiency which is still common in women, trace element deficiencies are usually seen only in patients on long-term total parenteral nutrition (TPN), and then only rarely now that trace element fortification of TPN solutions is the rule. Overnutrition is common in the Western world where it is a major cause of premature death.

All animals require an intake of water and energy. Also, during the course of evolution complex animals and humans have developed dietary needs for complex molecules whose synthetic pathways have been lost. Some minerals are also essential for humans. A list of dietary requirements is shown in Table 1. Table 2 lists recommended daily dietary allowances for healthy people. However, it should be remembered that these data are not completely relevant to sick patients.

The following chapters present a summary of current knowledge on nutrition. For those who need more detailed information a list of standard texts follows.

Table 1
HUMAN DIETARY REQUIREMENTS OF MINERALS AND COMPLEX COMPOUNDS

Classification	Nutrient
Amino acids	Isoleucine, leucine, lysine, methionine, phenylalanine, threonine, tryptophan, valine, histidine[a]
Fatty acids	Linoleic acid
Vitamins	
Water soluble	Ascorbic acid, biotin, cobalamin, folic acid, niacin, pantothenic acid, pyridoxin, riboflavin, thiamin
Fat soluble	A, D, E,[b] K
Minerals	Calcium, chlorine, magnesium, phosphorus, potassium, sodium
Trace elements	Chromium, cobalt, copper, iodine, iron, manganese, molybdenum, selenium, zinc

[a] Essential for growth in infants.
[b] Human requirements not established.

Table 2
RECOMMENDED DAILY DIETARY ALLOWANCES[1] FOR HEALTHY PEOPLE

	Age (years)	Weight (kg)	Weight (lb)	Height (cm)	Height (in)	Protein (g)	Fat-Soluble Vitamins: Vitamin A (μg RE)[2]	Vitamin D (μg)[3]	Vitamin E (mg α-TE)[4]	Water-Soluble Vitamins: Vitamin C (mg)	Thiamin (mg)	Riboflavin (mg)	Niacin (mg NE)[5]	Vitamin B_6 (mg)	Folacin[6] (μg)	Vitamin B_{12} (μg)	Minerals: Calcium (mg)	Phosphorus (mg)	Magnesium (mg)	Iron (mg)	Zinc (mg)	Iodine (μg)
Infants	0.0–0.5	6	13	60	24	kg × 2.2	420	10	3	35	0.3	0.4	6	0.3	30	0.5[7]	360	240	50	10	3	40
	0.5–1.0	9	20	71	28	kg × 2.0	400	10	4	35	0.5	0.6	8	0.6	45	1.5	540	360	70	15	5	50
Children	1–3	13	29	90	35	23	400	10	5	45	0.7	0.8	9	0.9	100	2.0	800	800	150	15	10	70
	4–6	20	44	112	44	30	500	10	6	45	0.9	1.0	11	1.3	200	2.5	800	800	200	10	10	90
	7–10	28	62	132	52	34	700	10	7	45	1.2	1.4	16	1.6	300	3.0	800	800	250	10	10	120
Males	11–14	45	99	157	62	45.	1000	10	8	50	1.4	1.6	18	1.8	400	3.0	1200	1200	350	18	15	150
	15–18	66	145	176	69	56	1000	10	10	60	1.4	1.7	18	2.0	400	3.0	1200	1200	400	18	15	150
	19–22	70	154	177	70	56	1000	7.5	10	60	1.5	1.7	19	2.2	400	3.0	800	800	350	10	15	150
	23–50	70	154	178	70	56	1000	5	10	60	1.4	1.6	18	2.2	400	3.0	800	800	350	10	15	150
	51+	70	154	178	70	56	1000	5	10	60	1.2	1.4	16	2.2	400	3.0	800	800	350	10	15	150
Females	11–14	46	101	157	62	46	800	10	8	50	1.1	1.3	15	1.8	400	3.0	1200	1200	300	18	15	150
	15–18	55	120	163	64	46	800	10	8	60	1.1	1.3	14	2.0	400	3.0	1200	1200	300	18	15	150
	19–22	55	120	163	64	44	800	7.5	8	60	1.1	1.3	14	2.0	400	3.0	800	800	300	18	15	150
	23–50	55	120	163	64	44	800	5	8	60	1.0	1.2	13	2.0	400	3.0	800	800	300	18	15	150
	51+	55	120	163	64	44	800	5	8	60	1.0	1.2	13	2.0	400	3.0	800	800	300	10	15	150
Pregnant						+30	+200	+5	+2	+20	+0.4	+0.3	+2	+0.6	+400	+1.0	+400	+400	+150	[8]	+5	+25
Lactating						+20	+400	+5	+3	+40	+0.5	+0.5	+5	+0.5	+100	+1.0	+400	+400	+150	[8]	+10	+50

	Age (years)	Vitamins (continued): Vitamin K (μg)	Biotin (μg)	Pantothenic Acid (mg)	Trace Elements: Copper (mg)	Manganese (mg)	Fluoride (mg)	Chromium (mg)	Selenium (mg)	Molybdenum (mg)	Electrolytes: Sodium (mg)	Potassium (mg)	Chloride (mg)
Infants	0–0.5	12	35	2	0.5–0.7	0.5–0.7	0.1–0.5	0.01–0.04	0.01–0.04	0.03–0.06	115–350	350–925	275–700
	0.5–1	10–20	50	3	0.7–1.0	0.7–1.0	0.2–1.0	0.02–0.06	0.02–0.06	0.04–0.08	250–750	425–1275	400–1200
Children and adolescents	1–3	15–30	65	3	1.0–1.5	1.0–1.5	0.5–1.5	0.02–0.08	0.02–0.08	0.05–0.1	325–975	550–1650	500–1500
	4–6	20–40	85	3–4	1.5–2.0	1.5–2.0	1.0–2.5	0.03–0.12	0.03–0.12	0.06–0.15	450–1350	775–2325	700–2100
	7–10	30–60	120	4–5	2.0–2.5	2.0–3.0	1.5–2.5	0.05–0.2	0.05–0.2	0.10–0.3	600–1800	1000–3000	925–2775
	11+	50–100	100–200	4–7	2.0–3.0	2.5–5.0	1.5–2.5	0.05–0.2	0.05–0.2	0.15–0.5	900–2700	1525–4575	1400–4200
Adults		70–140	100–200	4–7	2.0–3.0	2.5–5.0	1.5–4.0	0.05–0.2	0.05–0.2	0.15–0.5	1100–3300	1875–5625	1700–5100

[1] The allowances are intended to provide for individual variations among most normal persons as they live in the United States under usual environmental stresses. Diets should be based on a variety of common foods in order to provide other nutrients for which human requirements have been less well defined.

[2] Retinol equivalents. 1 retinol equivalent = 1 μg retinol or 6 μg β-carotene.

[3] As cholecalciferol. 10 μg cholecalciferol = 400 IU of vitamin D.

[4] α-Tocopherol equivalents. 1 mg α-tocopherol = 1 α-TE.

[5] 1 NE (niacin equivalent) is equal to 1 mg of niacin or 60 mg of dietary tryptophan.

[6] The folacin allowances refer to dietary sources as determined by *Lactobacillus casei* assay after treatment with enzymes (conjugases) to make polyglutamyl forms of the vitamin available to the test organism.

[7] The recommended dietary allowance for vitamin B_{12} in infants is based on average concentration of the vitamin in human milk. The allowances after weaning are based on energy intake (as recommended by the American Academy of Pediatrics) and consideration of other factors, such as intestinal absorption.

[8] The increased requirement during pregnancy cannot be met by the iron content of habitual North American diets or by the existing iron stores of many women; therefore, the use of 30—60 mg of supplemental iron is recommended. Iron needs during lactation are not substantially different from those of nonpregnant women, but continued supplementation of the mother for 2—3 months after parturition is advisable in order to replenish stores depleted by pregnancy.

From *Recommended Dietary Allowances*, 9th ed., Food and Nutrition Board, National Research Council, National Academy of Sciences, U.S.A., 1980. With permission.

GENERAL REFERENCES

1. **Shils, M. E. and Young, V. R.,** *Modern Nutrition in Health and Disease,* 7th ed., Lea & Febiger, Philadelphia, 1987.
2. **Passmoe, R. and Eastwood, M. A.,** *Davidson and Passmore — Human Nutrition and Dietetics,* 8th ed., Churchill Livingstone, London, 1986.
3. **Truswell, A. S.,** A.B.C. of Nutrition, British Medical Association, 1986.
4. **Paul, A. A. and Southgate, D. A. T.,** McCance and Widdowson's Composition of Foods, 4th ed., Her Majesty's Stationery Office, London, 1978.
5. U.S.D.A. Nutritive Value of Foods, Bull. #72, U.S. Department of Agriculture, U.S. Government Printing Office, Washington, D.C., 1981.

DIET COMPOSITION

Jo Ellen Welsh

INTRODUCTION

The typical North American diet consists of approximately 46% carbohydrate, 40% fat, and 12% protein (as % of total calories). Compared to dietary guidelines published by various expert scientific committees, consumption of fat, cholesterol, alcohol, sodium, and simple sugars, as well as total energy, is, in general, too high.[1,2] In contrast, increased consumption of fiber and other complex carbohydrates is currently recommended for the general population.[3] Table 1 compares the "ideal" diet (based on published dietary guidelines) with the typical daily intakes. It should be emphasized that in industrialized societies, problems arise more often from overconsumption and associated obesity, heart disease, and diabetes, than from nutrient deficiencies per se. However, a diet high in calories (refined sugars and fat) can often be suboptimal in various micronutrients, particularly the trace minerals.

In the following sections, some dietary components will be discussed separately in order to define clearly their metabolism and physiological effects. It is important to keep in mind, however, that nutrient interactions in a mixed meal are complex and remain pooorly characterized in many cases (i.e., fiber, calcium). In addition, complicating factors such as disease or drug treatment can have profound effects on nutrient metabolism.

CARBOHYDRATES

Dietary carbohydrates, which supply 4 cal/g, include monosaccharides (glucose, fructose, galactose), disaccharides (sucrose, maltose, lactose), and polysaccharides (starch, fiber). Carbohydrates must be broken down into monosaccharide units for absorption into the mucosal cell. The products of complete digestion are glucose, fructose, and galactose, all of which are absorbed into the portal blood via carrier-mediated transport mechanisms.

FIBER

Fiber is, by definition, that portion of dietary carbohydrate which is "resistant to digestion by enzymes produced by man".[3] Because dietary fiber is not hydrolyzed into absorbable monosaccharide units, it does not serve as an energy source. Its potential role in human health lies in its physiological effects within the gut lumen. Fiber is a heterogeneous group of compounds which includes cellulose, hemicellulose, pectin, lignin, gums, and mucilages. Due to its heterogeneous nature, the effects of fiber in the gastrointestinal tract, which are largely based on physicochemical properties, are extremely diverse and complex. In addition, modifications of the fiber by processing and/or cooking, as well as fermentation by gut bacterial enzymes, can radically alter the physiological effect.

The primary physicochemical effects of fiber in the gut are water retention, adsorption of organic compounds (bile acids, steroids), and cation exchange. Various fiber classes exhibit different abilities to elicit these effects, i.e., insoluble fibers (cellulose) exert effects primarily in the colon, whereas soluble fibers (pectin, gums) exert major effects in the stomach.

Five potential physiologic effects for different types of dietary fibers have been identified and extensively studied:

Table 1
COMPARISON OF TYPICAL NORTH AMERICAN DIET AND "IDEAL" DIET

Nutrient	Typical North American diet	Desirable intake and/or guidelines	Ref.
Total calories	Often overconsumption	Maintain ideal weight	1,2
Carbohydrates			
Total calories	46%	50—55%	1,2
Complex	47%	Increase consumption	1
Simple	53%	Decrease consumption	1
Fat			
Total calories	40%	<30—35%	1,2
Saturated	13—15%	10%	1,2
Polyunsaturated: saturated ratio	0.43	1—1.5	2
Protein			
Total calories	100 g/d 12%	2 × the requirement, however, no formal recommendations have been issued regarding protein intake	
Cholesterol	600—700 mg/d	300 mg/d	1,2
Fiber	15—20 g/d	Increase intake 2×	1,3
NaCl	10—12 g/d	Decrease discretionary use	1,2
Sugar	Varies	Decrease intake	2
Alcohol	Varies	Moderation	1,2

1. Laxation
2. Normalization of serum lipid levels
3. Attenuation of the postprandial glucose response
4. Suppression of appetite
5. Reduction in bioavailability of cations due to binding

Although mechanisms for these effects have been tentatively identified, the results are far from conclusive (Table 2). For details on the physiological effects of dietary fiber, see References 3 to 6.

The estimated daily intake of fiber is about 15 to 20 g/d. A twofold increase is currently recommended (see Table 1). Recommendations by the American, Canadian, and British diabetic associations now include increased fiber intake for the diabetic population as well. The increase should come from an increased consumption of high fiber foods, rather than fiber supplements, as purified fiber seems to exert different effects *in vivo* than does food fiber.[3,5]

Table 2
DIETARY FIBER AND ITS EFFECTS

Effect	Suggested mechanism	Fiber class
Laxation	Water adsorption	Nonfermentable, insoluble fibers (cellulose, lignin)
Normalization of serum lipid levels	Bind cholesterol and bile acids, decrease absorption (other unknown mechanisms likely contribute, also)	Lignin, hemicellulose
Attenuation of post-prandial glucose response	Delay gastric emptying, slow absorption and transit time	Soluble, viscous, gel-forming fibers (pectins, gums, hemicellulose)
Suppression of appetite	Bulk effect elicits satiety(?)	Data limited and very inconsistent
Binding of minerals (Ca, Fe, Zn)	Cations bind to negatively charged groups on fiber molecules; excreted	Insoluble fibers (phytate)

REFERENCES

1. Nutrition and Your Health: Dietary Guidelines for Americans, U.S. Department of Agriculture and U.S. Department of Health and Human Services, Washington, D.C.,
2. American Heart Association Nutrition Committee, Rationale of the diet-heart statement of the American Heart Association, *Arteriosclerosis,* 4, 177, 1982.
3. Health and Welfare Canada, report of the Advisory Committee on Dietary Fibre, Health Protection Branch, Canada, 1985.
4. The Clinical Role of Fibre, Symp. Proc., Medical Education Services, Mississauga, Canada, 1985.
5. **Crapo, P. A.,** Simple versus complex carbohydrate use in the diabetic diet, *Annu. Rev. Nutr.*, 5, 95, 1985.
6. **Vahouny, G. V.,** Conclusions and recommendations of the symposium on "Dietary fibers in health and disease", Washington, D.C., *Am. J. Clin. Nutr.*, 35, 152, 1982.

PROTEIN-ENERGY MALNUTRITION IN HOSPITAL

John Patrick

INTRODUCTION

Remarkably little of the day-to-day activity in a hospital biochemistry laboratory is concerned directly with nutrition. This is all the more surprising when one considers that almost all surveys of nutritional status in hospital patients show a malnutrition prevalency of at least 20%.[1] There are no routinely available reliable measures of energy status, protein status, or body composition. Some of the acute-phase responses to surgery and serious infection can be measured in some hospitals, but this is by no means universal.

The reasons for this deplorable deficiency in the biochemical armamentarium are not hard to find. Most physicians received less than 10 h of nutritional training during the years in medical school and are likely to look upon nutrition as applied common sense not rising to the level of modern, sophisticated, high-technology medicine. The question is, to what extent is this patronizing attitude justified? Clearly, a routine description of the biochemistry of proteins, carbohydrates, lipids, vitamins, and minerals will do little to change the status quo, since that is precisely what was taught in medical school.

The ''skeleton in the cupboard'' which needs rattling is that of the malnourished hospital patient. How do they come to be like that? What are the metabolic and morbidity consequences? What needs to be done and will biochemistry help? Also, would a better understanding of nutrition modify the biochemist's interpretation of the data produced by the laboratory?

The major cause of malnutrition in hospital is the failure to provide sufficient quantities of food; quality is not a major problem, except for the palate! The reason the issue is neglected is because medical students are not taught to be quantitative about tissue synthesis and because none of the currently available biochemistry data directs their attention to body composition. Most of the output of the biochemistry laboratory is expressed in concentration terms, whereas in many situations what is needed is an understanding of the changes in body composition which are occurring. For example, a patient with mild hyponatremia (approximately 125 to 130 mmol/l) and edema to the knees has an increased total body sodium and the hyponatremia is due to some combination of water retention and/or potassium deficiency. The fact that most medical students do not follow the logic of the statement shows that the implications of Edelman's work relating plasma osmolality to total body water, total exchangeable sodium, and total exchangeable potassium are still not widely enough appreciated.[2]

THE QUANTITATION OF TISSUE LOSS AND TISSUE SYNTHESIS

When disease induces catabolism, it is difficult to reverse immediately. It seems we are programmed to go into negative nitrogen balance under situations of stress;[3] consequently, the requirement for catch-up growth will be ongoing in hospital. It is also possible to become malnourished because of increased demands, e.g., hyperthyroidism, cystic fibrosis (energy requirements may increase to 130 to 150% of normal), decreased efficiency of feeding, as with chronic neurological diseases which impair swallowing, or because of decreased excretory function, e.g., renal or respiratory failure where limited capacities to excrete nitrogenous compounds or to clear carbon dioxide induce anorexia. All these patients need catch-up growth if they are to regain normal body composition. Despite the reflex responses of most medical students, the primary limiting substrate for catch-up growth is energy. Every

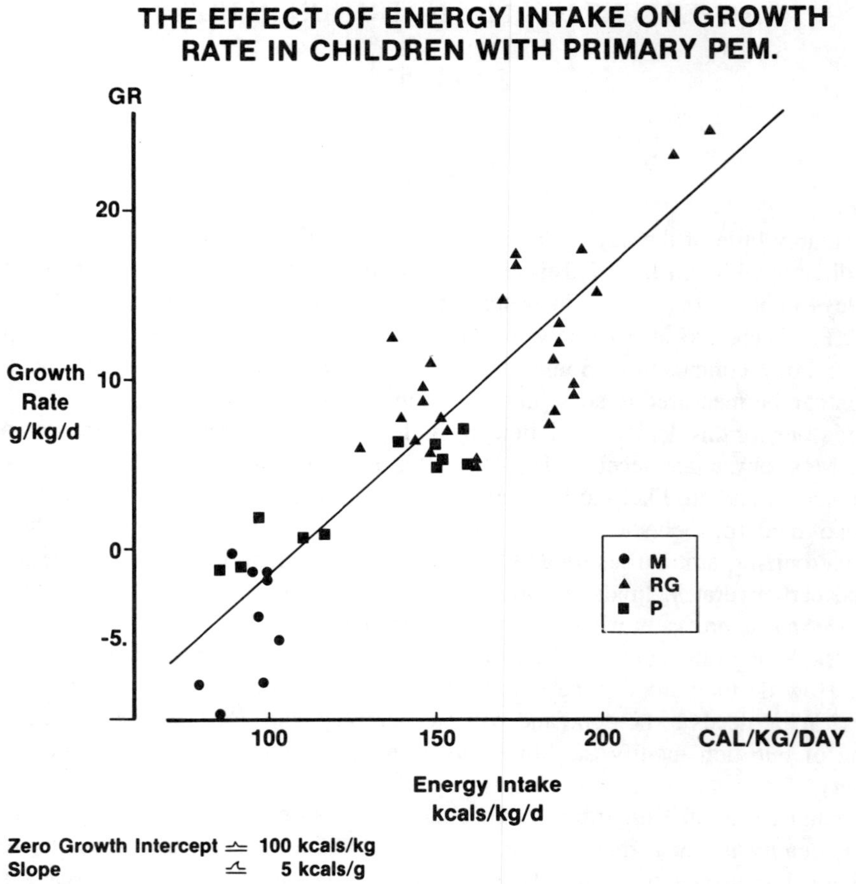

FIGURE 1. The graph displays the relationship between daily energy intake (kcal/kg/d) and daily weight gain (g/kg/d) in children aged 1 to 3 years, who are either malnourished and in the early stages of resuscitation (M) or are growing rapidly on a high-fat diet (RG) or on a high carbohydrate diet (P). The zero intercept is a measure of the average maintenance-energy requirements and the slope of the line indicates the energy requirement per gram of tissue laid down; in this case largely lean tissue.

gram of lean tissue requires 5 kcal for its deposition and every gram of adipose tissue around 10 kcal. Figure 1 illustrates this relationship with data obtained during the rapid nutritional resuscitation of severely malnourished children who are predominantly synthesizing muscle. The synthesis of 1 g of lean tissue includes about 250 mg of protein, whereas 1 g of adipose tissue contains only a negligible amount of protein. Thus, the protein energy ratio for lean tissue deposition is approximately 250 mg protein to 5 kcal energy or 1 kcal from protein in every 5 kcal, i.e., 20% of energy from protein. Because maintenance energy requirements for energy must be provided before net tissue synthesis can occur, this 20% energy from protein is an upper limit. The maintenance energy requirement in normal adult man is approximately 35 kcal/kg and the minimum protein requirement is around 0.5 g (2 kcal)/kg (i.e., about 6% energy from protein). Even less will serve if ketoacids and essential amino acid supplements are used. The most recent WHO/FAO report on protein and energy requirements no longer considers protein quality to be an important factor in human nutrition, because even with the most limited range of foods it is almost impossible to consume enough energy without necessarily also consuming enough protein.[4] The use of cassava as a staple may be exception. All this points out that energy should be the first consideration after weight loss, and body compositional measurements the most useful guide to the extent of

Table 1
QUANTITATION OF WEIGHT GAIN (10-kg CHILD, 3-kg DEFICIT)

Intake (kcal/kg/d)	Maintenance (kcal/kg/d)	Growth energy (kcal/d)	Weight gain (g/d)	Days to target	Cost ($)
100	100	0	0	∞	∞
105	100	50	10	300	240,000
110	100	100	20	150	120,000
120	100	200	40	75	60,000
140	100	400	80	32.5	30,000
180	100	800	160	16.25	15,000

Note: This table shows the theoretical relationship between energy intake and the correction of a 3-kg weight deficit in an infant. The cost of growth, 5 kcal/g, is taken from Figure 1 as is the maintenance energy requirement. For the purposes of the cost calculation a bed charge of $800/d has been utilized. In adults or older children the principles would be similar, but the maintenance energy costs would be smaller on a per kilogram basis.

the problem and the success of treatment. The impact of this insight on the management of weight loss is shown in Table 1. In these days of cost containment the economic facts are worth a second glance; attention to the provision of energy has the potential to dramatically reduce hospitalization times. Malnourished subjects have been repeatedly shown to have higher mortality and higher morbidity rates than normally nourished patients[5] (Table 1).

THE BIOCHEMICAL ASSESSMENT OF NUTRITIONAL STATUS

Because cellular energy status is so important one would like to be able to measure it, but currently there are no biochemical measurements directed towards assessment of ATP levels or ATP:ADP ratios *in vivo*. Leukocytes are probably the only cells that could be routinely sampled, but they present technical difficulties with handling. Indirect calorimetry is all that is likely to be available in the near future, and since quite long sampling periods are necessary for reliable data, only a few patients can be studied in this way. Stable isotope techniques for the measurement of energy expenditure are attractive for defined groups of patients with chronic nutritional problems,[6] but they are difficult to use in patients with acute illnesses. It is, of course, easier to follow body weight, and provided there are no abnormalities of sodium and/or water balance, this must represent changes in energy balance; a patient without an abnormality of sodium and/or water balance, who loses weight, is obviously in negative energy balance and vice versa. However, there are some practical *caveats* to the interpretation of a stable weight. In the chronically wasted patient, although weight is stable this has been achieved by diminishing metabolism, and almost every physiological process that has been examined is abnormal.[7] Although there are consequences in terms of diminished homeostatic capability, it may still be more appropriate to look upon these abnormalities as adaptations rather than primary pathological processes.[8] Adaptation to reduced energy supply, with continued, although slower growth, has been elegantly demonstrated in experimental animals.[9] Table 2 lists some of the responses to chronic protein-energy malnutrition that have been observed in the Third World. In most cases, the change can be viewed as having an energy conservation benefit purchased at the cost of diminished flexibility. The response of the Western internist to malnutrition must be to recognize what is happening and to approach correction with appropriate caution. Too rapid correction of malnutrition can be fatal.[10] Up to 3 weeks may be required before metabolism returns to normal. (The principles of treatment are summarized in Table 3). This is probably why short periods of total parenteral nutrition have not been shown to be of great benefit in improving ultimate outcome.

Table 2
RESPONSES TO PROTEIN-ENERGY MALNUTRITION

Reduction in resting metabolic rate
Reduction in protein synthesis rates
Reduction in membrane transport of sodium
Impaired thermogenic response to cold
Reduced insulin secretion in response to glucose
Reduced resting cardiac output and stoke volume
Reduced glomerular filtration rate and renal plasma flow

Table 3
NUTRITIONAL PRINCIPLES FOR "CATCH-UP" GROWTH

Energy is the most important requirement
  5—10 kcal(20—40 kJ) required for each gram of weight gain
Protein is rarely limiting
  Maintenance requirements of <5% total energy from protein
  Catch-up will not require more than 20% protein
Minerals can become limiting
Intercurrent infection inhibits growth

Now, given this background, what attitude should the biochemist take to the numerous schemes of biochemical nutritional assessment that are currently in vogue? Most are mainly dependent upon measurement of serum albumin.[5] Some even go so far as to equate a low serum albumin with kwashiorkor and to assume it equates with protein deficiency. The syndrome of kwashiorkor has a *sine qua non* of edema, but since children with analbuminemia do not have edema as a major problem and marasmic children without edema have a greater reduction in body mass than children with kwashiorkor, i.e., more protein deficiency than children with kwashiorkor, the relationship between hypoalbuminemia and edema is not as direct as the standard textbooks imply.[11,12] This simplistic approach to nutritional assessment is one of the major problems to be attacked if nutritional biochemistry is to become really useful. In this particular instance, it is likely that the patients with low serum albumin levels are actually manifesting evidence of salt retention and diluting their albumin mass, rather than protein depletion. Their subsequent poor prognosis is related to disordered sodium metabolism.[10,13] This is not to deny that sodium retention in these patients may be related to poor nutrition, but the sequence is undoubtedly much more complex than the proposed sequence: inanition — protein deficiency — hypoalbuminemia — edema — death or complications. What is needed is measurement of changes in body composition. The best that can currently be done is attempt to measure the balance status for important nutrients. Twenty-four-hour urine electrolytes, for example, allow reasonable estimates of sodium balance. If very little sodium is being excreted and the usual i.v. infusions are running, then salt retention leading ultimately to edema is easy to predict. Simultaneous measurement of urine potassium and nitrogen (or urea — approximately 75% of nitrogen derived from protein breakdown appears in the urine as urea) will allow at least an educated estimate of what is happening to body lean tissue composition. Although technically very simple, this represents a major conceptual step forward in patient management. Every gram of nitrogen lost or retained should be accompanied by 3 to 5 mmol of potassium, a fact which allows some checking for experimental errors both with weight loss and weight gain. Continued positive nitrogen balance in the absence of positive potassium balance is a biological nonsense. For the biochemist interested in improving the usefulness of the laboratory's output for nutritional assessment, it is measurements that can be extrapolated to body composition that are most likely to be helpful. Thus, if height and weight can be added to the data in the laboratory computer, it becomes possible to pick out patients who are likely to be malnourished, i.e., have a diminished weight for height, and to look at 24-h creatinine output (a crude estimate of muscle mass), urea output to distinguish the catabolic from the chronically starved, and sodium to determine who is at risk of the complications ensuing from salt retention. It is amazing how often one sees a patient receiving isotonic saline, because the serum sodium is slightly low without any consideration of the urinary sodium excretion, which is often very low. The consequences of this failure to think are entirely predictable. Equations relating serum albumin, skinfold thicknesses, and delayed hypersensitivity reactions to outcome need more thought. The usefulness of multivariate analyses, which relate particular biochemical measurements to patient outcome, should not be confused with the process of thinking about the nutritional changes occurring in particular patients. The former are useful policy tools; the latter should be the daily habit of individual patient care. The reason why a low serum albumin is an ominous prognostic sign is the critical intellectual problem. The fact that better nutrition which includes protein causes the albumin to revert to normal does not prove that the original hypoalbuminemia was caused by protein deficiency.

Despite this rather bleak review of the utility of currently available biochemical measurements for nutritional assessment, most clinical chemists will be asked to perform many biochemical tests on patients receiving nutritional support. In the view of the author, many of them are unnecessary, as can easily be demonstrated by asking the question, which abnormal value for the requested test will lead to a change in management? Table 4 lists a fairly typical recommended schedule which will be exceeded in many institutions; the comments are my editorial addition. Further, it should be noted that most studies of the usefulness of preoperative total parenteral nutrition have been inadequate in that they have not been randomized controlled trials, they have not had adequate criteria for the diagnosis of malnutrition, the surgery has not carried serious risk, and the period of parenteral feeding has not been long enough (7 d). However, combining the results from several studies which avoid most of these pitfalls yields the following figures; morbidity, 13% in total parenteral nutrition (TPN) group and 26% in controls ($p = 0.001$); and mortality, 6 vs. 15% ($p = 0.001$). Thus, TPN appears to be worthwhile provided it is used on appropriate (i.e., wasted or unable to eat) patients and for appropriate lengths of time (i.e., more than 7 d). If a laboratory is doing a lot of work for patients not fulfilling these criteria, then the protocols for the provision of TPN should be revised.

Thus, the main conclusion to be drawn from this introduction is that there are major nutritional problems in hospitals of which the biggest is undoubtedly malnutrition, and current biochemical tests are not useful for the determination of energy status which is the main requirement. The other major component of nutritional concern in hospital is TPN where biochemical monitoring is necessary in order to adjust rates of delivery to rates of utilization. The occasional need for more specific measurement will be discussed in the appropriate sections. Vitamin deficiencies, mineral deficiencies and other specific deficiencies, as, for example, of carnitine, undoubtedly occur but they are quantitatively small problems compared to protein energy malnutrition.

Table 4
RECOMMENDED MONITORING OF PATIENTS RECEIVING NUTRITIONAL SUPPORT

Measurement	Frequency during initial stabilization	Frequency after initial stabilization	Comment
		General	
Weight	Daily	Daily	Cheap and, if graphed properly, extremely useful
Height	Once (adults)	Once/week (children)	Only useful if measured carefully in children
HC (head circumference)	Not useful in adults	Weekly in children	
Skinfolds			
MAC[a]		Only necessary with chronic patients	
MAMC[b]			
		Urine	
Glucose	With each micturition initially	Daily	In the absence of glycosuria with IV glucose, running plasma glucose is unnecessary once stable levels have been acheived
		Blood	
Plasma electrolyte	3—4 ×/week	Weekly	To interpret plasma electrolytes without urine electrolyte and osmolality is inappropriate
BUN and creatinine	2 ×/week	Weekly	Useful to distinguish catabolic states from chronic states and to monitor renal function
Albumin	2 ×/week	Weekly	Does not alter treatment except, perhaps, rates of sodium administration
CBC and platelets	Weekly	Weekly	
Liver function tests	Weekly	Weekly	Only if the option to stop TPN is real, do these tests influence management
Cholesterol, triglyceride	Check clearance with each increment in lipid dosage	Weekly	The reliability of nephelometry has been questioned, but is still widely used. It is the clearance of the artificial chylomicras rather than cholesterol or triglyceride levels that are of critical importance
		Short Half-Life	
Proteins, e.g., retinal binding protein transferrin	Weekly	Weekly	If the patient is gaining weight without salt retention there is no benefit from these tests
Nitrogen balance	Weekly	Weekly	Comment as above
Trace elements	Monthly	Monthly	Plasma values are not good indices of deficiency
Vitamins	As indicated	As indicated	

[a] MAC = midarm circumference.
[b] MAMC = midarm muscle circumference.

REFERENCES

1. **Bistrian, B. R., Blackburn, M. D., Hallowell, E., and Meddle, R.,** Protein status of general surgical patients, *J.A.M.A.*, 230, 858, 1974.
2. **Elelman, I. S., Liebman, J., O'Meara, M. P., and Birkenfeld, L. W.,** Interrelations between serum sodium concentrations, serum osmolality and total exchangeable Na^+ and K^+ and TBW, *J. Clin. Invest.*, 37, 1256, 1958.
3. **Beisel, W. R.,** Mediators of fever & muscle proteolysis, *N. Engl. J. Med.*, 308, 586, 1983.
4. Energy and Protein Requirements, FAO/WHO/UNU Tech. Rep. Series, 1985, 724.
5. **Buzby, G. P., Muller, J. L., Matthews, D. C., Hobbs, C. L., and Rusato, E. F.,** Prognostic nutritional index in gastrointestinal surgery, *Am. J. Surg.*, 139, 160, 1980.
6. **Schoeller, D. A., Kushner, R. F., and Jones, P. J. H.,** Validation of doubly labeled water for measuring energy expenditure during parenteral nutrition, *Am. J. Clin. Nutr.*, 44, 291, 1986.
7. **Jackson, A. A. and Golden, M. H. N.,** Protein energy malnutrition, in *The Oxford Textbook of Medicine*, Weatherall, D. J., Ledingham, J. G. G., and Warrell, D. A., Eds., Oxford University Press, New York, 1983, 8.12.
8. **Waterlow, J. C.,** Metabolic adaptation to low intakes of energy and protein, *Annu. Rev. Nutr.*, 6, 495, 1986.
9. **Samonds, K. W. and Hegsted, D. M.,** Protein deficiency and energy restriction in young cebus monkeys, *Proc. Natl. Acad. Sci.*, 75, 1600, 1978.
10. **Patrick, J.,** Death during recovery from severe malnutrition and its possible relationship to sodium pump activity in the leukocyte, *Br. Med. J.*, 1, 1051, 1977.
11. **Patrick, J.,** Oedema in protein energy malnutrition: the role of sodium pump, *Proc. Nutr. Soc.*, 38, 61, 1979.
12. **Golden, M. H. N.,** Protein deficiency, energy deficiency and the oedema of malnutrition, *Lancet*, 1, 1261, 1982.
13. **Patrick, J. and Golden, M. H. N.,** Leucocyte electrolytes and sodium transport in protein energy malnutrition, *Am. J. Clin. Nutr.*, 30, 1478, 1977.

NUTRITION AND ELECTROLYTES

John Patrick and Jo Ellen Welsh

SODIUM AND POTASSIUM

Clinical chemists are often asked to consult on electrolyte problems; they should, therefore, be aware of the frequent interactions between nutritional state and electrolyte abnormalities. Reference to the example of potassium deficiency may be helpful.

Potassium deficiency can occur in three ways. First, it may occur because of straightforward negative balance due, for example, to diarrhea without adequate replacement; replacement is easy and effective. Second, it may occur because of altered membrane function leading to a diminished capacity to retain potassium within the cell. This happens in severe malnutrition[1] and it is not immediately corrected by potassium supplements. The correct treatment is nutritional rehabilitation, the principles of which have been well worked out.[2,3] It is interesting to look at a list of generalized diseases known to be associated with abnormalities of sodium transport (Table 1) and to note that many of them have major disturbances in energy metabolism and body composition, often with considerable degrees of malnutrition. The correction of the electrolyte problem is achieved by correction of the nutritional state. The third mechanism for potassium deficiency was beautifully illustrated by the experiments of Muntwyler and Griffin,[4] who induced hypokalemia and a drop of muscle potassium without loss of potassium from the body. Their experimental animals were pretreated by a regime which induced a loss of muscle mass and associated potassium; then, as soon as they were given enough energy and protein to rebuild their muscle, they attempted to do so. However, because every gram of nitrogen retained required 3 to 5 mmol of potassium and this was not provided, they rapidly became hypokalemic. Their muscle potassium content was also low. This phenomenon has also been demonstrated for zinc and for phosphate, although in the latter case it is the rebuilding of ATP levels rather than muscle that is the critical step. Nevertheless, it is important to appreciate that relative deficiency can be induced in the presence of positive balance.

Particularly in children, electrolyte problems are often seen together with a rapidly altering energy balance. The correction of the electrolyte problems requires that energy and other nutrients be in appropriate balance. The provision of recommended daily allowances has no relevance to this problem. Such recommendations were designed for normal people in good health, not for people recovering from severely catabolic illnesses. Ideally, energy expenditure and nitrogen excretion should be measured and thought given to the nutrient content of the tissues being synthesized. In the absence of sophisticated indirect calorimetry equipment the 1985 WHO/FAO/UNU approach to energy requirements should be used.[5]

CALCIUM

Calcium (Ca), phosphorus (P), and magnesium (Mg) are found predominantly in bone, where these minerals provide structure and density. Also, all three minerals are essential for normal bone cell metabolism. The large quantities of Ca, P, and Mg present in bone are in dynamic equilibrium with other body pools of these minerals and can be released when intakes are low. However, continued reliance on skeletal stores to maintain plasma mineral homeostasis will ultimately have an adverse effect on bone mass. For a detailed discussion of Ca, P, and Mg metabolism see the pertinent sections in this volume.

Ca is the most abundant mineral in the body (1200 g/70-kg man) and 99% is in bone; the remaining 1% of extraskeletal Ca is involved in such diverse functions as nerve impulse

Table 1
DISEASES ASSOCIATED WITH SODIUM TRANSPORT DEFECTS

7	
8	Chronic renal failure
9	Liver failure
10	Hyperthyroidism
11	Carcinomatosis
12	Protein-energy malnutrition
13	Hypertension

transmission, muscle contraction, blood clotting, and enzyme activation. The maintenance of plasma-ionized Ca levels is crucial to these extraskeletal functions of Ca, as these actions are dependent on a critical concentration gradient between extra- and intracellular Ca concentrations.

Plasma Ca is maintained at approximately 10 mg/dl (2.5 mmol/l) through coordinated hormonal processes which balance the entry of dietary Ca with fluxes into/out of bone and losses via urinary excretion. Parathyroid hormone (PTH) and vitamin D act in concert to increase plasma Ca, whereas the actions of calcitonin are antagonistic and result in reduction in plasma Ca (for an update on the hormonal regulation of extracellular Ca, see Reference 6). Total plasma Ca consists of three fractions: free ionized (47%), complexed (7%), and protein-bound (46%). Free ionized Ca is the functionally important fraction which is feed-back-controlled by the parathyroid gland.

Intestinal absorption of dietary Ca averages about 30%, but interactions in the gut can affect solubilization (for example, presence of fibers, P, acid, etc.). For a review on the bioavailability of dietary Ca, see Reference 7). The mechanism of intestinal Ca absorption has been divided into four phases: entry across the brush border, binding within subcellular sites, movement across the mucosal cell, and extrusion.[8] All four phases appear to be enhanced by vitamin D. PTH indirectly increases Ca absorption by stimulating activation of vitamin D.

Because of the large skeletal reserve of Ca and the hormonal processes for controlling Ca homeostasis, hypocalcemia does not occur during periods of low dietary Ca, but rather is secondary to other conditions (Mg deficiency, vitamin D deficiency, hypoparathyroidism). Clinical symptoms include tetany and neuromuscular irritability. Hypercalcemia (12 to 15 mg/dl) similarly is usually secondary to either hyperparathyroidism or malignancy. Symptoms are nonspecific and include Central nervous system dysfunction, arrhythmias, and hyper-tension. Chronic hypercalcemia can result in soft tissue calcification, particularly of the kidneys, pancreas, and blood vessels.

The concept of ''Ca deficiency'' is extremely difficult to define. Although it is well recognized that deposition of bone mineral during periods of active growth depends on the availability of Ca, the amount of Ca necessary for maintenance of the adult skeleton remains controversial. The U.S. FDA recommended daily allowance for adults of 800 mg is deemed too high by some who stress the ability of the body to adapt to habitually low Ca intakes by increasing the efficiency of intestinal Ca absorption.[9,10] Conversely, elderly subjects may need 1.0 to 1.5 g/d for balance, as both Ca absorption and the ability to adapt to low dietary Ca decline significantly with age.[11,13] In addition, many other physiologic, dietary, and lifestyle factors affect Ca balance, either negatively (high protein, fiber, and sodium intakes, alcohol, glucocorticoids) or positively (exercise, sun exposure, estrogens).[7] The relative contributions of these factors to age-related bone loss and development of osteoporosis (decreased bone mass) are currently under intense investigation (for a review, see References 12 and 14).

Concern about osteoporosis has led to the increased availability of a wide variety of Ca supplements. Although Ca is relatively nontoxic, chronic ingestion of large amounts (>1500 mg/d) can predispose to the formation of kidney stones in susceptible individuals. Those with a legitimate requirement for supplementation are advised to take $CaCO_3$, as its utilization is comparable to the more expensive organic preparations and it contains a higher percent Ca by weight. Small, divided doses throughout the day are better utilized than one large dose.[15]

PHOSPHORUS

Eighty-five percent of the body's P is contained as inorganic mineral salts of the skeleton; the remainder exists primarily as organic esters in soft tissues. Plasma P concentration is not as tightly controlled as that of Ca, because it reflects fluxes of P from intermediary metabolism as well as contributions from bone and soft tissue. Important extraskeletal roles of P include participation in carbohydrate, lipid, and protein metabolism (due to presence in intermediary metabolites, phospholipids, nucleic acids, and high energy phosphate compounds), acid-base balance, and constituent of cell membranes.[16]

P is absorbed largely as free phosphate, by complex mechanisms which remain poorly understood. The fact that vitamin D-deficient animals retain their ability to absorb significant quantities of P suggests that at least part of the mechanism is vitamin D independent.[17] The increase in P absorption seen in animals on low dietary P, however, does appear to be dependent on adequate vitamin D status.[17] Organic phosphate esters are, in general, not well absorbed, as they are not hydrolyzed by the human gut (for example, phytates). Overall, about 60 to 70% of total dietary P is absorbed.

Urinary phosphate excretion depends primarily on how much is absorbed from the diet; usually about 2/3 of absorbed P is excreted. Urinary P is largely inorganic phosphate with a HPO_4/H_2PO_4 normally of 4:1. Tubular reabsorption (about 85 to 95% of the filtered load) is increased by short-term cortisol therapy and growth hormone, and decreased by estrogen, thyroid hormone, PTH, long-term cortisol therapy, and hypercalcemia.

Dietary requirements for P are, with few exceptions, identical to those for Ca. Although it has been accepted that a Ca:P ratio close to 1 is optimal for skeletal health, adverse effects of lower ratios have not been demonstrated in humans[18,19] whose diets often contain much more P than Ca. one exception appears to be the development of "idiopathic hypocalcemia" in infants fed cow milk-based formulas,[20] which has been linked to their lower Ca:P ratio compared to human milk.

The homeostatic mechanisms for phosphate conservation coupled with its universal dietary occurrence prevent isolated P deficiency, even though the amount of P needed for growth and development is quite large. Hypophosphatemia, thus, is more often the result of metabolic disturbances than dietary lack. Some examples are excessive PTH secretion (PTH inhibits P reabsorption in kidney), metabolic acidosis (H^+ also inhibits P reabsorption in kidney), and excessive chronic use of P-binding antacids (render P unabsorbable). Attention to plasma P is also critical when refeeding patients (diabetics after ketoacidosis, alcoholics, starvation), as the sudden uptake of P by cells for phosphorylation of carbohydrates or buildup of bone or muscle mass may precipitate hypophosphatemia. Hypophosphatemia is also seen in genetic defects of phosphate transport (X-linked hypophosphatemia rickets). Symptoms of hypophosphatemia include weakness, anorexia, malaise, and skeletal pain.

Hyperphosphatemia is rare in the absence of renal disease, because the kidney can excrete 600 to 900 mg of P daily. However, during periods of severe muscle catabolism, cellular release of P can precipitate hyperphosphatemia. There are no specific symptoms of hyperphosphatemia, although resulting hypocalcemia can lead to neuroexcitability, tetany, etc. In severe renal failure, hyperphosphatemia can present a significant problem, leading to renal osteodystrophy, necessitating dietary P restriction.

Normal values for plasma P range from 2.5 to 5.0 mg/dl (0.8 to 1.6 mmol/l) in adults and from 4.0 to 7.0 mg/dl (1.3 to 2.3 mmol/l) in children. Because of the effect of acidosis on P metabolism, measurement of plasma pH and total CO_2 is necessary when evaluating hypophosphatemia.

MAGNESIUM

In contrast to Ca, specific and unique endocrine control mechanisms for Mg homeostasis have not been identified, yet plasma Mg is fairly constant at 1.8 to 2.5 mg/100 ml (0.75 to 1.00 mmol/l) in healthy adults. Intestinal absorption of dietary Mg varies widely among individuals (range 25 to 65%) and is generally inversely proportional to the amount present. Dietary Mg in the U.S. averages about 300 to 360 mg/d, the principal sources being cereal products, nuts, dairy products, and green vegetables.

Plasma Mg consists of three fractions: free (55%), complexed (13%), and protein bound (32%). Maintenance of extracellular Mg levels is primarily achieved through adjustments in urinary excretion. Normally, 95 to 97% of filtered Mg is reabsorbed; however, a renal threshold for Mg excretion operates to prevent large increases in plasma Mg. Factors that increase Mg excretion include extracellular volume expansion, hypercalcemia or hypercalciuria, diuretics, and phosphate depletion, while factors which decrease Mg excretion include Mg deficiency and PTH.[21]

The total amount of Mg in a 70-kg adult ranges from 20 to 28 g, 55% of which is located in bone. The major portion (27%) of extraskeletal Mg is in muscle. The biochemical functions of Mg are diverse and include activation of many key enzymes in intermediary metabolism as well as roles in neuromuscular transmission, nucleic acid metabolism, and protein synthesis.

The similarities in body distribution, homeostasis, and chemistry of Mg and Ca result in complex interactions between these two divalent cations.[21,22] In many instances, Mg acts antagonistically to Ca (for example, in activation of adenylate cyclase), whereas in other cases, Mg can mimic the effects of Ca (for example, in stimulation of PTH secretion).

Although primary Mg deficiency is quite rare, secondary deficiency is frequently seen clinically. Any conditions which decrease absorption (i.e., malabsorption syndromes) or increase losses (chronic diarrhea, vomiting, alcoholism) can lead to negative balance and depletion. As with P, refeeding after protein-calorie malnutrition, diabetic ketoacidosis, or alcoholism can lead to rapid uptake of Mg by cells, precipitating hypomagnesemia. Clinical symptoms of Mg depletion include hypocalcemia (which does not respond to Ca supplementation), hypokalemia, tetany, convulsions, nausea, weakness, and cardiac arrythmias. Many of the clinical symptoms appear to be secondary to hypocalcemia rather than unique to hypomagnesemia (see Reference 23 for complete description of sequelae of deficiency symptoms).

The crucial intracellular Mg level remains relatively stable even in deficiency states; however, hypomagnesemia induces profound changes in the metabolism of Ca, P, Na, and K. Intracellular levels of K and P decline, while those of Ca and Na increase during Mg depletion.[2] The relationships between these biochemical changes and deficiency symptoms and/or possible human disease remain to be elucidated.

Alterations in plasma lipoprotein profiles, membrane fluidity, platelet aggregation, and blood pressure in Mg depletion[24,25] support a role for Mg in cardiovascular functions. Although a relationship between Mg (as well as Ca) intake and heart disease in humans has long been postulated (the hard vs. soft water hypothesis), direct evidence remains lacking.

Abnormalities in bone structure/function and derangements in Ca homeostasis in experimental Mg deficiency[26] strongly suggest an important role for Mg in skeletal metabolism. The development of hypocalcemia during Mg deficiency has been extensively studied and seems to be related to failure of Ca release from bone and deposition of Ca in soft tissues.

Equilibrium normally exists between bone Mg and extracellular Mg; thus, bone Mg decreases with plasma Mg in deficiency states, especially during growth. Loss of bone Mg appears to affect the sensitivity of the skeleton to resorbing agents, particularly PTH. In addition, functional failure of the parathyroid gland may develop in severe hypomagnesemia.[23]

Decline in serum Mg correlates well with progressive Mg deficiency, except when complicating factors (for example, alcoholism) occur. Determination of 24-h urinary Mg is also useful when evaluating Mg status. A combination of low serum Mg (<1.2 mg/100 ml) and urinary excretion <1.2 mg/d) strongly indicates Mg deficiency. Total body Mg can be assessed accurately by bone or muscle biopsy, bone being the better indicator.[22]

Treatment of mild deficiency is simply increasing dietary Mg intake. More severe deficiency states can be treated with parenteral Mg, which is both safe and effective in the absence of renal disease. Treatment of related hypocalcemia with Ca supplementation should definitely be avoided because of the predisposition for soft tissue calcification in Mg deficiency.

REFERENCES

1. **Patrick, J. and Golden, M. H. N.,** Leucocyte electrolytes and sodium transport in protein energy malnutrition, *Am. J. Clin. Nutr.*, 30, 1478, 1977.
2. **Waterlow, J. C., Golden, M. H. N., and Patrick, J.,** Protein energy malnutrition: treatment, in *Nutrition in the Clinical Management of Disease,* Dickenson, J. W. T. and Lee, H. A., Eds., Edward Arnold, London, 1978.
3. **Jackson, A. A. and Golden, M. H. N.,** Protein energy malnutrition, in *The Oxford Textbook of Medicine,* Weatherall, D. J., Ledingham, J. G. G., and Warrell, D. A., Eds., Oxford University Press, New York, 1983, 8.12.
4. **Muntwyler, E. and Griffin, G. E.,** *Proc. Soc. Exp. Biol. Med.*, 89, 349, 1955.
5. Energy and Protein Requirements, FAO/WHO/UNU Tech. Rep. Series, 1985, 724.
6. **MacIntyre, I.,** The hormonal regulation of extracellular calcium, *Br. Med. Bull.*, 42, 343, 1986.
7. **Allen, L. H.,** Calcium bioavailability and absorption: a review, *Am. J. Clin. Nutr.*, 35, 783, 1982.
8. **Bronner, F., Pansu, D., and Stein, W. D.,** An analysis of intestinal calcium transport across the rat intestine, *Am. J. Physiol.*, 250, G561, 1986.
9. **Hegsted, D. M., Moscoso, I., and Collazos, C.,** A study of the minimum calcium requirements of adult men, *J. Nutr.*, 46, 181, 1952.
10. **Malm, O. J.,** Calcium requirements and adaptation in adult men, *Scand. J. Clin. Lab. Invest. Suppl.*, 10(36), 1, 1958.
11. **Heaney, R. P., Recker, R. R., and Saville, P. D.,** Menopausal changes in calcium balance performance, *J. Lab. Clin. Med.*, 92, 953, 1978.
12. **Slovik, D. M.,** The vitamin D endocrine system, calcium metabolism and osteoporosis, *Spec. Top. Endocr. Metab.*, 5, 83, 1983.
13. **Spencer, H., Kramer, L., Lesniak, M., DeBartolo, M., Norris, C., and Osis, D.,** Calcium requirements in humans, *Clin. Orthop. Relat. Res.*, 184, 270, 1984.
14. **Riggs, B. L. and Melton, L. J.,** Involutional osteoporosis, *N. Engl. J. Med.*, 314, 1676, 1986.
15. National Institutes of Health, Consensus Development Conference statement, *Osteoporosis,* 5(3), 1984.
16. **Pike, M. L. and Brown, R. L.,** *Nutrition: An Integrated Approach,* John Wiley & Sons, New York, 1984, 170.
17. **Lee, D. B. N., Walling, M. W., and Brautbar, N.,** Intestinal phosphate absorption: influence of vitamin D and non-vitamin D factors, *Am. J. Physiol.*, 250, G369, 1986.
18. **Zemel, M. B. and Linkswiler, H. M.,** Calcium metabolism in the young adult male as affected by level and form of phosphorus intake and level of calcium intake, *J. Nutr.*, 111, 315, 1981.
19. **Spencer, H., Kramer, L., Rubio, N., and Osis, D.,** The effect of phosphorus on endogenous fecal calcium excretion in man, *Am. J. Clin. Nutr.*, 43, 844, 1986.
20. **Venkataraman, P. S., Jsang, R. C., Greer, J. R., Noguchi, A., Laskarzewski, P., and Steichem, J. J.,** Late infantile tetany and secondary hyperparathyroidism in infants fed humanized cow milk formula, *Am. J. Dis. Child.*, 139, 664, 1985.
21. **Levine, B. S. and Coburn, J. W.,** Magnesium, the mimic/antagonist of calcium, *N. Engl. J. Med.*, 310, 1253, 1984.

22. **Cronin, R. E. and Knochel, J. P.,** Magnesium deficiency, *Adv. Intern. Med.*, 28, 509, 1983.
23. **Shils, M. E.,** Magnesium, in *Modern Nutrition in Health and Disease*, Goodhart, R. S. and Shils, M. E., Eds., Lea & Febiger, Philadelphia, 1980, 310.
24. **Rayssiguier, Y.,** Lipoprotein metabolism: importance of magnesium, *Magnesium Bull.*, 8, 186, 1986.
25. **Weinsier, R. L. and Norris, D.,** Recent developments in the etiology and treatment of hypertension: dietary calcium, fat and magnesium, *Am. J. Clin. Nutr.*, 42, 1331, 1985.
26. **Anders, G.,** The effects of magnesium on calcergic reactions and on the bones, *Magnesium Bull.*, 8, 5, 1986.

ETHANOL

Jo Ellen Welsh

Although ethanol is not considered a nutrient, it does contribute calories (7 cal/g) to the diet and, thus, may displace nutrient-containing foods. Ethanol also affects the metabolism of many nutrients (i.e., induces renal wasting of Mg and Zn) and has direct toxic effects on several tissues, primarily the liver. The spectrum of alcohol-induced liver disease includes fat accumulation, hepatitis, and cirrhosis.[1,2]

After absorption from the stomach and small intestine, alcohol is taken up by the liver (80%) and peripheral tissues (20%). Conversion to acetaldehyde (the intermediate which has been implicated in the pathogenesis of alcohol addiction) is via alcohol dehydrogenase and NAD. A second enzyme, acetaldehyde dehydrogenase, converts acetaldehyde to acetate, also with NAD as cofactor. Acetate (via acetyl coA) can be utilized for energy production or fatty acid synthesis, with subsequent storage as triglycerides. Competition for NAD between ethanol-metabolizing enzymes and lactate dehydrogenase prevents the conversion of lactate to pyruvate, and decreases gluconeogenesis from lactate.

In chronic alcoholics with high blood alcohol concentrations, the enzymes of the microsomal-ethanol oxidizing system (MEOS) are induced. This system catalyzes the oxidation of ethanol to acetaldehyde using NADPH as co-factor; thus, it does not contribute reducing equivalents for potential ATP synthesis. Therefore, at high intakes of ethanol, the caloric equivalent may be less than 7 cal/g.

The following abnormalities are usually present in chronic alcoholics: imbalance in NAD/NADH, elevated plasma and tissue triglycerides, acetate circulating as fuel source, and elevated blood lactate. The alcoholic may be under- or overweight, depending on whether alcohol is consumed instead of, or in addition to, the usual diet. In either case, malnutrition may be clinically evident.

Moderate alcohol intake has been associated with a decreased incidence of coronary artery disease, although the underlying mechanism has yet to be determined. Potential benefits include elevation of HDL cholesterol, alterations in certain apolipoprotein fractions, lowering of blood pressure, or an anticoagulant effect.[3]

REFERENCES

1. **Mitchell, M. C. and Herlong, H. F.,** Alcohol and nutrition: caloric value, bioenergetics and relationship to liver damage, *Annu. Rev. Nutr.*, 6, 457, 1986.
2. **Korsten, M. A. and Lieber, C. S.,** Nutrition in the alcoholic, *Med. Clin. North Am.*, 63, 963, 1979.
3. **Moore, R. D. and Pearson, T. A.,** Moderate alcohol consumption and coronary artery disease. A review, *Medicine*, 65, 242, 1986.

ESSENTIAL FATTY ACIDS

John Patrick

INTRODUCTION

Perhaps the most important point to make about dietary fats is that it is difficult to prepare food in Western society without including at least 20% fat; consequently, deficiency in any simple sense is unlikely, even in Aboriginal communities with reported fat consumption as low as 2.4 to 3.8 g per person per day; i.e., in approximately 1 g of essential fatty acids (EFA), overt EFA deficiency was not seen.[1] This intake of dietary EFA is considerably below the 2 to 5% of energy as EFA recommended by some authors today.[2]

Despite this initial *caveat* there is a rapidly growing literature about the importance of dietary fat in many diseases and important physiological systems (for review see References 10 and 11). The major roles of dietary fatty acids are their structural role in membranes, their reserve fuel function, and their role as precursors of the prostaglandins and leukotrienes. From the nutritional point of view there are only two essential dietary fatty acids, linoleic (n-6) and linolenic (n-3), where the digit indicates the first carbon with a double bond counting from the methyl terminus.

FATTY ACIDS IN MEMBRANES

The role of fatty acids in membrane function is still being studied. A truly adequate model of membrane structure has yet to be proposed. It is quite clear that the Singer-Nicholson model[3] must now be replaced by models which recognize the presence of microdomains within the membrane,[4] and a range of other factors including acyl chain length and saturation, which are known to alter membrane functions. Studies with liposomes constructed of single phospholipids and purified transport proteins have shown that there are extremely complex lipid-protein interactions even in this simplified model.[5] It seems that many of the classical features of EFA deficiency first described in the 1920s by Burr and Burr[6] are dependent upon the structural functions of fatty acids rather than on their role in prostaglandin function, because an analogue of arachidonic acid-columbinic acid cures these classical features but is not incorporated into eicosanoids.[7]

BRAIN FATTY ACIDS

The distribution of n-6 and n-3 fatty acids within the brain is interesting. The predominant n-6 fatty acid is arachidonic acid and the predominant n-3 fatty acid is docosahexaenoic acid (DHA) 22:6n-3. DHA is found mainly in phosphatidyl ethanolamine and phosphatidyl serine, particularly in the photoreceptor outer segments in the eye. In the rest of the brain, these DHA-rich phospholipids are found in the metabolically active regions like mitochondria and synaptic membranes, as opposed to myelin. Depletion of DHA has been shown in experimental animals to be associated with both reduced visual acuity and diminishing learning capacity, but the effects are hard to induce (for review see Reference 21).

FATTY ACIDS AND PROSTAGLANDINS

Nevertheless, it is the role of fatty acids as precursors of prostaglandins which is attracting most attention at present, and particularly the effects of different proportions of n-3 and n-6 fatty acids. The nomenclature of prostaglandins is based on a C20 chain which forms a

Table 1
FUNCTIONS OF PROSTAGLANDINS

Hemostasis — thromboxanes and prostacyclins
Regulation of blood flow — PGI2, PGE2
Renin release — PGI2
Vasopressin activity — PGE2
Renal blood flow — PGE2, PGI2
Closure of the ductus arteriosis — PGI2 PGE2
Gastric mucosal protection — PGE2
Contraction of the pregnant uterus — PGF2

Table 2
DISEASE PROCESSES THOUGHT TO BE MODULATED BY PROSTAGLANDINS

Inflammatory disease — PGE2 LTs (leukotrienes)
Bartter's syndrome — PGI2
Dysmenorrhea — PGF2
Hypercalcemia — PGE2
Associated with cancer

5-membered ring between C3 and C12. This is prostanoic acid. From this structure, the E series prostaglandins are derived by addition of a ketogroup at C9 and a hydroxyl group at C11, whereas the F series have hydroxyl groups at both sites. The suffixes 1, 2, and 3 refer to parent fatty acids dihomo-y-linolenic acid C20:3 (n-3), arachidonic acid C20:4 (n-6), and eicosapentanoic acid C20:5 (n-3). Oleic acid can also be similarly metabolized, although it is not an EFA. It is known as the n-9 series. Some of the functions are shown in Table 1 and some of the diseases thought to be due to increased production are shown in Table 2.

The control of prostaglandin production is still incompletely understood. The precursor pool is in the fatty acids esterified in the Sn-2 position of phospholipids. The total amount of prostaglandin production per day is of the order of 1 mg. This should be compared with a daily dietary intake in North America of EFAs of 100 to 200 mg. Nevertheless, EFA-deficient diets do eventually lead to a lowering of prostaglandin production, but skin changes precede changes in prostaglandin excretion.[8] The process of conversion of the fatty acid precursors to the prostaglandins involves desaturase and elongase enzymes with preferential affinities for n-3, n-6, n-9 fatty acids. Consequently, n-9 metabolites only accumulate when EFA deficiency is present and it is this fact which led to the use of 20:3 (n-9)/20:4 (n-6) ratios as an indicator of EFA deficiency. Currently, the upper limit of normality is 0.2.[9]

Although prostaglandins will be dealt with elsewhere in this volume, it is impossible to complete a section on dietary fatty acids without mentioning the prudent diet and the potential role of fish oils.

There is a continuing vigorous debate over the validity of the current governmental recommendations that everyone adhere to a prudent diet of which the features relevant to this section are the recommendation that fat constitutes no more than 30% of total intake, and that saturated, monounsaturated and polyunsaturated fats be distributed in the ratio 1:1:1.[12,13] There is no question that in appropriately selected extreme cases plasma cholesterol can be reduced by dietary means.[14,15] There is considerable argument as to whether this will alter the long-term risk of heart disease and stroke (for review see Reference 16). Furthermore, much of the public will not distinguish in their food-buying habits between saturated, polyunsaturated, and partially hydrogenated polyunsaturates (i.e., high *trans* fatty acids); nevertheless, the public is voting with its feet at present and buying more margarine and less butter. The argument is far from settled.

The role of fish oils is also in a state of rapid change,[20] with a considerable amount of evidence accumulating to indicate that fish oils are hypocholesterolemic and hypotriglyceridemic and that habitual consumption is associated with lower heart disease rates.[15] The mechanism is not settled but it may lie in the competition between n-3 and n-6 fatty acids for prostaglandin and leukotriene synthetic pathways.[17] There is also evidence that dietary manipulation of fat intake can alter the expression of immune diseases in animals.[18,19] We have not yet reached the stage where analysis of lipoprotein or membrane fatty acid profiles is a necessary laboratory test, but that possibility is not so remote as it once seemed.

REFERENCES

1. **Mitra, K.,** Observations on the diet and nutritional state of an Aboriginal (HOS) tribe, *Ind. J. Med. Res.*, 30, 91, 1942.
2. **Houtsmuller, V. and Vergroesen, A. J.,** Proc. 2nd Int. Conf. on Essential Fatty Acids, London, 1985.
3. **Singer, S. J. and Nicholson, G. L.,** The fluid mosaic model of the structure of cell membranes, *Science,* 175, 720, 1972.
4. **Karnovsky, M. J., Kleinfeld, A. M., Hoover, R. L., and Klausner, R. D.,** The concept of lipid domains in membranes, *J. Cell Biol.*, 94, 1, 1982.
5. **Carruthers, A. and Melchior, D. L.,** How bilayer lipids affect membrane protein activity, *Trends Biochem. Sci.*, 11, 331, 1986.
6. **Burr, G. O. and Burr, M. M.,** A new deficiency disease produced by the rigid exclusion of fat from the diet, *J. Biol. Chem.*, 82, 345, 1929.
7. **Houtsmuller, U. M. T.,** Columbinic acid: a new type of essential fatty acid, *Prog. Lipid Res.*, 20, 889, 1981.
8. **Hansen, H. S. and Jensen, B.,** Urinary prostaglandin E_2 and vasopressin excretion in essential fatty-acid deficient rats: effect of linoleic acid supplementation, *Lipids,* 18, 682, 1983.
9. **Holman, R. T.,** Essential fatty acid deficiency in humans, in *CRC Handbook Series in Nutrition and Food, Section E: Nutritional Disorders,* Vol. 3, Rechcigl, M., Jr., Ed., CRC Press, Boca Raton, FL, 1978, 335.
10. **Wahle, K. W. J.,** Fatty acid modification and membrane lipids, *Proc. Nutr. Soc.*, 42, 273, 1983.
11. **Clandinin, M. T., Field, C. J., Hargreaves, K., Morson, L., and Zsigmond, E.,** Role of diet fat in subcellular structure and function, *Can. J. Physiol. Pharmacol.*, 65, 546, 1985.
12. Consensus conference on lowering blood cholesterol to prevent heart disease, *JAMA,* 253, 2080, 1985.
13. Committee on Medical Aspects of Food Policy, Diet and Cardiovascular Disease, DHSS Report on Health and Social Subjects, No. 28, Her Majesty's Stationery Office, London, 1984.
14. **Ehnholm, C., Huttunen, J. K., Pietinen, P., Leino, U., Mutanen, M., Kostianen, E., Pikkarainen, J., Dougherty, R., Iacono, J., and Paska, P.,** Effect of diet on serum lipoproteins in a population with a high risk of coronary heart disease, *N. Engl. J. Med.*, 307, 850, 1982.
15. **Phillipson, B. E., Rothrock, D. W., Connor, W. E., Harris, W. S., and Illingworth, D. R.,** Reduction of plasma lipids, lipoproteins and apoproteins by dietary fish oils in patients with hyperglyceridemia, *N. Engl. J. Med.*, 312, 1210, 1985.
16. **Oliver, M. F.,** Strategies for preventing coronary heart disease, *Nutr. Rev.*, 43, 257, 1985.
17. **Glomset, J. A.,** Fish, fatty acids and human health, *N. Engl. J. Med.*, 312, 1253, 1985.
18. **Gorman, L. C.,** The role of diet in animal models of systemic lupus erythematosus, possible implications for human lupus, *Semin. Arthritis Rheum.*, 15, 61, 1985.
19. **Yoo, T. J., Chiu, H. C., Spector, A. A., Whiteaker, R. S., Denning, G. M., and Lee, N. F.,** Effect of fatty acid modifications of cultured hepatoma cells on susceptibility to complement mediated cytolysis, *Cancer Res.*, 40, 1084, 1984.
20. **Lands, W. E. M.,** *Fish and Human Health,* Academic Press, New York, 1986.
21. **Neuringer, M. and Connor, W. E.,** n-3 fatty acids in the brain and retina: evidence of their essentiality, *Nutr. Rev.*, 44, 285, 1986.

VITAMIN A

John Patrick

INTRODUCTION

An excellent and extensive review of vitamin A was published in 1985.[1] Vitamin A should be referred to as retinol. In practice, the term vitamin A is widely used and may or may not include carotenoids as well as retinol derivatives (aldehyde, ester, and acid). The vitamin A activity of these various compounds should now be expressed in retinol equivalents (1 equivalent = 1 μg retinol). A FAO/WHO committee[2] has published conversion factors of which the most important is for the conversion of β-carotene to retinol equivalents: RE = μg β-carotene/6. It is an arbitrary figure which should be altered depending upon source (oil or plant tissue) and dose (large or small, the latter being more efficiently used).

Vitamin A is susceptible to oxidation, particularly in the presence of trace elements, but food chemists have found ways to make it more stable in processed food; thus, deficiency from nutritional deprivation is unlikely in the first world.

METHODS OF DETERMINATION

There are still considerable technical problems with the analysis of vitamin A. The older spectrophotometric and fluorometric methods are now being replaced by high-performance liquid chromatographic (HPLC) methods. The large number of these methods indicates that the ideal assay has yet to be described (for review see Reference 3). The great asset of HPLC methods is their capacity to separate *cis* and *trans* isomers.

DIETARY SOURCES

Vitamin A is derived from long-chain fatty acid retinyl esters in fish oils, liver, kidney, eggs, and dairy products and from carotenoids in plants plus the ubiquitous enriched processed foods. (Carotenoids are split by a dioxygenase in the intestinal cells to produce two molecules of retinaldehyde.)

ABSORPTION AND METABOLISM

Vitamin A is esterified with long-chain fatty acids in epithelial cells and transported in chylomicrons via the lymph to the liver, for storage as fatty acid esters. This reserve is reduced to 10% before any changes occur in plasma retinol.[4] Secretion of vitamin A from the stored esters requires hydrolysis and attachment of retinol to retinol-binding protein (RBP) (1 mol retinol/mol protein). RBP normally associates on a molar basis with prealbumin.

RBP is unusual for a transport protein in that synthesis continues in the absence of an adequate supply of retinol; however, it is not secreted until retinol is available. Repletion with retinol can then lead to rapid secretion with transiently supranormal levels. A significant amount of catabolism occurs in the kidney because RBP is filtered and catabolized by the tubules. Renal failure limits this process and can cause a high plasma RBP; oral contraceptives also increase RBP. Protein deficiency and liver disease can produce low plasma RBP levels. Zinc deficiency and infections can also cause a low RBP. Because of the short half-life of RBP (7 h), it has been proposed as a marker protein for malnutrition.[5] The coexistence of abnormal renal function in severe protein and energy malnutrition will limit its usefulness.

Uptake by tissues depends upon RBP binding with a specific receptor. This, in turn, delivers retinol to an intracellular binding protein different in molecular weight and amino acid sequence from RBP. The properties of the currently known vitamin A binding proteins were reviewed in 1985.[11]

HYPERVITAMINOSIS A

Increasing numbers of reports of hypervitaminosis A are occurring, probably because of the availability of highly potent preparations and the great enthusiasm of many laymen and, regrettably, some physicians for megadosing. Vitamin A is membranolytic, lysing red cells *in vitro*. The main features of toxicity are erythema, eczema, alopecia, bone exostoses, hemorrhages, bone fractures, hepatic damage, and death. Dosages in excess of 60 mg are necessary for acute toxicity.

VITAMIN A AND VISION

This is still the only function for which a reasonable understanding can be claimed. Deficiency of vitamin A continues to be the major cause of blindness in Third World children.

The mechanism of vitamin A action in the retina is as follows: 11-*cis* retinaldehyde combines with opsin to form rhodopsin and light causes isomerization so that most of the retinaldehyde is in the all-*trans* form. This process straightens the polyene chain separating it from the protein. It is assumed that this conformational change leads to stimulation of optic neurons. An alteration in membrane permeability to sodium is involved in this process.

The regeneration of rhodopsin and the scavenging of 11-*cis* retinaldehyde by retinol dehydrogenase (a zinc metalloenzyme) are less well understood. Zinc deficiency can cause night blindness probably via impairment of this mechanism.

OTHER FUNCTIONS OF VITAMIN A

Unlike vision which requires retinaldehyde, other functions except reproduction seem to proceed equally well from retinoic acid which, *in vivo,* is irreversibly produced from retinaldehyde. Molecular mechanisms for the role of retinoic acid are not clear, but it does influence RNA synthesis. Culture of keratinocytes (in the presence of vitamin A) alters the pattern of proteins produced.[6] The effect of vitamin A on gene expression may be modified by any of the several binding proteins which have been discovered.

Vitamin A also acts as a sugar carrier in the synthesis of glycoproteins,[7] taking, for example, mannose from GDP mannose to donate it to membrane receptors. Vitamin A deficiency, therefore, leads to decreased glycoprotein biosynthesis such as gobletin in intestinal goblet cells. How essential the sugar carrier function is, remains unresolved, because animals given retionic acid, which will not form the sugar carrier, grow perfectly well.[8]

ASSESSMENT OF VITAMIN A STATUS

Ideally, since vitamin A is stored in the liver, a measure of liver vitamin A is required. A value of 20 μg RE per gram of liver has been suggested as safe.[9] Isotopic tracer approaches have been used; the problems, as always, are with the underlying assumptions about equilibration times, mixing, and compartmentation. Rietz found good correlation between liver measurements and isotopic measurements in rats.[10]

Plasma retinol does not fall until most of the liver stores are gone. A threshold figure is 20 μg/dl.[11] Impaired dark adaptation is often used to confirm a suspicion of vitamin A deficiency, especially when combined with a therapeutic trial. In small children, the best early sign is a history of stumbling and difficulty in reaching for objects at dusk. McLaren

Table 1
CLINICAL FEATURES OF VITAMIN A DEFICIENCY

General	Liver
Anorexia	Metaplasia of bile ducts
Growth failure and weight loss	Degeneration of Kuppfer cells
Dry and keratinized membranes	Nervous system
Increased infection rate	Incoordination
Eyes	Nerve degeneration
Bitot spots	Constriction at foramina
Night blindness	Hydrocephalus
Xerophthalmia	Bone
Keratomalacia	Defective modeling
Perforation of the eye	Restriction of brain and spinal cord cavities
Blindness	Cancellous bone abnormalities
Lungs	Reproduction
Pneumonia	Cessation of spermatogenesis
Renal	Menstrual irregularities
Cystitis	Fetal development abnormalities
Calculi	Blood
Nephrosis	Anemia

has described the clinical features in Jordanian children with malnutrition, emphasizing its high mortality.[12] Cases also occur in the first world, usually in the presence of malabsorption.[13]

The main clinical features of vitamin A deficiency are given in Table 1.

REQUIREMENTS

Many years ago, Hans Krebs showed that 390 μg of retinol was the minimum protective intake and most committees now recommend between 800 and 1000 μg/d. Breast milk provides 40 to 65 μg per retinol deciliter. In experimental human vitamin A depletion, 600 μg/d was required to correct all abnormalities.[14]

VITAMIN A AND CANCER

The squamous metaplasia seen in vitamin A deficiency is similar to some precancerous conditions. Vitamin A can reverse some cancerous changes *in vitro* and deficiency of vitamin A enhances susceptibility to chemical carcinogens. Peto et al.[15] found that in seven out of eight studies comparing retinol levels in controls and cancer patients, lower levels were found in the cancer patients.

RETINOIDS AND SKIN DISEASE

Synthetic retinoids have been used successfully in psoriasis, abnormal keratinization, and severe acne. They are, however, toxic and can only be used with very strict monitoring of liver function.

CONCLUSION

Although there are few situations in which vitamin A measurements are mandatory, more widespread measurement in clinical practice might be expected to produce interesting new data.

REFERENCES

1. **Pitt, G. A. J.,** Vitamin A, in *Fat-Soluble Vitamins,* Diplock, A. T., Ed., Heineman, London, 1985.
2. FAO/WHO Joint Expert Group, *WHO Tech. Rep. Ser.*, p. 362, 1967.
3. **Taylor, R. F.,** Chromatography of carotenoids and retinoids, *Adv. Chromatogr.*, 22, 157, 1983.
4. **Dowling, J. E. and Wald, G.,** *Proc. Natl. Acad. Sci. U.S.A.*, 44, 648, 1958.
5. **Venkataswamy, G., Glover, J., Cobby, M., and Pirie, A.,** Retinol-binding protein in serum of xerophthalmic malnourished children before and after treatment at a nutrition centre, *Am. J. Clin. Nutr.*, 30, p. 1968, 1977.
6. **Fuchs, E. and Green, H. J.,** Regulation of terminal differentiation of cultured human keratinocytes by vitamin A, *Cell,* 25, 617, 1981.
7. **De Luca, L. M.,** Direct involvement of vitamin A in glycosyl transfer reactions of mammalian membranes, *Vitam. Horm.*, 35, 1, 1977.
8. **Sporn, M. B. and Roberts, A. B.,** Role of retinoids in differentiation and carcinogenesis, *Cancer Res.*, 43, 3034, 1983.
9. **Olson, J. A.,** New approaches to methods for the assessment of nutritional status of the individual, *Am. J. Clin. Nutr.*, 35, 1166, 1982.
10. **Rietz, P., Wiss, O., and Weber, F.,** Metabolism of vitamin A and the determination of vitamin A status, *Vitam. Horm.*, 32, 237, 1974.
11. **Ong, D. E.,** Vitamin A-binding proteins, *Nutr. Rev.*, 43, 225, 1985.
12. **McLaren, D. S., Shirajian, E., Tchalian, M., and Khoury, G.,** Xerophthalmia in Jordan, *Am. J. Clin. Nutr.*, 17, 117, 1985.
13. **Anon.,** Vitamin A deficiency — a global disease, *Nutr. Rev.*, 43, 240, 1985.
14. **Sauberlich, H. E., Hodges, R. E., Wallace, D. L., Kolder, H., Canham, J. E., Hood, J., Raica, N., and Lowry, L. K.,** Vitamin A metabolism and requirements in the human studied with the use of labeled retinol, *Vitam. Horm.*, 32, 251, 1974.
15. **Peto, R., Doll, R., Buckley, J. D., and Sporn, M. B.,** Can dietary beta-carotene materially alter human cancer rates, *Nature,* 290, 201, 1981.

VITAMIN D

Jo Ellen Welsh

Vitamin D refers to a group of steroid compounds ("calciferols") with antirachitic properties which can be formed in both plants and animals by UV irradiation of precursors. Ergocalciferol (vitamin D_2) is derived from the plant sterol, ergosterol, and cholecalciferol (vitamin D_3) is derived from 7-dehydrocholesterol in animal epidermis. Vitamin D written without a subscript refers to either D_2 or D_3, as the metabolism and biological effects of these two compounds in humans appear to be equivalent.

Very few natural foods contain significant amounts of preformed calciferols (fish oils, liver, egg yolk), therefore, fortification of foods such as milk and cereals is now common. However, dietary vitamin D is only necessary if exposure to sunlight is extremely restricted, as adequate amounts can be generated *in vivo* under normal physiologic conditions.[1]

Dietary vitamin D is absorbed in the small intestine with other food lipids and, therefore, bile salts are necessary for complete absorption. Thus, vitamin D deficiency may be present in persons with chronic lipid malabsorption who have limited sun exposure.

Vitamin D (either from intestinal absorption or epidermal synthesis) is bound to a vitamin-D-binding protein (DBP) in plasma and transported to the liver. About 25% of the available vitamin D is then hydroxylated at the 25-position, the remainder being excreted (via bile) or stored (in fat, muscle, and liver). The product of liver hydroxylation, 25(OH) vitamin D, is the major circulating form of the vitamin whose plasma level reflects body stores. Circulating 25(OH) vitamin D, bound to DBP, is taken up by the kidney and further hydroxylated to yield either $1,25(OH)_2$ vitamin D or $24,25(OH)_2$ vitamin D. $1,25(OH)_2$ vitamin D is the biologically active vitamin D metabolite whose actions include stimulation of intestinal calcium and phosphorus absorption, renal calcium conservation, and bone mineral mobilization. Extensive evidence supports the view that $1,25(OH)_2$ vitamin D production is controlled by strict physiological signals related to the calcium needs of the individual. The major regulators of the renal 1-hydroxylase include peptide hormones (parathyroid hormone, calcitonin), ions (Ca^{2+}, Pi), and the vitamin D metabolites themselves. In fact, any agent that affects demand for, or handling of, calcium will indirectly influence $1,25(OH)_2$ vitamin D production. Such agents include estrogen, insulin, glucocorticoids, calmodulin, and prostaglandins; for an excellent review on the vitamin D endocrine system, see Reference 2.

The production of $24,25(OH)_2$ vitamin D is inversely proportional to that of $1,25(OH)_2$ vitamin D, however, the physiological significance of this pathway is unclear at present. Although it is traditionally considered to be a catabolic pathway,[3] possible functional roles for $24,25(OH)_2$ vitamin D in parathyroid hormone secretion and normal bone development have also been suggested.

The mode of action of $1,25(OH)_2$ vitamin D in its primary target tissues (intestine, kidney, and bone) parallels that of the classic steroid hormones (i.e., it binds to a subcellular receptor, interacts with DNA, initiates mRNA production and protein synthesis). The most widely studied protein induced by $1,25(OH)_2$ vitamin D is a cytosolic calcium-binding protein (CaBP) found in intestinal mucosal cells. The induction of CaBP is positively correlated with vitamin D-dependent active transport of calcium.[4] Although not as well characterized as CaBP, vitamin-D-dependent proteins induced in other target tissues include the 24-hydroxylase enzyme (kidney), and vitamin K-dependent GLA protein and alkaline phosphatase (bone).

Recently, specific receptors for $1,25(OH)_2$ have also been identified in tissues not previously considered to be target tissues for vitamin D action (thymus, adrenal, pancreas, pituitary, cancer cell lines). Thus, the biological actions of vitamin D appear to be as diverse as cell differentiation, insulin secretion, and host defense; for an update on the biochemistry of vitamin D receptors, see Reference 3.

Although vitamin D deficiency can simply be due to inadequate intake and/or low sun exposure, disturbances in any of the steps in absorption, activation, or action of the vitamin also occur clinically.[2] These include lipid malabsorption (disease- or drug-induced), liver or kidney disease of sufficient magnitude to interfere with activation of the parent compound, and genetic defects in $1,25(OH)_2$ vitamin D receptor binding or action (for example, vitamin D-resistant rickets).

Lack of vitamin D action results in chronically low plasma calcium and phosphorus due to impaired intestinal absorption and impaired release from bone. Low calcium $\times$ phosphorus product in plasma, in turn, cannot support normal bone formation, resulting in poorly mineralized (soft) bone (rickets in children, osteomalacia in adults).

Circulating level of 25(OH) vitamin D reflects body stores, and, thus, a low level will indicate either lack of the parent compound (i.e., either low skin synthesis, low dietary intake, or impaired absorption) or impaired activation in the liver. Because the renal production of $1,25(OH)_2$ vitamin D is geared to the calcium needs of the individual, the circulating level must be interpreted in light of the prevailing PTH, calcium, and phosphorus concentrations. For example, in a patient with hypocalcemia and elevated PTH, an elevation in $1,25(OH)_2$ vitamin D would be expected (assuming 25[OH] vitamin D is in the normal range). If the actual value for $1,25(OH)_2$ vitamin D is normal or below, a defect in the regulation or activity of the renal 1-hydroxylase should be suspected (for example, kidney disease, vitamin D-resistant rickets, type I). If, however, hypocalcemia persists despite elevations in PTH and $1,25(OH)_2$ vitamin D, target organ resistance (i.e., receptor defect) would be suspected (for example, vitamin D-resistant rickets, type II).

Measurement of the vitamin D metabolites is not routinely done in the clinical biochemistry lab, as the methods are expensive and time consuming. The most common procedures call for lipid extraction of plasma, separation of the various metabolites by chromatography (high performance liquid chromatography is optimal), and subsequent quantitation by individual competitive binding assays.[5]

Clinical detection of overt vitamin D deficiency, therefore, is usually based on characteristic skeletal changes and biochemical alterations. In rickets, flaring of the epiphyses of long bones, rachitic rosary, bowing of the legs, and spinal deformities occur. Plasma alkaline phosphatase is usually elevated, serum phosphorus is low, and calcium is frequently in the normal range. In adults, osteomalacia most often presents as bone pain, fractures, and skeletal deformities in chronic cases. Extensive skeletal demineralization is evident on X-ray, alkaline phosphatase is elevated, and plasma calcium is often low.

Since vitamin D can be stored, excessive intakes (2000 to 3000 IU/d) can lead to toxicity symptoms (nausea, vomiting, hypercalcemia, kidney damage, bone pain, and soft tissue calcification), although individual tolerances vary widely.

REFERENCES

1. **Fraser, D. R.,** The physiological economy of vitamin D, *Lancet,* 2, 969, 1983.
2. **Norman, A. W.,** The vitamin D endocrine system, *Physiologist,* 28, 219, 1985.
3. **Haussler, M. R.,** Vitamin D receptors: nature and function, *Annu. Rev. Nutr.,* 6, 527, 1986.
4. **Bronner, F., Pansu, D., and Stein, W. D.,** An analysis of intestinal calcium transport across the rat intestine, *Am. J. Physiol.,* 250, G561, 1986.
5. **Shepard, R. M. and DeLuca, H. F.,** Determination of vitamin D and its metabolites in plasma, *Methods Enzymol.,* 67, 393, 1980.

VITAMIN E

John Patrick

INTRODUCTION

The features of vitamin E deficiency, which are quite strikingly species dependent, were described in the 1920s. However, it was not until 1938 that the structure of tocopherol was published and only in 1974 that the IUPAC-IUB Commission on Biochemical Nomenclature sorted out the descriptive confusion. The term vitamin E is now used generically to include both tocols and tocotrienols having the biological activity of tocopherol. The only known biochemical function of vitamin E, which may be the mechanism of its action, is its capacity to react with oxidants which might otherwise cause lipid peroxidation. Kasparek has reviewed the chemistry.[1] There have been many reviews of vitamin E of which the recent one by Diplock is an excellent example[4] (see Table 1).

METHODS OF ANALYSIS

The analysis of vitamin E, once purified, can be achieved by spectrophotometric, colorimetric, and fluorometric methods. HPLC has provided accurate and rapid analyses from biological tissues. The demonstration that a given tocol or tocotrienol has vitamin E activity requires the use of a bioassay, usually rat sterility or erythrocyte hemolysis.

DIETARY SOURCES

Vitamin E deficiency should be rare in the first world because we consume a range of vegetable oils in which vitamin E activity is found. Some oils, like coconut oil (largely saturated), do not contain significant vitamin E activity. Infant formula in North America now contains from 120 to 200 μg/g of lipid, which appears to be adequate except in very low-birth-weight infants also receiving iron.[2] In the past it was low to inadequate.

ABSORPTION AND METABOLISM

Absorption of vitamin E is not well characterized because of the technical difficulties. A reasonable estimate for absorption from the normal diet is probably 30%. Esters of tocopherol are hydrolyzed in the gut and transport to the body involves carriage in the lymphatics as a lipoprotein (mainly VLDL complex). Bile salts and pancreatic secretions are necessary for absorption.[3] Linoleic acid inhibits vitamin E absorption. Metabolic inhibitors have little effect, suggesting that uptake is by nonsaturable passive diffusion, i.e., not carrier mediated. The tissue levels of vitamin E increase with age especially in adipose tissue. Liver tocopherol is around 2 μg/g at 3 months gestation, 5 μg/g at birth, and approximately 20 μg/g at maturity. Adipose tissue contains 5 to 10 times more, and the adrenal up to 500 times more. The role of vitamin E in the adrenal is unknown. The literature on subcellular distribution of vitamin E is contradictory.

The most common metabolic fate of vitamin E is oxidation to the tocopheroxy radical which can then be reduced to tocopherol. This occurs in a lipid environment. In a polar environment, like water, irreversible oxidation to a quinone can occur. The ease with which these reactions occur makes it extremely difficult to be certain of what happens *in vivo,* simply because of separation artifacts.

Excretion of vitamin E is predominantly via the gut — only 1% via the kidney; significant amounts may be lost through the skin.

Table 1
EFFECTS OF VITAMIN E DEFICIENCY

Species	Pathological changes	Ref.
Rat	Fetal resorption failure of ectodermal cavity formation	5
Mouse	Fetal resorption failure of ectodermal cavity formation	6
Guinea pig	Fetal resorption failure of ectodermal cavity formation	7
Rat	Atrophy of germinal epithelium in the testis	8
Hamster	Atrophy of germinal epithelium in the testis	8
Guinea pig	Atrophy of germinal epithelium in the testis	9
Mouse	Resistant to germinal atrophy for a long time	10
Most species	Necrotizing myopathies	
Growing chicks	Encephalomalacia, not simply due to lipid peroxidation (requires high linoleic acid intake)	11
	Exudative diathesis	12
Ruminants, pig, and rabbit	Myocardial damage	13
Rat	Hepatic necrosis (requires vitamin E deficiency, low sulfur amino acid intake, and Se deficiency)	15

REQUIREMENT FOR VITAMIN E

The requirement for vitamin E is difficult to specify because of multiple nutrient interactions. The recommendation is 3 to 10 mg/d, modified by age, sex, and pregnancy. The major nutrient interactions are: with polyunsaturated fatty acids (PUFA) the more PUFA the more vitamin E is required, but these two usually go together in food; with selenium, to some extent selenium can fulfill the role of vitamin E, thereby reducing the requirement for vitamin E; with other antioxidants in food, they also reduce vitamin E requirements in some experimental situations; finally, with vitamin A. For ill-understood reasons not related to antioxidant mechanisms, vitamin E reduces the rate at which vitamin A is lost from the liver in subjects consuming a vitamin A-deficient diet.

VITAMIN E DEFICIENCY — PATHOLOGY

The subject is complicated by species differences and by the interactions with selenium. Table 1 summarizes the most important points.

The pathogenic mechanisms are not completely understood. Many, but not all, investigators believe that the effects of vitamin E deficiency can be explained by lipid peroxidation. Part of the problem has been the difficulty of reliably measuring lipid peroxidation *in vivo*. It now seems that the measurement of pentane and ethane in expired air may solve this problem.[16] The technical difficulties are all secondary to the lability of the free radicals which it is presumed; Vitamin E destroys. There must also be a vitamin E regeneration mechanism, but to date, this has not been definitively described.

If the generally held view that the principal function of vitamin E is the protection of membranes from peroxidation damage is correct, then the intracellular locus of vitamin E ought to reflect this function. Mitochondrial and endoplasmic reticulum membranes have much higher degrees of membrane unsaturation than plasma membranes; therefore, vitamin E deficiency should affect these membranes most. This has been shown to be the case;[17] however, Giasuddin and Diplock found that when membrane lipids were fractionated into phospholipid categories there was no effect of tocopherol on the PUFA of the neutral lipids, but only on the polar lipids, suggesting that a more complex model of actions of vitamin E is required.[18]

VITAMIN E AND HUMAN HEALTH

For humans without gastrointestinal disease or the rare absence of lipoprotein carriers for vitamin E, it is extremely unlikely that vitamin E deficiency will occur. In experimental animals, dietary deficiency is difficult to achieve. Even the definition of a diet which is inadequate in vitamin E is difficult, because of the complex interactions with other factors such as amount and type of dietary fat, amounts of other antioxidants such as selenium, vitamin A, ambient oxidants like ozone and nitrous oxide, etc. In addition, each of these factors has its own train of modifying factors such as amounts of trace elements. Under these circumstances, controlled clinical trials with defined outcome measurements, in specified disease states, are the only way to demonstrate benefits. This has been achieved in a few instances.

Abetalipoproteinemia

Muller et al.[19] have reviewed the evidence that for this rare disease and for some cases of severe malabsorption, vitamin E supplementation is effective for the prevention of the degeneration of the nervous system which usually occurs in abetalipoproteinemia. In order to achieve appropriate results biochemical monitoring of vitamin E is necessary.

Retrolental Fibroplasia of Premature Neonates

Vitamin E supplementation reduces the severity of oxygen damage in these infants.[20]

Cystic Fibrosis

There is some evidence that there may be neurological problems in this disease consequent upon inadequate vitamin E levels. For review see Reference 21.

Intraventricular Hemorrhage (IVH)

There is suggestive evidence that the incidence of IVH in premature babies (less than 32 weeks) may be reduced by vitamin supplementation.[22] Monitoring of blood levels appears to be necessary.

Intermittent Claudication

There is one trial indicating that vitamin E supplementation may be helpful in intermittent claudication.[23]

RECOMMENDED INTAKES

In view of the foregoing description of the problems of describing a diet which is deficient in vitamin E, it is not surprising that the committees responsible for dietary recommendations write lengthy *caveats* before agreeing that the current intake of 7 to 9 mg of D-α-tocopherol is adequate, especially considering the 20% contribution of other vitamins.

CONCLUSION

For the clinical biochemist, it is apparent that reliable assays of vitamin E would be worthwhile additions to the management of some important groups of patients, particularly neonates and patients with cystic fibrosis.

REFERENCES

1. **Kasparek, S.,** *Vitamin E: A Comprehensive treatise,* Machlin, L. J., Ed., Marcel Dekker, New York, 1980.
2. **Gross, S. J. and Gabriel, E.,** Vitamin E status in preterm infants fed human milk or infant formula, *J. Pediatr.,* 106, 635, 1985.
3. **Gallo-Torres, H. E.,** Studies on the intestinal lymphatic absorption, tissue distribution and storage of vitamin E, *Acta Agric. Scand. Suppl.,* 19, 97, 1973.
4. **Diplock, A. T.,** *Vitamin E: In Fat Soluble Vitamins,* Ed., Heinemann, London, 1985.
5. **Evans, H. M. and Burr, G. O.,** New dietary deficiency with highly purified diets; supplementary requirement of diet of pure casein, sucrose and salt, *Proc. Soc. Exp. Biol. Med.,* 25, 41, 1927.
6. **Bryan, W. L. and Mason, K. E.,** Vitamin E deficiency in mouse, *Am. J. Physiol.,* 131, 263, 1940.
7. **Pappenheimer, A. M. and Goettsch, M.,** Death of embryos in guinea pigs on diets low in vitamin E, *Proc. Soc. Exp. Biol. Med.,* 47, 268, 1941.
8. **Mason, K. E.,** Differences in testis injury and repair after vitamin A deficiency, vitamin E deficiency and inanition, *Am. J. Anat.,* 52, 153, 1933.
9. **Pappenheimer, A. M. and Schogoleff, C.,** Testis in vitamin E deficient guineapigs, *Am. J. Pathol.,* 20, 239, 1944.
10. **Menschik, Z., Munk, M. K., Rogalski, T., Tymaszewski, O., and Szezesniak, T. J.,** Vitamin E studies on mice with special references to distribution and metabolism of lipids, *Ann. N.Y. Acad. Sci.,* 52, 94, 1949.
11. **Pappenheimer, A. M. and Goettsch, M.,** A cerebellar disorder in chicks apparently of nutritional origin, *J. Exp. Med.,* 53, 11, 1931.
12. **Dam, H. and Glavind, J.,** Alimentary exudative diathesis, *Nature,* 142, 1077, 1938.
13. **Gullickson, T. W.,** The relation of vitamin E to reproduction in dairy cattle, *Ann. N.Y. Acad. Sci.,* 52, 256, 1949.
14. **Schwarz, K.,** Tocopherol Als Leberschutzstoff, *Z. Physiol. Chem.,* 281, 101, 1944.
15. **Gyorgy, P. and Goldblatt, H.,** Further observations on production and prevention of dietary hepatic injury in rats, *J. Exp. Med.,* 89, 245, 1949.
16. **Hafeman, D. G. and Hoekstra, W. G.,** Lipid peroxidation in vivo during vitamin E and selenium deficiency in the rat as monitored by ethane evolution, *J. Nutr.,* 107, 656, 1977.
17. **Molenaar, I., Vos, J., Jager, F. C., and Hommes, F. A.,** The influence of vitamin E deficiency on biological membranes: an ultrastructural study on the intestinal epithelial cells of duckling, *Nutr. Metab.,* 12, 858, 1970.
18. **Giasuddin, A. S. M. and Diplock, A. T.,** The influence of vitamin E on membrane lipids of mouse fibroblasts in culture, *Arch. Biochem. Biophys.,* 210, 348, 1981.
19. **Muller, D. P. R., Lloyd, J. K., and Wolff, O. H.,** Vitamin E and neurological function, *Lancet,* 1, 225, 1983.
20. **Hittner, H. M. et al.,** Retrolental fibroplasia: efficacy of vitamin E in a double blind clinical study of preterm infants, *N. Engl. J. Med.,* 305, 1365, 1981.
21. **Roy, C. C. and Weber, A. M.,** Nutrition of the child with cystic fibrosis, in *Nutrition in Pediatrics,* Walker, W. A. and Watkins, J. B., Eds., Little, Brown, Boston, 1985, 463.
22. **Chiswick, M. I., Wynn, J., and Toner, N.,** Vitamin E and intraventricular hemorrhage in the newborn, *Ann. N.Y. Acad. Sci.,* 393, 109, 1982.

VITAMIN K

Jo Ellen Welsh

Vitamin K, a fat-soluble factor essential for normal blood coagulation, exists in two natural forms: phylloquinone (K1), derived from plants, and menaquinone (K2), derived from bacteria. Additionally, a water-soluble, synthetic form (menadione) exhibits potent vitamin K activity.

The only known biochemical function of vitamin K is as a co-factor for glutamyl carboxylases, enzymes which catalyze the carboxylation of glutamic acid residues in proteins. Proteins containing carboxylated glutamic acid residues (GLA proteins), which all appear to be involved in calcium binding, include four blood-clotting factors (prothrombin and factors VII, IX, X),[1] two bone-matrix proteins (osteocalcin and matrix GLA protein),[2,3] and a renal GLA protein.[1] As vitamin K-dependent glutamylcarboxylase activity has been detected in many other tissues (lung, testes, spleen, tumors),[1] it is likely that additional GLA proteins exist.

Primary vitamin K deficiency is extremely rare, except in newborn and premature infants, whose sterile gastrointestinal tract and minimal stores make them particularly susceptible to depletion (this condition is known as hemorrhagic disease of the newborn). Most documented cases of vitamin K deficiency are disease- or drug-induced. Because the absorption of vitamin K (from either diet or bacterial synthesis) occurs as for other food lipids, lipid malabsorption[2] may result in biochemical evidence of vitamin K deficiency. Additionally, since 50 to 100% of the daily requirement can be obtained from gut bacterial synthesis, chronic antibiotic use can result in vitamin K depletion.

Although prothrombin time is always increased in severe vitamin K deficiency, detection of the vitamin K-deficient state is ideally based on a radioimmunoassay for abnormal prothrombin molecules which lack the normal GLA residues.[5] This assay is 1000 times more sensitive than prothrombin time in detecting deficiency. The anticoagulant effect of dicoumeral and warfarin results from their potent inhibition of vitamin K action, presumably due to structural similarities between these compounds.

Although fat-soluble and presumably stored, little is known about the tissues involved. Metabolites of vitamin K appear in both feces and urine, thus, excretion is probably via both routes. There are no known reports of toxicity for any of the natural vitamin K substances. The synthetic, water-soluble menadione, however, is readily oxidized to toxic metabolites which cause hemolysis when given in excess.

REFERENCES

1. **Olson, R. E.,** The function and metabolism of vitamin K, *Annu. Rev. Nutr.*, 4, 281, 1984.
2. **Krasinski, S. D., Russel, R. M., Furie, B. C., Kruger, S. F., Jacques, P. F., and Furie, B.,** The prevalence of vitamin K deficiency in chronic gastrointestinal disorders, *Am. J. Clin. Nutr.*, 41, 639, 1985.
3. **Price, P. A. and Williamson, M. K.,** Primary structure of bovine matrix GLA protein, a new vitamin K-dependent bone protein, *J. Biol. Chem.*, 260, 14971, 1985.
4. **Price, P. A., Poser, J. W., and Raman, N.,** Primary structure of the carboxyglutamic acid-containing protein from bovine bone, *Proc. Natl. Acad. Sci.* U.S.A., 73, 3374, 1976.
5. **Blanchard, R. A., Furie, B. C., Kruger, S. F., Waneck, G., Jorgensen, M. J., and Furie, B.,** Immunoassay of human prothrombin species which correlate with functional coagulant activities, *J. Lab. Clin. Med.*, 101, 242, 1983.

CARNITINE

John Patrick

Carnitine is thought to be primarily required for the mitochondrial oxidation of fatty acids. Two carnitine acyl-transferase enzymes are involved (CAT I and CAT II). CAT I catalyzes the rate-limiting inward transfer of acyl groups, while CAT II transfers the acyl group to mitochondrial CoA, releasing carnitine for reutilization. A translocase is required for the movement of carnitine across the inner mitochondrial membrane.

L-Carnitine is synthesized from lysine and methionine via trimethyllysine, which is hydroxylated and converted, first to methylaminobutyraldehyde, then to butyrobetaine. This last step does not occur in either cardiac or skeletal muscle, but does occur in kidney and brain as well as liver. Carnitine is recycled and undergoes little degradation, usually being excreted intact. The metabolism of carnitine has been extensively reviewed recently.[1-5] In addition to its role in the transport of long-chain fatty acids into the mitochondrion, it is possible that carnitine is also involved in branch-chained α-keto acid oxidation, the removal of chain-shortened fatty acids from liver peroxisomes and adjustment of acyl-CoA/CoASH ratios. It has also been proposed that it buffers excess acyl groups in the mitochondrion by allowing their transport out of the mitochondrion and eventual excretion.[6] This has been documented in starvation.[7] Such a mechanism would also allow excretion of other toxic organic acids, if they have appropriate affinities for carnitine. All acyl CoA compounds are potentially toxic; thus, Stumpf et al.[6] have made a valuable contribution by focusing attention on the turnover of carnitine and on the ratio of acyl carnitine to free carnitine as opposed to the absolute levels of free and total carnitine, which have been previously used to diagnose carnitine deficiency. Thus, for instance, in Reye's syndrome, the elevated serum levels of carnitine found initially[8] are not so important as the facts that urinary excretion rates of both free and acyl carnitine are increased 40-fold over normal levels, and carnitine supplementation appears to halt clinical progression.[9] Although carnitine is synthesized in the body the requirement in pathological states may exceed the synthetic capacity.

The total body carnitine content of a normal 70-kg man is approximately 100 mmol, 98% of which is in muscle. The renal threshold is approximately 50 μmol/l. In dogs, extracellular fluid turnover time is 0.4 h, muscle turnover time 232 h, and whole-body turnover time 63 d. Data from man are comparable.[10]

Table 1 lists some of the main groups at risk of carnitine deficiency. Primary carnitine deficiency is usually classified into three groups, depending upon clinical presentation. The myopathic variety usually develops in the second decade with progressive muscular weakness and myoglobinuria. Pathologically, there is deposition of lipid in the muscle fibers and muscle carnitine is low, but levels in liver and blood are normal. Replacement of dietary long-chain lipids with short-chain lipids is effective therapy. The disease is rare.

Systemic carnitine deficiency is also rare. It presents early in life with muscle weakness, cardiomyopathy, and episodes of encephalopathy that resemble Reye's syndrome, except that plasma and tissue carnitine levels are low. The encephalopathy responds to i.v. glucose and the cardiomyopathy is improved by carnitine therapy. It is likely that defects in carnitine synthesis and transport at both the gastrointestinal and the renal epithelium can lead to the syndrome of carnitine deficiency.

A mixed picture unresponsive of carnitine therapy has also been described. For review of primary carnitine deficiency see Reference 4.

Secondary carnitine deficiency is likely to be of much more general importance. It occurs with a whole range of rare metabolic disorders of the mitochondrion, but also in chronic renal failure treated with hemodialysis, where carnitine both free and bound is lost in the

Table 1
CARNITINE DEFICIENCY STATES

Classical carnitine deficiency syndromes
Associated with a variety of rare metabolic defects (see Reference 4)
- Duchenne dystrophy
- Myocardial ischemia
- Liver disease
- Reye's syndrome
- Hyperlipidemia
- Dialysis

Total parenteral nutrition, especially neonate
Associated with drug toxicity, e.g., valproic acid[14]

Table 2
NORMAL CARNITINE LEVELS

	Carnitine		
	Total	**Free**	**Esterified**
Plasma (μmol/l)	10—50	10—45	0—5
Skeletal muscle (mmol/mgnoncollagen protein)	10—18	4—17	1—6
24-h urinary excretion (mmol)	300—350	140—200	140—200

dialysate; and also in Valproate-induced Reye's syndrome, probably because carnitine acts as a ''buffer'' for valproic acid, combining with it and being excreted in the urine like acyl carnitine.[4] In total parenteral nutrition (TPN) in neonates, especially the premature, carnitine deficiency has also been described,[11] probably because the capacity for its synthesis only develops late in gestation and appears to be triggered by the fat in breast milk.[12] Normally, breast milk supplies carnitine early in postnatal life. TPN and some soy-based formulas have not in the past provided carnitine. Most formula companies are now adding carnitine to their non-cow's milk-based formula and, no doubt, it will also be added to TPN solutions.

MEASUREMENT OF CARNITINE

The current radiometric assay for the measurement of carnitine is not ideal for clinical purposes.[13] The methods are, however, sensitive and specific. Normal values are given in Table 2, but it should always be remembered that plasma, liver, muscle, and urine values can move independently of one another depending upon the nature of the defect leading to suboptimal carnitine status.

CONCLUSION

The importance of carnitine in clinical medicine is yet to be clearly delineated. The advent of a more convenient assay will be the necessary prerequisite to an adequate data base on which to decide the issue. In the meantime it seems likely that carnitine will be added to TPN, non-cow's milk-based infant formulas, and enteral feeds.

REFERENCES

1. **Bremer, J.,** Carnitine metabolism and functions, *Physiol. Rev.,* 63, 1420, 1983.
2. **Borum, P. R.,** Carnitine, *Annu. Rev. Nutr.,* 3, 233, 1983.
3. **Rebouche, C. J. and Paulson, D. J.,** Carnitine metabolism and function in humans, *Annu. Rev. Nutr.,* 6, 41, 1986.
4. **Gilbert, E. F.,** Carnitine deficiency, *Pathology,* 17, 161, 1985.
5. **Engel, A. G. and Rebouche, C. J.,** Carnitine metabolism and inborn errors, *J. Inher. Metab. Dis.,* 7 (Suppl. 1), 38, 1984.
6. **Stumpf, D. A., Parker, W. D., and Angelini, C.,** Carnitine deficiency, organic acidemias and Reye's syndrome, *Neurology,* 35, 1041, 1985.
7. **Suzuki, M., Tokuyama, K., and Kinoshita, M.,** Urinary profile of L-carnitine and its derivatives in starved normal persons and ACTH injected patients with myopathy, *J. Nutr. Sci. Vitaminol. (Tokyo),* 29, 303, 1983.
8. **Broquist, H. P. and Borum, P. R.,** Carnitine, in *Advances in Nutritional Research, Vol. 4,* Draper, H. H., Ed., Plenum Press, New York, 1982, 181.
9. **Roe, C. R., Millington, D. S., Maltby, D., and Bohan, T. P.,** Status and function of L-carnitine in Reye's syndrome (RS) and related metabolic disorders, *J. Nutr. Reyes Syndr. Found.,* 4, 58, 1983.
10. **Rebouche, C. J. and Engel, A. G.,** Kinetic compartment analysis of carnitine metabolism in the human carnitine deficiency syndromes, *J. Clin. Invest.,* 73, 857, 1984.
11. **Schmidt-Somerfeld, E., Penn, D., Novak, M., and Wolf, H.,** Carnitine in human perinatal fat metabolism, *J. Perinat. Med.,* 13, 107, 1985.
12. **Hahn, P. and Novak, M.,** How important are carnitine and ketones for the newborn infant, *Fed. Proc.,* 44, 2369, 1985.
13. **Cedarblad, G. and Landstedt, S.,** A method for determination of carnitine in the picomole range, *Clin. Chim. Acta,* 37, 235, 1972.
14. **Murphy, J. V., Marquardt, K. M., and Shug, A. L.,** Valproic acid associated abnormalities of carnitine function, *Lancet,* 1, 820, 1985.

WATER-SOLUBLE VITAMINS

J. Thomas Hindmarsh

INTRODUCTION

In general, water-soluble vitamins function as co-factors or coenzymes in a wide variety of enzymatic reactions. Except for cobalamin, all can be synthesized by plants which provide good sources of them. Body storage is limited and, therefore, a continuous dietary supply is necessary (except for cobalamin). Also, urine excretion is rapid (except for cobalamin); therefore, cumulative toxicity is uncommon.

THIAMIN (VITAMIN B_1)

Physiology

Thiamin is a pyramidyl-substituted thiazole. The enzyme thiamin pyrophosphokinase present in brain and liver converts thiamin to its active form, thiamin pyrophosphate. Thiamin pyrophosphate is a coenzyme involved in the transfer of activated aldehydes. Two sites where this occurs are the transamination of α-ketoacids (the conversion of α-ketoglutarate to glutamate and of pyruvate to alanine) and transacetylase reactions. In the deficient subject, the substrates for these reactions accumulate (pyruvate, pentose sugars, α-ketocarboxylate derivates of leucine, isoleucine, and valine). The most abundant sources of thiamin are unrefined cereal grains, and organ meats, e.g., liver, heart, and kidney. The enrichment of flour now corrects the deficiency produced by its refinement. Thiamin is readily absorbed in the small intestine. Some fish contain a heat-labile enzyme (thiaminase) that destroys thiamin. Feeding a diet containing 10% or more of uncooked fish to foxes can produce thiamin deficiency (Chastek paralysis).[1]

Deficiency

The clinical features of thiamin deficiency are not clearly delineated, as the circumstances that produce dietary shortages usually result in multiple deficiencies of the B vitamins. In general terms, beriberi is the largest clinical manifestation of thiamin deficiency. It can be wet (associated with high output cardiac failure) or dry (with polyneuropathy). An infantile form has been described. In wet beriberi the clinical features are congestive cardiac failure. In dry beriberi the symptoms are those of a polyneuropathy which commonly produces symmetrical lesions involving the nerves of the lower limbs more than those of the upper limbs. Both sensory and motor nerves are involved. Signs include foot drop, muscle wasting, impaired knee and ankle jerks, paresthesia, burning sensations, loss of sensation, numbness, and loss of position sense.

Beriberi will develop in patients surviving largely on a rice diet. Others at risk are those living mostly on highly milled wheat flour, although flour enrichment in Western countries has largely corrected this problem. A review of thiamin intake in old people in the U.S. showed it to be adequate unless the subject had an associated alcohol problem.[2]

Alcoholics who have eaten little food for many weeks may develop dry beriberi. They may also develop Wernicke-Korsakoff syndrome which is an association of ophthalmoplegia, nystagmus, and ataxia, together with a severe memory and learning defect often accompanied by extensive confabulation. Mild thiamin deficiency in old people may manifest itself as cardiomyopathy. 5-Fluorouracil may produce subclinical thiamin deficiency possibly due to interference with thiamin phosphorylation. A recent Australian study has reported a relationship between sudden infant death syndrome and thiamin deficiency (detected by impaired erythrocyte transketolase activity[3]).

Assessment of Body Status

Methods for measuring the free vitamin level and thiamin pryophosphate in serum and urine are available and both have been used to assess body status. Urine excretion after a test load of thiamin has also been used to assess if body tissues are deficient, in which case it would be retained. Serum pyruvate levels may be elevated after a glucose load in thiamin deficiency. The most useful method for assessing thiamin status at the present time probably is the measurement of erythrocyte transketolase activity which is reduced in deficiency.[4] Measurements of enzyme activity with and without the addition of thiamin pyrophosphate to the incubation medium are compared.[5]

RIBOFLAVIN (VITAMIN B_2)

Riboflavin is a component of flavin nucleotides which are widely distributed in the animal and plant world and are essential coenzymes in many oxidation-reduction reactions in the respiratory chain and other metabolic pathways. It is readily absorbed in the small intestine and is naturally found in many vegetables, milk, liver, kidney, and heart. Deficiency produces angular stomatitis, cheilosis (fissuring and scaling of the lips), glossitis, and a normochromic normocytic anemia.

Erythrocyte and urinary riboflavin levels can be measured, but the test usually performed in investigating deficiency is based on determining the effect of added flavin adenine dinucleotide (FAD) or erythrocyte glutathione reductase activity. An increase in activity with FAD addition suggests riboflavin deficiency.[4-6]

PYRIDOXINE (VITAMIN B_6)

Pyridoxine, pyridoxal, and pyridoxamine are all equally effective precursors of the coenzymes pyridoxal phosphate and pyridoxamine phosphate. Meats, poultry, fish, yeast, and some seeds are all good sources. Aqueous solutions decompose in light, particularly at neutral or alkaline pH. They are readily absorbed in the gastrointestinal tract. Pyridoxal phosphate functions as a coenzyme in transamination and decarboxylation reactions and is, therefore, involved in amino acid metabolism and heme synthesis.

An isolated deficiency of vitamin B_6 is uncommon, however, deficiency may occur from the use of some drugs. Isoniazid (used in the treatment of tuberculosis) forms a hydrazone with pyridoxal and the complex is lost in the urine. Isoniazid is acetylated in the liver, but a fraction of humans are "slow acetylators" and only this group forms the hydrazone with isoniazid and is susceptible to pyridoxal deficiency from this therapy. Penicillamine (used in the treatment of Wilson's disease and heavy-metal poisoning) inactivates the coenzyme complex. There are several inherited diseases which respond to increased pyridoxine. Several of these appear to be the consequence of an inherited reduced affinity of a particular enzyme for the coenzyme.

Vitamin B_6 requirement is increased by the contraceptive pill. This appears to be the consequence of activation by estrogens of the enzymes converting tryptophan to nicotinic acid in the liver, some of which require pyridoxal phosphate as coenzyme. Tryptophan load tests on these subjects may show increased excretion of intermediary metabolites of the tryptophan-nicotinic acid pathway.[7] This can usually be prevented by large doses of pyridoxine. Pregnant women also may excrete increased amounts of tryptophan metabolites.[8] Pyridoxine toxicity (sensory neuropathy) has been reported in subjects taking 2 g or possibly less of the vitamin daily.[9,10]

A variety of methods can be used to assess vitamin B_6 status. Blood levels of pyridoxal phosphate can be measured directly; also, erythrocyte levels of aspartate and alanine aminotransferase before and after the addition of pyridoxal phosphate can be used.[5] The usual method for assessing vitamin B_6 status is the measurement of the urinary excretion of xanthurenic acid following an oral load of L-tryptophan. Because pyridoxal phosphate is necessary for the optimal conversion of tryptophan to niacin, urine xanthurenic acid is increased when patients with vitamin B_6 deficiency are given tryptophan by mouth. A similar test using an oral dose of methionine and measuring the ratio of urine cystathionine to cysteine sulfinic acid may also be used.[4]

BIOTIN

Much of our daily requirement for biotin is supplied by our intestinal bacteria. Egg yolk, animal tissues, tomatoes, and yeast are good sources of this vitamin. Egg white contains avidin, a heat-labile protein which combines with biotin to prevent its intestinal absorption. Biotin is a co-factor in several carboxylation reactions.

Deficiency must be rare, but has been reported due to the ingestion of large amounts of raw egg white.[11] Deficiency has also been reported in subjects maintained on total parenteral nutrition.[12] Clinical features include thinning of the hair with loss of color, eczema, paresthesia, depression, and lethargy. At least two inherited disorders of biotin metabolism have been reported which respond favorably to large doses of biotin.[13] Concentrations in biological samples can be measured by microbiological methods.[14]

PANTOTHENIC ACID

Pantothenic acid is widely distributed in animal tissues, whole grain cereals, and vegetables. Deficiency in humans is usually associated with deficiency of other B vitamins. It is readily absorbed from the gastrointestinal (GI) tract and is an essential component of coenzyme A, and as such is essential for the optimum metabolism of carbohydrates, fats, and amino acids. Coenzyme A acts as a carrier of acyl groups in fatty acid metabolism, in acetylation reactions, and in oxidative decarboxylations. Deficiency is reported to produce paresthesia of the hands and feet; in animals it produces adrenal cortical failure. Blood levels can be measured by radioimmunoassay or microbiological methods.[15]

NIACIN AND NIACINAMIDE

Niacin (nicotinic acid) or its corresponding amide, niacinamide (nicotinamide), is an essential dietary component. Plants and most animals can synthesize niacin from tryptophan by a pathway requiring pyridoxal phosphate as coenzyme. However, in man, about 60 mg of tryptophan are required to synthesize 1 mg of niacin, and, therefore, some preformed niacin (or niacinamide) is essential unless our diet is especially rich in tryptophan. A more useful nutritional parameter is the dietary niacin equivalent (niacin content added to 1/60 of the tryptophan content) rather than the niacin content alone.[16] Milk and eggs are poor sources of niacin, however, they are beneficial to patients with pellagra as they are rich in tryptophan.

Niacin and niacinamide are readily absorbed from the GI tract and are transported to the tissues where they are incorporated into NAD (nicotinamide adenine dinucleotide) and NADP (nicotinamide dinucleotide phosphate). These compounds serve as coenzymes in many reversible oxidation-reduction reactions. Corn (maize) contains very little tryptophan and niacin, and when this is the major source of dietary protein, the disease pellagra results. Patients with pellagra suffer from the triad of dermatitis, dementia, and diarrhea. Pellagra also may occur as a complication of the carcinoid syndrome, because tryptophan is catabolyzed at an excessive rate. It may also occur in Hartnup disease, an autosomal recessive disorder in which the neutral amino acids are poorly absorbed from the GI tract and poorly reabsorbed from the proximal renal tubule. Excessive dosing with niacin produces vascular flushing together with nausea, vomiting, and diarrhea. Hyperpigmentation of the skin may occur as may signs of chronic liver disease.

The diagnosis of pellagra is usually made from the typical clinical features and a dietary history. However, niacin status can also be tested by measuring the urinary excretion of *N*-1-methyl-nicotinamide and *N*-1-methyl-3-carboxamide-6-pyridone. Normally, adults excrete 20 to 30% of their niacin as the former and 40 to 60% as the latter; an excretion rate of pyridone/methyl-nicotinamide of 1.3 to 4 is acceptable. However, latent deficiency is present when the value falls below 1, and, as depletion occurs, pyridone disappears from the urine followed by methyl-nicotinamide. Also, pyridine nucleotides can be assayed directly in serum.[4]

ASCORBIC ACID

Chemistry

Ascorbic acid is similar in structure to a monosaccharide, but contains an enediol group which can be dehydrogenated to produce dehydroascorbic acid. The former is spontaneously oxidized to the latter and both forms are physiologically active; the acid has a pKa of 4.2 and is a relatively strong reductant. The dehydro form is more readily broken down than the reduced form; the resulting metabolites do not have antiascorbutic activity. Ascorbic acid is a white crystalline powder readily soluble in water. It is readily oxidized particularly in an alkaline medium and on exposure to heat. It is stable in cold acid solution.

Physiology

Good dietary sources of ascorbic acid are citrus fruits; canning and cooking produce losses of around 60%. The exact dietary requirement of ascorbic acid is not known. The recommended daily intake to maintain saturation of tissues in the human adult is around 60 mg/d (and additional 20 mg/d is recommended for pregnant women, and an additional 40 mg/d for lactating women). This 60-mg daily requirement is based upon the following findings: an intake of 77.5 mg/d is necessary to maintain the normal body pool of approximately 1500 mg and metabolic studies have shown that the catabolic rate of this pool is between 34 and 61.5 mg/d.[17] This pool is sufficient to prevent the development of scurvy for 1 to 1 $^1/_2$ months in volunteers ingesting a scorbutic diet. In patients with scurvy given isotope-labeled ascorbic acid, excretion of the labeled vitamin does not occur until the body pool reaches 1500 mg.

Clinical scurvy occurs when the body pool falls below 300 mg and a daily intake of 10 mg is sufficient to maintain this reduced pool and prevent the development of frank scurvy.

In common with all the water-soluble vitamins, ascorbic acid is readily absorbed from the GI tract possibly by some specific transport mechanism.[18] The greatest concentrations of ascorbic acid are found in the adrenal cortex, pituitary gland, corpus luteum, pancreas, liver, thymus, retina, and spleen. Higher concentrations are found in leukocytes, platelets, and erythrocytes than in the plasma.

The half-life of ascorbic acid in the human is only about 16 d. Daily urine excretion amounts to about half the dietary intake and usually amounts to between 20 and 30 mg for subjects on a normal diet. Ingested ascorbic acid is excreted in the urine as unchanged ascorbic acid, dehydroascorbic acid, and some is metabolized to oxalic acid. Megadoses of ascorbic acid (9 g daily by mouth) increase urine oxalic acid excretion from 20 to 40 mg daily.[1]

Ascorbic acid is an antioxidant and is employed in a series of biochemical conversions. One of its best-defined roles is as a co-factor for the enzymes responsible for hydroxylation of prolyl and lysyl residues in the formation of collagen and related tissues.[17] It is also involved in carnitine and catecholamine synthesis.

Dietary deficiency produces scurvy which is characterized by swelling of the gums and loosening of the teeth together with perifollicular and petechial hemorrhages on the buttocks, abdomen, legs, and arms. Later, larger bruises may occur spontaneously in the skin and subperiosteally (over the shins). Arthralgias may occur from joint effusions. Other features include small subconjunctival hemorrhages, drying of salivary and lacrimal glands, and edema of the legs. Anemia may occur as may congestive heart failure. Ascorbic acid requirements increase in stress and during surgery; also, steroid therapy increases the requirement for the vitamin. Old people commonly have inadequate intakes.[19] Megadoses of ascorbic acid appear to be non-toxic, however, increased production of oxalate might represent a hazard to persons predisposed to forming calcium oxalate kidney stones.

Assessment of Ascorbic Acid Status

A clinical diagnosis of frank scurvy is easy to make and laboratory tests are needed only for confirmation of the diagnosis. Subclinical scurvy is more difficult to detect. In these circumstances the biochemical tests are of more help. Obviously, it is important to ensure that no ascorbic acid is given before the tests are performed. Plasma levels reflect recent intake, but values below 0.2 mg/dl indicate deficiency. Leukocyte levels are more reflective of tissue content and their measurement is the preferred test.

REFERENCES

1. **Martin, D. W.,** Water soluble vitamins, in *Harper's Review of Biochemistry,* Martin, D. W., Mayes, P. A., Rodwell, V. W., and Granner, D. K., Eds., 20th ed., Lange, Los Altos, CA, 1985, 101.
2. **Rivlin, R. S.,** Nutrition and the health of the elderly, *Arch. Intern. Med.,* 143, 1200, 1983.
3. **Jeffrey, H. E., McCleary, B. V., Hensley, W. J., and Read, D. J. C.,** Thiamine deficiency — a neglected problem of infants and mothers — possible relationships to sudden infant death syndrome, *Aust. N.Z. J. Obstet. Gynaecol.,* 25, 198, 1985.
4. **McCormic, D. B.,** Vitamins, in *Textbook of Clinical Chemistry,* Tietz, N. B., Ed., W. B. Saunders, Philadelphia, 1986, 927.
5. **Bayoumi, R. A. and Rosalki, S. B.,** Evaluation of methods of coenzyme activation of erythrocyte enzymes for detection of deficiency of vitamin B_1, B_2, and B_6, *Clin. Chem.,* 22, 327, 1976.
6. **Sauberlich, H. E., Dowdy, R. P., and Skala, J. H.,** Riboflavin, in *Laboratory Tests for the Assessment of Nutritional Status,* CRC Press, Boca Raton, FL, 1974, 30.
7. **Rose, D. P.,** The influence of estrogens of tryptophan metabolism in man, *Clin. Sci,* 31, 265, 1966.
8. **Sauberlich, H. E., Dowdy, R. P., and Skala, J. H.,** Vitamin B_6, in *Laboratory Tests for the Assessment of Nutritional Status,* CRC Press, Boca Raton, FL, 1974, 37.
9. **Schaumburg, H., Kaplan, J., Windebank, A., Vick, N., Rasmus, S., Pleasure, D., and Brown, M. J.,** Sensory neuropathy from pyridoxine abuse, *N. Engl. J. Med.,* 309, 445, 1983.
10. **Berger, A. and Schaumburg, H.,** More on neuropathy from pyridoxine abuse, *N. Engl. J. Med.,* 311, 986, 1984.
11. **Davidson, S., Passmore, R., Brock, J. R., and Truswell, A. S.,** *Human Nutrition and Dietetics,* 7th ed., Churchill Livingstone, New York, 1979, 150.
12. **Lebenthal, E.,** *Total Parenteral Nutrition,* Raven Press, New York, 1986, 94.
13. **Bartlett, K., Ghneim, H. K., Stirk, H. J., and Wastell, H.,** Enzyme studies in biotin-responsive disorders, *J. Inher. Metab. Dis.,* 8 (Suppl. 1), 46, 1985.
14. **Sauberlich, H. E., Dowdy, R. P., and Skala, J. H.,** Biotin, in *Laboratory Tests for the Assessment of Nutritional Status,* CRC Press, Boca Raton, FL, 1974, 91.
15. **Wyse, B. W., Wittwer, C., and Hansen, R. G.,** Radioimmunoassay for pantothenic acid in blood and other tissues, *Clin. Chem.,* 25, 108, 1979.
16. **Davidson, S., Passmore, R., Brock, J. F., and Truswell, A. S.,** *Human Nutrition and Dietetics,* 7th ed., Churchill Livingstone, New York, 1979, 138.
17. **Levine, M.,** New concepts in the biology and biochemistry of ascorbic acid, *N. Engl. J. Med.,* 314, 892, 1986.
18. **Rose, R. C.,** Intestinal transport of vitamins, *J. Inher. Metab. Dis.,* 8 (Suppl. 1), 13, 1985.
19. **Milne, J. S., Lonergan, M. E., Williamson, J., Moore, F. M. L., McMaster, R., and Percy, N.,** Leucocyte ascorbic acid levels and vitamin C intake in older people, *Br. Med. J.,* 2, 383, 1971.

FOLATE AND VITAMIN B_{12}

Jo Ellen Welsh

INTRODUCTION

The metabolism of folate and vitamin B_{12} (cobalamin) is closely interrelated, as both are involved in "one carbon metabolism"; that is, they serve as acceptors or donors of one carbon units in a variety of reactions of amino acid and nucleotide metabolism.[1] Deficiency of either vitamin results in megaloblastic anemia,[2] a condition characterized by the presence of oval macrocytes in the peripheral circulation. Macrocytes (also called megaloblasts) are erythrocyte precursors which are released into the blood in B_{12} or folate deficiency because of derangements in DNA synthesis during red blood cell maturation (see details below). Although macrocytes are capable of carrying oxygen, they are fragile, easily hemolyzed, and, thus, have shorter life spans than mature erythrocytes. The ultimate effect is reduced oxygen transport to the tissues. Derangements of cell maturation are not restricted to the red cell line, as evidenced by effects of either deficiency on white cells and other rapidly dividing cell lines (particularly intestinal mucosal cells).

The metabolism of each vitamin will now be further discussed separately, with emphasis on differential diagnosis of the deficiency syndromes.

FOLATE (FOLIC ACID)

Folate consists of a pteridine nucleus and *p*-aminobenzoic acid, to which is attached from 1 to 7 glutamic acid residues. Folate is present in foods primarily as folylpolyglutamates, which are split into folylmonoglutamates by intestinal deconjugase enzymes prior to absorption. Plasma folate (folylmonoglutamate) circulates free or bound to plasma proteins.[3] Folate is delivered to the liver, bone marrow, CSF, and kidney where it is taken up against a concentration gradient. The folylpolyglutamates appear to be the major intracellular form, and, as this form cannot cross cell membranes as readily as the folylmonoglutamate form, folate concentration within cells is higher than in the extracellular fluid.[1]

The coenzyme form of folic acid is tetrahydrofolic acid (THFA), which is generated from folic acid by enzymatic reductions which utilize NADPH as co-factor.[3] THFA serves as a carrier for single carbon groups (formyl, formaldehyde, and methyl), which are important in amino acid interconversions, biosynthesis of purines and pyrimidines, and methylation reactions. The biochemistry of folate-requiring reactions is complex and beyond the scope of this review (see Reference 1). The end-result of folate deficiency is inhibition of DNA synthesis due to decreased availability of purines and thymidine, and, ultimately, impairment in cell division. Rapidly proliferating cells are most affected; bone marrow, intestinal mucosal cells, and tumor cells. Cancer chemotherapeutic agents such as methotrexate are folate antagonists which competitively inhibit folate-requiring reactions and, thus, block synthesis of thymidine and inhibit tumor cell growth.

Deficiency of folate can be related to low intakes, increased demands, malabsorption, or drug-induced metabolic interferences.[2] Although folates are present in most foods, they are easily destroyed by cooking and processing. The most common situation in which folate deficiency occurs is pregnancy; megaloblastic anemia is estimated to occur in one third of all pregnancies worldwide.[3] Even in developed countries, folate deficiency occurs in 3 to 5% of pregnancies.[4] Many drugs bind to and inhibit the dihydrofolate reductase enzyme which keeps folate in the active THFA form. Folate deficiency is the most important etiologic factor in the development of anemia in alcoholics due to a combination of a direct effect of alcohol on folate utilization, poor intake, and liver impairment.

Although serum folate decreases in folate deficiency, it does not always reflect tissue stores. Red cell folate levels are more indicative of tissue folate, but are also low in B_{12} deficiency. Thus, diagnosis of folate deficiency requires decreases in both serum folate (<3 ng/ml) and red cell folate (<150 ng/ml).[2] Urinary excretion of formiminoglutamate (FIGLU) is increased in folate deficiency, as this histidine catabolite is normally converted to glutamate by a folate-requiring enzyme system. Blood smear and/or marrow biopsy can detect megaloblastic changes, but, again, will not distinguish deficiency of folate from that of B_{12}. An *in vitro* measurement of conversion of deoxyuridine to thymidine (a THFA-requiring reaction) by peripheral lymphocytes (the dU suppression test) can indicate if thymidine synthesis is impaired. The test can then be repeated with added folate or B_{12} to distinguish which vitamin is lacking.

VITAMIN B_{12}

Vitamin B_{12} (cobalamin) has a complex structure which contains the metal cobalt surrounded by four nitrogen-containing five-membered rings. Two coenzyme forms of the vitamin have been identified; 5-deoxyadenosyl cobalamin and methyl cobalamin. The following three mammalian enzyme systems have been shown to require cobalamin as coenzyme:[1]

1. Methylmalonyl coA mutase (converts methylmalonyl coA to succinyl coA; key step in metabolism of branched chain fatty and amino acids)
2. Leucine 2,3-aminomutase (metabolic significance of this reaction is uncertain)
3. Methionine synthetase (generates methionine from homocysteine, with resynthesis of THFA from 5-methyl THFA)

It is reaction 3 that constitutes the link between B_{12} and folate metabolism. If B_{12} supply is limiting, folate becomes trapped in the inactive 5-methyl THFA form, resulting in functional folate deficiency and megaloblastic anemia. The anemia is responsive to treatment with folate. However, the neurological damage characteristic of B_{12} deficiency (progressive demyelination), which is presumably due to the role of B_{12} in reaction 1, will continue unless therapy with B_{12} is instituted.

Cobalamin is only present in foods of animal origin (meat, poultry, fish, eggs, dairy products) and is stringently conserved by the body. The daily requirement is extremely low (0.5 to 4 μg/d) and easily met, with the notable exception of strict vegetarians. Deficiency is more often the result of inadequate absorption, as cobalamin is absorbed by a complex process. The vitamin must be released from dietary protein by gastric acid and proteases, bound to "intrinsic factor (IF)", a 50,000 mol. wt. glycoprotein secreted by gastric parietal cells and transported to the ileum where the IF-B_{12} complex binds to the brush border in the presence of calcium. It has not been established when or where the complex becomes dissociated, but only B_{12} appears in plasma.[5] Transport of B_{12} in plasma is by a heterogeneous group of binding proteins labeled transcobalamin I, II, and III.[3] Transcobalamin II appears to be the major transporter of the vitamin to liver and bone marrow.

Interference with any of the steps in absorption or transport may result in deficiency, although symptoms usually do not appear for several years. Examples are gastrectomy (loss of IF-secreting cell mass), ileal resection (decreased surface area for absorption), or bacterial overgrowth of small bowel (bacteria use B_{12} for their own growth). The condition known as pernicious anemia is characterized by total or relative lack of IF, and may be congenital (genetic defect in IF production) or age dependent (genetically determined decline in IF secretion with age). If B_{12} deficiency is due to insufficient IF or to a noncorrectable defect in ileal structure or function, parenteral administration of maintenance doses (30 μg monthly) is necessary for life.[2]

The diagnosis of B_{12} deficiency is crucial to prevent the neurological damage which results if B_{12} deficiency is erroneously treated with folate. In contrast to some other water-soluble vitamins, serum B_{12} levels accurately reflect body stores, thus, a low level (<100 pg/ml) can be diagnostic for deficiency.[2] Additionally, increased urinary excretion of methylmalonate occurs in B_{12} deficiency due to a metabolic block. Because of the requirement of B_{12} for maintenance of folate in its active coenzyme form, evidence of folate deficiency (low red cell folate, increased urinary FIGLU, megaloblastic anemia) will be present in B_{12} deficiency as well. Serum folate, however, is usually normal or raised as its tissue utilization is impaired in B_{12} deficiency. The dU suppression test can help to distinguish between the two deficiencies (see discussion of folate), and tests for B_{12} absorption (Shilling test) can help pinpoint the problem. Caution in interpretation of gastrointestinal tests is necessary, however, as chronic deficiency of either vitamin can lead to impaired gut function due to effects on rapidly dividing mucosal cells.

REFERENCES

1. **Shane, B. and Stokstad, E. L. R.,** Vitamin B 12-folate interrelationships, *Annu. Rev. Nutr.,* 5, 115, 1985.
2. **Herbert, V.,** The nutritional anemias, *Hosp. Pract.,* 15, 65, 1980.
3. **Herbert, V., Colman, N., and Jacob, E.,** Folic acid and vitamin B 12, in *Modern Nutrition in Health and Disease,,* Goodhart, R. S. and Shils, M. E., Eds., Lea & Febiger, Philadelphia, 1980, 229.
4. **Taylor, K. B. and Anthony, L. E.,** Nutritional aspects of hematopoiesis, in *Clinical Nutrition,* McGraw-Hill, New York, 1983, 294.
5. **Jacob, E., Baker, S. J., and Herbert, V.,** Vitamin B 12-binding proteins, *Physiol. Rev.,* 60, 918, 1980.

Index

INDEX

A

B

H

N

O

P

X

Y

Z